Neurobiology of the Trace Amines

Experimental and Clinical Neuroscience

Neurobiology of the Trace Amines, edited by ***A. A. Boulton, G. B. Baker, W. G. Dewhurst,*** and ***M. Sandler,*** 1984
Neural Membranes, edited by ***Grace Y. Sun, Nicolas Bazan, Jang-Yen Wu, Giuseppe Porcellati,*** and ***Albert Y. Sun,*** 1983

NEUROBIOLOGY OF THE TRACE AMINES

Analytical, Physiological, Pharmacological, Behavioral, and Clinical Aspects

Edited by

A. A. BOULTON, G. B. BAKER, W. G. DEWHURST, and M. SANDLER

Humana Press · Clifton, New Jersey

Library of Congress Cataloging in Publication Data

Main entry under title:

Neurobiology of the trace amines.

"Based on the proceedings of Trace amines and the neurosciences, a meeting held at the University of Alberta, Edmonton, July 19–21, 1983 . . . . organized as a satellite meeting of the ninth meeting of the International Society for Neurochemistry, held in Vancouver, July 10–15, 1983"--Pref.

Includes index.

1. Biogenic amines--Congresses. 2. Neurochemistry--Congresses. I. Boulton, A. A. II. International Society for Neurochemistry. Meeting (9th : 1983: Vancouver, B.C.) [DNLM: 1. Amines--Congresses. QU 60 N494 1983]
QP801.B66N48 1984 612'.8042 84-626
ISBN 0-89603-063-6

Crescent Manor
PO Box 2148
Clifton, NJ 07015

Printed in the United States of America

Proceedings of a meeting held at the University of Alberta, Edmonton, July 19–21, 1983.

PREFACE

Many of the trace amines—more correctly called biogenic amines—have been known for decades, but because of their tiny concentrations (0.01–100 ng/g) in brain, it was only after the development of sophisticated analytical techniques (such as mass spectrometry) that they could be identified and quantitated in nervous tissue.

There are now more than 20 of them and most are related to the catecholamines and 5-hydroxytryptamine both structurally and metabolically. Their pharmacological and physiological properties make them prime candidates for a transmitter or neuromodulator role and many of them elicit profound behavioral syndromes after injection—one of them, phenylethylamine, has even been referred to as nature's amphetamine.

In the clinical sphere several have been shown to be involved in: Parkinsonism, schizophrenia, depression, agoraphobia, aggression, hyperkinesis, migraine, hypertensive crises, hypertyrosinemia, hepatic encephalopathy, epilepsy, and cystic fibrosis. Thus the research reported here on these intriguing "new" substances will be of great interest to psychiatrists, neurologists, biochemists, pharmacologists, physiologists, psychologists, behaviorists and indeed to all those working in the neurosciences and related fields today.

ACKNOWLEDGMENTS

This book is based on the proceedings of *Trace Amines and the Neurosciences,* a meeting held at the University of Alberta, Edmonton, July 19–21, 1983. This meeting was organized as a Satellite Meeting of the Ninth Meeting of the International Society for Neurochemistry, held in Vancouver, July 10–15, 1983. International organizers of the satellite meeting were Drs. A. A. Boulton (Saskatoon), W. G. Dewhurst (Edmonton), G. B. Baker (Edmonton), and M. Sandler (London). Members of the local organizing committee in Edmonton were G. B. Baker, W. G. Dewhurst, M. M. Ferguson, R. T. Coutts, and R. A. Locock.

The meeting was made possible by major conference grants from the Alberta Heritage Foundation for Medical Research, the Medical Research Council of Canada, and the University of Alberta, and by grants-in-aid from (in alphabetical order) Amersham Corporation, Beckman Instruments Inc., Bristol-Myers Inc., Ciba-Geigy Canada Ltd., Hewlett-Packard (Canada) Ltd., Mandel Scientific Co. Ltd., Novopharm Ltd., Pfizer Canada Inc., Rhône-Poulenc Pharma Inc., Smith Kline & French Canada Ltd., Squibb Canada Inc., Upjohn Company of Canada, and Wyeth Ltd.

We acknowledge with pleasure the presence of Dr. Lionel McLeod, President of the Alberta Heritage Foundation for Medical Research, who opened the conference.

CONTENTS

C. PHYSIOLOGY AND PHARMACOLOGY

Invited Communications

Free Communications

D. BEHAVIOR

Invited Communications

Free Communications

E. CLINICAL STUDIES

Invited Communications

Free Communications

CONTRIBUTORS

D. ADDINGTON • Regional Psychiatric Center, Saskatoon, Saskatchewan, Canada

A. ADELL • Instituto de Quimica Bio-Organica, Consejo Superior de Investigaciones Cientificas, Barcelona, Spain

I. ALTERMAN • National Institute of Mental Health, Bethesda, Maryland

**F. ARTIGAS • Instituto de Quimica Bio-Organica, Consejo Superior de Investigaciones Cientificas, Barcelona, Spain

B. A. BAILEY • Department of Biology, University of Waterloo, Waterloo, Ontario, Canada

*G. B. BAKER • Department of Psychiatry, University of Alberta, Edmonton, Alberta, Canada

*P. R. BIECK • Human Pharmacology Institute, CIBA-GEIGY GmBH, Tuebingen, Federal Republic of Germany

S. M. BONHAM-CARTER • The Bernhard Baron Memorial Research Laboratories and Institute of Obstetrics and Gynaecology, Queen Charlotte's Maternity Hospital, London, England

R. L. BORISON • Department of Psychiatry, Medical College of Georgia and Augusta V. A. Medical Center, Augusta, Georgia

*A. A. BOULTON • Psychiatric Research Division, University of Saskatchewan, Saskatoon, Sakastchewan, Canada

R. BOWEN • Department of Psychiatry, University of Saskatchewan, Saskatoon, Saskatchewan, Canada

J. BROADBENT • Psychiatric Research Division, University of Saskatchewan, Saskatoon, Saskatchewan, Canada

**O. H. CALLAGHAN • Department of Biochemistry, Chicago Medical School, University of Health Sciences, North Chicago, Illinois

J. CHEVESICH • Department of Biochemistry, Chicago Medical School, University of Health Sciences, North Chicago, Illinois

S. T. CHRISTIAN • Neurosciences Program, University of Alabama in Birmingham, Birmingham, Alabama

**S. J. COOPER • Department of Psychology, University of Birmingham, Birmingham, England

R. T. COUTTS • Neurochemical Research Unit, University of Alberta, Edmonton, Alberta, Canada

M. W. COUCH • Veterans Administration Medical Center and Department of Radiology, University of Florida College of Medicine, Gainesville, Florida

S. O. DAVIES • New York State Psychiatric Institute, New York, NY

B. A. DAVIS • Psychiatric Research Division, University of Saskatchewan, Saskatoon, Saskatchewan, Canada

*W. G. DEWHURST • Neurochemical Research Unit, Department of Psychiatry, University of Alberta, Edmonton, Alberta, Canada

*B. I. DIAMOND • Department of Psychiatry, Medical College of Georgia and Augusta V. A. Medical Center, Augusta, Georgia

S. DIAMOND • Diamond Headache Clinic, Chicago, Illinois

*C. T. DOURISH • Psychiatric Research Division, University of Saskatchewan, Saskatoon, Saskatchewan, Canada

**R. G. H. DOWNER • Department of Biology, University of Waterloo, Waterloo, Ontario, Canada

S. D. DUNLOP • Institute of Psychiatric Research, Department of Psychiatry, Indiana University School of Medicine, Indianapolis, Indiana

*D. A. DURDEN • Psychiatric Research Division, University of Saskatchewan, Saskatoon, Saskatchewan, Canada

*L. E. DYCK • Psychiatric Research Division, University of Saskatchewan, Saskatoon, Saskatchewan, Canada

J. P. EDSTROM • Department of Medical Physiology, University of Calgary, Calgary, Alberta, Canada

**D. M. FERGUSON • Department of Pharmacology, University of Health Sciences, The Chicago Medical School, North Chicago, Illinois

N. A. GARRICK • Clinical Neuropharmacology Branch, National Institute of Mental Health, Bethesda, Maryland

E. GELPÍ • Instituto de Quimica Bio-Organica, Consejo Superior de Investigaciones Cientificas, Barcelona, Spain

*A. J. GREENSHAW • Psychiatric Research Division, University of Saskatchewan, Saskatoon, Saskatchewan, Canada

S. J. HAQUE • Department of Biochemistry, University of Calcutta, Calcutta, India

**R. E. W. HARRISON • Department of Physiology, Rush Medical College, Chicago, Illinois

W. HARRISON • Depression Evaluation Services, New York State Psychiatric Institute, New York, NY

H. C. HENDRIE • Institute of Psychiatric Research, Department of Psychiatry, Indiana University School of Medicine, Indianapolis, Indiana

*T. P. HICKS • Department of Medical Physiology and Anatomy, University of Calgary, Calgary, Alberta, Canada

A. HITRI • Department of Psychiatry, Medical College of Georgia and Augusta V. A. Medical Center, Augusta, Georgia

Y. IMAZU • Department of Clinical Neurochemistry, Okayama University Medical School, Shikata-cho, Okayama, Japan

J. IRONS • Department of Physiology and Pharmacology, Queen's Medical Centre, Nottingham, England

*D. M. JACKSON • Department of Pharmacology, University of Sydney, Sydney, New South Wales, Australia

K. JELLINGER • Ludwig Boltzmann Institut fur Klinische Neurobiologie, Krankenhaus Wein-Laing, Wien, Austria

O. F. JENKINS • Department of Pharmacology, University of Sydney, Sydney, New South Wales, Australia

R. S. G. JONES • Biology Research Laboratories, CIBA-GEIGY AG, Basle, Switzerland

*A. V. JUORIO • Psychiatric Research Division, University of Saskatchewan, Saskatoon, Saskatchewan, Canada

*F. KAROUM • National Institute of Mental Health Research, St. Elizabeth's Hospital, Washington, DC

*E. KIENZL • Ludwig Boltzmann Institut fur Klinische Neurobiologie, Arbeitsgruppe Neurochemie, Krankenhaus Wein-Laing, Wien, Austria

D. F. KLEIN • New York State Psychiatric Institute, New York, NY

**K. KOBAYASHI • Department of Clinical Neurochemistry, Institute for Neurobiology, Okayama University Medical School, Shikata-cho, Okayama, Japan

J. LAUBER • Research Department, CIBA-GEIGY Ltd., Basle, Switzerland

D. F. LeGATT • Department of Laboratory Medicine, University of Alberta Hospital, Edmonton, Alberta, Canada

M. R. LIEBOWITZ • Anxiety Disorders Clinic, New York State Psychiatric Institute, New York, NY

S. LIPPER • National Institute of Mental Health, Bethesda, Maryland

K. LUKOWIAK • Department of Medical Physiology, University of Calgary, Calgary, Alberta, Canada

**C. A. MARSDEN • Department of Physiology and Pharmacology, Queen's Medical Centre, Nottingham, England

I. L. MARTIN • MRC Neurochemical Pharmacology Unit, Medical Research Council Centre, Cambridge, England

R. J. MARTIN • Department of Biology, University of Waterloo, Waterloo, Ontario, Canada

E. MARTINEZ • Instituto de Quimica Bio-Organica, Consejo Superior de Investigaciones Cientificas, Barcelona, Spain

P. McGRATH • New York State Psychiatric Institute, New York, NY

J. M. MIDGELEY • School of Pharmacy, University of London, London, England

*A. D. MOSNAIM • Department of Pharmacology, University of Health Sciences, The Chicago Medical School, North Chicago, Illinois

*D. L. MURPHY • Clinical Neuropharmacology Branch, National Institute of Mental Health, Bethesda, Maryland

E. NILSSON • Human Pharmacology Institute, CIBA-GEIGY GmbH, Tuebingen, Federal Republic of Germany

H. NOLLER • Institute for Physical Chemistry, Technical University, Vienna, Austria

D. PARKINSON • Department of Medical Physiology, Faculty of Medicine, University of Calgary, Calgary, Alberta Canada

**M. PAULOS • DuPont Co., Wilmington, Delaware

*S. R. PHILIPS • Psychiatric Research Division, University of Saskatchewan, Saskatoon, Saskatchewan, Canada

**M. K. PODDAR • Department of Biochemistry, University of Calcutta, Calcutta, India

F. M. QUITKIN • Depression Evaluation Service, New York State Psychiatric Institute, New York, NY

*G. P. REYNOLDS • Department of Neurological Surgery and Neurology, Addenbrooke's Hospital, Cambridge, England

P. RIEDERER • Ludwig Boltzmann Institut fur Klinische Neurobiologie, Arbeitsgruppe Neurochemie, Krankenhaus Wien-Laing, Wien, Austria

C. M. ROBINSON • Department of Physiology and Pharmacology, Queen's Medical Centre, Nottingham, England

*J. M. SAAVEDRA • Section on Pharmacology, Laboratory of Clinical Science, National Institute of Mental Health, Bethesda, Maryland

*M. SANDLER • The Bernhard Baron Memorial Research, Laboratories and Institute of Obstetrics and Gynaecology, Queen Charlotte's Maternity Hospital, London, England

C. SCHICK • Human Pharmacology Institute, CIBA-GEIGY GmbH, Tuebingen, Federal Republic of Germany

C. SHAH • Department of Psychiatry and Augusta V. A. Medical Center, Augusta, Georgia

**P. A. SHEA • Institute of Psychiatric Research, Department of Psychiatry, Indiana University School of Medicine, Indianapolis, Indiana

T. SHOHMORI • Department of Clinical Neurochemistry, Institute for Neurobiology, Okayama University Medical School, Shikata-cho, Okayama Japan

J. STEWART • Anxiety Disorders Clinic, New York State Psychiatric Institute, New York, NY

C. SUÑOL • Instituto de Quimica Bio-Organica, Consejo Superior de Investigaciones Cientificas, Barcelona, Spain

R. E. TESSEL • Department of Pharmacology and Toxicology, University of Kansas, Lawrence, Kansas

E. F. TORREY • National Institute of Mental Health Research, St. Elizabeth's Hospital, WAW Building, Washington, DC

S. TURKISH • Department of Psychology, University of Birmingham, Birmingham, England

J. M. TUSSELL • Instituto de Quimica Bio-Organica, Consejo Superior de Investigaciones Cientificas, Barcelona, Spain

*C. H. VANDERWOLF • Department of Psychology, University of Western Ontario, London, Ontario, Canada

A. J. VAZQUEZ • Department of Pharmacology, University of Health Sciences, The Chicago Medical School, North Chicago, Illinois

S. E. WADE • Institute of Psychiatric Research, Department of Psychiatry, Indiana University School of Medicine, Indianapolis, Indiana

P. C. WALDMEIER • Pharmaceuticals Division, CIBA-GEIGY Ltd., Basle, Switzerland

P. L. WALKER • The Bernhard Baron Memorial Research Laboratories and Institute of Obstetrics and Gynaecology, Queen Charlotte's Maternity Hosptial, London, England

**C. M. WILLIAMS • Veterans Administration Medical Center and Department of Radiology, University of Florida College of Medicine Gainesville, Florida

**M. E. WOLF • Tardive Dyskinesia Program, Veterans Administration Medical Center, North Chicago, Illinois

S. WORMITH • Regional Psychiatric Center, Saskatoon, Saskatchewan, Canada

R. J. WYATT • National Institute of Mental Health Research, Saint Elizabeth's Hospital, WAW Building, Washington, DC

*S. N. YOUNG • Department of Psychiatry, McGill University, Montreal, Quebec Canada

*P. H. YU • Cancer and Medical Research Building, University of Saskatchewan, Saskatoon, Saskatchewan, Canada

E. A. ZELLER • Chicago Medical School and Shriners Hospital, North Chicago, Illinois

* = Principal author, invited lecture
** = Principal author, free communication

A. INTRODUCTION

TRACE AMINES: THE EARLY YEARS

William G. Dewhurst

Co-Director, Neurochemical Research Unit, and Professor and Chairman, Department of Psychiatry, University of Alberta, Edmonton, Alberta, Canada

In preparing this historical introduction I must first make comment on the term "trace" amines. It came into existence at a meeting of the ACNP in Puerto Rico in 1975. However, I really find the phrase very awkward. To categorize amines (or any other biologically active chemical) by quantity seems almost meaningless without conditions being specified. If I wanted to consult a doctor I would find it much more helpful to know whether he was an obstetrician or proctologist rather than whether he was fat or thin. Hopefully, and eventually, we will find a classification based on function and perhaps during the course of this conference some useful generalizations might emerge. Until then I think we have to be content with chemical designations as the most precise.

One purpose of this historical introduction is to illustrate a functional approach which I have advocated and which I think still has to be extended in this field.

My own interest in these substances and their relation to affective disorder started in an oblique way and derived from the observation of Crane in 1956 that iproniazid given to tuberculosis patients could produce elevation of mood without necessarily producing an improvement in physical status. It soon became apparent that a group of drugs with the common property of inhibiting monoamine oxidase all had euphoriant properties of varying degrees. It seemed logical, therefore, to suppose that an examination of the processes by which these drugs acted might give us a clue about the possible biochemical basis of mood.

MONOAMINE OXIDASE INHIBITION AND DEPRESSION

At that time (the mid 1950's) three main substrates for monoamine oxidase were known: first, serotonin (or 5-hydroxytryptamine) which had been established in mammalian brain by Twarog and Page in 1953; second, norepinephrine which had been established in mammalian brain by Martha Vogt 29 years ago in 1954 with a characteristic distribution in the reticular system and hypothalamus; and, finally, a third substrate tryptamine which had been shown to exist in human urine by my erstwhile tutor and colleague, Dr. Richard Rodnight at the Maudsley Hospital and the Institute of Psychiatry of London University in 1956.

So in an initial study at the same Institute in collaboration with Dr. Michael Pare (now of St. Bartholomew's Hospital) we decided to examine a possible correlation between serotonin metabolism and the effects of monoamine oxidase inhibitors (Dewhurst and Pare, 1961[a,b]). In this work we used nialamide and as a measure of monoamine oxidase inhibition gave each patient a capsule of serotonin creatinine sulphate (which is metabolized normally by monoamine oxidase to 5-hydroxy indole acetic acid [5-HIAA]) and measured the 5-HIAA in the patient's urine after such an oral dose at 2, 4, 6, and 8 hour intervals before and two weeks after treatment.

In the 14 patients in whom there was a statistically significant reduction of monoamine oxidase activity on 5-HT as shown by the reduction of 5-hydroxy indole acetic acid excretion, none improved clinically, whereas the one patient who improved during the period of the study showed no statistically significant monoamine oxidase inhibition as judged by 5-HT metabolism.

Of course, such studies can be criticized because they examine, presumably, mainly peripheral metabolism, but it seems, nonetheless, fair to say that if the monoamine oxidase inhibitor can be shown to penetrate the blood-brain barrier, then similar effects on cerebral metabolism might be inferred. However, a direct effect on cerebral metabolism should be sought, if possible, and to this end I made a study of carcinoid tumors (Dewhurst, 1961).

These tumors mostly arise in the small intestine and stomach and in their characteristic form produce large amounts of serotonin. But an examination of the cerebral effects in such patients is of little use because serotonin does not pass the blood-brain barrier and even before it reaches it much is destroyed by monoamine oxidase in the liver, lungs and elsewhere.

However, there is a small group of atypical carcinoid tumors which produce 5-hydroxytryptophan (the precursor of 5-hydroxytryptamine) which can pass through the blood-brain barrier and then be decarboxylated in the brain to form 5-HT. Patients with such carcinoid tumors secrete large amounts of 5-HTP for many months or years and it thus seemed that here, surely, must lie the touchstone for testing whether or not an excess of serotonin in the brain can cause psychiatric disturbance in man. I managed to collect 8 such cases from the world literature and where sufficient detail about neuropsychiatric sequelae was lacking I enquired personally of the authors and was able to obtain information on all 8. In 6 of the patients no neuropsychiatric sequelae whatsoever were observed. In one patient (who was in terminal heart failure) a confusional state was noted and I think serotonin need not be adduced to explain this. The other had been one of the patients on whom Swedish workers had identified the carcinoid syndrome (with its right-sided heart lesions, flushing, diarrhea and the rest) and who had been in and out of hospital many times suffering with metastases in lungs and liver, and who had had many biopsies. She had been labelled 'cancerophobic.' Do we need to invoke 5-HT? I believe not.

These studies, to me at least, pointed rather strongly against a pathogenetic effect of serotonin in affective disorder.

So I turned to an examination of the effects of a monoamine oxidase inhibitor (phenelzine) on all the three substrates mentioned previously, in the urine of depressed patients (Dewhurst, 1961).

The most notable effect was on the excretion of tryptamine, which was increased some fourfold; 5-HT was intermediate, and the metanephrines (the quantitatively preponderant metabolite of the catecholamines) were affected least. And the differences between the differences were statistically significant.

This left me in a quandary because at that time (the early 1960's) no one really knew what the functions of tryptamine were although it showed the most significant changes.

At this stage, I was fortunate to join Professor E. Marley (again of the Institute of Psychiatry) to pursue the problem.

THE BIOLOGICAL EFFECTS OF TRYPTAMINE

It was clear from reading the literature, that much of the confusion arising about amine function from earlier studies had arisen from methodological errors such as the lack of appreciation that amines in solution are unstable, particularly at an alkaline pH; failure to control the environment when studying the effects of amines in conscious animals; and problems in ensuring that certain amines reached brain sites (because both serotonin and norepinephrine do not pass the blood-brain barrier). To this end, therefore, we designed a standardized environment for our animal studies in which sound, light, heat, and social stimuli were carefully controlled, and by the use of a one-way screen in the front of the experimental box the animal saw only its own reflection and not the face of the experimenter. Social stimuli could be permitted from a companion box at the side. We also wished to quantify continuously both behavioural and physiological stimuli and did this, by using electrocorticographic and electromyographic leads appropriately inserted as well as various measures of behaviour which were all converted by transducers into voltage signals and thence mathematically integrated continuously. Finally amongst various strategies that have been used to ensure that certain amines reach the brain (such as intraventricular cannulation) we chose in particular the study of young chicks as seemingly more physiological than cannulation and who between the ages of 1 to 4 weeks have an immature blood-brain barrier. This permits the passage of substances such as norepinephrine and serotonin but fortunately the receptors at that stage are sufficiently adult to give adult responses. We also, of course, used other species such as rats, kittens, and cats (Dewhurst and Marley, 1965[a,b,c]).

A BIOLOGICAL TYPOLOGY OF AMINES AND THEIR RECEPTORS

With such techniques and an examination of over 50 different amines, basically variants of the amines we know are present in the brain today, our results may be summarized for present purposes by dividing their effects into two groups (Dewhurst, 1968[a]).

An "excitant" group termed Type A produced all the effects of the alert state behaviourally and physiologically in terms of the ECoG and EMG: and a "depressant" group (termed Type C) which produced all the physiological and behavioural changes of physiological sleep. Because the terms "excitant" and "depressant" have so many meanings, I introduced the noncommittal terms "Type A" and "Type C" to designate these precise amines and their effects.

As might be expected these two groups are physiological antagonists.

But, more important than physiological antagonism is specific antagonism which defines a specific receptor. Such properties for Type A amines were shown by methysergide. Further identification of this Type A receptor came from other standard pharmacological techniques such as the application of Ahlquist's criteria. By these means we concluded that the receptor for Type A amines was a cerebral tryptamine receptor entirely similar to that identified in the stomach some years previously by Gaddum.

The receptor on which Type C amines (such as norepinephrine) act appears to be similar (judging by the effects of the specific antagonist dibenzyline) to peripheral α-receptors although identification was less certain than in the case of the tryptamine receptor. Of the Type A (excitant) amines, tryptamine and phenylethylamine (now called trace amines) were found to be the most potent. This group also included the synthetic substances amphetamine and methylphenidate.

MODELS FOR HUMAN MOOD CHANGE

Now the importance to me in establishing, as I believe, the Type A or tryptamine receptor in brain is the fact that amphetamine acts on it and that this substance is known to be euphoriant in man. Further, as this action is also blocked by methysergide, an extrapolation may properly be made from the animal receptor model to a basis of mood change in man without being accused of anthropomorphism.

One should also note that another conclusion from our work in the mid-sixties was that social aspects were also found empirically to be important. For example, one response of the chick to amphetamine-like substances has been described as a peculiar twittering which ethologists have termed "pleasure calls." Yet when the chick was placed in the controlled environment described previously, this twittering did not occur. However, when a psychosocial stimulus was exhibited the twittering did occur and this was shown to be highly significant in cross-over experiments. In the case of the chick the psychosocial stimulus was shown to be the presence of another chick or even the experimenter's hand at the conclusion of the experiment.

TRACE AMINES AND AFFECTIVE DISORDERS

The extrapolation to disease states in man is the postulate that a deficiency of the trace amines, phenylethylamine and tryptamine may be associated with and perhaps causative of depressed states in man, whereas an excess may be responsible for manic states. But in both states psychosocial stimuli are important. Remember those pleasure calls in the chick? (Dewhurst, 1968[a], 1969)

Further data in man initially came from urine studies and it was shown by the late Professor Fischer and others that phenylethylamine excretion is diminished in depression in man (Fischer et al., 1968). Similarly others have shown that tryptamine excretion is diminished in depression in man and returns to normal levels after recovery.

However, these are, admittedly, peripheral studies, as noted earlier. Nonetheless there are also other pieces of more indirect evidence which I think implicate the trace amines, particularly tryptamine. One example is the fact that tryptophan, especially when given with a monoamine oxidase inhibitor or nicotinamide, has been shown in a number of studies (not all, to be sure) to be euphoriant, whereas 5-HTP (the precursor of serotonin) as well as dopa (the precursor of the catecholamines) have not (Smith and Prockop, 1962; Dewhurst, 1969). If tryptophan (in some studies at least) can be shown to be euphoriant and 5-HTP is not, what is the logical answer? Surely it must be that this is due to the formation of something other than serotonin and why not tryptamine?

But in the late 1960's and early 1970's when these views were put forward, the very existence of tryptamine in brain was denied despite the fact that the substrate for its formation (tryptophan) was present as well as the necessary enzyme (amino acid decarboxylase). But even if present it was argued by others that the amounts would be so small that tryptamine could have little significance. And this, despite a half-life of ten minutes. Fortunately, shortly after this, two groups of workers were able to establish the presence of tryptamine and phenylethylamine unequivocally in brain. These were Drs. Axelrod and Saavedra working in Bethesda and Dr. Boulton and his group from Saskatoon using, respectively, radioenzymatic techniques and mass spectrometry. It is appropriate therefore for me to stop this early sketch at this point and hand over to Dr. Boulton who will take us through the 1970's. But before doing so I would like to illustrate the dramatic effects of a monoamine oxidase inhibitor on some trace amines revealed by the newer techniques of measurement. Figure 1 shows the change in rat diencephalon one and one-half hours after administration of tranylcypromine. The log scale must be noted. The effects of this monoamine oxidase inhibitor on the trace amines phenylethylamine and tryptamine are greater by several orders of magnitude than the changes on the established neurotransmitters. For those who prescribe such drugs such changes surely cannot be ignored.

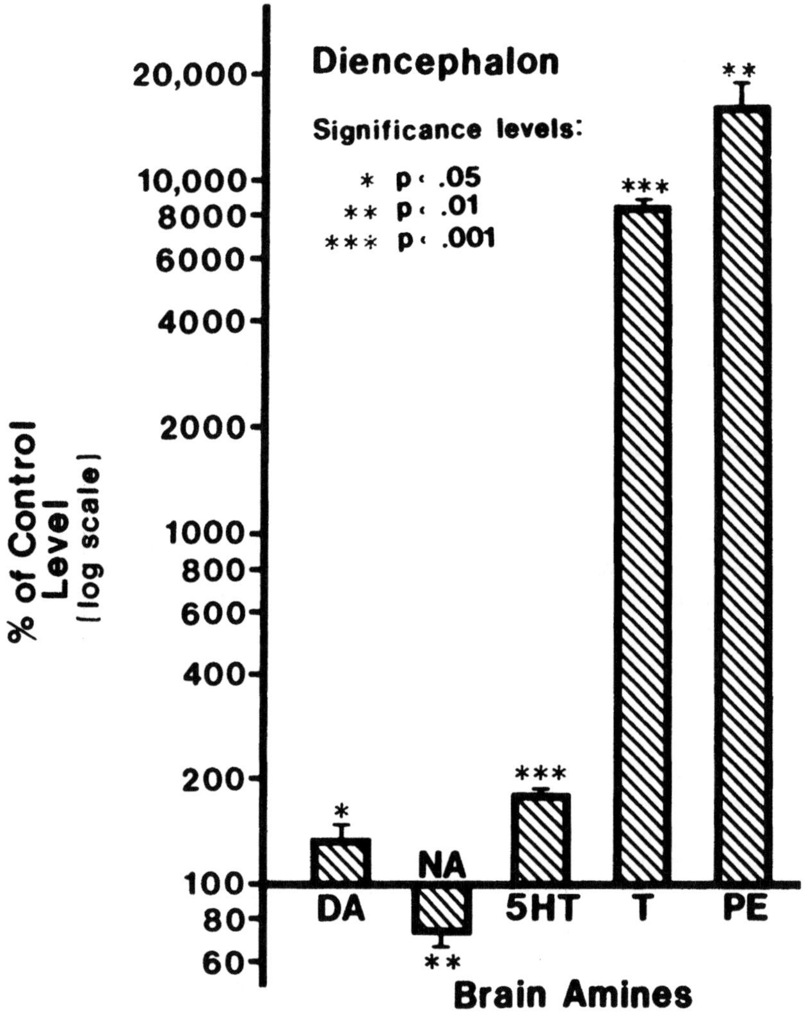

Figure 1. Levels of dopamine (DA), noradrenaline (NA), 5-hydroxytryptamine (5-HT), tryptamine (T), and phenylethylamine (PE), relative to control levels, in rat diencephalon, 1.5 hr. after a single 20 mg/kg i.p. injection of tranylcypromine. Bars represent S.E.M. Control values (ng/g) are: NA, 250 ± 26; DA, 960 ± 34; 5-HT, 692 ± 59; T, 0.7 ± 0.2; PE, 1.2 ± .15. Significance values determined using Student's _t_-test.

REFERENCES

Crane G. E. (1979) Further studies on iproniazid. J. Nerv. Ment. Dis. **124**, 322.

Curzon G. (1969) Tryptophan pyrrolase - a biochemical factor in depressive illness? Brit. J. Psychiat. **115**, 1367-1374.

Curzon G. (1970) Tryptophan pyrrolase - a biochemical factor in depressive illness? Brit. J. Psychiat. **116**, 571-572.

Dewhurst W. G. (1961) Amine metabolism in depressive illness. Dissertation, London University.

Dewhurst W. G. and Pare, C. M. B. (1961[a]) A clinical and biochemical study of monoamine oxidase inhibition in depressed patients. I. A clinical trial of nialamide.J. Ment. Sci. **107**, 239-243.

Dewhurst W. G. and Pare C. M. B. (1961[b]) A clinical and biochemical study of monoamine oxidase inhibition in depressed patients. II. 5-hydroxytryptamine tolerance before and after nialamide. J. Ment. Sci. **107**, 244-249.

Dewhurst W. G. and Marley E. (1965[a]) Methods for quantifying behaviour and cerebral activity and the effect of drugs under controlled conditions. Brit. J. Pharmacol. **25**, 671-681.

Dewhurst W. G. and Marley, E. (1965[b]) The effect of α-methyl derivatives of noradrenaline, phenylethylamine and tryptamine on the central nervous system of the chicken. Brit. J. Pharmacol. **25**, 682-704.

Dewhurst W. G. and Marley E. (1965[c]) Action of sympathomimetic and allied amines on the central nervous system of the chicken. Brit. J. Pharmacol. **25**, 705-727.

Dewhurst W. G. (1968[a]) New theory of cerebral amine function and its clinical application. Nature **218**, 1130-1133.

Dewhurst W. G. (1968[b]) Cerebral amines and behaviour. Lancet **ii**, 514-515.

Dewhurst W. G. (1969) Amines and abnormal mood. Proc. of the Royal Society of Medicine **102**, 1102-1107.

Dewhurst W. G. (1970) Tryptophan pyrrolase - a biochemical factor in depressive illness? Brit. J. Psychiat. **116**, 569-575.

Fischer E., Heller B., and Miro A. H. (1968) β-phenylethylamine in human urine. Arzneimittel-Forschung (Drug. Res. 18), 1486.

McKim H. R., Calverley D. G., Philips S. R., Baker G. B., and Dewhurst W. G. (1980) The effects of tranylcypromine on the levels of some cerebral amines in rat diencephalon, in Recent Advances in Canadian Neuropsychopharmacology (Grof, E. and Saxena, B., eds) pp. 7-13. S. Karger, Basel.

Philips S. R., Durden D. A., and Boulton, A. A. (1974) Identification and distribution of tryptamine in the rat. Can. J. Biochem. **52**, 447-451.

Rodnight R. (1956) Separation and characterization of urinary indoles resembling 5-hydroxytryptamine and tryptamine. Biochem. J. **64**, 621-626.

Saavedra J. M. and Axelrod J. (1972) A specific and sensitive enzymatic assay for tryptamine in tissues. J. Pharmacol. Exp. Ther. **182**, 363-369.

Smith B. and Prockop D. J. (1962) Central-nervous-system effects of ingestion of L-tryptophan by normal subjects. New England J. Med. **267**, 1338.

Twarog B. M. and Page I. H. (1953) Serotonin content of some mammalian tissues and urine and a method for its determination. Amer. J. Physiol. **175**, 157-161.

Vogt M. (1954) The concentration of sympathin in different parts of the central nervous system under normal conditions and after administration of drugs. J. Physiol. (Lond.) **123**, 451-481.

Weil-Malherbe H. (1968) Amines, alerting, and affect. Lancet **ii**, 219-220.

TRACE AMINES AND THE NEUROSCIENCES: AN OVERVIEW

Alan A. Boulton

Psychiatric Research Division
Cancer and Medical Research Building
University of Saskatchewan
Saskatoon, Saskatchewan S7N OWO

First I wish to join with Professors Dewhurst, Baker and Sandler in welcoming everyone to Edmonton, Alberta and to this the first international conference on the 'Trace Amines.' The trace amines are a group of substances (and there are now more than twenty of them) that, over the past decade, have emerged not as metabolic accidents, as was once proposed, but as functionally relevant and intriguing substances.

In 1971, in drawing attention to these biogenic amines, present in tissues and brain in such tiny quantities, I referred to them as microamines (Boulton, 1971), a term meant to imply that their presence, primarily in the brain, was in microamounts (i.e. 0.1-10 nanograms) as compared with the classical putative monoamine neurotransmitters [dopamine (DA), noradrenaline (NA) and 5-hydroxytryptamine (5-HT) - usually referred to as the biogenic amines] that are present in microgram amounts. In 1971 controversy still existed concerning the actual levels of these microamines; this was more or less settled in December 1975, however, during a study group at the American College of Neuropsychopharmacology meeting in Puerto Rico when agreement was reached on the likely brain levels. It was at this meeting that the neologism "Trace Amine' was adopted (Usdin and Sandler, 1976). Some of the trace amines, however, are present in microgram amounts in some species and in some tissues and certainly not in trace amounts after treatment with monoamine oxidase inhibitors (Philips and Boulton, 1979). Some of the classical monoamines, on the other hand, are sometimes present in certain tissues or brain regions in trace amounts; nevertheless the misnomer has remained operative and it seems destined to continue for some time yet. The result of this, of course, is that even the metabolites of the trace amines (i.e. the 'trace acids') are now encumbered with the same misleading title (Davis and Boulton, 1981). One only has to look at the urinary concentrations of phenylacetic acid, m- and p-hydroxyphenylacetic acid and p-hydroxymandelic acid to realise just how misleading such a title can be.

It is not my intention in this brief overview to describe what is going to be discussed in much more detail during the various sessions to follow at this conference or what has been written in several recent reviews (Boulton, 1976; 1978a,b; 1979a; Boulton and Juorio, 1982; Durden and Boulton, 1982); rather I am going to skip over the methodological, analytical, distribution, pharmacological, physiological and bioclinical studies in order to draw attention to the emerging field of the behavioural pharmacology of the trace amines, particularly in relation to certain components of spontaneous activity, their probable involvement in the mediation of action of some of the classical monoamines and their possible role as neuromodulators.

That the trace amines were not recently identified follows from a perusal of the literature; p-tyramine (p-TA) for example, was first identified and synthesized by Schmitt and Nasse in 1865 whilst phenylethylamine (PE), tryptamine (T) and octopamine (OA) were first described, and aspects of their properties investigated, in the first half of the 20th century. They have been taken less seriously than they deserved to be for several reasons; first, because of their low tissue levels methods for their unambiguous analysis and quantitation were not available; only the advent of mass spectrometry (Durden and Karoum, this book), gas chromatography (Baker et al., this book), radioenzymatic assay (Saavedra, this book), novel derivatives and sophisticated chromatography enabled their identification and quantitation to become routine. Secondly they have been described as indirectly acting sympathomimetic amines not possessing appreciable effects in their own right; because of this they have been used quite simply as drugs to manipulate the levels of other biogenic amines. In addition histochemical procedures to visualize them have not been available. Finally, they share in common with the classical amines apparently the same synthetic and catabolic enzymes. Because of all this the trace amines have been referred to as either 'metabolic accidents' [despite the fact that the trace amines are phylogenetically older than the catecholamines (CA's)], or co- and/or false transmitters; furthermore there has been a tendency to dismiss them precisely because of their tiny concentrations. Only recently have the trace amines,and PE and T in particular, been shown to possess turnover rates that are so fast that in unit time quantities similar to those existing for DA, NA and 5-HT are synthesized (Wu and Boulton, 1973, 1974, 1975; Durden and Philips, 1980).

Although it is now possible to quantitate the trace amines in relatively small amounts of tissue, it will be necessary to develop more refined techniques in order to map their detailed anatomical distributions in the brain, locate trace aminergic pathways and identify and quantitate their *in vivo* release. To this end recent developments in negative ion chemical ionisation (Wood, 1982) and the use of metastable ions in the mass spectrometer (Durden, 1982, 1983) utilizing newly developed derivatives (Davis, 1979) are promising. For routine analyses in larger tissue samples or body fluids the use of gas chromatography and high pressure liquid chromatography seems destined to be adopted (see Baker *et al.*, Young, Bailey *et al.*, this book). The pharmacological investigation of the trace amines and acids has revealed that whilst they possess some properties in common with the CA's and 5-HT they also may be distinguished from them. In uptake and release studies for example (see Dyck, this book) it is now possible to release selectively the TA's in the presence of DA by using stimulant drugs and aminotetralins. Other differential effects on concentration and turnover (of TA's and DA) are found in the presence of stimulants, neuroleptics, DA agonists and antagonists, CA precursors and metabolic inhibitors and stress; in fact a reciprocal relationship between *p*-TA and *m*-TA and DA turnover has now been established (see Juorio, this book). Using the predictive properties of this reciprocal relationship along with the administration of suitable precursors and decarboxylase inhibitors it has proved possible to conclude that although hydroxylation and dehydroxylation reactions occur in the synthesis of the TA's, their principal biosynthesis is by decarboxylation of *p*- and *m*-tyrosine. These studies have, in addition, indicated that there may be more than one aromatic amino acid decarboxylase, that maybe the TA's are involved in the control of DA synthesis and that *m*-tyrosine must be a normal tissue constituent (Juorio and Boulton, 1982; Boulton and Juorio, 1983; Juorio, 1983).

Although Trendelenburg (1972) in his benchmark review pointed out that terms such as 'directly' and 'indirectly' acting as applied to the trace amines were inappropriate, it was not until the late 70's that iontophoretic studies indicated that at least some of the trace amines possessed inhibitory neurotransmitter properties. Tryptamine, for example, is a more potent inhibitory neurotransmitter than 5-HT and both T and OA have now virtually been established as neurotransmitters in their own right (see Jones, Hicks this book). Although the TA's and PE are inhibitory when iontophoresed in conventional amounts (Henwood *et al.*, 1979) they are less potent than DA and NA. When they are iontophoresed along with DA and NA, however, but in amounts insufficient to cause post-synaptic effects, they vastly potentiate the effects of the CA's, thus lending credence to the suggestion that the trace amines may be neuromodulators (Boulton, 1976, 1979b, 1980). The recent claim that there are receptors for PE and T (Kellar and Cascio, 1982; Hauger *et al.*, 1982 and Kienzl *et al.*, this book) is further evidence of a synaptic role and probably heralds a flurry of activity in this field.

As will be seen from the session on the behavioural aspects of the trace amines (Jackson; Greenshaw; Diamond; Dourish, this book) this is probably now the most active field in trace amine research. To date most attention has been paid to PE, really because of its structural similarity to amphetamine and the suggestion that it is involved in the aetiology of schizophrenia, depression, agoraphobia and aggression (see also Yu *et al.*; Sandler *et al.*; Murphy *et al.*; Reynolds and Bieck, this book). It is a fact, however, that many of the trace amines (PE, T, *m*-TA, *p*-TA, OA) are behaviourally active and likely to be involved in at least some aspects of spontaneous behaviour, stereotypy, locomotor activity, motivated behaviour, motor control, sexual behaviour, eating and drinking, self stimulation and reward, sleep, aggression, etc. A few months ago (Boulton, 1982), with the help of Dr. Dourish I was able to indicate which monoamines seemed to be involved in which particular aspects of motivated behaviour and motor control. The question marks in Table 1 reveal just how much is lacking in our knowledge of the role of the trace amines in behaviour.

Table 1

The Possible Role of Various Biogenic Amines in Motivated Behaviour and Motor Control

Amine	Locomotor Activity	Stereotypy	Feeding and Drinking	Sexual Behaviour	Aggression	Sleep	Reward
DA	+	+	±	+	+	−	+
NA	+	−	+	−	±	−	+
5-HT	−	+	+	−	−	+	−
PE	+	+	−	+	+	?	+
p-TA	±	+	?	?	?	?	−
m-TA	+	+	?	?	?	?	?
T	−	+	?	?	?	+	?

− Evidence for an inhibitory role; + evidence for an excitatory role; ± evidence for an excitatory or inhibitory role; ? unknown at present.

Table 2
The Role of Some Biogenic Amines in the Mediation of Various Components of Spontaneous Locomotor Activity and Stereotypy

Behavioural Components	DA	NA	5-HT	PE	p-TA	m-TA	T
Forward Walking	+	+	−	+	?	+	−
Rearing	+	−	−	+	?	?	?
Sniffing	+	+	0	+	?	?	?
Reactivity	+	+	0	+	?	+	?
Headbobbing	+	0	0	+	?	?	?
Headweaving	+	0	+	+	?	+	+
Forepaw padding	+	0	+	+	?	+	+
Grooming and licking	+	±	−	+	?	?	?
Abortive grooming	+	?	−	+	?	?	?
Backward walking/circling	+	−	+	+	?	+	?
Wet dog shakes	−	0	+	0	?	+	?
Head twitch	0	0	+	0	+	?	−

+ Evidence for an excitatory role; − evidence for an inhibitory role; ± evidence for an excitatory or inhibitory role; 0 available evidence suggests no involvement; ? unkown at present.

In a similar manner the involvement (when known) of the various biogenic amines in components of spontaneous locomotor activity and stereotypy are shown in Table 2.

I entertain little doubt that by utilising peripheral and central injections in the presence and absence of suitable drugs, suitable paradigms and newly developed techniques and applications that it will soon be possible to fill in many of the gaps in Tables 1 and 2 and surely the time has come, with respect to the putative monoamine neurotransmitters, to challenge existing concepts and beliefs and our persisting but oddly simplistic view that the CA's and 5-HT can explain all things.

Just how the trace amines act is not yet known; certainly some of them, particularly T and OA, may be neurotransmitters in their own right. Because of their very fast turnover rates, however, their ability to potentiate the actions of conventional monoamine neurotransmitters when they are co-iontophoresed with them and because of

Table 3

Suggested definition for a substance involved in the process of neurotransmission

NEUROHUMOR	- source, site, type, mode.
Source	= neuronal, glial, cytosol, granules, etc.
Site of action	= presynaptic, postsynaptic, blood vessel, muscle, etc.
Type of action	= synaptic, nonsynaptic, etc.
mode of action	= primary (similar to acetylcholine in peripheral junctions eliciting propagation of the impulse), secondary (eliciting subthreshold response), direct (action caused by the transmitter itself), indirect (action mediated by a different substance, etc.)

their well known ability to release CA's, I have proposed that they may act as neuromodulators or synaptic activators (Boulton, 1976, 1979, 1980) which I define as shown in Table 3 following Dismuke's (1979) concepts of neurohumors. A trace amine could thus be defined as: -
Neurohumor - cytosolic, postsynaptic, secondary (in)direct.

In particular I envisage that hyperpolarisation or depolarisation of the synaptic membrane occurs in such a way as to not create significant post-synaptic effects. The activation would be spatially and temporally discrete but continuous and it is suggested that it be a requirement for conventional neurotransmission [that is the activators (or neuromodulators) maintain the synapse in a state of readiness]. Anything that upsets this synaptic homeostasis would lead to aberrant signal propagation and the processes subserving mentation. I further envisage that the activation or homeostasis could occur by the trace amine acting alone (i.e. the left side of Fig. 1) or by involving (releasing) another neurotransmitter [probably one of those monoamines present in larger quantities (Fig. 1 right)].

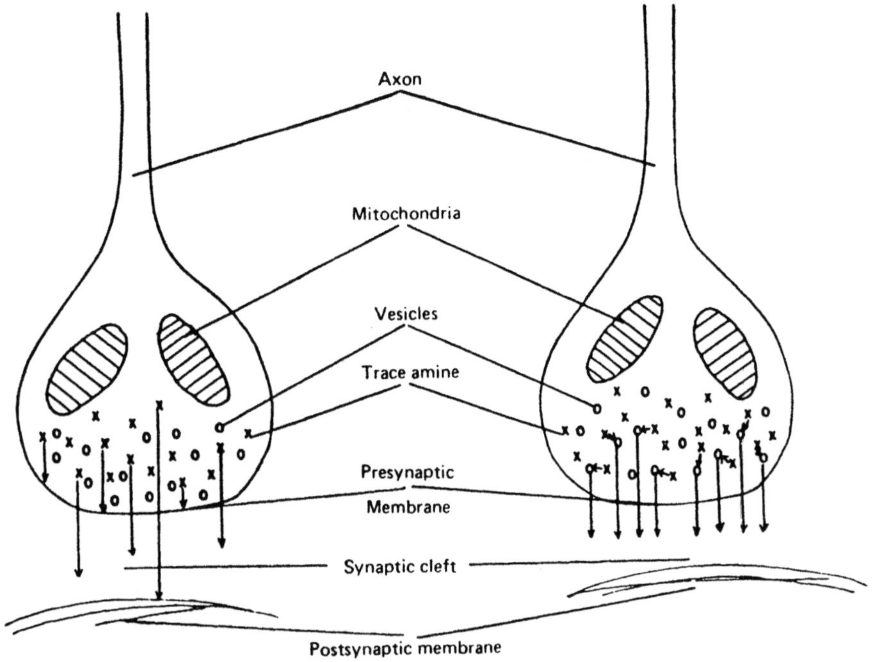

Fig. 1. Possible synaptic mechanisms of action of the trace amines.

Several of the trace amines and/or their metabolites and/or associated enzymes have been implicated in schizophrenia, depression, aggression, migraine, Parkinsonism, hypertension, hyperkinesis, epilepsy, hepatic coma and phenylketonuria. The evidence to support these involvements (both bioclinical and pharmacological) in schizophrenia, depression, agoraphobia and aggression will be discussed by Yu, Sandler, Murphy, Reynolds and Bieck during this meeting.

For the future it seems to me that developments will take place in several areas. Clearly in order to map trace aminergic pathways and identify nuclei, refinements and improvements in quantitation and localisation will be required; such developed methods can then be applied to an examination of the in vivo release of the trace amines and thus a direct appraisal of whether or not they are co- and/or false transmitters. In neuro-physiological/pharmacological terms the way in which the trace amines modulate the actions of other putative neurotransmitters (peptides and amino acids as well as amines) must be investigated as well as their specificity and mechanisms of action. Searches for, and investigation of, the properties of trace amine receptors will also continue. In metabolic terms a more detailed understanding of the synthesis and catabolism of the trace amines, especially the contribution from the CNS, is required, as well as investigation of the specificity and properties of the synthetic and catabolic enzymes.

Another most important area for study and one possessing possible pharmaceutical industrial implications concerns the effects of deuterium substitution; already it has been shown that α,α-dideutero substitution in the alkyl side chain of the trace amines reduces their oxidative deamination by monoamine oxidase (Yu et al., 1981, 1982) as well as vastly potentiating their behavioural effects (Dyck et al., 1983; Dourish et al., 1983a,b). Already it is possible to manipulate differentially the levels of the trace amines by judicious selection of enzyme inhibitors, precursors, neuroleptic and stimulant drugs, etc. If particular trace amines are involved in the regulation of specific aspects of behaviour or in the aetiology of certain mental disorders, the presence of strategically placed deuterium atoms either in the amine, its precursor, or a drug could further increase or decrease these levels. Now that some of the trace amines have been implicated in some mental disorders it becomes necessary to undertake more carefully designed studies, in carefully selected patients, in order to establish firmly that they are indeed involved and in what way. Even this partial list of future studies leaves me with a sense of optimism and excitement and it is with this feeling that I look forward to our conference and the future.

ACKNOWLEDGEMENT

I thank my colleagues, past and present, who have contributed so substantially to the work of the Research Division in Saskatoon and I thank Saskatchewan Health and the Medical Research Council of Canada for continuing financial support.

REFERENCES

Boulton A.A. (1971) Amines and theories in psychiatry. Lancet 2, p.7871.

Boulton A.A. (1976) Cerebral aryl alkyl aminergic mechanisms, in Trace Amines and the Brain (E. Usdin and M. Sandler, eds) pp 22-39, M. Dekker Inc., New York.

Boulton A.A. (1978a) Cerebral trace amines: their function and relevance, in Neuropsychopharmacology 1, (P. Deniker, C. Radouco-Thomas and A. Villeneuve, eds) pp 1405-1412, Pergamon Press, New York.

Boulton A.A. (1978b) The tyramines: functionally significant amines or metabolic accidents? Life Sci. 23, 659-672.

Boulton A.A. (1979a) Trace amines in the central nervous system, in International Rev. Biochem. 26 (K. Tipton ed.) pp 179-206, University Park Press, Baltimore.

Boulton A.A. (1979b) The trace amines: Neurohumors (cytosolic, pre- and/or post-synaptic, secondary, indirect?). Behav. Brain Sci. 2, p. 418.

Boulton A.A. (1980) The properties and potential function of some brain trace amines in neurochemistry and clinical neurology, in Progress in Clinical Biological Research (L. Battistin, G. Hashim and A. Lajtha, eds), pp 291-303, Allan R. Liss Inc., New York.

Boulton A.A. (1982) Some aspects of basic psychopharmacology: the trace amines. Prog. Neuro-psychopharmacol. Biol. Psychiat. 6, 563-570.

Boulton A.A. and Juorio A.V. (1982) Brain trace amines, in Handbook of Neurochemistry (A. Lajtha ed) 1, 189-222, Plenum Press, New York.

Boulton A.A. and Juorio A.V. (1983) Cerebral decarboxylation of meta- and para-tyrosine. Experientia 39, 130-134.

Davis B.A. (1979) Some new fluorescent derivatives for the mass spectrometric quantitation of biogenic amines. Biomed. Mass Spec. 6, 149-156.

Davis B.A. and Boulton A.A. (1981) Longitudinal urinary excretion of some 'trace acids' in a human male. J. Chromatog. Biomed. Applns. 222, 161-169.

Dismukes R.K. (1979) New concepts of molecular communication among neurones. Behav. Brain Sci. 2, 409-448.

Dourish C.T., Dewar K.M., Dyck L.E. and Boulton A.A. (1983a) Potentiation of the behavioural effects of the antidepressant phenelzine by deuterium substitution. Psychopharmacol. (in press).

Dourish C.T., Greenshaw A.J. and Boulton A.A. (1983b) Deuterium substitution enhances the effects of β-phenylethylamine on spontaneous motor activity in the rat. Pharmacol. Biochem. Behav. (in press).

Durden D.A. (1982) Selected metastable peak monitoring in mass spectrometry with stable isotope labelled internal standards and linked magnetic and electric sector fields. Anal. Chem. 54, 666-670.

Durden D.A. (1983) Determination of meta- and para-hydroxyphenylacetic acid levels in single caudate nuclei by selected metastable peak monitoring: a new sensitive gas chromatographic-mass spectrometric procedure. J. Neurosci. Meth. 7, 61-66.

Durden D.A. and Boulton A.A. (1982) Mass spectrometric analysis of some neurotransmitters and their precursors and metabolites, in Handbook of Neurochemistry (A. Lajtha, ed) 2, 103-132.

Durden D.A. and Philips S.R. (1980) Kinetic measurements of the turnover rates of phenylethyalmine and tryptamine in vivo in rat brain. J. Neurochem. 34, 1725-1732.

Dyck L.E., Durden D.A., Yu P.H., Davis B.A. and Boulton A.A. (1983) Potentiation of the biochemical effects of β-phenylethylhydrazine by deuterium substitution. Biochem. Pharmacol. 32, 1519-1522.

Hauger R.L., Skolnick P. and Paul S.M. (1982) Specific [^{3}H]-β-phenylethylamine binding sites in rat brain. Eur. J. Pharm. 83, 147-148.

Henwood R.W., Boulton A.A. and Phillis J.W. (1979) Iontophoretic studies of some trace amines in the mammalian CNS. Brain Res. 164, 347-351.

Juorio A.V. (1983) The effect of some decarboxylase inhibitors on mouse striatal tyramines. Neuropharmacol. 22, 71-73.

Juorio A.V. and Boulton A.A. (1982) The effect of some amino acid precursors and enzyme inhibitors on the mouse striatal concentration of tyramines and HVA. J. Neurochem. 39, 859-863.

Kellar K.J. and Cascio C.S. (1982) [^{3}H]-Tryptamine: high affinity binding sites in rat brain. Eur. J. Pharm. 78, 475-478.

Philips S.R. and Boulton A.A. (1979) The effects of monoamine oxidase inhibitors on some arylalkylamines in rat striatum. J. Neurochem. 33, 159-167.

Schmitt R. and Nasse O. (1865) Beitrag zur Keuntnifs des tyrosins. Liebigs Ann. Chem. 133, 211-216.

Trendelenburg U. (1972) Classification of sympathomimetic amines. Handbook Exp. Pharmacol. 33, 336-362.

Usdin E. and Sandler M. (eds) (1976) Trace Amines and the Brain, Marcel Dekker, New York.

Wood P. (1982) A selected ion monitoring assay for dopamine and its metabolites using negative chemical ionization. Biomed. Mass Spec. 9, 302-306.

Wu P.H. and Boulton A.A. (1973) Distribution and metabolism of tryptamine in rat brain. Can. J. Biochem. 51, 1104-1112.

Wu P.H. and Boulton A.A. (1974) Distribution, metabolism, and disappearance of intraventricularly injected p-tyramine in the rat. Can. J. Biochem. 52, 374-381.

Wu P.H. and Boulton A.A. (1975) Metabolism, distribution and disappearance of injected β-phenylethylamine in the rat. Can. J. Biochem. 53, 42-50.

Yu P.H., Barclay S., Davis B.A. and Boulton A.A. (1981) Deuterium isotope effects on the enzymatic oxidative deamination of trace amines. Biochem. Pharmacol. 30, 3089-3094.

Yu P.H., Kazakoff C.W., Davis B.A. and Boulton A.A. (1982) Deuterium isotope effect on the enzymatic oxidation of dopamine and serotonin. Biochem. Pharmacol. 31, 3697-3698.

B. ANALYSIS

QUANTIFICATION OF TRACE AMINES AND THEIR METABOLITES BY HIGH RESOLUTION OR METASTABLE ANALYSIS USING DOUBLE FOCUSSING MASS SPECTROMETRY

David A. Durden

Psychiatric Research Division
Cancer and Medical Research Building
University of Saskatchewan
Saskatoon, Saskatchewan S7N OWO

INTRODUCTION

During the past 15 years, many methods have been developed for the detection of the trace amines and their metabolites. The trace amines that have been studied include phenylethylamine (PE), *meta*- and *para*-tyramine (*m*-TA and *p*-TA), tryptamine (T), phenylethanolamine (PEOH), *meta*- and *para*-octopamine (*m*-OA and *p*-OA) and *meta*- and *para*-synephrine (*m*-Syn and *p*-Syn). The trace metabolites for which methods have been developed are phenylacetic acid (PAA), *meta*- and *para*-hydroxyphenylacetic acids (*m*-HPAA and *p*-HPAA), *meta*- and *para*-hydroxymandelic acids (*m*-HMA and *p*-HMA), and *para*-hydroxyphenylethanol (*p*-HPE). Although attempts have been made to extend chemical and bioassay procedures which were successful for determinations of catecholamines, of these procedures, spectrofluorimetric or radioenzymatic assays were of mixed success and appear to have fallen into disuse. By far the greater number, and apparently universally successful procedures, involve mass spectrometry.

Mass spectrometry can be used in conjunction with chromatographic separation to provide very specific and sensitive methods for the analysis of trace amines and their metabolites. It may be used in its low or high resolution mode of operation. With low resolution mass spectrometry (LRMS) one can detect ions characteristic of the compound and measure their nominal mass to charge ratios (m/z) and relative intensities which are used in conjunction with chromatographic data to identify compounds. High resolution mass spectrometry (HRMS) provides precise m/z values of the ions and hence their elemental compositions, which can be used to verify the compound identity. The chromatographic methods used have been either thin layer chromatography (TLC) on silica gel thin layers or gas chromatography (GC) using packed or high resolution capillary columns. In this laboratory we have determined trace amines as their 1-dimethylaminonaptha-lene-5-sulphonyl (dansyl) derivatives separated by TLC and identified and quantified by HRMS with a double focussing mass spectrometer using deuterium labelled internal standards (Durden *et al*, 1973; Boulton *et al*, 1973; Philips, *et al*, 1974, 1974a, 1975; Durden, 1978; Durden and Boulton, 1979; Durden *et al* 1978, 1980; Durden and Philips, 1980) and trace acids as their perfluorinated acyls, perfluorinated esters using high resolution capillary gas chromatography (HRGC) and the double focussing mass spectrometer operated in high resolution or selected metastable peak monitoring mode again with deuterated internal standards (Durden and Boulton, 1981, 1982; Davis and Boulton, 1982; Davis *et al*, 1982; Durden 1982, 1983).

AMINES

Methods

For the analysis of PE, *m*-TA, *p*-TA and T, the following procedure may be used. Tissue (up to 200 mg) is homogenized with a "Polytron" homogenizer in 2 ml 0.05 N $HClO_4$ in 16x100 mm glass culture test tubes and appropriate amounts (usually 25 to 50 ng) of the deuterated internal standards are added. A mixture of Na_2CO_3 and $NaHCO_3$ (1:1 w/w), approximately 300 mg, is added to the homogenate. This usually dissolves completely. Two ml of acetone containing 20 mg dansyl chloride are then added with mixing (Vortex mixer). Some of the carbonate will precipitate in the presence of acetone. The tubes are sealed and left overnight in the dark at room temperature to permit the reaction to go to completion and excess dansyl chloride to hydrolyse. Next day, the tubes are centrifuged at low speed (1600 rpm) in a clinical centrifuge and the supernatant transferred to a second 16x100 mm test tube. The supernatant is extracted twice with 2 ml hexane and the organic layer saved in test tubes. The tubes are then placed in a water bath at 40-45°C and the organic layer evaporated in a stream of nitrogen. The dansyl derivatives are dissolved in a few drops (100-200 μl) of the mixture, acetone-toluene (1:1 v/v) and applied to the silica gel layers (Merck silica gel 60, 20x20 cm, 0.25 mm thick, No. 5763, Brinkmann Instruments, Rexdale, Ontario, Canada) in a 2 cm streak. Six samples plus a concentrated dansyl standard may be accommodated on one plate. After development in the first solvent system (toluene-ethyl acetate, 5:2) the zones are located under long-wave UV light, and scraped into test tubes and extracted twice with acetone-methanol (1:1 v/v) and the samples transferred to new tubes in which the organic layer is again removed under a nitrogen stream and the dansyl derivative redissolved in acetone-toluene. The samples are then spotted on the second TL plate and developed with a second solvent system [PE, toluene-triethylamine, 5:1 (v/v); m-TA, p-TA, toluene-triethylamine-methanol, 50:5:1 (v/v/v); T, toluene-triethylamine-methanol, 50:5:7 (v/v/v)]. The zones are again visualized under UV and circumscribed with a metal stylus. The silica gel is carefully powdered in situ and then sucked into one half of a melting point capillary which is constricted mid-length to hold a small piece (1 mm x 5 mm) of glass fibre filter paper (Whatman GFA). The dansyl amine is then eluted from the silica by drawing 25 ul of ethyl acetate through the silica and the capillary is sealed with hematocrit clay so that there is an air gap adjacent to the ethyl acetate (Philips et al, 1974). Samples are stored in a freezer (-20°C) until analysed by mass spectrometry.

The beta-hydroxyphenylalkylamines, OA and Syn may be determined using the mixed dansyl-acetyl derivative (Durden et al,. 1978,1980). After extraction and removal of the solvent, the dansyl derivatives are acetylated at the beta-hydroxy group in ethyl acetate-pyridine-acetic anhydride (10:1:1 v/v/v) for 1 hr at room temperature in the dark and then separated unidimensionally in two or three TLC systems (Durden et al, 1980).

All organic solvents are HPLC grade (Caledon Laboratories, Georgetown, Ontario, Canada) or distilled from glass in the laboratory. Water is purified by reverse osmosis and then distilled in glass. Deuterium labelled internal standards are either synthesised in the laboratory or purchased from Merck, Sharpe and Dohme Canada Ltd.

The amines are quantified by observing the evaporation profile of the molecular ions of the dansyl amine and the dansyl amine deuterium labelled standard as they evaporate from the direct insertion probe (at temperatures ranging from 250°C to 310°C). The mass spectrometer is operated at high resolution; 10,000 for the mono-dansyl derivatives of PE and T and 7,000 for *bis*- and *tris*-derivatives. The peak profiles are automatically integrated by a minicomputer (Durden, 1978).

Discussion

When this method was developed in 1972 it became apparent that it would be necessary to derivatize the trace amines as the chemistry of free base amines is not very suitable for mass spectrometric analysis at the nanogram level. Several derivatives were considered. Unfortunately the derivatives suitable for GC analysis, e.g. perfluoroacyls, do not give very intense molecular ions and we felt that the base peak ions were not specific and too low in mass to permit unequivocal identification at this level. The dansyl derivatives on the other hand exhibited intense molecular ions (especially for PE and T) and were easily separated by an existing TLC technology. Later, we observed that for the *beta*-hydroxyamines, OA and Syn, dansylation alone was not sufficient as the alcoholic group was not derivatized and appeared to cause decomposition during MS analysis and that acetylation of the group to make the mixed dansyl-acetyl derivative improved the MS sensitivity by a factor of 50 (Durden *et al*, 1980). Separation of the *m*- and *p*-isomers, however, is more difficult but can be undertaken as the dansyl derivative before the acetylation step. PEOH can be analysed as either the dansyl (Durden, 1978) or mixed dansyl-acetyl derivative as there is little difference in sensitivity with either derivative. It is important, however, that dansyl PEOH is isolated from mono-dansyl tyramines as their molecular ions are identical.

It was also evident that a suitable internal standard would be essential. Homologs of the amines are not suitable for TLC analysis as their dansyl derivatives separate from the amine of interest, whereas stable isotope labelled analogs do not. For the early analysis of PE, p-TA and T (Durden et al., 1973; Philips et al., 1974, 1974a), 1,1-dideutero analogs were used. We have since found that the 1,1,2,2-tetra-deutero analogs give greater sensitivity with lower blank values, and greater dynamic range due to the reduced signal of the ions at m/z values greater than the most abundant of the unlabelled dansyl compound caused by the less abundant isotopes (C^{13}, N^{15}, O^{18} and S^{34}).

The use of high resolution mass analysis gives an additional advantage over low resolution mass spectrometry. In addition to providing a precise m/z value and hence elemental composition of the ion, high resolution separates ions of the same nominal m/z but different elemental compositions thus reducing blank values.

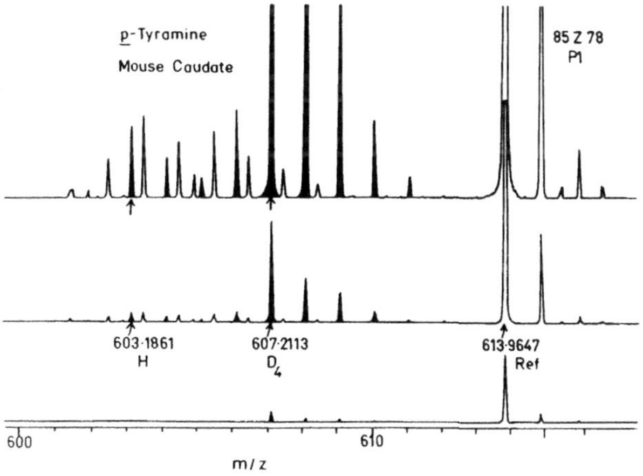

Figure 1. Dansyl-p-TA extract from a mouse caudate nucleus at 7,000 resolution. Ions in black are due to bis dansyl tyramines; m/z 603.1861 is dansyl-p-TA, and 607.2113 dansyl-p-TA-d_4. The mass reference ion $C_{12}F_{24}N$ of perfluoro-tri-n-butylamine is 613.9647.

This is demonstrated in Fig. 1 which shows the p-TA extract from a mouse caudate nucleus. Note that adjacent to m/z 603.1861, the molecular ion of dansyl-p-TA, there is another peak of similar intensity at about m/z 603.4. Under low resolution conditions, the two peaks would be detected as one as m/z 603 and the value for p-TA would be considerably larger. Results for m-TA and p-TA for three rat brain tissues obtained at low and high resolution settings are shown in Table 1. The blanks mean values are shown for comparison. These mean values were subtracted from the tissue measured values to obtain the corrected values. In this experiment, low resolution mass analysis gave blanks of several nanograms which in the case of m-TA were larger than the tissue measured values and thus negative values would be calculated. High resolution gave much lower and more consistent blank values, and hence greater sensitivity.

We have found that for analysis of the dansyl amines following TLC separation, a resolution of at least 5,000 is necessary to obtain the desired sensitivity and specificity.

ACIDS

Method

The tissue is homogenized in 2 ml 0.1 M $ZnSO_4$ and the deuterated standards PAA-d_2, m-HPAA-d_2 and p-HPAA-d_2 (100 ng each) are added and thoroughly mixed. 2 ml 0.1 M Ba(OH_2) is added to precipitate the protein which is removed by centrifugation (10,000 g for 15 min). The supernatant is acidifed to pH 1 with HCl and the acids extracted into ethyl acetate (2x4 ml). The volume of the ethyl acetate is reduced to 100-200 μl and 50 μl triethylamine is added. The solution is transferred to conical reaction vials and taken to dryness. Benzene (50 μl) is added to remove any traces of water azeotropically as the samples are dried again. The sample is reacted with 50 μl pentafluoropropionic anhydride (PFPA) and 100 μl of 1,1,1-trifluoroethanol (TFEOH) for 1 hr at 70°C. The volume is then reduced to about 20 μl under a stream of nitrogen and a second amount of PFPA (100 μl) added with 150 μl hexane and the vials reheated for 1 hr at 70°C. The volume is

Table 1

The Effect of Mass Spectrometer Resolution on the Analysis of _meta_- and _para_-Tyramine as Their Dansyl Derivatives

Sample	Mass Spectrometer Resolution: 7,000		1,000	
	Amount of amine in sample			
_m-TA_	Measured	Less blank	Measured	Less blank
Blanks	0.14±0.01 (3)		4.20±1.0 (3)	
Tissue 1	2.66	2.52	4.29	0.09
Tissue 2	2.13	1.99	3.82	-0.38
Tissue 3	1.12	0.98	3.16	-1.04
_p-TA_				
Blanks	0.25±0.12 (3)		2.60±0.23 (3)	
Tissue 1	2.93	2.68	4.62	2.02
Tissue 2	3.42	3.17	5.45	2.85
Tissue 3	1.79	1.54	3.14	0.54

Values are in nanograms ± S.E.M. and 'n' for Blanks. Internal standards were 44 ng _m_-TA-d_4 and 50 ng _p_-TA-d_4.

then reduced to 20 μl hexane (200 μl) is added and shaken for 30 s with 100 μl phosphate buffer (pH 6). The phosphate buffer layer is removed and a second 100 μl added and the vial shaken. The hexane layer is removed into a 200 μl vial and reduced to 20 μl in volume. Samples 0.5 to 1.0 μl are injected into the GC-MS.

GC separation of PAA-TFE derivative and _m_-HPAA and _p_-HPAA-PFP-TFE derivatives occurs on an SGE OV-101 SCOT capillary column (Mandel Scientific, Rockwood, Ontario, Canada) which is connected directly to the ion source of a VG-70-70F double focussing mass spectrometer with a short length of 0.25 mm id silica capillary. The column is operated isothermally at 170°C with helium carrier gas flow at 29 cm/sec.

For the HRMS analysis the mass spectrometer is operated at 5,000 resolution and is adjusted in selected ion monitoring (SIM) mode to detect the molecular ion of PAA and PAA-d_2 at m/z 218.0555 and 220.0680 and the fragment ions of _m_-HPAA, _p_-HPAA, _m_-HPAA-d_2 and _p_-HPAA-d_2 at m/z 253.0288 and 255.0413 (Durden and Boulton 1982; Davis _et al_, 1982).

To attain greater sensitivity for the analysis of _m_-HPAA and _p_-HPAA, the mass spectrometer is operated in the selected metastable peak monitoring (SMPM) mode (Durden, 1982, 1983). The PFP-TFE derivatives and HRGC conditions above are used. The amount of internal standard, however, may be reduced to 1 or 2 ng (Durden, 1983). The mass spectrometer is adjusted, at low resolution, with linked magnetic (B) and electric sector (E) fields and constant accelerating field (V), to detect the "metastable" decompositions of the molecular ion m/z 380 to the base peak m/z 253 (and 382 to 255 of the d_2 standards) which occur in the first field free region of the mass spectrometer. This procedure has enabled quantitation of _m_-HPAA and _p_-HPAA in a single rat caudate nucleus (Durden, 1983).

DISCUSSION

Several derivatives are suitable for making the trace acids amenable to GC-MS analysis. The phenolic group may be acylated conveniently as the trifluoroacetate (TFA), pentafluoropropionate (PFP), or heptafluorobutyrate (HFB), and the carboxyl group esterified as the methyl (Me), propyl (Pr), butyl (Bu), isobutyl (i-but), trifluoroethyl (TFE), pentafluoro-n-propyl (PF-n-P), hexafluoroisopropyl (HFIP) or pentafluorobenzoyl (PFB) esters. Our early procedures for the analysis of m-HPAA and p-HPAA (Durden and Boulton, 1981) and m-HMA (Davis and Boulton, 1981) used the HFB-Me derivative. The Me ester of PAA is not very suitable for GC-HRMS analysis due to its low mass and lack of a hetero atom (such as fluorine) which would permit high resolution separation of its ions from ions due to other endogenous background material. Consequently, since we wished to determine PAA and m- and p-HPAA in the same sample, we now use PFP acyls and TFE or PF-n-P ester derivatives (Durden and Boulton, 1982; Davis et al., 1982) in conjunction with OV-101 or OV-17 SCOT capillary columns which provide more than adequate resolution. The PFB ester, although suitable for analysis of PAA (Fellows et al., 1978) increases unacceptably the GC retention time of m-HPAA and p-HPAA and the catecholic acids.

Since fluorine is mono-isotopic, the signal from ions of two mass units greater than the most abundant, due

Table 2

Determination of Phenylacetic Acid in a Pooled CSF Sample by High and Low Resolution GC-MS

Method	MS Resolution	GC Column	PAA Concentration
HRGC-HRMS	5,000	Capillary	69.6±5.2 (8)
HRGC-LRMS	1,000	Capillary	66.6±1.8 (5)
LRGC-HRMS	5,000	Packed	115.0±4.8 (3)*
LRGC-LRMS	1,000	Packed	101.2±1.8 (4)*

Concentrations are in ng/ml ± S.D. (n)
Capillary Column - OV101 SCOT, 50 m x 0.5 mm i.d.
Packed Column - 1.5% SP2250/1.95% SP2401 on Supelcoport 100/120. 1.8 m x 2 mm i.d.
*These values are different from the HRGC-HRMS method at $p < .001$.
Derivative - TFE detected at masses 218.0555 and 220.0680.

to the natural abundance of ^{13}C, ^{17}O and ^{18}O are of low intensity and hence the di-deutero acids internal standards are acceptable. We have used higher numbers of deuterium labels, but find that they are chromatographed at somewhat different retention times from the proteo compound (Durden and Boulton, 1982). The deuterium labelled analogs elute before the proteo compounds with the difference in retention time increasing with the number of deuterium atoms.

We have also examined the use of high and low resolution mass spectrometry and compared capillary columns versus packed columns. Some results obtained for PAA and p-HPAA in human csf under various GC-MS conditions are shown in Tables 2 and 3. It can be seen that for analysis of PAA, the use of a capillary column gave lower values but that increasing resolution only affected the standard deviation. On the packed column PAA is not fully resolved from other material which gives a signal at m/z 218. In the analysis of p-HPAA using the PFP-TFE derivative, the more important factor appears to be the mass spectrometer resolution. A similar trend was observed for m-HPAA. However, it is not fully resolved from p-HPAA using the packed column and because of the low levels in CSF

Table 3

Determination of *para*-Hydroxyphenylacetic Acid in a Pooled CSF Sample by High and Low Resolution GC-MS

Method	MS Resolution	GC Column	p-HPAA Concentration
HRGC-HRMS	5,000	Capillary	68.8±4.8 (8)
HRGC-LRMS	1,000	Capillary	86.6±1.4 (5)*
LRGC-HRMS	5,000	Packed	69.4±1.6 (4)
LRGC-LRMS	1,000	Packed	81.0±0.9 (4)*

Concentrations are in ng/ml ± S.D. (n)
Capillary Column - OV101 SCOT, 50 m x 0.5 mm i.d.
Packed Column - 1.5% SP2250/1.95% SP2401 on Supelcoport 100/120, 1.8 m x 2 mm i.d.
*These values are different from the HRGC-HRMS method at $p < .001$.
Derivative - PFP-TFE at m/z 253.0288 and 255.0413.

reliable results were not obtainable. Thus we may conclude that the combination of high resolution mass spectrometry with high resolution gas chromatography (HRMS-HRGC) can be expected to give consistently reliable results when one is determining a variety of compounds.

The effect of analysis using SMPM is demonstrated in Table 4. Rat caudate nuclei were determined by the two methods of normal HRGC-HRMS-SIM as above and HRGC-SMPM with different amounts of tissue. Essentially the same results were obtained from the three procedures, but because of the greater sensitivity of SMPM, it was possible to get the same standard error using one caudate nucleus and 1 or 2.5 ng internal standard as was obtained using HRMS-SIM of a pool of 10 caudate nuclei and 100 ng of internal standard. HRMS-SIM analysis of the single tissue gave a much larger (and unacceptable) standard error. SMPM appears to be a method of obtaining increased sensitivity for certain compounds. Unfortunatey it is not universally usable, as is normal HRMS-SIM as it is very dependent upon

Table 4

Concentrations of _m_- and _p_-Hydroxyphenylacetic Acids in Rat Caudate Nucleus Determined by SMPM or HRMS-SIM

Method	_m_-HPAA	_p_-HPAA
SMPM Single caudate Resolution 800 1.0 or 2.5 ng d_2 I.S.	9.2±1.5 (9)	43.0±3.7 (10)
HRMS-SIM Pooled 10 caudates Resolution 5,000 100 ng d_2 I.S.	9.4±1.2 (4)	43.0±3.9 (4)
HRMS-SIM Single caudate Resolution 5,000 1.0 or 2.5 ng I.S.	8.9±4.5 (9)	47.0±9.4 (10)

Values are ng/g, mean ± S.E.M. (number of determinations)

the physical chemistry of the compound. For example the metastable yield for the fragmentation m/z 362 to m/z 303 of the HFB-Me derivative was much lower for _m_-HPAA compared to _p_-HPAA, and PAA-TFE gave virtually no signal for the main fragmentation pathway m/z 218 to m/z 91 (Durden, 1982).

CONCLUSION

From the above, it can be seen that the present methodology of using TLC-HRMS analysis of the dansyl derivative of the trace amines and HRGC-HRMS analysis of the perfluoroacyl and ester derivatives of the trace acids enables detection of the compounds in subnanogram quantities. In order to detect picogram quantities of the compounds, advances in derivatization and mass spectrometry will be required. The choice of derivative may be used to concentrate the ionization of the molecule into a few significant and specific ions. Such an approach has been examined in the development of the di-n-butylaminonapthalene sulphonyl (bansyl) and related derivatives and a significant gain in sensitivity is apparent (Davis, 1979; Davis _et al_, 1980). Mass spectrometric procedures which increase the signal to noise ratio such as the SMPM method described above and the use of negative ion chemical ionization (Hunt and Crow, 1978; Lewy and Markey, 1978) with electron capturing derivatives appear to be new procedures which may provide the necessary sensitivity and specificity for picogram analysis.

ACKNOWLEDGEMENTS

I thank Dr. A.A. Boulton for his advice and encouragement and the Medical Research Council of Canada and Saskatchewan Health for continuing financial support.

REFERENCES

Boulton A.A., Philips S.R. and Durden D.A. (1973) The analysis of certain amines in tissues and body fluids as their dansyl derivatives. _J. Chromatog._ _82_, 137-142.

Davis B.A. (1979) Some new fluorescent derivatives for the mass spectrometric quantitation of biogenic amines. Biomed. Mass Spectrom. 6, 149-156.

Davis B.A. and Boulton A.A. (1981) Excretion of m-hydroxymandelic acid in human urine. J. Chromatog. Biomed. Appli. 222, 271-275.

Davis B.A., Durden D.A. and Boulton A.A. (1980) Recent strategies for increasing the sensitivity of detection and quantitation of some biogenic amines by direct probe high resolution mass spectrometry, in Recent Developments in Mass Spectrometry in Biochemistry and Medicine 6 (Frigerio A., ed), pp 83-93. Elsevier Scientific Pub. Co., Amsterdam.

Davis B.A., Durden D.A. and Boulton A.A. (1982) Plasma concentrations of p- and m-hydroxyphenylacetic acid and phenylacetic acid in humans. Gas chromatographic-high resolution mass spectrometric analysis. J. Chromatog. 230, 219-230.

Durden D.A. (1978) Analysis of amines by mass spectrometry: identification and quantitation of trace amines at the picomole level, in Research Methods in Neurochemistry Vol. 4 (Marks, N. and Rodnight, R., eds), pp 205-250. Plenum Press, New York.

Durden D.A. (1982) Selected metastable peak monitoring in mass spectrometry with stable-isotope-labeled internal standards and linked magnetic and electric sector fields. Anal. Chem. 54, 666-670.

Durden D.A. (1983) Determination of meta- and para-hydroxyphenylacetic acid levels in single caudate nuclei by selected metastable peak monitoring: a new sensitive gas chromatographic-mass spectrometric procedure. J. Neurosci. Meth. 7, 61-66.

Durden D.A. and Boulton A.A. (1979) Mass spectrometric analysis of metabolites with specific reference to biogenic amines, in: Techniques in the Life Sciences, Vol. 2, Techniques in Metabolic Research (Kornberg H.L., Metcalfe J.C., Northcote D.H., Pogson C.I. and Tipton K.F., eds), pp 1-25. Elsevier-North Holland, Amsterdam.

Durden D.A. and Boulton A.A. (1981) Identification and distribution of m- and p-hydroxyphenylacetic acid in the brain of the rat. J. Neurochem. 36, 129-135.

Durden D.A. and Boulton A.A. (1982) Identification and distribution of phenylacetic acid in the brain of the rat. J. Neurochem. 38, 1532-1536.

Durden D.A., Juorio A.V. and Davis B.A. (1978) Analysis of p-synephrine and related β-hydroxyphenylethylamines by direct probe high resolution mass spectrometry, in Quantitative Mass Spectrometry in Life Sciences II (de Leenheer A.P., Roncucci, R.R. and van Peteghem C., eds), pp. 389-397. Elsevier Scientific Publishing Co., Amsterdam.

Durden D.A., Juorio A.V. and Davis B.A. (1980) Thin layer chromatographic and high resolution mass spectrometric determination of β-hydroxyphenylethylamines in tissues as their dansyl-acetyl derivatives. Anal. Chem. 52, 1815-1820.

Durden D.A. and Philips S.R. (1980) Kinetic measurement of the turnover rates of phenylethylamine and tryptamine *in vivo* in the rat brain. J. Neurochem. 34, 1725-1732.

Durden D.A., Philips S.R. and Boulton A.A. (1973) Identification and distribution of β-phenylethylamine in the rat. Can. J. Biochem. 51, 995-1002.

Fellows L.E., King G.S., Pettit B.R., Goodwin B.L., Ruthven C.R.J. and Sandler M. (1978) Phenylacetic acid in human cerebrospinal fluid and plasma: selected ion monitoring assay. Biomed. Mass Spectrom. 5, 508-511.

Hunt D.F. and Crow F.W. (1978) Electron capture negative ion chemical ionization mass spectrometry. Anal. Chem. 50, 1781-1784.

Lewy A.J. and Markey S.P. (1978) Analysis of melatonin in human plasma by gas chromatography negative chemical ionization mass spectrometry. Science 201,741-743.

Philips S.R., Durden D.A. and Boulton A.A. (1974) Identification and distribution of p-tyramine in the rat. Can. J. Biochem. 52, 366-373.

Philips S.R., Durden D.A. and Boulton A.A. (1974a) Identification and distribution of tryptamine in the rat. Can. J. Biochem. 52, 447-451.

Philips S.R., Davis B.A. Durden D.A. and Boulton A.A. (1975) Identification and distribution of meta-tyramine in the rat. Can. J. Biochem. 53, 65-79.

THE USE OF ENZYMATIC RADIOISOTOPIC MICROASSAYS FOR THE QUANTIFICATION OF β-PHENYLETHYLAMINE, PHENYLETHANOLAMINE, TYRAMINE AND OCTOPAMINE

Juan M. Saavedra

Section on Pharmacology, LCS, NIMH,

Bethesda, Maryland 20205

ABSTRACT

Enzymatic radioisotopic microassays have been used to quantitate those biogenic amines present in mammalian brain in small amounts. Methods were developed for β-phenylethylamine, phenylethanolamine, octopamine and tyramine. The use of partially purified and group-specific N-methyl and O-methyltransferases, combined with selective procedures for the complete separation of the reaction products, including chromatographic steps, resulted in specific and sensitive microassays. Similar procedures are used for the determination of the activities of enzymes which synthesize or metabolize these amines.

The use of such methods allows the study of levels and metabolism of amines in discrete areas of mammalian brain, plasma, and in single cells from invertebrates, where these amines might play a role as neurotransmitters.

INTRODUCTION

Enzymatic-isotopic methods are based on the incubation of an extract of the biological material with specific methyltransferase enzymes, together with a donor of methyl groups, S-adenosyl-L-methionine (SAME). The methyltransferases catalyze, in the test tube as well as normally in tissues, the transfer of the methyl group from SAME to the O- or N-terminal positions of the amines, resulting in the formation of the corresponding O- and N-methyl derivatives. The use of a donor of radioactively-labelled methyl groups ([^{3}H]methyl-SAME), of high specific activity, and the utilization of selective organic extraction and chromatographic procedures to separate and quantitate the radioactive O- or N-derivatives formed, allowed the development of a number of specific and sensitive micromethods, most of them with a sensitivity at the low picogram level (Table 1). Such a sensitivity was necessary to accurately detect low levels of biogenic amines in brain. The radioenzymatic methods utilizing N- or O-methyltransferases are the methods of choice for the quantification of small amounts of N-

Table 1. Sensitivity of the enzymatic-isotopic methods for the determination of biogenic amines in tissues

Amine	Enzyme	Sensitivity (picograms)
N-acetylserotonin	HIOMT	50
Serotonin	NAT - HIOMT	10
Phenylethanolamine	PNMT	25
Octopamine	PNMT	25
β-Phenylethylamine	DBH - PNMT	200
Tyramine	DBH - PNMT	200
Histamine	HMT	5
Dopamine	COMT	5
Noradrenaline	COMT	2
Adrenaline	COMT	2

HIOMT = Hydroxyindole O-methyltransferase (EC 2.1.1.4)
NAT = N-acetyltransferase (EC 2.3.1.5)
PNMT = Phenylethanolamine N-methyltransferase (EC 2.1.1.28)
DBH = Dopamine-β-hydroxylase
HMT = Histamine N-methyltransferase (EC 2.1.1.8)
COMT = Catechol-O-methyltransferase (EC 2.1.1.7)

acetylserotonin, serotonin, dopamine, noradrenaline, adrenaline, histamine, β-phenylethylamine, phenylethanolamine, octopamine and tyramine.

GENERAL PROCEDURE

A preliminary, simple extraction procedure is necessary to separate the endogenous amine from contaminating materials present in the tissues. The procedures involve acid precipitation and centrifugation with subsequent analysis of the supernatant or organic extraction of the material with evaporation of the organic phase prior to the enzymatic assay.

The extracted material is incubated in the presence of a radioactively-labelled donor of methyl groups ([^{3}H]methyl-SAME) with high-specific activity and partially purified O- or N-methyltransferases. In some cases a double-enzyme incubation is performed to convert the amine to a derivative with high affinity for the methylation process.

The use of specific methyltransferase enzymes results in the selective methylation of relatively small groups of endogenous compounds. A very simple purification procedure

Table 2. General procedure for the enzymatic-isotopic assay of biogenic amines and related substances

1. Extract the amine from the tissue or body fluid (acid or buffer extraction).
2. Incubate with N- or O-methyltransferases and radioactive methyl donor ([^{3}H]methyl-SAME).
3. Extract the radioactive N- or O-methylated products formed in the reaction with specific organic solvents.
4. Eliminate the radioactive contaminants by selective drying and/or chromatographic procedures.
5. Count the radioactivity by liquid scintillation spectroscopy.

is usually necessary to prepare the methyltransferases. The use of partially purified enzymes is sufficient for selectivity of methylation and allows the storage of relatively stable enzyme preparation for months.

Specificity of the assays is dependent on the separation of the selected derivatives from other possible products as well as from the unreacted labelled SAME and its radioactively labelled metabolites. This procedure is usually carried out by a preliminary extraction of the labelled material into specific organic solvents, taking advantage of the favorable partition coefficient of the N- or O-methylated amine derivates. This is followed by specific chromatographic procedures and the drying of the organic phase, to eliminate radioactively-labelled volatile contaminants (Table 2). Internal standards are carried through the entire procedure to correct for losses of material. A linear relationship between the amount of amine present in the tissues and the number of dpm obtained is found in most assays from 5 to 2,000 picograms with an intraassay variability of less than 5%.

ENZYME PURIFICATION

A partial purification of the methylating enzymes is necessary to provide adequate preparations for their use in the enzymatic-isotopic assays. These procedures generally involve simple ammonium sulfate precipitation and dialysis steps, and result in the production of highly stable enzyme preparations (Table 3).

PURIFICATION OF DOPAMINE-β-HYDROXYLASE

Chromaffin granules are isolated from fresh bovine adrenals (Smith and Winkler, 1967). The granules are lysed by freeze-thawing which is followed by hand homogenization in a Teflon-glass homogenizer in 4 mM Tris-HCl buffer (pH 7.5). After centrifugation at 100,000 g the DBH in the supernatant is separated from contaminating proteins by means of affinity chromatography on a concanavalin A-sepharose column (Pharmacia Fine Chemicals, Piscataway, NY) (Molinoff et al., 1971; Tallman et al., 1976).

Table 3. Enzymes used in the detection of biogenic amines

Enzyme	Source	Purification procedure	Amine measured
Histamine-N-methyl-transferase (EC 2.1.1.8)	guinea pig brain	1-2	Histamine
Catechol-O-methyl-transferase (EC 2.1.1.7)	rat liver	1-2	Dopamine Noradrenaline Adrenaline
Dopamine-β-hydroxylase (DBH) (EC 1.14.2.1)	bovine adrenal medulla	1-2-3	β-Phenylethylamine Tyramine
Hydroxyindole-O-methyl-transferase (HIOMT) (EC 2.1.1.4)	bovine pineal	1-2	N-acetylserotonin Serotonin
N-acetyltransferase (NAT) (EC 2.3.1.5)	rat liver	1-2	Serotonin
Phenylethanolamine-N-methyltransferase (PNMT) (EC 2.1.1.28)	bovine adrenal medulla	1-2-4	Phenylethanolamine Octopamine Phenylethylamine Tyramine

Purification procedures: 1. Ammonium sulphate fractionation. 2. Dialysis. 3. Concanavalin-A sepharose column. 4. Acid precipitation.

PURIFICATION OF PHENYLETHANOLAMINE-N-METHYLTRANSFERASE

Fresh bovine adrenal medullas are kept at 4°C until processed (less than six hours). Tissues are homogenized with four volumes of 0.25 M sucrose, centrifuged at 6,000 x g for 30 min to remove cell debris, and filtered through gauge. The supernatant is centrifuged at 78,000 x g for one hour, and then filtered through gauge to remove lipids. The supernatant is precipitated with ammonium sulfate, and the 30-60% precipitate is resuspended in a small volume of 0.05 M, potassium-phosphate buffer, pH 7.4 and dialyzed against 200 volumes of the same buffer, overnight (Axelrod, 1962).

The dialyzed preparation is adjusted to pH 5.5 by a dropwise addition of 1 N acetic acid, and the preparation is centrifuged for 20 minutes at 10,000 x g. The supernatant is adjusted to pH 7.4 with 1 N NaOH, and dialyzed against 200 volumes of 0.05 M potassium phosphate buffer, pH 7.4, overnight. The final protein concentration is 8 to 10 mg/ml with an activity of 500 to 800 units/ml with octopamine as substrate (Saavedra et al., 1974).

RADIOENZYMATIC ASSAYS: GENERAL CONCEPT

Amines such as phenylethanolamine and octopamine can be measured after incubation with phenylethanolamine-N-methyltransferase (PNMT) which specifically methylates the N-terminal group of β-hydroxylated amines. The products are later selectively extracted and quantitated after thin-layer chromatography.

β-Phenylethylamine and tyramine, however, should first be hydroxylated by incubation with dopamine-β-hydroxylase (DBH), to form phenylethanolamine and octopamine, respectively, and then incubated with PNMT (Saavedra and Axelrod, 1973; Saavedra, 1974a, 1974b; Tallman et al., 1976) (Fig. 1).

PHENYLETHANOLAMINE

Endogenous phenylethanolamine can be quantitated by incubation with phenylethanolamine-N-methyltransferase and

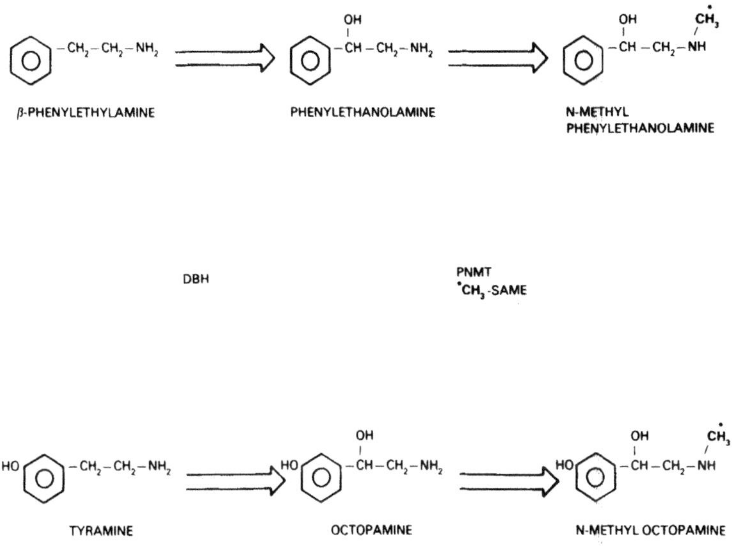

FIGURE 1

[^{3}H]methyl-SAME (Saavedra and Axelrod, 1973). The routine application of a thin-layer chromatographic step to further purify the radioactive product formed is necessary for adequate specificity. The method could be used for the simultaneous determination of phenylethanolamine and octopamine (Saavedra, 1974b) (Fig. 1).

Male Sprague-Dawley rats weighing 150-200 g are killed by decapitation. Organs are rapidly removed, frozen on dry-ice, weighted, and homogenized in 5-10 volumes of ice-cold 20 mM Tris-HCl buffer (pH 8.6) containing the monoamine oxidase inhibitor, iproniazid (50 µg/ml). The homogenates are heated at 90°C for three minutes and centrifuged at 12,000 rpm in a Sorvall refrigerated centrifuge. A 200 µl aliquot of the supernatant fluid is transferred to a 15-ml glass stoppered tube and incubated for 20 min at 37°C after addition of a mixture containing 10 µl of partially purified phenylethanolamine-N-methyltransferase, 5 µl (0.54 nmol) of [^{3}H]methyl-SAME (specific activity, 4.54 Ci/mmol) and 35 µl of 20 mM Tris-HCl buffer (pH 8.6). One ng of phenylethanolamine is added to another aliquot as an internal standard. The incubation is stopped by addition of 0.5 ml of 0.5 M borate buffer (pH 10) and the radioactive product is extracted with 6 ml of a mixture containing 95% heptane and 5% isoamylalcohol (v/v), by shaking for 15 sec. After the phases are separated by centrifugation, a 5 ml aliquot of the organic phase is transferred to counting vials and evaporated to dryness under reduced pressure at 40°C. One ml of ethanol and 10 ml of phosphor containing 40 ml of Liquifluor (New England Nuclear Corp., Boston, MA) per liter of toluene are added to the residue, and radioactivity is measured.

Normally occurring β-hydroxylated compounds such as metanephrine, octopamine, catecholamines, and other phenylethylamines (dopamine, phenylethylamine and tryamine), present at 100-times the concentration of phenylethanolamine, give negligible interference (less than 3%). Identification of N-[^{3}H]methylated phenylethanolamine in tissues is made by means of thin-layer chromatography on precoated Eastman chromagram sheets (Saavedra and Axelrod, 1973).

β-PHENYLETHYLAMINE

Rats are killed by decapitation, and their organs immediately removed, frozen in dry-ice, weighed while still frozen and homogenized in 4 volumes of ice-cold 0.1 N HCl, containing pargyline, 50 ml/l. After centrifugation at 10,000 g for 30 min in a refrigerated centrifuge, the supernatant is divided into 2 ml aliquots and transferred to 15 ml glass stoppered centrifuge tubes. To one of the aliquots, 10 ng of authentic β-phenylethylamine are added as an internal standard. The pH is adjusted to 11-11.5 with 1 N NaOH, and the aqueous phase is extracted with 8 ml of toluene by shaking for 30 min in a mechanical shaker. After centrifugation, 6 ml of the organic phase are transferred to a centrifuge tube containing 300 μl of 0.1 N HCl. These tubes are shaken for 30 min, and after centrifugation the organic phase is carefully aspirated and discarded. The tubes containing the HCl are placed in a vaccuum dessicator at room temperature for 20 min in order to remove traces of toluene, and 200 μl of the HCl transferred to a new centrifuge tube. After addition of 20 μl of 1 N NaOH, the centrifuge tubes are incubated at 37°C after the addition of (a) 50 μl of a mixture A (freshly prepared) containing 5 μl of 0.24 M ascorbic acid, pH 5.5, 25 μl of 0.5 M sodium fumarate, pH 5.5, 10 μl of 0.012 M pargyline, 2 μl (1,500 units) of catalase, and 8 μl of 1.0 M Tris-acetate buffer, pH 5.5; (b) 50 μl of partially purified DBH. After 60 min, 100 μl of a mixture B, containing 70 μl of 1 M Tris-HCl buffer, pH 8.6, 5 μl of [^{3}H]methyl-SAME (0.54 nmol), 5 μl of distilled water, and 20 μl of partially purified PNMT are added, and the incubation continued for 30 min. Blanks consist of the same incubation mixture, replacing the DBH by 50 μl of distilled water. The reaction is stopped with 0.5 ml of 0.5 M borate buffer, pH 10, and the product extracted with 6 ml of a mixture containing 97% toluene and 3% isoamyl alcohol (v/v). After centrifugation, 4 ml of the organic phase are transferred to a counting vial and dried under vacuum for 1 hour at 40°C. After drying, the activity is measured by liquid scintillation spectroscopy.

The identification of the [^{3}H]N-methylated products formed in the reaction is made with thin-layer chromatography (Saavedra, 1974a).

The assay for β-phenylethylamine is reasonably easy to perform, rapid, fairly inexpensive and about 50 times more sensitive than methods currently in use. The double enzymatic procedure allows tissue blanks to be used and endows the assay with marked specificity. The presence of endogenous phenylethanolamine is corrected by measurement of "blank" values when DBH is omitted from the incubation medium. The use of the present assay for the simultaneous determination of β-phenylethylamine and phenylethanolamine, however, is not recommended, due to the low extraction of the latter amine from strongly alkaline solutions.

OCTOPAMINE

The enzymatic-radioisotopic method for octopamine is based upon the N-methylation of octopamine by the use of a partially purified preparation of phenylethanolamine-N-methyltransferase in the presence of [^{3}H]methyl-SAME. The radioactive products formed, N-methyl and N,N, dimethyl-octopamine (synephrine and N-methylsynephrine) are isolated and identified by solvent extraction, evaporation procedures to remove volatile contaminants, and thin-layer chromatography. The routine separation of the radioactive synephrines formed in the reaction by a thin-layer chromatography step provides a necessary measure of specificity (Saavedra, 1974b). This assay can also be adapted for the simultaneous measurement of both octopamine and phenylethanolamine in the same sample (Saavedra, 1974b).

Rat organs are dissected in the cold, immediately frozen on dry-ice, weighed while still frozen, and homogenized in 25 volumes of ice-cold 0.02 M Tris-HCl buffer, pH 8.6, containing the monoamine oxidase inhibitor, iproniazid, 1 x 10^{-3}M. The homogenates are heated in a water bath at 90°C for 3 min. The proteins are removed by centrifugation and 200 µl aliquots of the clear supernatant are incubated in 15-ml glass-stoppered centrifuge tubes at 37°C for 20 min, after addition of a mixture containing 10 µl of partially purified phenylethanolamine-N-methyltransferase, 5 µl of SAME (0.54 nmole) and 35 µl of 0.02 M Tris-HCl buffer, pH 8.6. Blanks are prepared replacing the tissue by 200 µl of the same buffer. One nanogram of octopamine is added to another aliquot as an internal standard. The reaction is stopped with 0.5 ml of 0.5 M borate buffer, pH 10. Ten micrograms of authentic synephrine (N-methyloctopamine) (1 mg/ml in distilled water) are added as a carrier, and the products are extracted into 6 ml of a mixture containing three parts of toluene and two parts of isoamyl alcohol (v/v), by shaking in a Vortex mixer for 30 sec. After centrifugation, the organic phase is quantitatively transferred to 15 ml glass-stoppered tubes containing 1 ml of 0.5 M borate buffer, pH 10, shaken in a Vortex mixer for 30 sec, and centrifuged. A 4 ml aliquot of the organic phase is transferred to counting vials containing 2 ml of toluene: isoamyl alcohol mixture, and the contents are dried overnight in a chromatography oven at 80°C. The radioactivity is determined by liquid scintillation spectroscopy.

Identification of radioactive synephrine and N-methylsynephrine is made by means of thin-layer chromatography on precoated Eastman chromagram sheets (Saavedra, 1974b). Tissues are carried through the entire procedure as described above. After developing, the sheets are stained with diazotized p-nitroaniline, followed by ninhydrin. The sheets are then cut into 1 cm sections and transferred to counting vials containing 1 ml of ethanol. Ten milliliters of phosphor are added to each vial, and the radioactivity is counted. The percentage of authentic octopamine in the tissue is determined as the fraction of radioactivity found to be isographic with synephrine and N-methylsynephrine (Saavedra, 1974b).

The simultaneous determination of phenylethanolamine and octopamine in the same sample is performed as follows: the tissue supernatants are divided and processed as described above, with the exception that another aliquot is carried through the procedure, with the addition of 1 ng phenylethanolamine as internal standard. After stopping the reaction as described above, the [^{3}H]methyl-phenylethanolamine is separated from the [^{3}H]methyloctopamine by extraction with 6 ml of heptane containing 5% isoamyl alcohol (v/v). A 4 ml aliquot of the organic mixture is transferred to counting vials and dried at 40°C under vaccum for 1 hour. The remaining heptane: isoamyl alcohol mixture is carefully aspirated, the [^{3}H]-methyloctopamine is extracted with 6 ml of toluene: isoamyl alcohol (3:2, v/v) mixture, and processed as in the octopamine assay described above (Table 4).

TYRAMINE

Tyramine can be quantitated by a procedure similar to the one used for octopamine after being converted to octopamine by dopamine-β-hydroxylase. This method depends on the β-hydroxylation of tyramine by dopamine-β-hydroxylase (DBH) (EC 1.14.2.1) (Molinoff et al., 1971), followed by the transfer of a tritiated methyl group to the N-terminal of octopamine by phenylethanolamine-N-methyltransferase (EC 2.1.1.28) (Axelrod, 1962); the enzymatically formed

Table 4. Simultaneous determination of octopamine and phenylethanolamine

	Octopamine					Phenylethanolamine	
	Weight (mg)	cpm over blank	Apparent ng/g	% of Apparent	Autentic ng/g	cpm over blank	ng/g
Rat hypothalamus	40	840	26	80	20.8	692	20
Rat pineal gland	1	408	472	70	331	362	489
Fetal rat brain (16 days' gestation)	54	1463	28.8	70	20.2	1205	18

Tissues were processed after homogenization in 1 ml of ice-cold 0.02 M TrisHCl buffer, pH 8.6, containing iproniazid, 1×10^{-3}M. Blank values were 180 cpm for octopamine and 92 cpm for phenylethanolamine. Authentic amines added to tissue supernatants gave 4,000-4,700 cpm (octopamine) and 3,000-3,800 cpm (phenylethanolamine) per ng.

[^{3}H]synephrine is extracted into an organic solvent and the radioactivity determined (Saavedra, 1974b). The routine application of a thin-layer chromatographic step to all samples is necessary for adequate specificity (Tallman et al., 1976).

To measure tyramine, rats are decapitated, their organs removed immediately and frozen on dry-ice. Tissues are weighed while still frozen and are homogenized in either 5 or 10 volumes of ice-cold 0.1 N HCl. After centrifugation at 20,000 g for 20 min in a refrigerated centrifuge, the supernatant is divided into 1 ml samples and placed in 15 ml glass stoppered centrifuge tubes. To one of the samples, 10 ng of authentic tyramine are added as an internal standard. The mixture is neutralized with 100 µl 1.0 N NaOH and 0.5 ml of 1 M Tris-HCl buffer (pH 10.0) is added. The aqueous phase is extracted with 7 ml of methylacetate. After centrifugation 6.0 ml of the organic phase is transferred to a tube containing 500 µl of 0.1 N HCl. These tubes are mixed, centrifuged, and after centrifugation, the organic phase is aspirated. The water phase is washed with 4 ml of toluene to remove traces of the methylacetate and the tubes containing the HCl are placed in a vacuum dessicator at room temperature for 1 hr. They could then be frozen overnight. 300 µl of this solution are transferred to a new tube and neutralized with 30 µl of 1.0 N NaOH. The tubes are then incubated at 37°C for 1 hr after the addition of a solution containing 6 µl 0.3 M ascorbic acid, 25 µl 0.5 M sodium fumarate, pH 5.5, 3 µl (2,000 units) catalase (Boehringer Mannheim Biochemicals), 16 µl 1 M Tris-acetate buffer (pH 5.5), and 50 µl of DBH. Blanks either contain water instead of DBH or are not incubated for the 1 hr period. After this incubation, 100 µl of a mixture containing 85 µl 1 M Tris-HCl (pH 8.6), 10 µl PNMT and 5 µl [^{3}H]methyl-SAME are added and the incubation is continued for another 25 min. The reaction is stopped with 0.5 ml of 0.5 M borate buffer (pH 10.0) containing 10 µg nonradioactive synephrine as carrier and the product is extracted into 6 ml of toluene: isoamyl alcohol (3:2, v/v). After centrifugation, 5 ml of the organic phase is transferred to another fresh tube containing 1 ml of 0.5 M borate buffer (pH 10). After mixing and centrifugation, 4 ml of the washed organic phase are transferred to a counting vial and dried overnight at 80°C in a chromatography oven. The dried fraction is dissolved in 1 ml of ethanol and the radioactivity counted.

The supernatant fraction must be extracted into methylacetate at pH 10.0; the use of either a higher or lower pH markedly reduces the recovery of tyramine. The use of toluene to remove traces of methylacetate from the HCl fraction is essential, as it is necessary to remove traces of organic solvent under vacuum; these solvents, if not removed, drastically reduce the amount of subsequent enzymatic activity. The use of a monoamine oxidase inhibitor is unnecessary if the enzymes are purified as above; however, cruder enzyme fractions may contain this enzyme which will reduce the observed values for tyramine. The purified DBH and PNMT above are suitably stored at -15°C in small (3 ml) samples and would remain stable for several months; repeated freezing and thawing of either enzyme should be avoided. The tritiated SAME once thawed should be kept at 5°C and not refrozen.

The use of a tissue blank for each determination is indicated as several β-hydroxylated amines (octopamine and phenylethanolamine) are extracted to a minor extent and would otherwise lead to high values for tyramine. Internal standards serve to correct for losses during extraction procedures and typically about a 50-55% recovery is found from brain tissue. The additional borate buffer wash and final drying procedure at 80°C are necessary to remove various volatile and soluble contaminants which would increase the blank values. [^{3}H]Synephrine is identified by TLC on precoated Eastman chromagram sheets of silica gel (Saavedra, 1974b).

The enzymatic-radioisotopic assay has been modified including a dansylation procedure and three successive thin-layer chromatographic separations to isolate the meta- and para- isomers of octopamine (Danielson et al., 1977). The ortho isomer, however, can not be separated by this technique, which does not consider the formation of N-methylsynephrine derivatives during the enzymatic reaction (Molinoff et al., 1969) and might be therefore not specific for determination of position isomers.

REFERENCES

Axelrod, J. (1962) Purification and properties of phenylethanolamine N- methyl transferase. J. Biol. Chem. 237, 1657-1660.

Danielson, T.J., Boulton, A.A., Robertson, H.A. (1977) m-Octopamine, p-octopamine and phenylethanolamine in rat brain: A sensitive, specific assay and the effects of some drugs. J. Neurochem. 29, 1131-1135.

Molinoff, P.B., Landsberg, L. and Axelrod, J. (1969) An enzymatic assay for octopamine and other β-hydroxylated phenylethylamines. J. Pharm. Exp. Ther. 170, 253-261.

Molinoff, P.B., Weinshilboum, R. and Axelrod, J. (1971) A sensitive enzymatic assay for dopamine-β-hydroxylase. J. Pharm. Exp. Ther. 178, 425-431.

Saavedra, J.M. (1974a) Enzymatic isotopic assay for and presence of β-phenylethylamine in brain. J. Neurochem. 22, 211-216.

Saavedra, J.M. (1974b) Enzymatic-isotopic method for octopamine at the picogram level. Anal. Biochem. 59, 628-633.

Saavedra, J.M. (1979) Microquantitation of neurotransmitters in specific areas of the central nervous system. Int. Rev. Neurobiol. 21, 259-274.

Saavedra, J.M. and Axelrod, J. (1973) Demonstration and distribution of phenylethanolamine in brain and other tissues. Proc. Natl. Acad. Sci. USA 70, 769-772.

Saavedra, J.M., Palkovits, M., Brownstein, M.J. and Axelrod, J. (1974) Localisation of phenylethanolamine N-methyltransferase in the rat brain nuclei. Nature 248, 695-696.

Smith, A.D. and Winkler, H. (1967) A simple method for the isolation of adrenal chromaffin granules on a large scale. Biochem. J. 103, 480-482.

Tallman, J.F., Saavedra, J.M. and Axelrod, J. (1976) A sensitive enzymatic-isotopic method for the analysis of tyramine in brain and other tissues. J. Neurochem. 27, 465-469.

GAS CHROMATOGRAPHY FOR ANALYSIS OF TRACE AMINES IN TISSUES AND BODY FLUIDS

Glen B. Baker[1], Ronald T. Coutts[1] and Ian L. Martin[2]
[1]Neurochemical Research Unit, Dept. of Psychiatry and Faculty of Pharmacy and Pharmaceutical Sciences, University of Alberta, Edmonton, Alberta, and [2]MRC Neurochemical Pharmacology Unit, Medical Research Council Centre, The Medical School, Hills Road, Cambridge, England

ABSTRACT

The application of gas chromatography to analysis of trace amines and their metabolites in tissues and body fluids is reviewed. Studies using techniques developed by the authors, i.e. acetylation followed by perfluoroacylation for electron-capture gas chromatography, are described in detail. The advantages and disadvantages of gas chromatography relative to other analytical techniques are outlined, and the possible future directions of gas chromatographic analysis of trace amines are discussed.

INTRODUCTION

The low absolute concentrations of the trace amines in most tissues have made studies on the function of these compounds extremely difficult. It is only in the last decade that methods have been developed with adequate sensitivity and specificity for confident examination of these substances in nervous tissue. Methodologies now utilized extensively include mass spectrometry (alone or combined with gas chromatography), radioenzymatic procedures, high-pressure liquid chromatography and gas chromatography with electron-capture detection (GC-ECD). In the present report, we will review the use of GC for analysis of trace amines and their metabolites in tissues and body fluids and speculate on possible future directions which GC may take.

ANALYSIS OF TRACE AMINES AND THEIR METABOLITES BY GAS CHROMATOGRAPHY: AN OVERVIEW

Of the various detector systems available, ECD and flame-ionization detection (FID) have been utilized most extensively for analysis of trace amines and their metabolites in nervous tissue and body fluids. GC-FID has been employed in the analysis of 2-phenylethylamine (PE) and tryptamine (T) in putrefying human tissue (Oliver *et al.*, 1977) and for measurement of plasma levels of PE in dogs following infusion of this amine (Cone *et al.*, 1978). Reynolds and Gray (1976) analyzed PE in urine by GC-FID after conversion to its trifluoroacetyl derivative; these authors employed the same method for detection of N-methyl-PE in urine (Reynolds and Gray, 1978). Haeffner *et al.* (1976) utilized GC-FID for identification of *p*-tyramine (*p*-TA) in urine; a mixed trifluoroacetyl-trimethylsilyl derivative of the amine was formed for analysis. Borison *et al.* (1974) analyzed PE in rabbit brain by GC-FID after conversion of the amine to its dinitrophenylsulfonic acid derivative, but the values obtained were much higher than those reported by other workers using a high-resolution mass spectrometric procedure (Boulton *et al.*, 1975). In a study on urinary excretion of acid metabolites of amines in depressed patients, Sandler *et al.* (1979) employed a GC with a capillary column and FID.

Perfluoroacylating reagents have been used most commonly for derivatization of trace amines for analysis by GC-ECD, but nitro-containing reagents have also found some use. Edwards and Blau (1972, 1973) utilized dinitrophenyl derivatives for analysis of PE in tissues from drug-treated rats. Doshi and Edwards (1979, 1981) reported that 2,6-dinitro-4-trifluoromethylbenzenesulfonic acid is a very suitable reagent for GC-ECD quantification of catecholamines and histamine, and suggested that it may also be applicable to other amines.

Trifluoroacetic, pentafluoropropionic and heptafluorobutyric anhydride have been popular for derivatization of a wide variety of alcohols, phenols and amines, including trace amines. Schweitzer *et al.* (1975) extracted PE from human urine and derivatized it with heptafluorobutyric anhydride prior to analysis by GC-ECD. Martin and Baker (1977) in a GC-ECD assay, derivatized PE in aqueous solution with acetic anhydride, extracted the resultant N-acetyl PE with ethyl acetate, and reacted with pentafluoropropionic anhydride (PFPA); the procedure improved the extractability of the PE and was applicable to quantification of PE in control rat brain. This method was subsequently applied to the analysis of tryptamine (T) in brain and urine (Baker *et al.*, 1979, 1981; Calverley *et al.*, 1980). The method has now been modified to permit efficient extraction and quantification of *m*- and *p*-TA and related phenolic amines (Coutts *et al.*, 1980; Baker *et al.*, 1981, 1982; LeGatt *et al.*, 1981) and details will be given later in this report. Pentafluorobenzoyl chloride, a reagent which is known to react with and impart high sensitivity on ECD to amphetamine and related compounds (Midha *et al.*, 1979), has been utilized for analysis of PE in sheep brain and human urine (Reynolds *et al.*, 1980; Blau *et al.*, 1979).

Reports on the quantification of metabolites of the trace amines by gas chromatography are less frequent than for the amines themselves, although interest in this area is now increasing steadily. Sandler *et al.* (1979) employed GC-FID combined with a capillary column to quantitate phenylacetic acid (PAA), *p*-hydroxyphenylacetic acid (*p*-HPAA) and *p*-hydroxymandelic acid (*p*-HMA) in urine samples from humans. GC-ECD methods have been reported in which acid metabolites of trace amines have been reacted with fluorinated anhydrides (to derivatize phenol and alcohol groups) and/or fluorinated or non-fluorinated alcohols (to derivatize carboxylic acid groups). Such methods have been employed for GC analysis of *m*- and *p*-HPAA in brain (Davis *et al.*, 1977; McQuade *et al.*, 1981) and of PAA, *m*- and *p*-HPAA and *m*- and *p*-HMA in urine (Davis and Boulton, 1981a,b).

RESULTS OBTAINED IN BRAIN TISSUE AND BODY FLUIDS USING ACETYLATION FOLLOWED BY PERFLUOROACYLATION

In our laboratory, we have successfully employed acetylation (under aqueous conditions) followed by perfluoroacylation (under anhydrous conditions) for analysis of trace amines (Martin and Baker, 1977; Coutts et al., 1980; Baker et al., 1979). The various procedures have now been modified and combined into one to provide for simultaneous analysis of PE, m- and p-TA and T in a single sample (Baker et al., 1981; LeGatt et al., 1981). The details of the procedure are given below.

The brain sample is homogenized in ice-cold 0.1N perchloric acid containing 10 mg % EDTA. Any of the following amines can be added as internal standards: 5-methyltryptamine (for T only), benzylamine, 3-phenylpropylamine, 2-(4-chlorophenyl)ethylamine or tranylcypromine. After centrifugation at 12,000 x g for 15 minutes, the supernatant is retained for analysis. The supernatant or urine sample (to which internal standard has been added) is adjusted to pH 7.8. Sodium phosphate buffer (pH 7.8; 0.25 M) (1/10 the total volume of the brain supernatant or urine) is added, and the sample is shaken with the liquid ion-pairing compound di(2-ethylhexyl)phosphoric acid (DEHPA) (2.5% v/v in chloroform). After centrifugation, the lower phase is retained and back-extracted with 0.5N HCl. The HCl phase is neutralized with solid sodium bicarbonate and reacted with acetic anhydride (1/10 the volume of the HCl phase). After effervescence has ceased, the sample is shaken with ethyl acetate to extract the acetylated amines. At this point, the ethyl acetate layer is divided into two portions, one for analysis of T (portion A) and one for analysis of PE, m-TA and p-TA (portion B). Portion A is shaken briefly with 1/10 the volume of distilled water (to remove residues of sodium bicarbonate) and the organic layer is retained and evaporated to dryness under a stream of nitrogen. The residue is reacted with PFPA (75 μl) in the presence of 25 μl of ethyl acetate for 30 minutes at 60°. At the end of this time, the reaction mixture is partitioned between cyclohexane (300 μl) and saturated sodium tetraborate buffer (3.0 ml). A portion of the cyclohexane phase is injected on a GC-ECD equipped with a 10 m OV-101 fused silica capillary column. Portion B is shaken (40 minutes) with 1/10 the volume of 10 N NH_4OH solution to hydrolize the acetylated phenolic groups. The aqueous phase is neutralized with 6N HCl, and the organic phase is retained and taken to dryness under a stream of nitrogen. To the residue is added ethyl acetate (25 μl) and TFAA (75 μl). The mixture is left at room temperature for 30 minutes and then partitioned between cyclohexane

(300 μl) and saturated sodium tetraborate buffer (3.0 ml). An aliquot of the organic phase is used for GC-ECD analysis using the same column as described above for T. Under these conditions, the derivatives (Figure 2) are N-acetyl,N-trifluoroacetyl-PE (LeGatt *et al.*, 1981), O-trifluoroacetyl,N-acetyl,N-trifluoroacetyl-*m*- or *p*-TA (Coutts *et al.*, 1980; Baker *et al.*, 1982) and a spirocyclic derivative of T (Blau *et al.*, 1977). PE can be assayed using either procedure, but the method used for portion B has proved more satisfactory for urine.

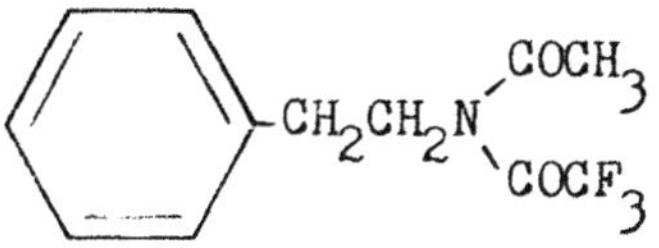

I

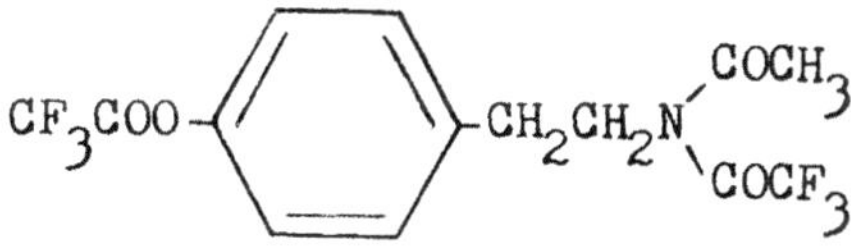

II

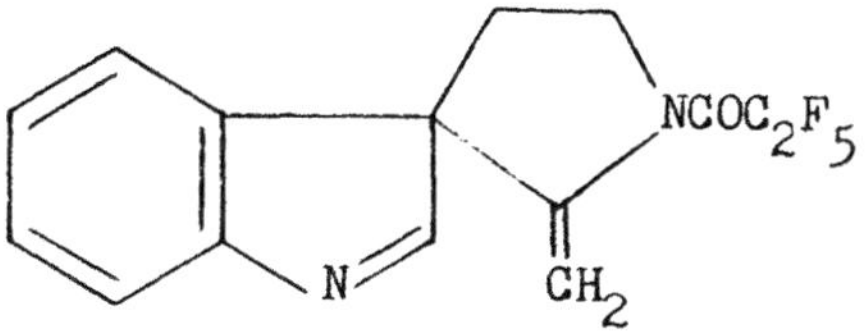

III

Figure 2. Derivatives of trace amines formed by acetylation/perfluoracylation procedures (Baker *et al.*, 1981). I = derivatized 2-phenlethylamine, II = derivatized *p*-tyramine, and III = derivatized tryptamine.

Mean normal values obtained using this technique are as follows: (a) brain (ng/g), PE 1.1 (Martin and Baker, 1977); *p*-TA, 1.8 (Baker *et al*., 1982) and T, <1 (Calverley *et al*., 1980) (b) human urine (μg/24 h), PE, 20; *m*-TA, 101; *p*-TA, 500; and T, 80 (Coutts *et al*., 1981). Except for PE in urine, these levels are in good agreement with values obtained using mass spectrometric techniques (Durden *et al*., 1973; Philips *et al*., 1974a,b; Slingsby and Boulton, 1976; Karoum *et al*., 1978). Urinary PE levels are somewhat higher than those reported using integrated-ion-current mass spectrometry (Slingsby and Boulton, 1976) and combined gas chromatography-mass spectrometry (Karoum *et al*., 1979). Control values of *m*-TA and T in brain could not be obtained because of background noise, but both can be quantitated in rats treated with monoamine oxidase inhibitors; both amines are readily measurable in urine samples.

The procedures described above provide derivatives which are stable and sensitive. In addition, 5-hydroxytryptamine, 3-methoxytyramine and normetanephrine can be assayed simultaneously (Baker *et al*., 1981).

FUTURE DIRECTIONS IN ANALYSIS OF TRACE AMINES BY GAS CHROMATOGRAPHY

Because of its relatively low cost compared to mass spectrometry, GC represents an attractive analytical alternative to researchers wishing to study the trace amines. Several trace amines (as well as other amines of biological importance) can be assayed simultaneously; this is an advantage over the radioenzymatic procedures available. Although radioenzymatic methods are generally more sensitive than GC procedures, expensive radiochemicals must be purchased and disposed of. Techniques utilizing HPLC are now popular for analysis of catecholamines, 5-HT and their metabolites; HPLC is not now employed extensively for analysis of trace amines or their metabolites, but, as indicated in the articles by Bailey *et al*. and Martin *et al*. elsewhere in this book, this situation may soon change.

Other derivatives and detectors may prove useful in future GC analysis of trace amines. Recent research in our laboratories has shown that acetylation in aqueous solution followed by pentafluorobenzoylation under anhydrous conditions is useful for analysis of PE in brain tissue (Hampson, Baker and Coutts, submitted). Heptafluorobutyrylation combined with GC with nitrogen-phosphorous detection (NPD) has been used for analysis of tranylcypromine, a monoamine oxidase inhibitor closely related structurally to PE (Bailey and Barron, 1980). The measurement of N-methyl-T and N,N-dimethyl-T in urine using trifluoroacetylation and GC-NPD has been reported (Oon and Rodnight, 1977). Javaid and Davis (1981) reported the properties of various derivatives of PE and related analogues with regard to GC-NPD analysis. It has been reported (Narasimachari and Friedel, 1981) that PEA and T can be measured in plasma, urine, cerebrospinal fluid and brain after formation of isothiocyanate derivatives and analysis by GC-NPD. Jacob et al. (1978) reported a sensitivity of 500 fg for N,N-dimethylthiophosphinylaniline on GC-NPD, and it is possible that reagents such as dimethylthiophosphinic acid (Jacob et al., 1978) and diethylchlorophosphate (Deo and Howard, 1978) may prove useful as reagents for derivatization of trace amines.

As mentioned earlier in this paper, pentafluorobenzoyl chloride (PFBC) has proved to be a useful derivatizing reagent for GC-ECD analysis of trace amines. These studies were carried out by reacting PFBC with the amine under anhydrous conditions. However, it is known that PFBC can react with phenols and amines under aqueous basic conditions if PFBC is added in the organic phase containing a mutually miscible solvent such as acetonitrile (Argauer, 1968; Cole et al., 1977; Singh et al., 1979). Such conditions have been used in our laboratories to assay PE and amphetamine analogues in brain tissue (Cristofoli et al., 1982) and to assay trace amines in foodstuffs (Baker, Coutts, Wong and Pasutto, unpublished). Further studies in brain tissue and urine are underway.

With the development of newer and more sensitive derivatizing reagents, the development of new methods of utilizing these reagents, and the advance in instrumentation (e.g. capillary columns and automatic samplers), the future of GC in analysis of trace amines seems secure.

ACKNOWLEDGEMENTS

Funding for research has been provided by the Medical Research Council of Canada, the Alberta Heritage

Foundation for Medical Research, the Alberta Mental Health Research Fund, and Health and Welfare Canada.

REFERENCES

Argauer R. J. (1968) Rapid procedure for the chloroacetylation of microgram quantities of phenols and detection by electron-capture gas chromatography. Anal. Chem. **40**, 122-124.

Bailey E. and Barron E. J. (1980) Determination of tranylcypromine in human plasma and urine using high-resolution gas-liquid chromatography with nitrogen-sensitive detection. J. Chromatogr. Biomed. Appl. **183**, 25-31.

Baker G. B., Calverley D. G., Dewhurst W. G., and Martin I. L. (1979) A sensitive gas chromatographic technique for quantification of urinary tryptamine. Br. J. Pharmac. **67**, 469P-470P.

Baker G. B., Coutts R. T., and Martin I. L. (1981) Analysis of amines in the central nervous system by gas chromatography with electron-capture detection. Progr. Neurobiol. **17**, 1-24.

Baker G. B., LeGatt D. F., and Coutts R. T. (1982) A gas chromatographic procedure for quantification of para-tyramine in rat brain. J. Neurosci. Methods **5**, 181-188.

Blau K., Claxton I. M., Ismahan G., and Sandler M. (1979) Urinary phenylethylamine excretion: gas chromatographic assay with electron-capture detection of the pentafluorobenzoyl derivative. J. Chromatogr. Biomed. Appl. **163**, 135-142.

Blau K., King G. S., and Sandler M. (1977) Mass spectrometric and nuclear magnetic resonance confirmation of a 3,3-spirocyclic indole derivative formed from melatonin and related acyl tryptamines. Biomed. Mass Spectrom. **4**, 232-236.

Borison R. L., Mosnaim A. D., and Sabelli H. C. (1974) Biosynthesis of brain 2-phenylethylamine: influence of decarboxylase inhibitors and D-amphetamine. Life Sci. **15**, 1837-1848.

Boulton A. A., Juorio A. V., Philips S. R., and Wu P. H. (1975) Some arylalkylamines in rabbit brain. Brain Res. **96**, 212-216.

Calverley D. G., Baker G. B., McKim H. R., and Dewhurst W. G. (1980) A gas chromatograph technique using electron-capture detection for simultaneous estimation of tryptamine and 5-hydroxytryptamine in biological tissue. Can. J. Neurol. Sci. **7**, 237.

Cole W. J., Parkhouse J., and Yousef Y. Y. (1977) Application of the extractive alkylation technique to the pentafluorobenzoylation of morphine (a heroin metabolite) and surrogates, with special reference to the quantitative determination of plasma morphine levels using mass fragmentography. J. Chromatogr. **136**, 409-416.

Cone E. J., Risner M. E., and Neidert G. L. (1978) Concentrations of phenethylamine in dog following single doses and during intravenous self-administration. Res. Commun. Chem. Path. Pharmacol. **22**, 211-232.

Coutts R. T., Baker G. B., and Calverley D. G. (1980) A rapid, sensitive method of measuring meta- and para-tyramine levels in urine using electron-capture gas chromatography. Res. Commun. Chem. Path. Pharmacol. **28**, 177-184.

Coutts R. T., Baker G. B., LeGatt D. F., McIntosh G. J., Hopkinson G., and Dewhurst W. G. (1981) Screening for amines of psychiatric interest in urine using gas chromatography with electron-capture detection. Progr. Neuro-Psychopharmacol. **5**, 565-568.

Cristofoli W. A., Baker G. B., Coutts R. T., and Benderly A. (1982) Analysis of a monofluorinated analogue of amphetamine in brain tissue using gas chromatography with electron-capture detection. Progr. Neuropsycho-Pharmacol. Biol. Psychiat. **6**, 373-376.

Davis B. A. and Boulton A. A. (1981a) Excretion of m-hydroxymandelic acid in human urine. J. Chromatogr. Biomed. Appl. **222**, 271-275.

Davis B. A. and Boulton A. A. (1981b) Longitudinal urinary excretion of some "trace" acids in a human male. J. Chromatogr. Biomed. Appl. **222**, 161-169.

Davis B. A., Durden D. A., Pun-Li P., and Boulton A. A. (1977) Gas chromatographic procedure for the determination of meta- and para-hydroxyphenylacetic acids. J. Chromatogr. **142**, 517-522.

Deo P. G. and Howard P. H. (1978) Phosphorylation of alcohols/phenols for gas-liquid chromatographic separation and flame photometric detection. J. Offic. Anal. Chem. **61**, 210-213.

Doshi P. S. and Edwards D. J. (1979) Use of 2,6-dinitro-4-trifluoromethylbenzenesulfonic acid as a novel derivatizing reagent for the analysis of catecholamines, histamines and related amines by gas chromatography with electron-capture detection. J. Chromatogr. **176**, 359-366.

Doshi P. S. and Edwards D. J. (1981) Effects of L-DOPA on dopamine and norepinephrine concentrations in rat brain assessed by gas chromatography. J. Chromatogr. **210**, 505-511.

Durden D. A., Philips S. R., and Boulton A. A. (1973) Identification and distribution of β-phenylethylamine in the rat. Can. J. Biochem. **51**, 995-1002.

Edwards D. J. and Blau K. (1972) Analysis of phenylethylamines in biological tissues by gas-liquid chromatography with electron-capture detection. Anal. Biochem. **45**, 387-402.

Edwards D. J. and Blau K. (1973) Phenylethylamine in brain and liver of rats with experimentally induced phenylketonuria-like characteristics. Biochem. J. **132**, 95-100.

Haeffner L. J., Magen J., and Kowlessar O. D. (1976) The gas-liquid chromatographic separation of selected catecholamines on polyamide A103. J. Chromatogr. **118**, 425-428.

Jacob K., Falkner C., and Vogt W. (1978) Derivatization method for the high-sensitivity determination of amines and amino acids as dimethylthiophosphinic amides with the alkali flame-ionization detector. J. Chromatogr. **167**, 67-75.

Javaid J. I. and Davis J. M. (1981) GLC analysis of phenylalkyl primary amines using nitrogen detector. J. Pharm. Sci. **70**, 813-815.

Karoum F., Nasrallah H., Potkin S., Chuang L., Moyer-Schwing J., Phillips I., and Wyatt R. J. (1979) Mass fragmentography of phenylethylamine, m- and p-tyramine and related amines in plasma, cerebrospinal fluid, urine and brain. J. Neurochem. **33**, 201-212.

LeGatt D. F., Baker G. B., and Coutts R. T. (1981) Simultaneous extraction and separation of trace amines of biological interest. J. Chromatogr. Biomed. Appl. **225**, 301-308.

Martin I. L. and Baker G. B. (1977) A gas-liquid chromatographic method for the estimation of 2-phenylethylamine in rat brain tissue. Biochem. Pharmac. **26**, 1513-1516.

McQuade P. L., Juorio A. V., and Boulton A. A. (1981) Estimation of p- and m-isomers of hydroxyphenylacetic acid in mouse brain by a gas chromatographic procedure: their regional distribution and the effects of some drugs. J. Neurochem. **37**, 735-739.

Midha K. K., McGilveray I. J., and Cooper J. K. (1979) A GLC-ECD assay for simultaneous determination of fenfluramine and norfenfluramine in human plasma and urine. Can. J. Pharm. Sci. **14**, 18-21.

Narasimhachari N. and Friedel R. O. (1981) Quantitation of biologically important primary amines as their isothiocyanate derivatives by gas chromatography using nitrogen detection and validation by selected ion monitoring. Clin. Chim. Acta. **110**, 235-243.

Oliver J. S., Smith H., and Williams D. J. (1977) The detection, identification and measurement of indole, tryptamine and 2-phenylethylamine in putrefying human tissue. Forensic Sci. **9**, 195-203.

Oon, M. C. H. and Rodnight R. (1977) A gas chromatographic procedure for determining N,N-dimethyltryptamine and N-monomethyltryptamine in urine using a nitrogen detection. Biochem. Med. **18**, 410-419.

Philips S. R., Durden D. A., and Boulton A. A. (1974a) Identification and distribution of p-tyramine in the rat. Can. J. Biochem. **52**, 336-373.

Philips S. R., Durden D. A., and Boulton A. A. (1974b) Identification and distribution of tryptamine in the rat. Can. J. Biochem. **52**, 447-451.

Reynolds G. P. and Gray D. O. (1976) A method for the estimation of 2-phenylethylamine in human urine by gas chromatography. Clin. Cim. Acta **70**, 213-217.

Reynolds G. P. and Gray D. O. (1978) Gas chromatographic detection of N-methyl-2-phenylethylamine: a new component of human urine. J. Chromatogr. Biomed. Appl. **145**, 137-140.

Reynolds G. P., Sandler M., Hardy J., and Bradford H. (1980) The determination and distribution of 2-phenylethylamine in sheep brain. J. Neurochem. **34**, 1123-1125.

Sandler M., Ruthven C. R. J., Goodwin B. L., and Reynolds G. P. (1979) Deficient production of tyramine and octopamine in cases of depression. Nature (Lond.) **278**, 357-358.

Schweitzer J. W., Friedhoff A. J., and Schwartz R. (1975) Phenethylamine in normal urine: failure to verify high values. Biol. Psychiat. **10**, 277-285.

Singh J., Cochrane W. P., and Scott J. (1979) Extractive acylation of ethylenethiourea from water. Bull. Environm. Contam. Toxicol. **23**, 470-474.

Slingsby J. M. and Boulton A. A. (1976) Separation and quantitation of some urinary arylalkylamines. J. Chromatogr. **123**, 51-56.

Free Communications

STUDIES ON TRYPTAMINE METABOLISM BY GC-MS AND HPLC TECHNIQUES

C. Suñol, J.M. Tusell, F. Artigas, E. Martinez, A. Adell and E. Gelpí
Analytical Neurochemistry Unit. Instituto de Química Bio-Orgánica. C.S.I.C.
c/ Jorge Girona Salgado, s/n. Barcelona-34
Spain.

INTRODUCTION

The indoleamine tryptamine (T) belongs to the group of the "trace amines". Like the rest of these amines it also has a presumed physiological role in accordance with the various actions it exerts in chemical neurotransmission. It has been shown that it plays a predominant role in the behavioural "5-HT syndrome" following administration of tryptophan (TP) and a monoaminooxidase inhibitor (MAOI) since suppression of T synthesis partially abolishes the stereotyped behaviour. This effect could be explained by T acting directly on postsynaptic serotonin (5-HT) receptors or by indirectly increasing the concentration of 5-HT at these sites (Marsden and Curzon, 1979). Along these lines iontophoretic studies have shown that T can modify the effects of 5-HT either by abolishing 5HT excitatory responses or by enhancing its depressant actions (Jones and Boulton, 1980). Also, T has been related to the firing inhibition following the stimulation of the nucleus raphe medianus, whereas the excitatory period was assigned to 5-HT (Jones, 1982a; Jones and Broadbent, 1982). Cox, Lee and Martin (1981) have shown that unilateral intrahypothalamic injection of T or 5-HT produces opposite effects on body temperature, likewise suggesting the existence of a raphe-hypothalamic tryptaminergic pathway (Cox et al.,1981). All of these data support the presumption that T and 5-HT act on different receptor populations. Recently, it has been described that [^{3}H]T binds to rat brain cortex in a saturable and high affinity fashion (Kellar and Cascio, 1982) and that these sites are distinct from the serotonergic receptors, this indicating the possible existence of a specific receptor for T.

Although the literature on T is rather scarce as compared to that on 5-HT, there is yet a great deal of speculation and work done on T as reflected in the extensive review by Jones (1982b).

The poor knowledge of the exact role played by T in the CNS is mainly a consequence of the difficulties associated with its detection, the same applying to its deaminated metabolite, indoleacetic acid (IAA). The fluorimetric determination based on the method of Hess and Udenfriend (1959) together with the radioenzymatic assay (Saavedra and Axelrod, 1972) give abnormally high values for T. Until now, only the mass spectrometric technique has been able to measure the extremely low amounts of T (<0.5 ng/g) (Philips and Boulton, 1974; Artigas et al., in press, a) and its metabolite IAA (10-15 ng/g) (Warsh et al., 1977a; Artigas and Gelpí, 1979; Tusell et al., in press) in rat brain tissue. Recently, HPLC with fluorimetric detection has been shown to be a suitable method to determine IAA in brain tissue and CSF (Anderson and Young, 1981; Tusell et al., in press), thus allowing the incorporation of this instrumentally simpler technique to the analysis of IAA and the other indoleamines in brain tissue.

An account is presented herein on our accumulated experience in the applicability of both GC-MS and HPLC procedures to the study of T and IAA in brain tissue and biological fluids.

METHODS

Materials

<u>Chemicals</u>. T hydrochloride and IAA come from Sigma Chem. Co. (St. Louis, MO, USA). $\alpha,\alpha,\beta,\beta$-T-$d_4$ (free base) was from Massanalysis (Bromma, Sweden) and IAA-d_2 was from Merck, Sharp and Dohme (Montreal, Canada). The isotopic purity of both was checked by SIM and was found to be better than 98%. Stock solutions of T and IAA are prepared in distilled methanol and stored at -30ºC. Working solutions for each experiment (approx. 1 ng/μL) can be prepared by dilution.

All solvents used for the extraction procedures are always analytical grade and are freshly distilled once before use. Water for the preparation of homogenization media is distilled over potassium permanganate. Other chemicals used are analytical grade.

GC-MS(SIM) instrumentation. A Hewlett Packard 5995 GC-MS system equipped with an open-split interface and a standard 9825A microprocessor, has been used for the work with capillary columns (25 m long, 0.3 mm i.d. bonded phase OV-101) also obtained from Hewlett Packard (Palo Alto, CA).

HPLC instrumentation. The HPLC system consists of a solvent delivery pump (model 6000) and an U6K injector from Waters Associates (Milford, Ma) fitted to a μ-Bondapak C_{18} reverse phase column. A guard column is placed between the injector and the chromatographic column. Column effluents are monitored with a 650-10S fluorimetric detector (Perkin Elmer, Norwalk, Ct). Fluorescence excitation and emission wavelengths are set at 280 and 340 nm, respectively.

Experimental Procedures

Some of the initial data shown in results were obtained using the GC-MS(SIM) technique with packed columns and have already been described in detail (Artigas and Gelpí, 1979). The latest analytical procedures that we are presently using have also been described elsewhere. Briefly, they can be summarized as follows:

T and IAA determination on rat brain tissue by GC-MS (Artigas et al., in press, a; Tusell et al., in press). Rat brains are homogenized in 10 vols of KCl 1 mol/L, 0.1% ascorbic acid, adjusted at pH 2 with HCl and containing the corresponding amounts of deuterated standards. After centrifugation at 43500 x g for 30 minutes, the supernatants are extracted with 1.5 vols of ethyl acetate. The organic layer is evaporated to dryness and the residue is then derivatized by combined methylation-perfluoroacylation. The content of IAA in the sample is determined by GC-MS. The aqueous layer is adjusted to pH 11.5, previously adding 100 μg of ascorbic acid, and extracted twice with 1.5 vols of dichloromethane. The organic layer is next evaporated to dryness under Helium flow and the residue can be perfluoroacylated for the GC-MS analysis of T. The usual chromatographic and spectrometric conditions are: injection in the splitless method and separation in a capillary column at 180-200ºC. The identity of endogenous T and IAA is determined by their coelution with the corresponding internal standards T-d_4 and IAA-c_2 and by their SIM response ratios at m/z 276, 289 and 276, 335, respectively. Routine assays can be performed by focusing on m/z 289 and 292 for T-2PFP and T-d_4-2PFP, and 276, 278 for IAA-Me-PFP and IAA-d_2-Me-PFP.

IAA determination on rat brain tissue by HPLC (Tusell et al., in press). The homogenate obtained as above (without adding the deuterated homologs) is divided into two aliquots and one of them is supplemented with 10-15 ng of IAA. After centrifugation at 43500 x g for 30 minutes, the corresponding supernates are extracted with 1.5 vols of ethyl acetate. The organic layers are evaporated to dryness and redissolved in 300 μL of chromatographic eluent. The HPLC separation is carried out isocratically with 0.01 mol/L sodium acetate at pH 4.0 and 40% methanol. The content of IAA in each sample can be calculated by the method of standard addition.

Determination of TP, IAA and 5-HIAA in CSF by HPLC (Adell et al., 1983). Samples of CSF (20-100 μL) are directly injected into the liquid chromatograph. Chromatographic separation of TP metabolites can be performed through a linear programm with a mixture of methanol/water and a 0.005 mol/L solution of 1-pentane sulfonic acid at pH 2.6, as the high and low concentration eluents, respectively. The content of IAA in each sample can be calculated by reference to a calibration curve.

IAA determination on plasma by HPLC (Martínez et al., 1983). Plasma is deproteinized with trichloroacetic acid in the presence of ascorbic acid. After vigorous shaking and centrifugation the clear supernatant is adjusted to pH 1.5-2 and aliquots of it are directly injected into the liquid chromatograph. The chromatographic separation is carried out isocratically by a slight modification of the procedure described above. The content of IAA is determined by the method of standard addition.

RESULTS AND DISCUSSION

Determination of T and IAA in CNS

Before the advent of on-line liquid chromatographic selective detectors (e.g. fluorimetric and electrochemical) mass spectrometry by itself or combined to a gas chromatograph (GC-MS) was the only method one could use to reliably analyze minute amounts of indoleamines in brain tissue. In our hands, the GC-MS technique coupled to a previous sample clean-up procedure by adsorption on a XAD-2 resin, allowed us to determine TP, T, IAA, 5-HT and 5-HIAA in single samples of rat brain tissue. The concentration found in this way for the trace amine T was below the detection limit of the system available to us at that time (about 0.4 ng/g), whereas that of its metabolite IAA was 13 ng/g (Artigas and Gelpí, 1979). The analyses of these compounds were carried out on GC packed columns and, since our aim was the concurrent determination of TP and the above four indole metabolites, the overall clean up procedure was optimized so that compromise conditions for all five metabolites were prevailing. Other authors (Warsh et al., 1977 b) had also failed to detect endogenous rat brain T using conventional GC-MS techniques, although Philips et al., (1974) had succesfully determined T in rat brain with a more sophisticated apparatus (direct probe-high resolution mass spectrometry). It was thus evident that the routine assay of T would require a higher degree of sensitivity.

Recently, we have achieved the detection of rat brain T by means of a selective extraction on CH_2Cl_2 and further analysis of the extract by capillary GC-MS. The high resolving power of GC capillary columns (as compared to packed columns) as well as their enhancement of the signal to noise ratio, resulting in much narrower peaks, allows the quantitative determination of T in rat brain tissue down to levels of about 100 pg/g (Artigas et al., in press, a). Fig. 1 shows the selected ion current profile (SICP) of brain T and IAA in a 5 day old rat by capillary GC-MS. The newer flexible, inactive fused silica capillary columns can be used in routine GC-MS work with enormous advantages over classical glass capillary columns. Thus, IAA can also be determined at a high sensitivity with this kind of methodology. For instance, detection limits of 5 pg of authentic IAA can be achieved (Fig. 2). However, we have lately shown that the low endogenous amount of IAA in rat brain can also be readily determined by HPLC coupled to fluorimetric detectors (HPLC-F) with an accuracy equivalent to that of GC-MS, as illustrated by the excellent correlation coefficient (r^2=0.991) that we obtained in a recent comparison of both techniques (capillary GC-MS and HPLC-F) (Tusell et al., in press). Table I shows the good concordance between the HPLC and GC-MS determinations within a wide range of concentrations. In this case, on account of the lower costs and wider availability of the HPLC equipment in many laboratories,

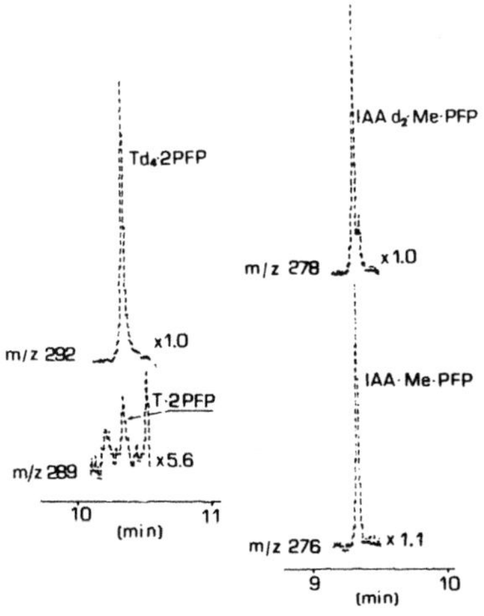

Figure 1.- T and IAA determination by capillary GC-MS. Chromatographic conditions: see methods.

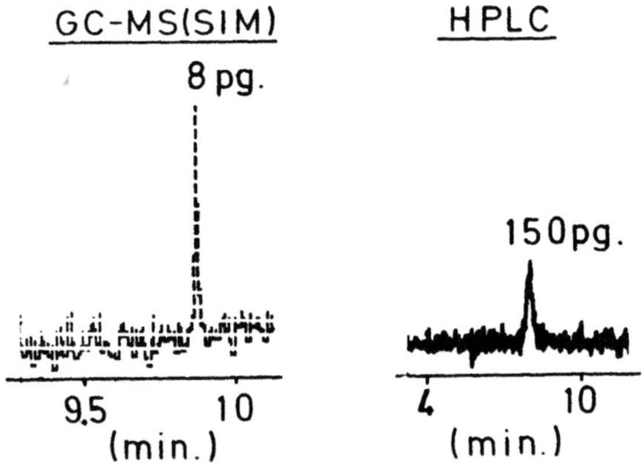

Figure 2.- Selected ion current and HPLC-F profiles of amounts of IAA close to the detection limits.

Table I

SAMPLES	HPLC (found)	GC-MS (found)	"THEORETICAL" VALUE
A	15.7 ng/g	18.2 ng/g	18.2 ng/g
B	35.7	36.7	37.6
C	79.2	73.8	79.4
D	114.2	118.1	120.2

GC-MS would be better used as a reference analytical technique. Also, as HPLC equipment is less sophisticated than GC-MS systems, in the long run, it is also less prone to breakdowns. Thus, HPLC would be preferred over GC-MS for the routine analysis of IAA. Moreover, the HPLC assay requires no more sample handling that a single step extraction of IAA into ethyl acetate, avoiding the sample derivatization step needed for GC-MS.

Likewise, CSF can also be analyzed by HPLC. In this case the sample, frozen immediately after lumbar puncture, is thawed and directly injected into the liquid chromatograph without further manipulation. Figure 3 illustrates this kind of determination for IAA, 5-HIAA and TP. As stated in methods, the fluorimetric detection of IAA, as well as of other indole compounds, depends on the natural fluorescence of the indole ring. Unfortunately and contrary to serotonin or the catecholamines (Tusell et al., 1982), both T and IAA lack easily oxidizable groups, so that the use of the highly sensitive electrochemical detection methods, cannot be presently applied to the measurement of traces of these compounds in a biological matrix.

The availability of both of these sensitive and accurate instrumental techniques (GC-MS and HPLC-F) has enabled us to undertake a thorough study of the ontogenetic development of tryptamine, this being an aspect on which nothing was known, in spite of the increasing body of data concerning various neurochemical and functional considerations. Probably this was because, as stated above, until recently only one research group (Philips et al., 1974) was able to confidently measure T in brain samples. Briefly, the results obtained in the studies of the ontogeneses of brain T and IAA (Artigas et al., in preparation) indicate that the highest values of this trace amine in rat brain are reached within 24 hours from birth, with concentrations decreasing

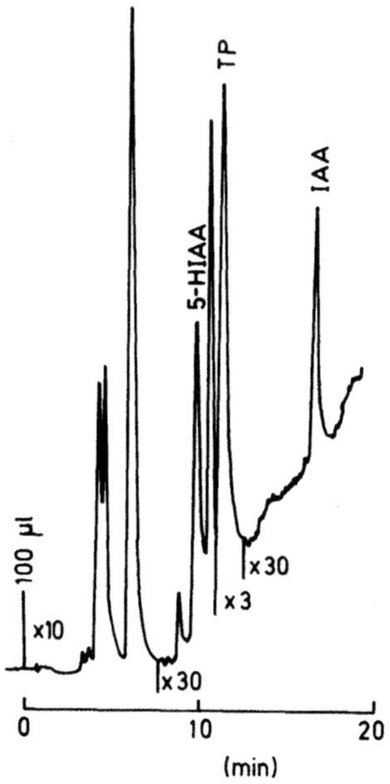

Figure 3.- HPLC-F determination of TP and its acidic metabolite on CSF samples. This sample was from a patient suffering neurological disorders (Gilles de la Tourette Syndrome). The content of IAA was estimated to be 11 ng/ml. Chromatographic conditions: see Methods.

to day 5 (108 pg/g) followed by a further increase up to adult values (494 pg/g). The ontogeneses of T and IAA parallel that of TP, with higher concentrations being found in the early stages of extrauterine life. This kind of development is clearly different from that of the classic neurotransmitter amines, such as the catecholamines or serotonin, whose cerebral concentrations are low at birth and increasingly grow to the adult values (Rastogi and Singhal, 1974). Interestingly, the T/TP ratio is significatively enhanced from day 16 on, whereas the IAA/T ratio diminishes, suggesting some kind of storage mechanism. This is interesting inasmuch as it relates to the unsolved problem of the existence or not of subcellular stores of T. In this regard we need to consider that according to our results the IAA/T ratio of 22 is very high relative to the 5-HIAA/5-HT ratio of 0.79 in the 45 day old rat. This could also be taken as an indication of poor neuronal storage mechanisms for T, so that the amine would not be efficiently protected against the action of MAO. However, Boulton and Baker (1975) have found a high percentage of T in the synaptosomal (P_2B) fraction, with a similar profile to that of 5-HT and catecholamines. In contrast, Juorio (1982) has reported that reserpine increases T levels in mouse striatum, suggesting that it could be protected by some kind of reserpine-resistant mechanism. Our preliminary results on the action of reserpine on the indoleamines are shown in Table II. Along the lines of the well known amine depleting action of reserpine, 5-HT is depleted to a 15% of the control value, 5-HIAA increases to a 283% and IAA concentration is also enhanced by a 50%. T was not determined in this study. It would be interesting to see whether the concentration of T as well as its amine/acid ratio, are affected by reserpine in different regions of the rodent brain. Also, since T concentration in brain is extremely sensitive to TP, the action of reserpine on the brain content of the latter should likewise be carefully checked. In this context, Marsden and Curzon (1974) found that a reserpine dose of 10 mg/Kg does not modify the TP content of cerebral cortex in rat, while Young et al. (1980), taking the concentration of TP in CSF as an index of its content in the CNS, showed a significant increase after the reserpine treatment. Work is in progress to resolve these discrepancies as well as the question of a possible storage mechanism for T.

Peripheral metabolism of T

T is considered a "trace" amine in the mammalian brain but probably this is not the case in the rest of the body. Unfortunately, the initial T data on lung, heart and brain, by Saavedra and Axelrod (1973), were obtained by the isotopic-enzymatic technique which has been shown to give abnormally high values in brain tissue. However, since it has also been shown that T crosses the blood brain barrier (Oldendorf and Braun, 1976), its peripheral metabolism can affect its availability to brain. To study this metabolism we have described an HPLC procedure to determine IAA as well as 5-HIAA in plasma. The HPLC-F analyses of IAA in plasma samples from healthy individuals show that the mean concentration of this acid is very high (270 ± 27 ng/ml; $\bar{x} \pm$S.E.M., n=15) as compared to that of 5-HIAA ($9.9 \pm 1,0$ ng/ml; $\bar{x} \pm$ S.E.M.,n=16). Thus, the plasmatic compartment of the human

Table II

	5-HT	5-HIAA	IAA	5-HIAA/5-HT
Control	589 ± 71 (6)	732 ± 109 (6)	16.4 ± 1.3 (3)	1.2 ± 0.1 (6)
Reserpinized	86 ± 17 (4)**	2068 ± 292 (4)**	25.2 ± 2.1 (4)*	24.9 ± 2.3 (4)°

- Values expressed as ng/g of wet tissue ($\bar{x} \pm$ S.E.M.)
- Male adult Sprague-Dawley rats were treated with saline (control animals) or 10 mg/Kg reserpine i.p., 20 hours before sacrifice
- * $p < 0.05$; ** $p < 0.005$; ° $p < 0.001$ (Student's t-test)

body shows a striking pattern: while 5-HIAA -the metabolite of the more concentrated amine, 5-HT- is in the low ng/mL range, the concentration of IAA -arising from the "trace" amine- is about 30 times greater (Martínez et al., 1983). In rat plasma, the concentration of IAA is also high (191± 64 ng/mL, $\bar{x}$±S.D.) (Artigas et al., in press,b). In absence of reliable data on the content of T in the various body tissues, these results could be interpreted as follows:

a) Most of the plasmatic IAA comes from transamination of TP to indolepyruvic acid (IPyA) and further decarboxylation of the latter (Weissbach et al., 1959)

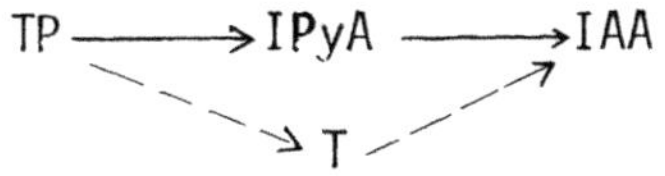

b) The high plasmatic content of IAA reflects either a high concentration or a very high metabolism of T in the body tissues, being the pathway through IPyA a minor one.

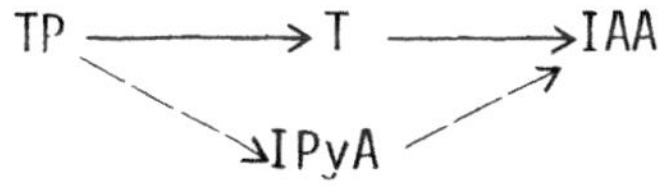

The obvious difference between both hypotheses is that only in b) the measurements of IAA would reflect T metabolism. The i.p. treatment of rats with either 100 mg/Kg benserazide (L-aromatic aminoacid decarboxylase inhibitor), 15 mg/Kg tranylcypromine or 100 mg/Kg iproniazid (MAO inhibitors) produces significative decreases of 69%, 57% and 73% respectively (Artigas et al., in press,b). These results indicate that the formation of IAA from TP through IPyA is indeed a minor pathway, since the inhibition of the enzymes involved in the synthesis of IAA via T clearly reduces the plasmatic content of this acid to about 30% of its control value.

In conclusion, this retrospective view of past and recent work, carried out both in our as well in other laboratories in the elusive field of tryptamine's neurochemical function shows that, after a difficult start, analytical techniques are now sufficiently refined for its routine measurement within a wide dynamic range. Evidently, the application of such techniques will facilitate the studies aimed at defining this "trace amine" true role in CNS.

Acknowledgements: This work was supported by research grants from the Comisión Asesora de Investigación Científica y Técnica and by a research grant from the Fondo de Investigaciones Sanitarias de la Seguridad Social.

We thank Productos Roche and Smith, Kline & French SAE (Barcelona, Spain) for the generous gift of benserazide and tranylcypromine.

REFERENCES

Adell, A., Tusell, J.M., Artigas, F., Martínez, E., Suñol, C. and Gelpí, E. (1983) Fluorometric determination of tryptophan and its brain indoleamine metabolites by ion-pair HPLC. *J. Liq. Chromatog.* 6, 527-541.

Anderson, G.M. and Young, J.G. (1981). Minireview: applications of liquid chromatographic-fluorimetric systems in neurochemistry. *Life Sci.* 28, 507-517.

Artigas, F. and Gelpí, E. (1979). A new mass fragmentographic method for the simultaneous analysis of Tryptophan, Tryptamine, Indole-3-acetic acid, Serotonin and 5-hydroxyindole-3-acetic acid in the same sample of rat brain. *Anal. Biochem.* 92, 233-242.

Artigas, F., Suñol, C., Tusell, J.M., Martínez, E. and Gelpí. (in press, a). Determination of tryptamine in brain tissue by capillary GC-MS (selected ion monitoring) *Biomed. Mass Spectrom.*

Artigas, F., Martínez, E., Tusell, J.M., Suñol, C. and Gelpí, E. (in press,b). On the metabolic origin of plasmatic indole-3-acetic acid in the rat. *Biochem. Pharmacol.*

Boulton, A.A. and Baker, G.B. (1975). The subcellular distribution of β-phenylethylamine, p-tyramine and tryptamine in rat brain. *J. Neurochem.* 25, 477-481.

Cox, B., Lee, T.F. and Martin, D. (1981). Different hypothalamic receptors mediate 5-hydroxytryptamine- and tryptamine- induced core temperature changes in the rat. *Br. J. Pharmac.* 72, 477-482.

Cox, B., Davis, A., Juxon, V., Lee, T.F. and Martin, D. (1981). Core temperature changes following electrical stimulation of the midbrain raphé nuclei and intrahypothalamic injection of tryptamine in the rat. *Br. J. Pharmac.* 74; 216 P.

Hess, S.M., Redfield, B.G. and Udenfriend, S. (1959). The effect of monoamine oxidase inhibitors on the tryptamine content of animal tissues and urine. *J. Pharm. Exp. Ther.* 127, 178-181.

Jones, R.S.G. and Boulton, A.A. (1980). Tryptamine and 5-hy-

droxytryptamine: actions and interactions on cortical neurones in the rat. *Life Sci.* 27, 1849-1856.
Jones, R.S.G. (1982 a). Responses of cortical neurones to stimulation of the nucleus raphé medianus: A pharmacological analysis of the role of indoleamines. *Neuropharm.* 21, 511-520.
Jones, R.S.G. (1982 b). Tryptamine: a neuromodulator of neurotransmitter in mammalian brain?. *Prog. Neurobiol.* 19 117-139.
Jones, R.S.G. and Broadbent, J. (1982). Further studies on the role of indoleamines in the responses of cortical neurones to stimulation of nucleus raphé medianus: effects of indoleamine precursor loading. *Neuropharm.* 21, 1273-1277.
Juorio, A.V. (1982). The effects of some antipsychotic drugs, D-amphetamine and reserpine on the concentration and rate of accumulation of tryptamine and 5-hydroxytryptamine in the mouse striatum. *Can. J. Physiol. & Pharmac.* 60, 376-380.
Kellar, K.J. and Cascio, C.C. (1982) [^{3}H]tryptamine: High affinity binding sites in rat brain. *Europ. J. Pharmac.* 78, 475-478.
Marsden, C.A. and Curzon, G. (1974). Effects of lesions and drugs on brain tryptamine. *J. Neurochem.* 23, 1171-1176.
Marsden, C.A. and Cruzon, G. (1979). The role of tryptamine in the behavioural effects of tranylcypromine + L-tryptophan. *Neuropharm.* 18, 159-164.
Martínez, E., Artigas, F., Suñol, C., Tusell, J.M. and Gelpí, E. (1983). High performance liquid chromatography determination of indole-3-acetic acid and 5-hydroxyindole-3-acetic-acid in human plasma. *Clin. Chem.* 29, 1354-1357.
Oldendorf, W.H. and Braun, L.D. (1976). [^{3}H]-tryptamine and [^{3}H]-water as diffusible internal standards for measuring brain extraction of radio- labeled substances following carotin injection. *Brain Res.* 113, 219-224.
Philips, S.R., Durden, D.A. and Boulton, A.A. (1974). Identification and distribution of tryptamine in the rat. *Can J. Biochem.* 52, 447-451.
Rastogi, R.B. and Singhal, R.L. (1974). Thyroid hormone control of 5-hydroxytryptamine metabolism in developing rat brain. *J. Pharm. Exp. Ther.* 191, 72-81.
Saavedra, J.M. and Axelrod, J. (1972). A specific and sensitive assay for tryptamine in tissues. *J. Pharm. Exp. Ther.* 182, 363-369.
Saavedra, J.M. and Axelrod, J. (1973). Effects of drugs on

the tryptamine content of rat tissues. *J. Pharm. Exp. Ther.* 185, 523-529.
Tusell, J.M., Suñol, C., Artigas, F., Martínez, E. and Gelpí, E. (1982). Neurochemical studies on catecholamines and indoleamines by high-performance liquid chromatography with electrochemical detection. *Chromatog.* 16, 112-116.
Tusell, J.M., Artigas, F., Suñol, C., Martínez, E. and Gelpí, E. (in press). Comparison of HPLC and GC-MS methods for the analysis of indole-3-acetic acid in brain tissue. *J. Chromatog. Biomed. Appl.*
Warsh, J.J., Chan, P.W., Godse, D.D., Coscina, D.V. and Stancer, H.C. (1977 a). Gas chromatography-mass fragmentographic determination of IAA in rat brain. *J. Neurochem.* 29, 955-958.
Warsh, J.J., Godse, D.D., Stancer, H.C., Chan, P.W. and Coscina, D.V. (1977 b). Brain tryptamine in rats by a new gas chromatography-mass fragmentographic method. *Biochem. Med.* 18, 10-20.
Young, S.N., Anderson, G.M. and Purdy, W.C. (1980). Indoleamine metabolism in rat brain studied through measurements of tryptophan, 5-HIAA and IAA in cerebrospinal fluid. *J. Neurochem.* 34, 309-315.

A RAPID AND SPECIFIC TECHNIQUE FOR THE EXTRACTION OF TYRAMINE AND OCTOPAMINE FROM BIOLOGICAL TISSUES FOR HPLC ANALYSIS

B.A. Bailey, R.J. Martin, R.G.H. Downer

Department of Biology
University of Waterloo
Waterloo, Ontario, Canada

High performance liquid chromatography (HPLC) with electrochemical detection (ElCD) is a widely accepted procedure for estimation of catecholamines (Shoup, 1982), however the technique has not been applied extensively to analysis of the monohydroxy analogues of catecholamines. The infrequent use of HPLC/ElCD for determination of phenolamines is due in part to the high electrode potentials that are required to effect electro-oxidation of these compounds (Bailey et al., 1982) and to the lack of a convenient technique for extracting the monoamines from biological tissues. The constraints due to detection have recently been resolved by the use of a dual coulometric detection system (Martin et al., 1983; see also paper by Martin et al. in this volume) and the present report summarises a rapid, specific extraction procedure that affords up to 75% recovery of tyramine and octopamine from a variety of biological tissues. A more detailed description of the extraction procedure including data on factors influencing recovery will be published elsewhere.

EXTRACTION PROCEDURE

Dissected brain and heart tissues were weighed and macerated in the frozen state before being added to cooled 15 ml Corex centrifuge tubes containing 2.5 ml 0.2 N perchloric acid for homogenisation in a Branson ultrasonic homogenizer. Insect tissues were placed in 1.5 ml polypropylene tubes containing 200 µl of 0.1 N perchloric acid. Following the addition of sodium heparin (brain and heart, 2000 units in 0.5 ml 0.2 N perchloric acid; insect nerve cord and hemolymph, 100 units in 50 µl 0.1 N perchloric acid) to facilitate lipoprotein precipitation the homogenate was centrifuged for 30 min (vertebrate tissues, 30,000X g; insect tissues, 40,000X g).

Phenolamines were extracted using 3.0 ml disposable C-18 extraction columns (J.T. Baker, Philipsburg, N.J.). The columns were first washed with 1 column-volume of methanol:acetonitrile (1:1, v:v) containing 15 mM trichloroacetic acid (TCA) followed by 1 volume methanol and then 2 volumes double distilled, deionised water (final distillation over potassium permanganate). Fluids were drawn through the column by vacuum pressure with care being taken to avoid drying of the column matrix.

The pH of the supernatant obtained from the perchloric acid homogenate was adjusted to 8.7 with 1 N ammonium hydroxide and the supernatant then applied to the extraction column and washed sequentially with 2 volumes water and 1 ml methanol. The methanol rinse was drawn completely through the column. Amines were eluted from the extraction column with 800 µl methanol containing 15 mM TCA. The eluant was dried at 65°C under a stream of nitrogen and reconstituted with a known volume of water so that a 50 µl aliquot provided an appropriate sample for injection onto the HPLC column.

CHROMATOGRAPHY

Extracted amines were separated and estimated by HPLC/ElCD using a dual coulometric detection system according to the procedure of Martin et al. (1983).

RESULTS AND DISCUSSION

Effect of pH on recovery. Catecholamines and phenolamines are ionized at acidic pH and, therefore, are poorly retained on C-18 reverse phase columns at low pH. By adjusting the pH towards the relatively high pK values of these compounds (Mack and Bönisch, 1979), a greater capacity for retention is achieved and the amines are adsorbed by the reverse phase matrix. However, extraction of the amines at basic pH demonstrates low recovery of catecholamines relative to phenolamines; furthermore, the recovery of catecholamines decreases as the pH is increased from 8.20 to 9.25. Catecholamines are more readily oxidized than phenolamines (Bailey et al., 1982) and the poor recovery at pHs above 8.0 is due to oxidation of catecholamines. Thus, at pH 9.25 the recovery of norepinephrine, epinephrine and dopamine is less than 20% while 60% octopamine and tyramine are recovered at this pH. The optimal pH for recovery of octopamine and tyramine was 8.75, therefore, selective extraction of these phenolamines was conducted at this pH.

Extraction of phenolamines from biological tissues. Typical chromatograms of samples extracted from rat brain and heart and insect haemolymph and nerve cord are illustrated in Fig. 1 (for standards see paper by Martin et al. in this volume). The extraction efficiency of octopamine from nerve cord (74%) and hemolymph (75%) is similar to that obtained for standard solutions (75%), however the octopamine recovery from heart and brain is 58% and 55% respectively. The decreased recovery from vertebrate tissues results from the presence in these tissues of unidentified compounds which compete with octopamine and tyramine for binding sites on the column matrix. The application of relatively small aliquots of the tissue supernatant to the extraction column is recommended in order to enhance recovery. The percentage recovery obtained for octopamine and tyramine from vertebrate tissues is superior to that provided by the extraction procedure employed in the radioenzymatic assay (Molinoff et al., 1969) that has been used to produce many estimates of octopamine levels; furthermore, the procedure is less complex than that reported by Baker et al. (1980). Thus, the recoveries obtained in the present study provide further vindication for the usefulness of the extraction procedure.

The octopamine and tyramine levels of several vertebrate and invertebrate tissues as obtained by HPLC/ElCD are indicated in Table 1 and compare

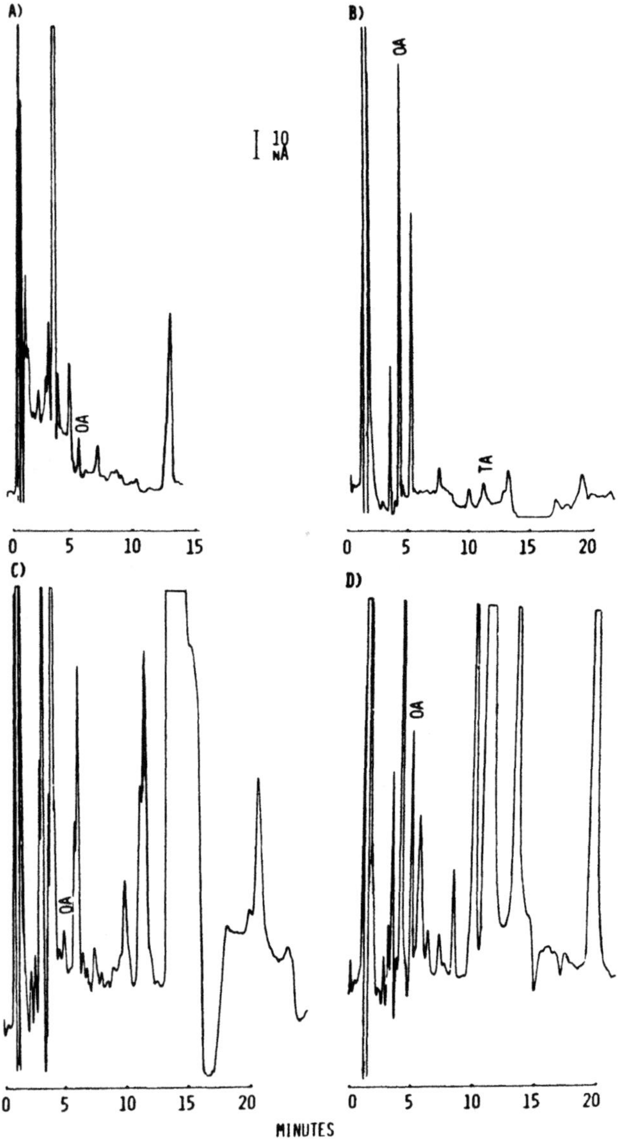

Fig. 1. Chromatograms of extracted samples of insect haemolymph. (A), insect nerve cord (B), rat brain (C), and rat heart (D). Octopamine was detected coulometrically with the detector electrode set at 0.75V after the eluant was passed through a screening electrode set at 0.50V.

Table 1. Octopamine and tyramine content of various biological tissues

	PHENOLAMINE CONTENT	
TISSUE	OCTOPAMINE	TYRAMINE
Brain	4.64±2.36 ng/g (6)*	n.d.
Heart	12.37±8.97 ng/g (6)	n.d.
Insect haemolymph	29.0±9.0 nanomolar (12)	317.5±45.7 nanomolar (3)
Insect nerve cord	778.65±87.24 ng/g (7)	113.50±23.41 ng/g (3)

*Values indicate mean ± standard deviation for number of determinations shown in parentheses.

favourably with values reported by other workers using a variety of analytical procedures (Axelrod and Saavedra, 1974; Karoum et al., 1979; Dymond and Evans, 1979).

ACKNOWLEDGEMENT

Supported by an operating grant from the Natural Sciences and Engineering Research Council of Canada to RGHD.

REFERENCES

Axelrod, J. and Saavedra, J.M. (1974) Octopamine, phenylethanolamine, phenylethylamine and tryptamine in the brain. Ciba Foundation Symposium 22 (new series), 51-59.

Baker, G.B., Coutts, R.T., and LeGatt, D.F. (1980) A procedure for extraction and separation of phenylethylamine, tyramine and octopamine. Biochem. Soc. Trans. 8, 622-623.

Bailey, B.A., Martin, R. J., and Downer, R.G.H. (1982) Simultaneous determination of dopamine, norepinephrine, tyramine and octopamine by reverse-phase high performance liquid chromatography with electrochemical detection. J. Liquid Chromatog. 5, 2435-2452.

Dymond, G.R., and Evans, P.D. (1979) Biogenic amines in the nervous system of the cockroach, Periplaneta americana: association of octopamine with mushroom bodies and dorsal unpaired median (DUM) neurones. Insect Biochem. 9, 535-545.

Karoum, F., Nasrallah, H., Potkin, S., Chuang, L., Moyer-Schwing, J., Phillips, I., and Wyatt, R.J. (1979) Mass fragmentography of phenylethylamine, m- and p-tyramine and related amines in plasma, cerebrospinal fluid, urine and brain. J. Neurochem. 33, 201-212.

Mack, F., and Bönisch, H. (1979) Dissociation constants and lipophilicity of catecholamines and related compounds. Arch. Pharmacol. 310, 1-9.

Martin, R.J., Bailey, B.A., and Downer, R.G.H. (1983) Rapid estimation of catecholamines, octopamine and 5-hydroxytryptamine from biological tissues using high performance liquid chromatography with coulometric detection. J. Chromatogr. (in press)

Molinoff, P.B., Landsberg, L., and Axelrod, J. (1969) An enzymatic assay for octopamine and other β-hydroxylated phenylethylamines. J. Pharmac. exp. therap. 170, 253-261.

Shoup, R.D. (ed.) (1982) Recent reports on liquid chromatography/electrochemistry, BAS Press, West Lafayette.

ANALYSIS OF OCTOPAMINE, DOPAMINE, 5-HYDROXYTRYPTAMINE AND TRYPTOPHAN IN THE BRAIN AND NERVE CORD OF THE AMERICAN COCKROACH

R.J. Martin, B.A. Bailey, R.G.H. Downer

Department of Biology
University of Waterloo
Waterloo, Ontario, Canada

The monohydroxy analogue of norepinephrine, octopamine (OA), appears as a trace amine in a variety of vertebrate tissues (Yui et al., 1980; Boulton and Juorio, 1981), and is a primary component of the insect nervous system (Evans, 1980). In insects, OA serves as a neurotransmitter, neuromodulator and neurohormone (Orchard, 1982) and has been proposed as the sympathomimetic effector of physiological responses to excitation (Hoyle, 1975; Downer, 1979). The demonstrated importance of OA in insects, together with its potential significance in vertebrates, indicates the need for a convenient, sensitive procedure for simultaneous determination of OA and other monoamines in small amounts of biological tissue. High performance liquid chromatography (HPLC) with electrochemical detection (ElCD) has been used successfully to estimate catecholamines and 5-hydroxytryptamine (5-HT) (Shoup, 1982), but the high potential required to electro-oxidize OA has generally excluded this compound from such analyses (Bailey et al., 1982). The problem of detection has recently been resolved by the use of two coulometric detectors in series (Martin et al. 1983). Eluant from the HPLC column is first exposed to a detector set at a relatively low potential which effects complete oxidation of catecholamines; the eluant then passes to a second detector which is set at the higher potential required for oxidation of phenolamines.

In the present study the principal catecholamines, phenolamines and indolalkylamines of the insect nervous system have been estimated in whole brain and individual ganglia of the nerve cord of the American cockroach, Periplaneta americana, using HPLC/ElCD.

ANALYSIS

The analytical procedure employed in the present study has been described previously (Bailey et al., 1982; Martin et al., 1983) and only salient features will be described in the present account.

Dissected tissues were homogenized by sonication over ice in 50 µl 0.2 N perchloric acid containing dihydroxybenzylamine (DHBA) as internal standard and sodium heparin to facilitate lipoprotein precipitation. Following centrifugation at 40,000X g for 30 min, 10 µl of the supernatant was injected onto an Ultrasphere IP, C18 column (Beckman, 150 x 4.6 mm ID, 5 µm particle size) using a sixport Rheodyne 7125 valve fitted with a 20 µl holding loop. The mobile phase comprising 1.5 µM EDTA, 2.0 µM SDS, 0.02 M TCA, 0.075 M monobasic sodium phosphate, 20% acetonitrile and 12% methanol was pumped at 0.8 ml/min using a Spectra-Physics model 740B solvent delivery system equipped with pulse dampening. Eluted peaks were detected using the ESA 5100A Coulometric Detection System which comprises a coulometric guard cell and a solid state analytical cell containing two coulometric working electrodes in series. The guard cell was located before the injector and, when operated at high potential (0.9V), oxidized mobile phase contaminants thereby reducing baseline current and noise at the analytical detectors. The first detector was set at 0.5V for detection of catecholamines, 5-HT, some tryptophan (TP) and DHBA while the second detector was set at 0.75V and detected OA, tyramine (TA), 5-HT and TP. Signals from the coulometric detector were integrated by computing integrators (SP4270 and SP System 1, Spectra-Physics).

RESULTS AND DISCUSSION

The efficacy of the analytical procedure is illustrated in Fig. 1 which demonstrates typical chromatograms of standards and direct injections of the whole brain supernatant. The chromatograms demonstrate the increased selectivity that can be achieved when the dual coulometric detectors are employed in the screen mode. For example, the similar retention times of DHBA and OA would make baseline resolution of these compounds difficult with conventional detector systems; however, the increased selectivity afforded by the dual detector results in complete oxidation and detection of DHBA at the first detector and a clean OA fraction available for oxidation at the second detector.

Most published analyses of OA have employed a radioenzymatic procedure in which a radiolabelled methyl group is transferred to OA in the presence of phenylethanolamine-N-methyltransferase and S-adenosylmethionine with the resulting radiolabelled synephrine then being extracted and counted (Molinoff et al., 1969). This procedure has been criticized for lack of specificity (Dougan and Wade, 1981) and the additional purification steps required to ensure specific estimation of OA render the procedure laborious for routine processing of large numbers of samples. By contrast, HPLC/ElCD combines convenience with sensitivity and enables estimation of a spectrum of biogenic amines.

Table 1 demonstrates application of the procedure to the analysis of monoamines in the insect nervous system. The values obtained for OA, dopamine (DA) and norepinephrine (NE) compare favourably with those reported by Evans (1980) for this species with most nerve cord ganglia containing about 2 times as much OA as DA. By contrast, in the brain, the DA levels are almost double those of OA. In addition, the results indicate variable amounts of 5-HT in the different ganglia and high levels of TP in all ganglia. TP and 5-HT levels are high in the brain whereas TA and NE are present in relatively low concentrations.

The availability of a convenient, sensitive analytical procedure for determining levels of a spectrum of biogenic amines in discrete regions of the insect nervous

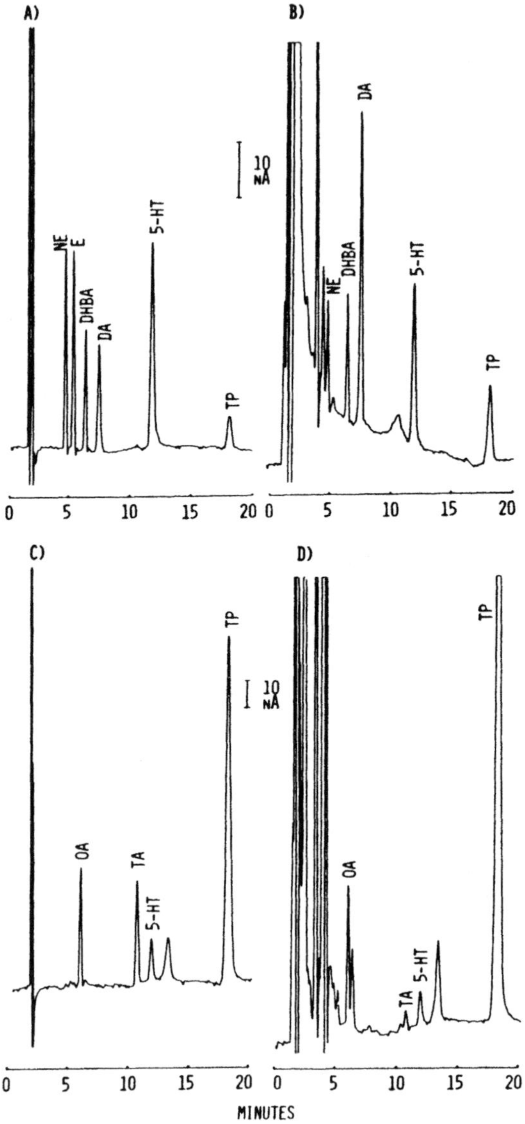

Fig. 1. Typical chromatograms obtained with detector 1 at 0.5V (A,B) and detector 2 at 0.75V (C,D).
(A), standard: 500 pg NE, E, DA, 5-HT; 250 pg DHBA; 1000 pg TP. (C) standard: 500 pg OA, TA, 5-HT; 1000 pg TP. (B) and (D) supernatant from whole brain of *Periplaneta americana* (1/5 brain equivalent). Chromatographic conditions as described in text.

Table 1. Biogenic amine levels in the central nervous system of *Periplaneta americana* as estimated by HPLC/ElCD

TISSUE	BIOGENIC AMINE CONTENT (pMole)					
	OA	DA	5-HT	TP	NE	TA
WHOLE BRAIN	21.63±3.39	40.83±2.67	15.02±2.64	42.51±3.56	6.33±0.80	2.24±0.31
PROTHORACIC GANGLION	4.79±0.30	2.44±0.20	1.31±0.39	16.65±3.85	N.D.	N.D.
MESOTHORACIC GANGLION	4.37±0.41	2.66±0.31	1.80±0.33	23.40±5.33	N.D.	N.D.
METATHORACIC GANGLION	4.68±0.33	2.68±0.23	2.30±0.17	24.07±6.29	N.D.	N.D.
ABDOMINAL GANGLIA 1-5 (PLUS INTERCONNECTIVES)	4.89±0.37	3.44±0.29	3.58±0.31	25.23±7.96	N.D.	N.D.
TERMINAL ABDOMINAL GANGLION	2.89±0.29	1.35±0.14	0.67±0.06	7.12±1.56	N.D.	N.D.

Values indicate mean ± standard error of the mean for 5 determinations.

system contributes further to the advantages of this system as a model for neurobiological investigation.

REFERENCES

Bailey, B.A., Martin, R.J. and Downer, R.G.H. (1982) Simultaneous determination of dopamine, norepinephrine, tyramine and octopamine by reverse phase liquid chromatography with electrochemical detection. J. Liquid Chromatog. 5, 2435-2452.

Boulton, A.A. and Juorio, A.V. (1981) Brain trace amines, in Handbook of Neurochemistry, Vol. 1 (Lajtha A., ed). Plenum Press, New York.

Dougan, D.F.H. and Wade, D.N. (1981) Aminergic neurons, phenolamine pathways and octopamine. TIPS, 1981, 113-116.

Downer, R.G.H. (1979) Induction of hypertrehalosemia by excitation in Periplaneta americana. J. Insect Physiol. 25, 59-63.

Evans, P.D. (1980) Biogenic amines in the insect nervous system. Adv. Insect Physiol. 15, 317-473.

Hoyle, G. (1975) Evidence that insect dorsal unpaired median (DUM) neurons are octopaminergic. J. exp. Zool. 193, 425-431.

Martin, R.J., Bailey, B.A. and Downer, R.G.H. (1983) Rapid estimation of catecholamines, octopamine and 5-hydroxytryptamine from biological tissues using high performance liquid chromatography with coulometric detection. J. Chromatogr. (in press).

Molinoff, P.B., Landsberg, L. and Axelrod, J. (1969) An enzymatic assay for octopamine and other β-hydroxylated phenylethylamines. J. Pharmac. exp. Therap. 170, 253-261.

Orchard, I. (1982) Octopamine in insects: neurotransmitter, neurohormone and neuromodulator. Can. J. Zool. 60, 659-669.

Shoup, R.E. (1982) Recent reports on liquid chromatography/electrochemistry, BAS Press, West Lafayette.

Yui, Y., Itokawa, Y. and Kawai, C. (1980) A rapid and highly sensitive method for determination of picogram levels of norepinephrine and epinephrine in tissues by high performance liquid chromatography. Analyt. Biochem. 108, 11-15.

NATURAL OCCURRENCE AND METABOLISM OF THE ISOMERIC OCTOPAMINES AND SYNEPHRINES

Clyde M. Williams, Margaret W. Couch and John M. Midgley

Veterans Administration Medical Center and Department of Radiology, University of Florida College of Medicine, Gainesville, Florida, U.S.A., 32610 and School of Pharmacy, University of London, London, U.K., WCIN IAX

Introduction: The development of a radiochemical enzyme assay for the quantitative determination of p-octopamine (Molinoff et al, 1969) led to its discovery in a wide variety of invertebrate nerve systems and in mammalian sympathetic nerves. Subsequent work led to the conclusion that p-octopamine is an invertebrate neurotransmitter and a cotransmitter with norepinephrine in mammalian sympathetic nerves (Axelrod and Saavedra, 1977). It was discovered in 1976 that the radiochemical enzyme assay was not specific for p-octopamine because the m- and p- isomers of octopamine could not be resolved. The use of a modification of the technique showed that both m- and p-octopamine are present in rat salivary gland (Robertson et al, 1977) and brain (Danielson et al, 1977). The radiochemical enzyme assay depends upon norepinephrine N-methyltransferase, which can accept all three positional isomers of octopamines as substrates with varying efficiency. The resultant products, the corresponding synephrines, are also substrates for the enzyme (Axelrod, 1962; Fuller et al, 1981). Consequently any one of these amines (or a mixture of them) would have been detected by the original unmodified assay but identified and quantified as p-octopamine. The advent of gas chromatography-mass spectrometry (GCMS) techniques has allowed the unequivocal resolution, identification and quantitative determination of all six octopamines and synephrines, together with their metabolites. This review summarizes recent work using GCMS methods which provide conclusive evidence that all three of the former and two of the latter amines occur naturally.

Occurrence in mammals: A negative chemical ionization (NCI) GCMS method has been developed recently for the quantitative determination of the octopamines and synephrines (Williams et al, 1983) and their distribution in various tissues of the rabbit is shown in Table 1.

TABLE 1
CONCENTRATION OF OCTOPAMINES AND SYNEPHRINES IN RABBIT TISSUE
(ng g^{-1})

Amine	Adrenal	Heart	Spleen	Vas Deferens	Brain
o-Octopamine	ND	ND	ND	ND	ND
m-Octopamine	10	1.8	1.2	4.5	ND
p-Octopamine	9	5.4	2.7	15	0.5
o-Synephrine	ND	ND	ND	ND	ND
m-Synephrine	88	ND	ND	4.6	ND
p-Synephrine	59	ND	ND	ND	ND

o-Octopamine: This amine was detected qualitatively in rat salivary and adrenal gland by a single ion electron-impact (EI) GCMS method (Williams and Couch, 1978), but could not be located subsequently in adrenal gland or other tissues using the more rigorous NCI GCMS technique (Table 1). Since the lower limit of detection of o-octopamine by this method is about 1-10 ng g^{-1}, it must be present in organs other than those examined or in a lower concentration in the organs investigated. When this method was employed for the analysis of rat urine, o-octopamine could not be detected in untreated animals (limit of detection of 10-20 ng day^{-1}), but after prior administration of a monamine oxidase (MAO) inhibitor rats excreted 300 ng day^{-1}: thus providing definitive evidence for the natural occurrence of o-octopamine (James et al, 1983).

m-Octopamine: m-Octopamine was initially identified in rat salivary gland (Robertson et al, 1977) and brain (Danielson et al, 1977) by a modified radiochemical enzyme assay. This result was confirmed qualitatively in rat salivary and adrenal gland and in bovine adrenal gland by a single ion EI GCMS method (Williams and Couch, 1978) and these results have been extended to other organs (in the rabbit) by a NCI GCMS method (Table 1). m-Octopamine was found in several sympathetically innervated organs and the concentrations reported in Table 1 were significantly increased (30 fold in heart and spleen) after MAO inhibition. m-Octopamine was not detected in whole rat brain but was found after MAO inhibition, thus confirming the results of Danielson et al (1977). The ratio of m-octopamine to p-octopamine varies from 1.1 in adrenal to 0.1 in monamine oxidase inhibited brain: corresponding values of 0.3 were observed in heart, spleen and vas deferens. These values are in good agreement with corresponding ratios found by Robertson et al (1977) (1.0 in rat salivary gland) and Danielson et al (1977) (0.06 in MAO-inhibited rat hypothalamus).

p-Octopamine: Robertson et al (1977) found that the concentration of p-octopamine in rat salivary gland was 260 ng g^{-1} when determined by the original radiochemical enzyme assay; but was found to be 8 ng g^{-1} when determined by the modified method. They concluded that the unmodified technique measured more than endogenous p-octopamine, and our work confirms this conclusion. Table 2 shows that the concentration of p-octopamine determined by the radiochemical enzyme assay (Molinoff and Axelrod, 1969) in 5 organs of the rat is significantly higher than the corresponding value obtained by the NCI GCMS method (Williams et al, 1983) for the same organs of the rabbit. Hence, the presence of p-octopamine in the tissues of sympathetically innervated organs is confirmed unequivocally but the concentrations of the amine are significantly less than those formerly reported.

TABLE 2
CONCENTRATION OF p-OCTOPAMINE DETERMINED BY DIFFERENT METHODS
(ng g^{-1})

	Radiochemical Enzyme Assay	NCI-GCMS
Adrenal	461	9
Brain	2.4	0.5
Heart	50	5.2
Spleen	14	2.7
Vas Deferens	30	15

o-Synephrine: o-Octopamine is a poor substrate for norepinephrine N-methyl transferase (Fuller et al, 1981) which may be the reason that o-synephrine has not yet been detected in any biological tissue (Table 1 - detection limit 10 ng g^{-1}) or rat urine (detection limit 200 ng day^{-1}) (James et al, 1983).

m-Synephrine: The first report of the natural occurrence of m-synephrine was in 1980 and the organ was adrenal gland (Midgley et al, 1980; Durden et al, 1980). These authors used different GCMS methods but similar results were obtained (Table 3). Norepinephrine N-methyl transferase is highly localized in adrenal gland but small amounts occur also in midbrain and heart (Axelrod, 1962). Since both m- and p-octopamine are found in these organs, it was anticipated that both m- and p-synephrine would also be present. Appreciable quantities of these amines were found in adrenal but not in any other tissue examined, except for vas deferens (in which m-synephrine was found) (Table 1).

TABLE 3
COMPARISON OF THE CONCENTRATIONS OF m- AND p-SYNEPHRINE OBTAINED BY DIFFERENT METHODS ($ng\ g^{-1}$)

Tissue	m-Synephrine	p-Synephrine	References
Bovine adrenal	83	-	Midgley et al (1980)
Bovine adrenal	160	310	Durden et al (1980)
Mouse adrenal	580	970	Durden et al (1980)
Rabbit adrenal	88	59	Williams et al (1983)

p-Synephrine: This amine is a constituent of oranges and its variable presence in human urine has been attributed to dietary factors. Durden et al (1980) first demonstrated that it occurred naturally (in bovine and mouse adrenal) and this observation has been confirmed in rabbit adrenal (Table 3) but it has not been detected in other tissues to date (Table 1).

Octopamines in the invertebrate nervous system: Farnham et al (1978) failed to find p-octopamine in the abdominal ganglia of Aplysia by EI GCMS despite an earlier report that it had been detected by the radiochemical enzyme assay (Saavedra et al, 1974). However, p-octopamine was located in other ganglia of Aplysia, but the possible presence of o- and m-octopamine and the synephrines in invertebrate nerves has not been investigated so far.

Metabolism of exogenous octopamines and synephrines: The metabolism of exogenous octopamines and synephrines in mammals is summarized in Table 4. In mammals aldehydes formed by deamination of primary and secondary biogenic catecholamines (such as norepinephrine and epinephrine) can be reduced to the corresponding glycol or oxidized to the corresponding acid. The monophenolic octopamines and synephrines also follow these two pathways and a variable amount of the amine is excreted as a conjugate. In rats the dominant pathway is reductive deamination to the corresponding conjugated hydroxyphenylglycol with a smaller amount being metabolized by oxidative deamination to the unconjugated hydroxymandelic acid. In humans, the reverse is true; the major pathway being oxidative.

Natural occurrence in urine of octopamines, synephrines and their metabolites: The natural occurrence of amines can be inferred from the presence in urine of their terminal metabolites. This information is summarized in Table 5. Neither o-octopamine, o-synephrine nor OHPG could be detected in the urine of normal humans or rats. However OHMA was easily identified in both species and also in patients with phenylketonuria where it is excreted in very large amounts (Crowley et al., 1980). Neither m-octopamine nor m-synephrine could be located in normal human

TABLE 4
METABOLISM OF EXOGENOUS OCTOPAMINES AND SYNEPHRINES
% administered dose

Amine	Species (route)	Ref.	Metabolite Acid*	Glycol*	Amine
o-octopamine	Rat (ip)	1	16(u)	33(c)	10(u), 4(c)
m-octopamine	Rat (ip)	2	nq(x)	nq(x)	nq(x)
	Human (iv)	3	57(u)	-	11(u),13(c)
p-octopamine	Human (iv)	4	66(u)	-	8(u), 8(c)
o-synephrine	Rat (ip)	1	14(u)	19(c)	10(u), 9(c)
m-synephrine	Human (oral)	5	35(u)	9(c)	60(c)
	Rat (ip)	5	6(u)	50(c)	9(c), 7(u)
p-synephrine	Human (oral)	6	67(u)	-	3(u)

* Refers to the corresponding hydroxymandelic acid and hydroxyphenylglycol; u-unconjugated; c-conjugated; x-conjugation not determined; nq-not quantitative; 1-James *et al*, 1983; 2-Maruyama *et al*, 1968; 3-Hengstmann *et al*, 1975; 4-Hengstmann *et al*, 1974; 5-Ibrahim *et al*, 1983; 6-Hengstmann and Aulepp, 1978.

or rat urine but both MHMA and MHPG were easily detected in both species and significant elevations of both MHMA and MHPG have been detected in 3 of 4 patients with neuroblastoma (Williams *et al.*, 1982). *p*-Octopamine and *p*-synephrine are both known to occur in human urine although the latter substance is believed to be of dietary origin. PHMA and PHPG occur in both humans and rats, in amounts much greater than the *o*- or *m*- analogs.

Physiology of the octopamines and synephrines: *m*-Synephrine is a powerful pressor and mydriatic agent but the *p*-isomer is a very much weaker pressor agent. *m*-Octopamine has significant pressor activity in the rat (1/100th that of norepinephrine) (Fregly *et al.*, 1979) but its activity on aortic strip reactivity is nearly as powerful as that of norepinephrine (Ress *et al.*, 1980). *p*-Octopamine is very much less active than *m*-octopamine in both these functions and neither *o*-octopamine nor *o*-synephrine has any known physiological or pharmacological activity at all.

Robertson (1981) has recently reviewed the evidence for *p*-octopamine as a neurotransmitter in invertebrates and vertebrates and has concluded that, while *p*-octopamine has a central role in invertebrate neurotransmission, it and other trace amines have only a minor function (if any) in the vertebrate nervous system. However there are alternative possibilities for a physiological role of the octopamines in mammalian sympathetic nerves. The two classical autonomic neurotransmitters are acetylcholine and norepinephrine but it is now clear that there are non-adrenergic, non-cholinergic nerves which also supply smooth muscle and the

TABLE 5
URINARY CONCENTRATIONS OF ENDOGENOUS OCTOPAMINES, SYNEPHRINES AND METABOLITES
ug day^{-1}*

Amine	Species	Ref.	Metabolite Acid	Glycol	Amine
o-octopamine	Human	1	11(u)	-	-
	Rat	2	1.1(u)	ND(< 0.2)	ND(< 0.2)
m-octopamine	Human	1	44(u)	-	-
	"	3	58(u)	-	-
	"	4	56(u)	11(c)	-
	Rat	4	0.8(u)	0.8(c)	-
p-octopamine	Human	5	-	-	0-30(c)
	Rat	6	60(u)	-	0.10(u)
	Human	6	2500(u)	-	-
	"	1	3300(u)		
	"	7	-	91(c)	-
	Rat	7	-	4.8(c)	
o-synephrine	Rat	2			ND(< 0.2)
m-synephrine	Human	8			ND
	Rat	8			ND
p-synephrine	Human	9			650

*Literature values in ng mg^{-1} creatinine converted to ug day^{-1} by assuming a daily excretion of 1.5 g creatinine in humans and 5 mg in rats; u-unconjugated; c-conjugated; ND-not detected; 1-Midgley et al, 1979; 2-James et al, 1983; 3-Davis and Boulton, 1981; 4-Crowley et al 1982; 5-Kakimoto and Armstrong, 1962a; 6-Kakimoto and Armstrong, 1962b; 7-Edwards et al, 1979; 8-Ibrahim et al, 1983; 9-Pisano et al, 1961.

presence of these nerves has been established in a wide variety of visceral and vascular organs (Burnstock et al., 1979). These non-adrenergic, non-cholinergic fibers to the gastrointestinal tract, lung, bladder and some blood vessels may be served by putative neurotransmitters such as ATP, serotonin, dopamine, GABA and a variety of polypeptides. Perhaps this list of possible autonomic neurotransmitters should be extended by the addition of m- and p-octopamine and possibly o-octopamine and m-synephrine.

Summary: Both m- and p-octopamine occur in several sympathetically innervated organs in a ratio varying from 1.0 (adrenal) to 0.2 (spleen). The total concentration ranges from 4 (spleen) to 20 (adrenal) ng g^{-1}. o-Octopamine has been identified unambiguously in the urine of rats previously treated with monamine oxidase inhibitors but cannot be detected in tissues (lower limits of detection 10 ng g^{-1}). o-Synephrine has not been detected in any tissue or fluid examined. m-Synephrine is present in adrenal gland in amounts of about 100 ng g^{-1} but not in any other tissue except vas deferens (1-5 ng g^{-1}). p-Synephrine is also present in adrenal gland in concentrations similar to that of m-synephrine but has not been detected elsewhere. These amines are metabolized by mammals largely to the corresponding hydroxyphenylglycol and hydroxymandelic acid.

References:

Axelrod J. (1962) Purification and properties of phenylethanolamine-N-methyl transferase. J. Biol. Chem. 237, 1657-1660.

Axelrod J. and Saavedra J. M. (1977) Octopamine. Nature 265, 501-504.

Burnstock G., Hokfelt T., Gershan M. D., Iversen L. L., Kosterlitz H. W. and Szurszewski J. H. (1979) Non-adrenergic, non-cholinergic autonomic neurotransmission mechanisms. Neurosciences Res. Prog. Bull. 17, 379-519.

Crowley J. R., Couch M. W., Williams C. M., James M. I., Ibrahim K. E. and Midgley J. M. (1982) Identification and quantitative determination of m-hydroxyphenylglycol in mammalian urine. Biomed. Mass Spect. 9, 146-152.

Crowley J. R., Midgley J. M., Couch M. W., Garnica A. and Williams C. M. (1980) Increased excretion of o-hydroxymandelic acid in phenylketonuria. Biomed. Mass Spect. 7, 349-353.

Danielson T. J., Boulton A. A. and Robertson H. A. (1977) m-Octopamine, p-octopamine and phenylethanolamine in rat brain. A sensitive, specific assay and the effects of some drugs. J. Neurochem. 29, 1131-1135.

Davis B. A. and Boulton A. A. (1981) Excretion of m-hydroxymandelic acid in human urine. J. Chromatog. 222, 271-275.

Durden D. A., Juorio A. V. and Davis B. A. (1980) Thin-layer chromatographic and high resolution mass spectrometric determination of β-hydroxyphenylethylamines in tissues as dansyl-acetyl derivatives. Anal. Chem. 52, 1815-1820.

Edwards D. J., Rizk M. and Neil J. (1979) Simultaneous analysis of phenylglycols and phenylethanols in human urine by gas chromatography-mass spectrometry. J. Chromatog. 164, 407-416.

Farnham P. J., Novak R. A. and McAdoo D. J. (1978) A re-examination of the distribution of octopamine and phenylethanolamine in the Aplysia nervous system. J. Neurochem. 30, 1173-1176.

Fregly M. J., Kelleher D. L. and Williams C. M. (1979) Adrenergic activity of ortho- meta- and para-octopamine. Pharmacol. 18, 180-187.

Fuller R. W., Hemrick-Luecke S. K. and Midgley J. M. (1981) Comparison of o-octopamine and related phenylethanolamines as substrates for norepinephrine N-methyl transferase. Res. Comm. Chem. Path. Pharmacol. 33, 207-213.

Hengstmann J. H. and Aulepp H. (1978) Pharmakokinetik und Metabolismus von ^{3}H-Synephrine. Arzneim-Forsch. Drug. Res. 28, 2326-2328.

Hengstmann J. H., Konen W., Konen C., Eichelbaum M. and Dengler H. J. (1974) The physiological disposition of p-octopamine in man. Naunyn-Schmiedeberg's Arch. Pharmacol. 283, 93-106.

Hengstmann J. H., Konen W., Konen C., Eichelbaum M. and Dengler H. J. (1975) Bioavailability of m-octopamine in man related to its metabolism. Europ. J. Clin. Pharmacol. 8, 33-39.

Ibrahim K. E., Midgley J. M., Crowley J. R. and Williams C. M. (1983) The mammalian metabolism of R-(-)-m-synephrine. J. Pharm. Pharmacol. 35, 144-147.

James M. I., Midgley J. M. and Williams C. M. (1983) Metabolism and biosynthesis of (+)-o-octopamine and (+)-o-synephrine in the rat. J. Pharm. Pharmacol. (in press)

Kakimoto Y. and Armstrong M. D. (1962a) The phenolic amines of human urine. J. Biol. Chem. 237, 208-214.

Kakimoto Y. and Armstrong M. D. (1962b) On the identification of octopamine in mammals. J. Biol. Chem. 237, 422-427.

Maruyama K., Tanaka A., Urakubo G., Irino O. and Fukawa K. (1968) Studies on norphenylephrine (NP). I. Metabolites of NP in animals. Yakugaku Zasshi 88, 1516-1522.

Midgley J. M., Couch M. W., Crowley J. R. and Williams C. M. (1979) Identification and quantitative determination of o- and m-hydroxymandelic acid in human urine. Biomed. Mass Spect. 6, 485-490.

Midgley J. M., Couch M. W., Crowley J. R. and Williams C. M. (1980) m-Synephrine: normal occurrence in adrenal gland. J. Neurochem. 34, 1225-1230.

Molinoff P. and Axelrod J. (1969) Octopamine: normal occurrence in sympathetic nerves of rats. Science 164, 428-429.

Molinoff P. B., Landsberg L. and Axelrod J. (1969) An enzymatic assay for octopamine and other β-hydroxylated phenylethyl amines. J. Pharmacol. Exp. Ther. 170, 253-261.

Pisano J. J., Oates J. A., Karmen A., Sjoerdsma A. and Udenfriend S. (1961) Identification of p-hydroxy-α-(methylamino-methyl)benzyl alcohol (synephrine) in human urine. J. Biol. Chem. 236, 898-901.

Ress R. J., Rahmani M. A., Fregly M. J., Field F. P. and Williams C. M. (1980) Effects of isomers of octopamine on in vitro reactivity of vascular smooth muscle of rats. Pharmacol. 21, 342-347.

Robertson H. A. (1981) Octopamine - after a decade as a putative neuroregulator in Essays in Neurochemistry and Neuropharmacology ed. by Youdim M. B. H., Lovenberg W., Sharman D. F. and Laguado J. R. 5, 47-73.

Robertson H. A., David J. C. and Danielson T. J. (1977) Effects of denervation on the levels of p-octopamine, m-octopamine (norphenylephrine) and phenylethanolamine in rat salivary gland. J. Neurochem. 29, 1137-1139.

Saavedra J. M., Brownstein M. J., Carpenter D. O. and Axelrod J. (1974) Octopamine: presence in single neurons of Aplysia suggests neurotransmitter function. Science 185, 364-365.

Williams C. M. and Couch M. W. (1978) Identification of ortho-octopamine and meta-octopamine in mammalian adrenal and salivary gland. Life Sci. 22, 2113-2120.

Williams C. M., Crowley J. R., Couch M. W. and Midgley J. M. (1982) Increased excretion of m-hdyroxyphenylglycol and m-hydroxymandelic acid in neuroblastoma. Biochem. Med. 28, 305-309.

Williams C. M., Couch M. W. and Crowley J. R. (1983) m-Octopamine: normal occurrence with p-octopamine in mammalian sympathetic nerves. (in preparation)

OCTOPAMINE IS PRESENT IN RETINAS OF VARIOUS MAMMALIAN SPECIES

T.P. Hicks and D. Parkinson

Departments of Medical Physiology and Anatomy,
Faculty of Medicine, The University of Calgary,
Calgary, Alberta. Canada T2N 4N1

It has become increasingly evident during the past decade that octopamine functions as a synaptic transmitter, or neuromodulator, within the nervous systems of many invertebrates. A similar role for this compound in the mammalian nervous system is less firmly established and remains, in some measure, a controversial issue. Certainly octopamine is present in mammalian CNS and the fact of its occurrence in trace quantities within the brain is compromised somewhat by a relatively high rate of turnover. There is evidence of post-synaptic receptors for octopamine being pharmacologically distinct from those of the catecholamines, but to date it has proved impossible to identify discrete populations of neurones in the mammalian CNS which contain, or are exclusively responsive to, octopamine.

Ehinger (1976) reported uptake of exogenous octopamine into the rabbit retina, but the measurement of endogenous octopamine in this tissue has not been reported. The retina has several advantages for examining the possible role of octopamine as a synaptic transmitter in mammalian nervous systems: (1) it is a part of the CNS, but is anatomically isolated and therefore can be subjected to local experimental manipulations while minimizing systemic side effects; (2) the structure and function of the retina have been well investigated with anatomical, immunochemical and electrophysiological techniques; (3) the level of activity of its constituent neurones can be modulated by employing natural stimuli (ie: light); and, (4) neuronal processing is directed predominantly centripetally, with no or little central input to complicate the interpretation of results. We therefore sought to establish whether or not octopamine occurs in this tissue in mammals, and if so, whether it is distributed evenly across a number of species.

METHODS

Animals were killed by cervical dislocation and both of the retinas quickly dissected out, pooled and frozen on dry ice. Tissues were stored at -70° until assayed. Octopamine levels were measured by the radioenzymatic method with phenylethanolamine-N-methyl transferase (PNMT) and ^{3}H-S-adenosyl methionine (SAM) as cofactor (Saavedra, 1974). Retinas were homogenized in 250 μl/retina of 20 mM TRIS/10^{-4}M pargyline pH 8.6, heated at 95° C for 10 min and centrifuged at 10,000 g for 15 min. Four equal aliquots of the supernatant were dispensed and 500 pg of octopamine was added to two of these to provide an external standard. The final incubation mixture (100 μl) contained 80 μl sample, 10 μl PNMT and 10 μl (1 μCi) ^{3}H-SAM. After incubation at 37° for 45 min, the reaction was stopped by the addition of 300 μl 0.5M sodium borate pH 10.0 containing 200 ng synephrine as carrier. The aqueous phase was extracted with 1 ml toluene/isoamyl alcohol. After a brief centrifugation, the organic layer was taken to a fresh tube and evaporated under vacuum. The tritium-labelled material which remained was dissolved in 1 ml methanol and estimated by liquid scintillation spectroscopy. A new standard curve was plotted for each occasion to demonstrate the linearity and sensitivity of the assay. Tissue octopamine levels were calculated by the external standard ratio method after correction for a reagent blank.

RESULTS

The radioenzymatic assay in our hands was linear from 100 pg to 2 ng. A typical standard curve is given in Figure 1. The sensitivity of the method, as defined by sufficient octopamine to give a response of twice the blank counts, was around 50 pg.

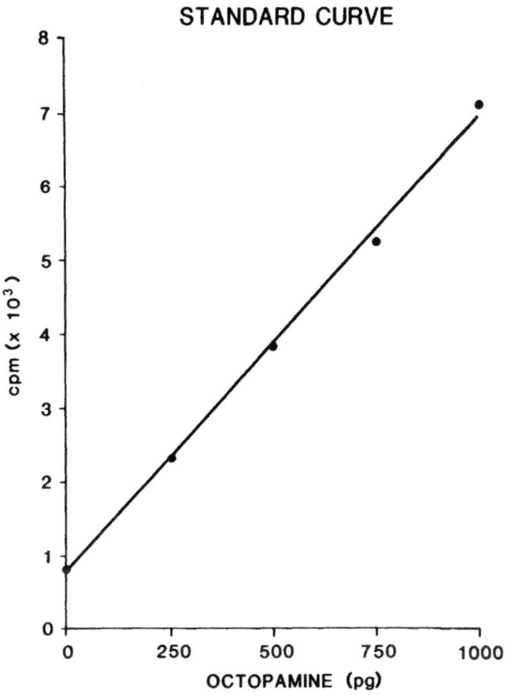

Fig. 1. A standard curve for octopamine, illustrating the linearity of the assay as well as its sensitivity. Ordinate: counts per minute, Abcissa: octopamine content.

Octopamine has been detected in all three of the animal species so far examined, namely rat, hamster and guinea pig. The results are given in Table 1. Retinal octopamine levels were similar in all three species.

Administration of the monoamine oxidase inhibitor pargyline, intraperitoneally (150 mg/kg) 6 h prior to eye removal, increased retinal octopamine levels by about 70% in the guinea pig. In one experiment, in which two rats were treated similarly with pargyline, retinal octopamine levels were also increased.

TABLE 1. RETINAL OCTOPAMINE LEVELS

SPECIES	[OCTOPAMINE](PG/MG PROTEIN)
Hamster	170 ± 8 (5)
Rat	277 ± 30 (5)
Guinea Pig (Control)	120 ± 15 (7)
Guinea Pig (150 mg/kg Pargyline, i.p. 6 h)	200 ± 25 (5)

DISCUSSION

The results indicate the existence of radioenzymatically detectible octopamine in various mammalian retinas. When expressed relative to tissue wet weight, these values are comparable to levels reported in other parts of the mammalian CNS (Buck et al., 1977, Danielson et al., 1977, Hicks, 1977). The cephalopod retina is morphologically similar to the vertebrate retina, the major difference being that the former is oriented with the photoreceptors proximal to the vitreal surface. *Octopus vulgaris* retina contains considerable amounts of octopamine (50 ng/g wet weight, Juorio and Molinoff, 1974).

The retina would be a useful model system for examining the possible transmitter role of octopamine, but first several basic criteria need to be met. An initial question which must be answered is: Is the material we identify as radio-labelled product, in fact, derived from endogenous octopamine? The high specificity of the radioenzymatic method provides some degree of confidence in the data. Chromatographic systems have been described for a more sophisticated variation of this assay technique (Danielson et al., 1977) and these methods would provide further evidence that our data reflect endogenous octopamine content.

Of all the principal biogenic amine transmitters now known, including noradrenaline, dopamine and 5-hydroxtryptamine, mammalian retina only contains significant amounts of dopamine (Häggendal & Malmfors, 1965). Dopamine is present in a subset of amacrine neurones in this tissue. Neurochemical evidence indicates that these neurones are activated when the retina is exposed to photic stimulation but do not respond in the dark (Iuvone et al., 1978;

Parkinson & Rando, 1983). It has been suggested that octopamine found in mammalian tissues may arise from a biosynthetic route involving conversion of dietary tyramine to octopamine by dopamine-β-hydroxylase in noradrenergic neurones. In the retina, such a pathway is less likely, due to the absence of noradrenergic neurones. The exception would be those retinas which are heavily vascularized, such as the rat retina, which would then contain vascular adrenergic neurones. Yet we could still find significant amounts of octopamine in the guinea pig retina, which is avascular.

Pretreatment of guinea pigs with pargyline increased retinal octopamine levels by about 70% after 6 h. This result indicates that retinal octopamine is metabolized by monoamine oxidase. The increased octopamine levels could arise not only from reduced catabolism of this amine, but also from an increased rate of synthesis arising from elevated systemic levels of tyramine. It would be interesting to examine the effect of an intraocular injection of pargyline (to inhibit retinal MAO without the possibility of accompanying systemic effects) on retinal octopamine levels. Such localized experiments could also be performed with inhibitors of various biosynthetic enzymes, to determine their roles in octopamine synthesis. In addition, it might then be demonstrated whether the octopamine was of retinal or systemic origin.

The retina has proved to be a good model system for studying the more established synaptic transmitters; this tissue may provide useful evidence to support or refute the notion that octopamine has a role as a mammalian central synaptic transmitter.

ACKNOWLEDGEMENTS

The present work was supported by the AHFMR and the Canadian MRC.

REFERENCES

Buck S.H., Murphy R.C. and Molinoff P.B. (1977) The normal occurrence of octopamine in the central nervous system of the rat. *Brain Res.* 122, 281-297.

Danielson T.J., Boulton A.A. and Robertson H.A. (1977) *m*-Octopamine, *p*-octopamine and phenylethanolamine in rat brain: a sensitive, specific assay and the effects of some drugs. *J. Neurochem.* 29, 1131-1135.

Ehinger B. (1976) Weak neuronal accumulation of octopamine in dopaminergic neurons of the rabbit retina. *Experientia* 32, 890-891.

Häggendal J. and Malmfors T. (1965) Identification and cellular localization of the catecholamines in the retina and the choroid of the rabbit. *Acta physiol. scand.* 64, 58-66.

Hicks T.P. (1977) The possible role of octopamine as a synaptic transmitter: a review. *Can. J. Physiol. Pharmacol.* 55, 137-152.

Iuvone P.M., Galli C.L., Garrison-Gund C.K. and Neff N.H. (1978) Light stimulates tyrosine hydroxylase activity and dopamine synthesis in retinal amacrine cells. *Science* 202, 901-902.

Juorio A.V. and Molinoff P.B. (1974) The normal occurrence of octopamine in neural tissues of the *octopus* and other cephalopods. *J. Neurochem.* 22, 271-280.

Parkinson D. and Rando R.R. (1983) Effect of light on dopamine turnover and metabolism in rabbit retina. *Invest. Ophthalmol. Vis. Sci.* 24, 384-388.

Saavedra J.M. (1974) Enzymatic-isotopic method for octopamine at the picogram level. *Anal. Biochem.* 59, 628-633.

C. PHYSIOLOGY AND PHARMACOLOGY

INVESTIGATION OF TRACE AMINE METABOLISM IN THE CENTRAL NERVOUS SYSTEM THROUGH MEASUREMENTS ON CEREBROSPINAL FLUID

Simon N. Young

Department of Psychiatry, McGill University,

1033 Pine Avenue West, Montreal, Quebec,
Canada, H3A 1A1

ABSTRACT

The metabolism of trace amines in the CNS can be studied through measurement of their product acids in the CSF. Animal studies suggest that an appreciable amount of tryptamine in the brain is derived from the periphery and diffuses into the brain. The same is probably true for phenylethylamine. Phenylethylamine is metabolized in human brain at about the same rate as the biogenic amines and p-tyramine at about half that rate. Tryptamine metabolism is only somewhat slower, while m-tyramine metabolism is a quantitatively minor pathway. Tryptophan availability is an important controlling factor for tryptamine turnover while aromatic amino acid decarboxylase levels are important for both tryptamine and phenylethylamine. Unlike biogenic amine metabolism, trace amine metabolism does not seem to be influenced in any important way by the age or sex of the subject. Human brain tryptamine metabolism is increased by melatonin administration, possibly due to its action on the activity of aromatic amino acid decarboxylase. In narcolepsy and idiopathic hypersomnia CSF levels of the tryptamine metabolite, indoleacetic acid, and of dopamine, are low. As both tryptamine and dopamine are present in brain at the highest levels in the basal ganglia, this brain region is implicated in the pathophysiology of hypersomnia. CSF indoleacetic acid was elevated in hepatic coma, the elevation being related to the grade of coma. This is consistent with the trace amine/false transmitter hypothesis of hepatic coma.

INTRODUCTION

Investigation of the validity of the theories that relate abnormal trace amine metabolism to neuropsychiatric disorders will ultimately depend on measurements of trace amine metabolism in human brain. For the biogenic amines two main methods have been used to study human brain metabolism: postmortem studies and measurements on cerebrospinal fluid (CSF). For the trace amines direct measurements on human brains may have limited applicability because of rapid postmortem changes of some of the trace amines (Philips et al., 1978). Thus, CSF studies are likely to be the main source of information on human trace amine metabolism for many years to come. This paper discusses the extent to which measurements of trace amine metabolites in the CSF reflect metabolism of the parent amines in the CNS and describes some of the factors influencing trace amine metabolism in the brain.

METHODS

In the author's studies the rat experiments were done on male Sprague-Dawley retired breeders. To obtain rat CSF, the animals were anesthetized with Nembutal, the atlanto-occipital membrane was exposed, and 75-150 μl of CSF was taken from the cisterna magna by a syringe, after puncturing the membrane. Human CSF was collected from patients undergoing diagnostic pneumoencephalography. The first and last CSF samples were collected. The first sample consists of CSF derived from the lumbar sac, while the last sample, taken after injection of oxygen, was originally in the basal cisterns and is referred to here as cisternal CSF. In other studies CSF was obtained by simple lumbar puncture from the fasting patients. Tryptophan, 5-hydroxyindoleacetic acid (5HIAA), indoleacetic acid (IAA) and indolpropionic acid were measured by direct injection of CSF into a high performance liquid chromatograph (HPLC) with a reverse-phase column and fluorometric detection (Anderson and Purdy, 1979). Phenylacetic acid (PAA), p-hydroxyphenylacetic acid (pHPAA) and m-hydroxyphenylacetic acid (mHPAA) in CSF were measured by gas chromatography-mass spectrometry (Davis et al., 1982).

ARE CSF AMINE METABOLITES DERIVED FROM THE BRAIN?

The common assumption that the level of any amine metabolite in the CSF reflects metabolism of the parent amine in the CNS is not necessarily valid. Compounds present in CSF at low levels might well be derived totally or partially from the periphery and diffuse into CSF. In a study of indoles in CSF we found indolepropionic acid in rat and human CSF at concentrations not far below that of IAA (Young et al., 1980a). Indolepropionic acid is formed by bacterial metabolism in the intestine, and diffuses into the CSF via the blood. We showed that IAA can also diffuse into CSF when its plasma level is very high. However, the normal blood-CSF gradient for IAA would not allow any appreciable peripheral contribution into the CSF (Young et al., 1980a). Animal experiments also show that PAA does not penetrate the blood-CSF barrier (Pedemonte et al., 1976). pHPAA and mHPAA, which are more polar than PAA and are also present in plasma at a lower concentration than PAA (Davis et al., 1982) are also unlikely to diffuse from blood to CSF in any important amounts.

METABOLIC ORIGINS OF TRACE AMINE METABOLITES IN CSF

It is necessary to consider not only the anatomical origin but also the metabolic origin of compounds in CSF. IAA can be found from tryptamine (T), but could it also be formed by transamination of tryptophan to indolepyruvic acid and decarboxylation and oxidation to IAA? If the formation of IAA from tryptophan is catalyzed by the same route and by the same enzymes that convert 5-hydroxytryptophan to 5HIAA, then inhibitors of these enzymes should affect CSF 5HIAA and IAA in the same manner. We found that the monoamine oxidase (MAO) inhibitor pargyline and the decarboxylase inhibitor brocresine decrease rat CSF 5HIAA and IAA to a similar extent (Young et al., 1980b).

mHPAA could be formed by transamination and decarboxylation of m-tyrosine but m-tyrosine is not thought to occur naturally. p-Tyrosine is transaminated to its keto-acid, which is metabolized further to homogentisic acid rather than to pHPAA. Similarly, phenylalanine can be transaminated but further metabolism produces other compounds rather than PAA (Kaufmann, 1977). The idea that CSF PAA comes from β-phenylethylamine (PE) in the CNS is also supported by the data of Karoum et al. (this volume). Thus IAA, PAA pHPAA and mHPAA all seem to be derived from their parent trace amines [T, PE, p-tyramine (p-TA) and m-tyramine (m-TA) respectively] rather than by transamination and decarboxylation of amino acids. This conclusion is supported by the fact that the regional distribution of all these acids in rat brain parallels that of their relevant parent trace amines (Warsh et al., 1977; Durden and Boulton, 1981, 1982).

DIFFERENCES IN THE METABOLISM OF TRYPTAMINE AND 5-HYDROXYTRYPTAMINE

Measurements of IAA and 5HIAA in rat CSF provide a simple method for studying the metabolism of the parent indoleamines in rat brain. Tryptophan loading experiments showed that metabolism of T is much more responsive to changes in precursor availability than is metabolism of 5-hydroxytryptamine (5HT) (Young et al., 1980b). Thus, CSF 5HIAA only doubled, while IAA could be increased 25-fold with a large tryptophan load. This is because tryptophan hydroxylase is much closer to saturation with tryptophan than is aromatic amino acid decarboxylase. Another difference in the synthesis of the two compounds concerns their source. A peripheral decarboxylase inhibitor (carbidopa or benserazide) lowers rat CSF IAA but not 5HIAA, suggesting that some of the T in brain is formed peripherally and then diffuses into brain (Young et al., 1980b). A peripheral origin does not, of course, exclude a central function.

The two indoleamines also showed differences in their catabolism. Thus, clorgyline, which inhibits MAO A preferentially, lowered rat CSF 5HIAA but not IAA (Young et al., 1980b). Deprenyl, an MAO B inhibitor, had the opposite effect. Thus the relative efficacies of the two forms of MAO which have been established *in vitro* (Houslay et al., 1976) seem also to apply to the catabolism of the indoleamines *in vivo*.

Catabolism of amines is affected by their release and here also the two indoleamines differ. Thus, reserpine increased rat CSF 5HIAA because of release of 5HT from storage vesicles. However, CSF IAA levels showed no evidence for release of T (Young et al., 1980b). Cold stress which causes release of 5HT and thus an increase of CSF 5HIAA (Isaac, 1975) also failed to increase IAA (Young et al., 1980b). The data from these two experiments support the idea that T is not released from neurons in the same way as 5HT (Hery et al., 1979), and is probably not stored in vesicles.

CSF MEASUREMENTS FOR THE STUDY OF TRACE AMINE METABOLISM IN HUMAN CNS

The Influence of Transport on Amine Metabolites in Lumbar CSF

A variety of significant positive correlations are seen between the levels of the acid metabolites of the biogenic amines and trace amines in human lumbar CSF. However, most of these correlations are not seen in cisternal CSF (Young et al., 1980c, 1982). Presumably the correlations in lumbar CSF are due in part to the transport system which removes these acids from the CSF. This probenecid-sensitive transport system is present in the choroid plexuses (Forn, 1972) and in the cortical and spinal subarachnoid spaces (Wolfson et al., 1974; van der Poel et al., 1977). There is no evidence for such a transport system in the basal cisterns. Thus, the significant positive correlation between, for example, 5HIAA and IAA in lumbar CSF could be due to a similar rate of removal of these two compounds from the CSF, and also possibly inhibition of IAA egress by the higher levels of 5HIAA. In the basal cisterns, where there is no transport system, no such correlation is seen (Young et al., 1980c). Obviously, if the levels of trace amine metabolites in lumbar CSF are influenced in part by the transport of the acids out of CSF, more information can be obtained from cisternal CSF or from measurements on lumbar CSF of patients given probenecid to block the transport system. Probenecid is known to increase the levels of the acid metabolites of trace amines in lumbar CSF (Karoum et al., 1977). Nonetheless, in spite of the disadvantage of lumbar CSF measurements without probenecid administration, as will be seen below useful data can be obtained from such measurements.

Concentration Gradients of Trace Amine Metabolites in Human CSF

The sharp concentration gradient for homovanillic acid between the ventricles, basal cisterns and lumbar sac reflects the fact that dopamine is metabolized very much more actively in the brain than in the spinal cord. Spinal cord metabolism of 5HT is more active than that of dopamine and 5HIAA shows less of a gradient than homovanillic acid

(Garelis et al.,1974). Among the trace amine metabolites IAA and pHPAA show a significant cisternal-lumbar gradient, while PAA and mHPAA do not (Young et al., 1980c, 1982). A significant ventricular-lumbar gradient has also been demonstrated for pHPAA and the p-octopamine metabolite, p-hydroxymandelic acid (Karoum et al., 1975; Kobayashi et al., this volume). These data indicate that spinal cord metabolism is relatively more important for PE and m-TA than it is for T, p-TA or p-octopamine.

The Influence of Sex and Age on Amine Metabolites

The mean levels of the biogenic amine metabolites, 5HIAA, homovanillic acid and 3-methoxy-4-hydroxyphenylethylene glycol are all higher in CSF from women than from men (Young et al., 1980c; Maas et al., 1982). However, the trace amines, T, PE, m-TA and p-TA show no sex differences in their rates of metabolism (Young et al., 1980c, 1982). IAA levels in lumbar CSF are higher in females than in males, but this is probably due to inhibition of the transport of IAA out of CSF by the higher levels of 5HIAA, which also show this sex difference. In cisternal CSF the same sex difference is seen for 5HIAA, but not for IAA (Young et al., 1980c).

Among the trace amines considered here only m-TA may show age-related changes. However as the significant positive correlation between mHPAA and age is seen in lumbar CSF and not cisternal CSF, the effect may not be a real one (Young et al., 1982).

Rates of Metabolism of Trace Amines in Human CNS

The concentration of IAA in rat CSF is about 5% of the 5HIAA level (Young et al., 1980a). Studies on rat brain show that the rate of T metabolism is about 5% of the rate of 5HT metabolism (Warsh et al., 1979). Thus, the relative amounts of the metabolites in CSF seem to reflect the relative rates of metabolism of the parent amine in the CNS. In human CSF the IAA content is 20% of the 5HIAA concentration (Young et al., 1980a), suggesting that metabolism of T is relatively more important in human brain than in rat brain.

CSF data indicate that PE is metabolized in human brain at about the same rate as the biogenic amines (Young et al., 1982), a situation that is known to be true for rat brain (Durden and Philips, 1980). CSF data also indicate that p-TA is metabolized in human brain at about half the rate of PE, while m-TA is a minor metabolite having a turnover of only 16% of the rate of p-TA (Young et al., 1982).

Precursors and Enzymes Controlling the Rate of Trace Amine Synthesis in Human CNS

In human CSF there is a significant positive correlation between the concentrations of tryptophan and IAA (Young et al., 1980c). One explanation for this is that tryptophan availability normally is one factor controlling the synthesis of T. Support for this idea comes from tryptophan loading experiments. In humans, IAA rises more than 5HIAA, and after a 6 g tryptophan load T metabolism in human brain can be as high as 40% of the rate of 5HT metabolism (Young and Gauthier, 1981). There is no correlation between phenylalanine and PAA in human CSF, a surprising conclusion as aromatic amino acid decarboxylase from the rat is as far from saturation with phenylalanine (Fellman et al., 1976) as with tryptophan (Ichiyama et al., 1970). Levels of PAA and IAA are significantly positively correlated in human cisternal CSF (Young et al., 1982). The parent amines, PE and T, are the only two formed exclusively by the action of aromatic amino acid decarboxylase. Thus, the decarboxylase, which is known to be very variable in human brain (Sacks et al., 1979), may be an important factor controlling metabolism of both PE and T in brain.

Effect of Melatonin on Tryptamine Metabolism in Human Brain

In rat brain, melatonin increases the level of pyridoxal phosphokinase (Anton-Tay et al., 1970). This should increase brain pyridoxal phosphate and thus might increase aromatic amino acid decarboxylase activity, as pyridoxal phosphate is its cofactor. Therefore we gave 0.5 g oral melatonin loads to patients undergoing pneumoencephalography 8 h before CSF was taken. IAA levels in both lumbar and cisternal CSF doubled due to the melatonin, indicating an increased rate of T metabolism (Young et al., 1983), although there was no effect on 5HIAA.

Trace Amine Metabolism in Neuropsychiatric Disorders

Numerous theories relate altered trace amine metabolism to neuropsychiatric disorders, but most of these theories have not yet been tested experimentally in any adequate way. Much of the clinical data comes from studies on plasma and urine, although there is some CSF data. Thus, CSF PAA has been reported to be low in depression and high in schizophrenia (Sandler et al., 1977, 1978). We have looked at T metabolism in two disorders. In narcolepsy and in idiopathic hypersomnia we found low levels of IAA in the CSF (Montplaisir et al., 1982). This was of special interest because CSF dopamine was also low and both dopamine and T are found in highest concentration in the caudate nucleus (Philips et al., 1978). This structure is therefore implicated in the pathophysiology of hypersomnia.

Liver failure can be associated with neuropsychiatric changes and coma. Baldessarini and Fisher (1977) have suggested that altered brain function in this condition may be due to the accumulation of trace amines in the CNS which interfere with the action of the biogenic amines. Although many studies have looked at trace amines in the brains of animals with liver failure or in the plasma of patients with hepatic coma little work has been done looking at CSF metabolites in the human condition. We found that lumbar CSF IAA was higher in patients with cirrhosis of the liver than in controls. It was also higher in the CSF of patients in coma than in those with hepatic cirrhosis but not in coma. There was a strong correlation between the grade of coma and CSF IAA (Young and Lal, 1980). How far these changes in the metabolism of T (or of other trace amines) are relevant to the induction of hepatic coma or are simply a reflection of advanced liver dysfunction is unclear.

CONCLUSIONS

CSF measurements are a suitable method for studying trace amine metabolism in the human CNS. At the moment data on the CSF levels of trace amine metabolites is fragmentary and there is little or no information for some compounds, for example, the octopamine metabolites. Many more studies are needed to complete our knowledge of the factors normally controlling trace amine metabolism in human brain and to test theories relating trace amines to neuropsychiatric disorders. Advances in methodology should make it possible to measure the amines themselves in CSF and this will be a valuable additional source of information in such studies.

ACKNOWLEDGEMENTS

Work in the author's laboratory is supported by the Medical Research Council of Canada.

REFERENCES

Anderson G.M. and Purdy W.C. (1979) A liquid chromatographic-fluorometric system for the analysis of indoles in physiological samples. Anal. Chem. 51, 283-286.

Anton-Tay F., Sepulveda J., and Gonzalez S. (1970) Increase of brain pyridoxal phosphokinase activity following melatonin administration. Life Sci. 9, 1283-1288.

Baldessarini R.J. and Fischer J.E. (1977) Substitute and alternative neurotransmitters in neuropsychiatric illness. Arch. Gen. Psychiat. 34, 958-964.

Davis B.A., Durden D.A., and Boulton A.A. (1982) Plasma concentrations of p and m-hydroxyphenylacetic acid and phenylacetic acid in humans: gas chromatographic-high resolution mass spectrometric analysis. J. Chromatogr. 230, 283-286.

Durden D.A. and Philips S.R. (1980) Kinetic measurements of the turnover rates of phenylethylamine and tryptamine in vivo in the rat brain. J. Neurochem. 34, 1725-1732.

Durden D.A. and Boulton A.A. (1981) Identification and distribution of m- and p-hydroxyphenylacetic acids in the brain of the rat. J. Neurochem. 36, 129-135.

Durden D.A. and Boulton A.A. (1982) Identification and distribution of phenylacetic acid in the brain of the rat. J. Neurochem. 38, 1532-1536.

Fellman J.H., Roth E.S., and Fujita T.S. (1976) Decarboxylation to tyramine is not a major route of tyrosine metabolism in mammals. Arch. Biochem. Biophys. 174, 562-567.

Forn J. (1972) Active transport of 5-hydroxyindolcacetic acid by the rabbit choroid plexus in vitro; blockade of probenecid and metabolic inhibitors. Biochem. Pharmacol. 21, 619-624.

Garelis E, Young S.N., Lal S. and Sourkes T.L. (1974) Monoamine metabolites in lumbar CSF: The question of their origin in relation to clinical studies. Brain Res. 79, 1-8.

Héry F., Simonnet G., Bourgoin S., Soubrié P., Artand F., Hamon M., and Glowinski J. (1979) Effect of nerve activity on the in vivo release of (^{3}H)serotonin continuously formed from L-(^{3}H)tryptophan in the caudate nucleus of the cat. Brain Res. 169, 317-334.

Houslay M.D., Tipton K.F., and Youdim M.B.H. (1976) Multiple forms of monoamine oxidase: Fact and artefact. Life Sci. 19, 467-478.

Ichiyama A., Nakamura S., Nishizuka Y., and Hayaishi O. (1970) Enzymic studies on the biosynthesis of serotonin in mammalian brain. J. Biol. Chem. 245, 1699-1709.

Isaac L. (1973) Temperature alteration of monoamine metabolites in cerebrospinal fluid. Nature (New Biol.) 243, 269-271.

Karoum F., Gillin J.C., Wyatt R.J., and Costa, E. (1975) Mass-fragmentography of nanogram quantities of biogenic amine metabolites in human cerebrospinal fluid and whole rat brain. Biomed. Mass Spectr. 2, 183-189.

Karoum F., Bunney W., Gillin J.C., Jimerson D., van Kammen D., and Wyatt R.J. (1977) Effect of probenecid on the concentration of the lumbar cerebrospinal fluid acidic metabolites of tyramine, octopamine, dopamine and norepinephrine. Biochem. Pharmacol. 26, 629-632.

Kaufman S. (1977) Phenylketonuria: biochemical mechanisms, in Advances in Neurochemistry, Vol. 2 (Agranoff B.W. and Aprison M.H., eds.) pp. 1-132. Plenum Press, New York.

Maas J.W., Kocsis J.H., Bowden C.L., Davis J.M., Redmond D.E., Hanin I., and Robin E. (1982) Pretreatment neurotransmitter metabolites and response to imipramine or amitriptyline treatment. Psychol. Med. 12, 37-43.

Montplaisir J., de Champlain J., Young S.N., Missala K., Sourkes T.L., Walsh J., and Rémillard G. (1982) Narcolepsy and idiopathic hypersomnia: biogenic amines and related compounds in CSF. Neurology 32, 1299-1302.

Pedemonte W.A., Mosnaim A.D., and Bulat M. (1976) Penetration of phenylacetic acid across the blood-cerebrospinal fluid barrier. Res. Commun. Chem. Path. Pharmacol. 14, 111-116.

Philips S.R., Rozdilsky B.G., and Boulton A.A. (1978) Evidence for the presence of m-tyramine, p-tyramine, tryptamine, and phenylethylamine in the rat brain and several areas of the human brain. Biol. Psychiat. 13, 51-57.

Sacks W., Vogel W.H., Nagatsu T., Lloyd K.G., and Sandler M. (1979) Is there dopa decarboxylase in human brain? in Catecholamines: Basic and Clinical Frontiers (Usdin E., Kopin I.J., and Barchas J., eds.) pp. 127-131. Pergamon Press, New York.

Sandler M., Ruthven C.R.J., Goodwin B.L., et al. (1978) Raised cerebrospinal fluid phenylacetic acid concentration: preliminary support for the phenylethylamine hypothesis of schizophrenia? Commun. Psychopharmacol. 2, 199-202.

Sandler M., Ruthven C.R.J., Goodwin B.L., and Coppen A. (1979) Decreased cerebrospinal fluid concentration of free phenylacetic acid in depressive illness. Clin. Chim. Acta 93, 169-171.

van der Poel F.W., van Praag H.M., and Korf J. (1977) Evidence for a probenecid-sensitive transport system of acid monoamine metabolites from the spinal subarachnoid space. Psychopharmacology 52, 35-40.

Warsh J.J., Chan P.Q., Dodse D.D., Coscina D.V., and Stancer H.C. (1977) Gas chromatographic-mass fragmentographic determination of indole-3-acetic acid in rat brain. J. Neurochem. 29, 955-958.

Warsh J.J., Coscina D.V., Godse D.D., and Chan P.W. (1979) Dependence of brain tryptamine formation on tryptophan availability. J. Neurochem. 32, 1191-1196.

Wolfson L.I., Katzman R., and Escriva A. (1974) Clearance of amine metabolites from the cerebrospinal fluid: the brain as a "sink". Neurology (Minneap.) 24, 772-779.

Young S.N. and Lal S. (1980) CNS tryptamine metabolism in hepatic coma. J. Neural Transm. 47, 153-161.

Young S.N., Anderson G.M., Gauthier S., and Purdy W.C. (1980a) The origin of indoleacetic acid and indolepropionic acid in rat and human cerebrospinal fluid. J. Neurochem. 34, 1087-1092.

Young S.N., Anderson G.M., and Purdy W.C. (1980b) Indoleamine metabolism in rat brain studied through measurements of tryptophan, 5-hydroxyindoleacetic acid and indoleacetic acid in cerebrospinal fluid. J. Neurochem. 34, 309-315.

Young S.N., Gauthier S., Anderson G.M., and Purdy W.C. (1980c) Tryptophan, 5-hydroxyindoleacetic acid and indoleacetic acid in human cerebrospinal fluid: interrelationships and the influence of age, sex, epilepsy and anticonvulsant drugs. J. Neurol. Neurosurg. Psychiat. 43, 438-445.

Young S.N. and Gauthier S. (1981) Effect of tryptophan administration on tryptophan, 5-hydroxyindoleacetic acid and indoleacetic acid in human lumbar and cisternal cerebrospinal fluid. J. Neurol. Neurosurg. Psychiat. 44, 323-327.

Young S.N., Davis B.A., and Gauthier S. (1982) Precursors and metabolites of phenylethylamine, *m* and *p*-tyramine and tryptamine in human lumbar and cisternal cerebrospinal fluid. J. Neurol. Neurosurg. Psychiat. 45, 633-639.

Young S.N., Gauthier S., Kiely M.E., Lal S., and Brown G.M. (1983) Effect of oral melatonin administration on melatonin, 5-hydroxyindoleacetic acid, indoleacetic acid and cyclic nucleotides in human cerebrospinal fluid. In Preparation.

ANALYSIS OF TRACE AMINES: ENDOGENOUS LEVELS AND THE EFFECTS OF VARIOUS DRUGS ON TISSUE CONCENTRATIONS IN THE RAT

S.R. Philips

Psychiatric Research Division,
Cancer and Medical Research Building,
University of Saskatchewan,
Saskatoon, Saskatchewan S7N OWO

ABSTRACT

The trace amines phenylethylamine (PE), p-tyramine (p-TA), m-tyramine (m-TA) and tryptamine (T) have been analyzed as their dansyl derivatives by the highly specific and sensitive integrated ion current mass spectrometric procedure. Of the rat tissues examined, the kidney contained the highest concentrations of the trace amines. In the brain, these substances were anatomically most concentrated in the caudate nucleus, while subcellularly, significant amounts were associated with particulate material, including the synaptosomes. PE and T, and to a lesser extent p-TA and m-TA, exhibited very high turnover rates and their tissue concentrations increased rapidly after inhibition of monoamine oxidase. The accumulation rates of PE and T in the brains of rats treated with pargyline were similar to those of the classical neurotransmitters dopamine (DA), noradrenaline (NA) and 5-hydroxytryptamine (5-HT), although these latter compounds are present endogenously in much greater amounts. The trace amines have also been identified in various regions of the human brain and appear to be most concentrated in the basal ganglia.

INTRODUCTION

As recently as the early 1970's, very little was known about either the amounts of the trace amines which were present in the body, or the distribution of these substances throughout the body. Nevertheless, a number of reports had appeared in the literature which suggested that some of these compounds, in particular PE, p-TA, m-TA and T, might be involved in some way in a variety of psychiatric and neurological disorders. Much of the evidence had been obtained from clinical urinary studies. PE had already been shown to be present in small quantities in human urine. It was known to be excreted in greater than normal amounts by patients suffering from phenylketonuria, and had been reported to be excreted in less than normal amounts by patients suffering from depression. It was also known to possess weak sympathomimetic properties, to stimulate motor activity, and to mimic the pharmacological response of amphetamine (see Durden et al, 1973). p-TA had been reported to precipitate migraine attacks in migraine patients and, following ingestion, to be excreted in conjugated form in smaller quantities by migraine patients than by normal subjects. The unconjugated amine had also been reported to be excreted in abnormal amounts by patients with Parkinson's disease and schizophrenia (see Philips et al, 1974a). It had been suggested that m-TA, another of the trace amines, might be responsible for the central excitatory actions of the amino acid m-tyrosine, since m-tyrosine was readily decarboxylated to the amine by aromatic amino acid decarboxylase. It was also believed that m-TA might play a role in Parkinson's disease because of its ability to reduce the depletion of DA stores by reserpine (see Philips et al, 1975). Finally, T, a trace amine believed to arise in mammalian tissues by decarboxylation of tryptophan, had been reported to be excreted in abnormally high amounts by some Parkinson patients. Furthermore, in some longitudinal studies involving schizophrenic patients, exacerbation of psychotic behavior appeared to be correlated with increased urinary excretion of T, while an improvement in the mental state was accompanied by a reduction in excreted T toward more normal levels (see Philips et al, 1974b).

In view of the possible clinical importance of the trace amines, it seemed essential to develop methods for their analysis which would allow the positive identification and determination of these substances, not only in whole brain, but in specific regions of the brain. Because of the very small amounts of the trace amines believed to be present in mammalian tissues, much more specific and sensitive procedures than those used in earlier urinary studies were required. The analytical procedures described here have been used to establish the endogenous trace amine levels in several peripheral tissues and brain regions of the rat, as well as in central nervous system tissue of a number of other species. They have also been used to examine the effects of a variety of metabolic inhibitors, antipsychotics and other substances on tissue trace amines.

DETERMINATION OF ENDOGENOUS TRACE AMINE CONCENTRATIONS IN THE RAT

PE was the first trace amine to be determined by the mass spectrometric integrated ion current procedure (Durden et al, 1973). In order to reduce the likelihood of interference from other substances during mass spectrometric analysis, the amine was first isolated from most other tissue constituents by means of a rather extensive purification procedure (Fig. 1). The tissue was homogenized in 0.4 N perchloric acid and a known amount of 1,1-dideutero-2-PE (deuterated PE) hydrochloride was added to act as an internal standard. The homogenate was centrifuged and the supernatant decanted and adjusted to pH 7. An amine fraction was isolated by percolating the solution through a column of Biorad AG 50W-X2 [H^+]. The amines were eluted and dansylated, and dansyl PE was isolated by thin layer chromatography.

To determine quantitatively the amount of PE in a sample, an aliquot of the separated dansyl PE was introduced into an unheated probe tip, then inserted into the vacuum system of the mass spectrometer to evaporate the solvent. The sample was then placed in the hot ion source and the ion currents produced by the molecule ions of dansyl PE (m/e 354.1402) and dansyl deuterated PE (m/e 356.1527) were recorded as the sample evaporated from the probe tip. Since the amount of deuterated PE in the sample was known, the amount of endogenous PE could be determined by measuring the areas under the two ion curves.

The TA's and T were measured by procedures similar to that used for PE, although there were some differences in

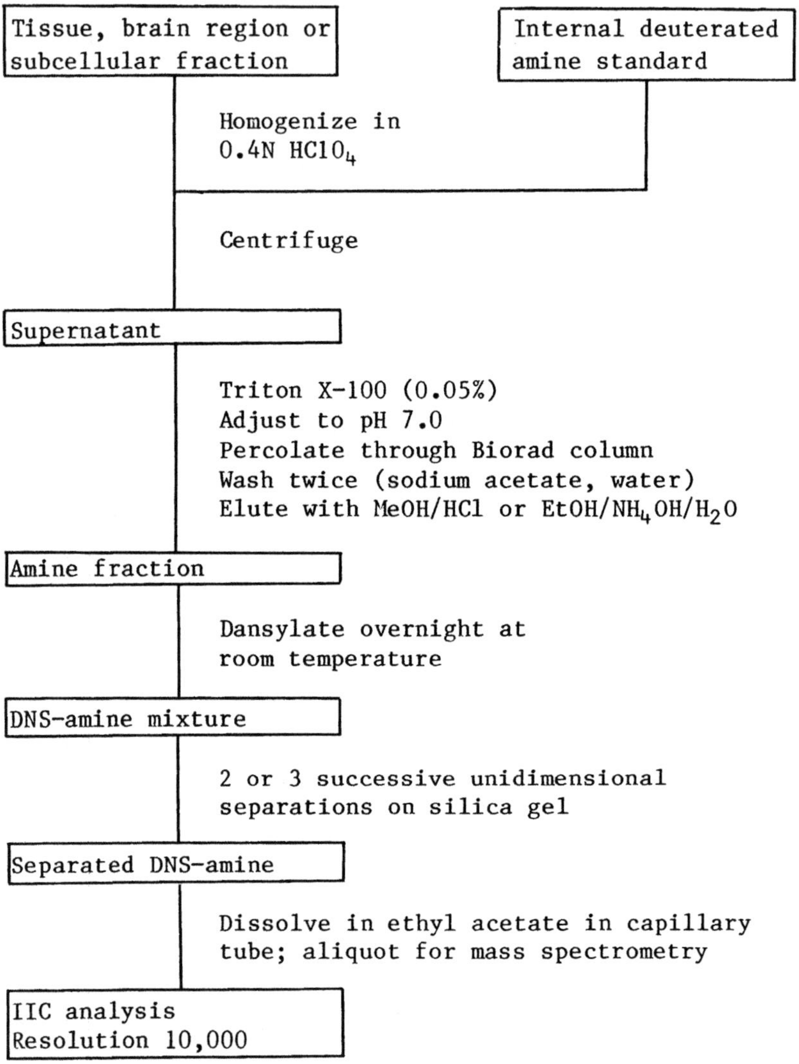

Figure 1 - Isolation and quantitative determination of the trace amines.

the solvent used to elute the former amines from the Biorad column, and in the solvents used to isolate the dansylated compounds by thin layer chromatography (Philips et al, 1974a, b, 1975). The molecule ions of the bis dansyl TA's

Table 1

Distribution of PE, p-TA, m-TA and T in the Rat

Tissue	PE	p-TA	m-TA	T
Brain	1.8	2.0	0.3	0.5
Heart	5.7	3.4	0.4	0.6
Kidney	20.5	32.7	12.6	8.0
Liver	2.0	1.5	0.3	0.7
Lung	4.0	3.0	0.3	0.5
Spleen	4.7	3.4	0.3	0.4
Blood	1.9	1.3	0.2	0.2

Values, in ng/g, are means of 6-9 samples. Amine concentrations in brain and peripheral tissues (except blood) are taken from Durden *et al*, 1973 (PE); Philips *et al*, 1974a (p-TA); Philips *et al*, 1975 (m-TA); and Philips *et al*, 1974b (T).

(m/e 603.1861) and dideuterated TA's (m/e 605.1987) and of monodansyl T (m/e 393.1511) and dideuterated T (m/e 395.1637) were used to determine the endogenous levels of these substances by the integrated ion current procedure.

Endogenous tissue and brain region concentrations of PE, p-TA, m-TA and T, shown in Tables 1 and 2, indicate that the kidney contained substantially greater amounts of each amine than did any of the other tissues examined. This finding may reflect the relatively high activity of L-aromatic amino acid decarboxylase in that tissue (Davis and Awapara, 1960). In the brain, the trace amines were most concentrated in the caudate nucleus, although PE was more evenly distributed throughout the rest of the brain than were the TA's or T.

Table 2

Distribution of PE, p-TA, m-TA and T in Rat Brain

Brain Region	PE	p-TA	m-TA	T
Hypothalamus	2.1	1.2	0.2	0.2
Caudate Nucleus	4.2	11.1	3.0	1.5
Cerebellum	2.1	0.1	<0.1	0.1
Stem	1.6	0.4	0.1	0.1
Rest	1.4	0.5	0.2	0.1

Values, in ng/g, are means of 5-6 samples.

SUBCELLULAR DISTRIBUTION OF THE TRACE AMINES IN RAT BRAIN

Once the analytical methods for measuring the trace amines had been established, it became possible to investigate a variety of other problems concerning these compounds. One of the first applications of the procedure was to determine the subcellular distributions of PE, *p*-TA and T in the rat brain (Boulton and Baker, 1975). Whole brains (minus the cerebellum) were homogenized in 0.32 M sucrose. Subcellular fractionation, essentially according to the procedure of Whittaker and Barker (1972), was carried out to yield a nuclear debris pellet, P_1, a crude mitochondrial pellet, P_2, and a microsomal pellet, P_3. P_2 was occasionally fractionated further to yield subfractions P_2A (myelin), P_2B (synaptosomes) and P_2C (mitochondria). The amines were then analyzed in each fraction.

Approximately 50% of recovered PE, 40% of *p*-TA and 60% of T was closely associated with the combined particulate fractions ($P_1 + P_2 + P_3$), while an additional 15-20% of recovered T was more loosely associated with the particulate material, perhaps as a noncovalent lipid-amine complex (Boulton *et al.*, 1972). The remaining amine was found to be soluble in the high speed supernatant, S_3. Among the particulate fractions, approximately 55% of the recovered particulate PE, 65% of the p-TA and 55% of the T was associated with the P_2 fraction. Subfractionation of the P_2 pellet indicated that for each amine, 50-55% of the recovered amine was associated with the P_2B, or synaptosomal fraction.

As a result of these experiments, in which it was established that the trace amines were at least partly associated with particulate subcellular material in the brain, and other studies in which the heterogeneous nature of their anatomical distribution in the brain was demonstrated, it was suggested that the trace amines might play a role in neurotransmission. Rather than acting as transmitters in the conventional sense, however, it was proposed (Boulton and Baker, 1975) that they might function as activators of regions of the synaptic area. As such, they would not cause the actual propagation of the nervous impulse, but might, instead, maintain the synapse in a state of readiness for firing by a larger quantity of a different neurotransmitter.

EFFECT OF MONOAMINE OXIDASE INHIBITORS ON TRACE AMINE CONCENTRATIONS IN TISSUES AND BRAIN REGIONS OF THE RAT

The very low tissue concentrations which we and others (Saavedra and Axelrod, 1972; Saavedra, 1974; Willner et al, 1974; Tallman et al, 1976) reported for the trace amines suggested that either these substances were formed very slowly or they were metabolized very rapidly. Earlier work by Wu and Boulton (1973, 1974, 1975), in which intraventricularly injected labeled amines had been found to disappear at a very rapid rate, suggested that the latter was more likely. To investigate further the metabolic breakdown of the trace amines, we examined in some detail the effects of inhibiting that breakdown on trace amine levels in various tissues and brain regions of the rat. The monoamine oxidase inhibitors pargyline (75 mg/kg) and iproniazid (100 mg/kg) were administered intraperitoneally, and the animals were killed 4 hours later.

The results of these experiments are presented in Table 3. Both inhibitors produced marked increases in PE

Table 3

The Effects of Pargyline (Pg) and Iproniazid (Ip) on Trace Amine Concentrations in Peripheral Tissues and Brain Regions of the Rat

	PE		p-TA		m-TA		T	
	Pg	Ip	Pg	Ip	Pg	Ip	Pg	Ip
Tissue								
Heart	91	89	6.5	2.8	3.1	2.5	7.6	12.0
Kidney	773	967	73.0	34.2	20.0	16.8	105	155
Liver	487	316	9.7	3.4	1.7	1.2	27.6	93.1
Lung	414	522	10.0	4.2	2.0	1.4	36.1	76.4
Spleen	270	224	8.2	2.5	1.6	0.8	12.6	11.0
Blood	33	36	2.9	1.0	0.3	0.3	2.0	5.3
Brain Region								
Whole Brain	185	114	13.2	3.9	1.9	1.6	9.4	5.6
Hypothalamus	139	150	2.8	1.6	1.3	0.7	11.2	6.8
Caudate Nucleus	213	227	123	49.2	15.4	12.5	40.4	9.5
Cerebellum	136	176	0.4	0.4	0.2	0.2	5.0	3.8
Stem	152	199	0.8	0.4	0.6	0.3	7.4	3.9
Rest	173	218	5.9	2.4	1.5	0.9	7.5	3.5

Values, in ng/g, are means of 3-7 samples.

in all peripheral tissues examined; in the liver, values were 160-240 fold greater than those observed in untreated animals (Table 1). T was also increased substantially in all tissues, especially by iproniazid. The TA's, however, were much less affected; p-TA, in fact, was apparently unchanged by iproniazid in most tissues. PE was increased to a similar extent in each of the brain regions tested and appeared to be relatively evenly distributed throughout the brain. The TA's and T were increased more by pargyline than by iproniazid, and were most concentrated in the caudate nucleus.

From these experiments, a number of general features of the trace amines became apparent: (1) the tissue containing the highest concentration of trace amines, in both normal and monoamine oxidase-inhibited animals, appeared to be the kidney; (2) the largest increases in trace amine concentrations following monoamine oxidase inhibition occurred in the liver; (3) blood levels were as low as, or lower than those in any of the other tissues tested; (4) among the brain regions examined, the caudate nucleus contained the highest concentration of the trace amines in both normal and monoamine oxidase-inhibited animals. Their heterogeneous anatomical distribution was most marked in the case of the TA's and T, while PE was more evenly distributed throughout the brain, perhaps reflecting the more non-polar nature of its molecule and its greater ability to cross the blood-brain barrier. It is likely that much of the PE (and perhaps T) found in the brains of monoamine oxidase-inhibited animals was produced elsewhere in the body and was carried to the brain by the blood, while the TA's, which are synthesized from their precursor amino acids, were probably retained at or near their sites of synthesis.

The effects of the monoamine oxidase inhibitors clorgyline, deprenyl, tranylcypromine and phenelzine on trace amine levels in the rat caudate nucleus were also studied (Philips and Boulton, 1979). Clorgyline and deprenyl have been reported to inhibit preferentially the A and B forms of monoamine oxidase, respectively (Johnston, 1968; Knoll et al, 1965), while tranylcypromine and phenelzine have been used clinically to treat depression. Dose-response curves obtained using these inhibitors showed clearly that PE was unaffected by up to 50 mg/kg of clorgyline while *p*-TA, *m*-TA and T were increased by doses as low as 0.5, 1.0 and 5 mg/kg, respectively. Deprenyl, on the other hand, increased PE at a dose of 1 mg/kg, but had little effect on the TA's and T at doses up to 10 or 50 mg/kg, respectively. Tranylcypromine increased all the trace amines at 1 mg/kg, and caused very large increases in PE and T at higher doses. Phenelzine at 0.1 and 1 mg/kg caused increases in *p*-TA and T, and in *m*-TA, respectively, but did not affect PE. At higher doses, both PE and T increased sharply.

The results of these experiments suggested that while PE was oxidized by monoamine oxidase B, the TA's and T were oxidized by monoamine oxidase A, and perhaps to a lesser extent by the B form. They further suggested that by administering the proper monoamine oxidase inhibitor, it might be possible to manipulate selectively trace amine concentrations in the brain.

KINETIC MEASUREMENTS OF THE TURNOVER RATES OF PE AND T IN RAT BRAIN

The turnover rates of PE and T were estimated in rat brain by measuring their accumulation rates in the presence of a high dose of pargyline (Durden and Philips, 1980). When the animals were given pargyline in doses ranging from 19 to 450 mg/kg, then killed 2 hours later, the greatest brain concentration of PE was found after a 225 mg/kg dose of the drug. The results of these experiments indicated that the maximum accumulation rate of PE could therefore be determined by injecting that dose and measuring the amine concentration at various times after drug administration. Under those conditions, the PE concentration increased linearly for the first two hours after pargyline injection, then appeared to stabilize. Its rate of accumulation, determined from the initial slope of the plot of amine concentration versus time, is indicated in Table 4, and is compared with literature values for the synthesis rates of DA and NA.

The brain concentration of T exhibited a clearly defined two-step increase as the pargyline dose was increased, presumably reflecting the sequential inhibition of type B, then type A monoamine oxidase. The accumulation rates measured after 225 and 450 mg/kg pargyline were linear for

Table 4

Comparison of Some Kinetic Parameters for PE, DA and NA in Rat Brain

Amine	Accumulation Rate (nmol/g/h)	Endogenous Concentration (nmol/g)	Half-life (min)
PE	1.5	0.015[a]	0.4
DA	2.8[b]	7.5	110
NA	0.4[b]	2.5	250

[a]Endogenous concentration taken from Durden *et al* (1973).
[b]Calculated from Table 1 of Brodie *et al* (1966).

Table 5

Comparison of Some Kinetic Parameters for T and 5-HT in Rat Brain

Amine	Accumulation Rate (nmol/g/h)	Endogenous Concentration (nmol/g)	Half-life (min)
T	0.24[a]	0.0016	0.28
T	0.16[b]	0.0016	0.42
5-HT	1.55[c]	2.85	77

[a]Data obtained after administration of 450 mg/kg pargyline.
[b]Data obtained after administration of 225 mg/kg pargyline.
[c]Calculated from data in Table 2 of Beck *et al* (1977).

two hours after drug administration. They are shown in Table 5 and compared to the synthesis rate for 5-HT.

From Tables 4 and 5 it is clear that the accumulation rates of PE and T in the brain are comparable to those of the classical neurotransmitters, in spite of the much lower endogenous concentrations of the trace amines. These experiments provided clear evidence of the very rapid metabolic turnover rates of PE and T. By measuring their accumulation rates, we have implied nothing about the origin of these substances. They may be synthesized in the brain itself or they may be produced elsewhere and transported to the brain by the blood. Since both amines are able to cross the blood-brain barrier quite readily (Oldendorf, 1971), it is likely that a portion of the observed concentration increase occurs as a result of synthesis in and transport from another part of the body.

TRACE AMINES IN THE HUMAN BRAIN

By the late 1970's, the trace amines had been clearly identified and measured in mammalian brain. They had been shown to display a very active metabolism and were known to respond with rapid concentration changes to monoamine oxidase inhibitors, some of which had occasionally been used clinically in treating depression. Because earlier urinary studies had suggested that these substances might play a role in psychiatric disorders, it seemed useful to attempt to establish their presence in the human brain (Philips *et al*, 1978).

Analyses were performed on post-mortem brains obtained from four nonpsychiatric patients, aged 59-70 years. The tissue was obtained 4 or 10 hours after death, dissected, and analyzed. The same amines were also measured in post-mortem rat brain in an attempt to determine the extent of changes which might occur post-mortem in human brain. To simulate the post-mortem temperature conditions experienced by a human body, the animals were killed by injecting air into the heart, allowed to remain at room temperature for one hour after death, then transferred to a cold room (4°C) and stored for up to 11 hours before the brains were removed and the trace amines analyzed.

Table 6

Trace Amines in Post-Mortem Human Brain

Brain Region	PE	*p*-TA	*m*-TA	T
Frontal Cortex	0.8	0.1	<0.1	<0.1
Cerebellar Cortex	1.1	0.5	<0.1	0.1
Dentate Nucleus	-	0.1	<0.1	<0.1
Substantia Nigra	-	1.0	2.2	<0.1
Parietal Cortex	0.6	0.1	<0.1	0.1
Cerebellar White	1.0	0.1	<0.1	<0.1
Subcortical White	0.6	0.1	<0.1	0.1
Caudate Nucleus	1.5	1.2	3.0	0.2
Putamen	0.9	1.4	3.7	0.2
Int. and Ext. Pallidum	1.7	1.0	1.7	0.2
Anterior Internal Capsule	2.0	0.4	0.3	0.1
Posterior Internal Capsule	0.8	0.2	0.7	0.1
Dorsal Thalamus	0.3	0.8	0.2	0.1
Hypothalamus	-	0.2	0.2	<0.1
Ventral Thalamus	0.7	0.4	0.2	0.1
Mid Thalamus	0.9	0.3	<0.1	0.2
Hippocampus	0.4	<0.1	<0.1	0.1

Values, in ng/g, represent means of 2-4 samples, except for hippocampus, and PE levels in the pallidum and dorsal thalamus, which are single values.

Table 7

Trace Amine Concentrations in Post-Mortem Rat Brain

Time (h)	PE	p-TA	m-TA	T
0	2.1	1.0	0.2	0.2
0.5	2.8	1.4	0.2	0.8
1	3.8	1.6	0.2	1.6
2	8.9	2.3	0.2	3.2
4	5.6	2.8	0.2	3.4
9	8.5	3.3	0.2	4.5
12	4.1	2.2	0.3	3.5

Values, in ng/g, are means of 3-8 samples.

The results of these experiments (Table 6) indicated that, in the human brain, PE, the TA's and T were most concentrated in the basal ganglia, but that PE and T were somewhat more evenly distributed throughout the brain than were the TA's. While the presence of the trace amines and their apparent heterogeneous distribution has therefore been established, little can be inferred from these results about their in vivo concentrations. Changes might have occurred during the post-mortem period which caused substantial changes in the amine concentrations before the analyses could be performed. Furthermore, the values shown represent mean values of at most four specimens, and in some brain regions, a wide range of amine concentrations was observed.

Post-mortem changes in rat brain trace amine levels suggested that a gradual increase occurred in all amines except m-TA, with the greatest increase occurring in T (Table 7). Whether these changes are representative of those occurring in the human brain will be difficult to determine.

CONCLUSION

The trace amines have been shown to occur in several peripheral tissues in the mammalian body and to be distributed heterogenously throughout the brain. Their localized concentrations in the brain, and their association with the particulate fractions of the brain, especially the synaptosomal fraction, suggest that they may have some functional significance. The rapid metabolic turnover rates of these substances, at least of PE and T, and the fact that their tissue concentrations increase very rapidly when their degradation is impeded, suggest that if the trace amines function either as transmitter substances or as synaptic activators, their effects are likely to be strongly accentuated in the event of a disturbance in their metabolism. Finally, preliminary evidence from post-mortem tissue has indicated that the trace amines are present and distributed heterogeneously throughout the human brain. It is, therefore, quite possible that if the concentration changes oberved in the rat apply as well to the human, then trace amines may indeed play a role in some psychiatric and neurological disorders.

ACKNOWLEDGMENT

I thank Dr. A.A. Boulton for his advice and encouragement and the MRC of Canada and Saskatchewan Health for financial support.

REFERENCES

Beck O., Wiesel F.-A. and Sedvall G. (1977) Mass fragmentographic determination of 5-hydroxytryptamine and 5-hydroxyindole-3-acetic acid in brain tissue using deuterated internal standards. J. Chromatogr. 134, 407-414.

Boulton A.A. and Baker G.B. (1975) The subcellular distribution of β-phenylethylamine, p-tyramine and tryptamine in rat brain. J. Neurochem. 25, 477-481.

Boulton A.A., Wu P.H. and Philips S. (1972) Binding of some primary aromatic amines to certain rat brain particulate fractions. Can. J. Biochem. 50, 1210-1218.

Brodie B.B., Costa E., Dlabac A., Neff N.H. and Smookler H.H. (1966) Application of steady state kinetics to the estimation of synthesis rate and turnover time of tissue catecholamines. J. Pharmacol. Exp. Ther. 154, 493-498.

Davis V.E. and Awapara J. (1960) A method for the determination of some amino acid decarboxylases. J. Biol. Chem. 235, 124-127.

Durden D.A. and Philips S.R. (1980) Kinetic measurements of the turnover rates of phenylethylamine and tryptamine *in vivo* in the rat brain. J. Neurochem. 34, 1725-1732.

Durden D.A., Philips S.R. and Boulton A.A. (1973) Identification and distribution of β-phenylethylamine in the rat. Can. J. Biochem. 51, 995-1002.

Johnston J.P. (1968) Some observations upon a new inhibitor of monoamine oxidase in brain tissue. Biochem. Pharmacol. 17, 1285-1297.

Knoll J., Ecseri Z., Kelemen K., Nievel J. and Knoll B. (1965) Phenylisopropylmethylpropinylamine (E-250), a new spectrum psychic energizer. Archs. int. Pharmacodyn. Ther. 155, 154-164.

Oldendorf W.H. (1971) Brain uptake of radiolabeled amino acids, amines, and hexoses after arterial injection. Am. J. Physiol. 221, 1629-1639.

Philips S.R. and Boulton A.A. (1979) The effect of monoamine oxidase inhibitors on some arylalkylamines in rat striatum. J. Neurochem. 33, 159-167.

Philips S.R., Durden D.A. and Boulton A.A. (1974a) Identification and distribution of *p*-tyramine in the rat. Can. J. Biochem. 52, 366-373.

Philips S.R., Durden D.A. and Boulton A.A. (1974b) Identification and distribution of tryptamine in the rat. Can. J. Biochem. 52, 447-451.

Philips S.R., Davis B.A., Durden D.A. and Boulton A.A. (1975) Identification and distribution of *m*-tyramine in the rat. Can. J. Biochem. 53, 65-69.

Philips S.R., Rozdilsky B. and Boulton A.A. (1978) Evidence for the presence of *m*-tyramine, *p*-tyramine, tryptamine and phenylethylamine in the rat brain and several areas of the human brain. Biol. Psychiat. 13, 51-57.

Saavedra J.M. (1974) Enzymatic isotopic assay for and presence of β-phenylethylamine in brain. J. Neurochem. 22, 211-216.

Saavedra J.M. and Axelrod J. (1972) A specific and sensitive enzymatic assay for tryptamine in tissues. J. Pharmacol. Exp. Ther. 182, 363-369.

Tallman J.F., Saavedra J.M. and Axelrod J. (1976) A sensitive enzymatic-isotopic method for the analysis of tyramine in brain and other tissues. J. Neurochem. 27,465-469.

Whittaker V.P. and Barker L.A. (1972) The subcellular fractionation of brain tissue with special reference to the preparation of synaptosomes and their component organelles, in, Methods of Neurochemistry, Vol. 2, (Fried, R., ed), pp 1-52. Marcel Dekker, New York.

Willner J., LeFevre H.F. and Costa E. (1974) Assay by multiple ion detection of phenylethylamine and phenylethanolamine in rat brain. J. Neurochem. 23, 857-859.

Wu P.H. and Boulton A.A. (1973) Distribution and metabolism of tryptamine in rat brain. Can. J. Biochem. 51, 1104-1112.

Wu P.H. and Boulton A.A. (1974) Distribution, metabolism and disappearance of intraventricularly injected p-tyramine in the rat. Can. J. Biochem. 52, 374-381.

Wu P.H. and Boulton A.A. (1975) Metabolism, distribution and disappearance of injected β-phenylethylamine in the rat. Can. J. Biochem. 53, 42-50.

DRUG-INDUCED CHANGES IN THE CENTRAL METABOLISM OF TYRAMINE AND OTHER TRACE MONOAMINES: THEIR POSSIBLE ROLE IN BRAIN FUNCTIONS

A. V. Juorio

Psychiatric Research Division,
Cancer and Medical Research Building,
University of Saskatchewan,
Saskatoon, Saskatchewan, S7N OWO

ABSTRACT

It has been known for quite a number of years that PE, _p_-TA, _m_-TA or T belong to a group of monoamines that are present in neural tissue in small concentrations and for this they have been named trace amines. The trace amines are present in discrete areas of the mammalian brain and possess a high metabolic rate as determined by their accumulation after MAO inhibition. Their concentrations are increased after the administration of their respective precursor amino acids or in some instances by specific activation of l-aromatic amino acid decarboxylase. In contrast, l-aromatic amino acid decarboxylase inhibitors reduce their brain concentrations. The administration of antipsychotics or stimulant drugs have unraveled intriguing interrelations between the concentrations of some trace amines and the metabolism of the classical transmitters. No changes in neuronal firing rate were observed after the iontophoretic administration of small amounts of trace amines. In contrast, these small amounts potentiate the firing rate changes induced by iontophoresing some of the classical transmitters such as DA, NA or 5-HT. These findings suggest that the trace amines possess a role in neural transmission; this may be a regulatory effect on the action of the classical transmitters or as neurotransmitters in some pathways. Future work will have to decide whether any of these occur.

DEFINITION

Trace amines is an all embracing term that refers to biogenic amines that are present in neural tissues in concentrations ranging from 0.1 to 100 ng/g (Boulton, 1976). The amines included within these concentrations are PE, PEOH, p-TA, m-TA, o-TA, p-OA, m-OA, o-OA, p-synephrine, m-synephrine, o-synephrine and T. Some classical transmitters such as DA, NA, A and 5-HT can in some anatomical locations fall within the terms of the definition of trace amines. The name trace amines is an inadequate one because the definition is based on tissue levels and not on their functional role. The lack of a better term, however, makes it necessary to continue with the use of the name trace amines, at least for the time being, for this group of

Table 1

The concentration of some trace amines and classical transmitters in the rat or fowl whole brain, octopus ganglia and starfish arm nerve

	Rat (ng/g)	Fowl (ng/g)	Octopus (ng/g)	Starfish (ng/g)
PE	1.8[1]	0.7[7]	3.0[8]	4.4[9]
p-TA	2.0[2]	0.5[7]	80[8]	2.2[9]
m-TA	0.3[3]	0.3[7]	0.6[8]	<0.8[9]
DA	600[4]	290[7]	4690[8]	5950[9]
T	0.5[5]	<0.3[7]	<0.6[8]	1250[9]
5-HT	400[6]	700[7]	4180[8]	<3.0[9]

The superscripts indicate the source of reference.
[1]Durden et al, 1973; [2]Philips et al, 1974a; [3]Philips et al, 1975; [4]Bertler and Rosengren, 1959; [5]Philips et al, 1974b; [6]Bogdanski et al, 1956; [7]Juorio, 1976; [8]Juorio and Philips, 1976; [9]Juorio and Robertson, 1977.

compounds. In this short review I shall discuss only PE, p-TA, m-TA and T.

PRESENCE

The p- and m- isomers of TA, PE and T are present in the mammalian or avian brain (Table 1); their concentrations, in comparison with that of DA or 5-HT are quite small (Table 1) but show an uneven regional distribution (Table 2). The highest concentrations of PE, p-TA, m-TA and T are present in the caudate nucleus (Table 2). A portion of PE, p-TA and T were found associated with a subcellular particulate fraction (Boulton and Baker, 1975). In contrast, neural tissues of some invertebrates contain relatively high concentrations of p-TA or T (Table 1) and sometimes fall outside the definition of a trace amine.

SYNTHESIS

PE and T are formed by decarboxylation of phenylalanine or tryptophan respectively (Lovenberg, Weissback and Udenfriend, 1962; Saavedra and Axelrod, 1973; Saavedra, 1974). The case for p- and m-TA is more complex since p-tyrosine decarboxylation is very slow (Fellman, Roth and Fujita, 1976) and m-tyrosine has not yet been detected in the brain (Durden, unpublished). Also, both p- and m-TA can be formed by hydroxylation of PE and this has been shown to occur in the periphery (Boulton, Dyck and Durden, 1974) and in the brain (Silkaitis and Mosnaim, 1976; McQuade and Juorio, 1982). Recent experiments have shown that the parenteral administration of L-phenylalanine or p-tyrosine increases the mouse striatal concentration of p-TA, an effect that is enhanced by monoamine oxidase inhibition (Juorio and Boulton, 1982) and reduced by inhibition of aromatic-l-amino acid decarboxylase inhibitors. The parenteral administration of L-phenylalanine or m-tyrosine increase striatal m-TA concentrations and this was also further enhanced by monoamine oxidase inhibition (Juorio and Boulton, 1982). Also, it was observed that after *in vivo* administration, m-tyrosine is a better substrate for decarboxylation than p-tyrosine (Juorio and Boulton, 1982). Quite surprisingly, the administration of various aromaticl-aminoacid decarboxylase inhibitors (NSD 1055, NSD 1034, α-fluoromethyldopa), at low or moderate

Table 2

The rat brain regional distribution of some trace amines and trace acids

Regions	PE (ng/g)	PAA (ng/g)	*p*-TA (ng/g)	*p*-HPAA (ng/g)	*m*-TA (ng/g)	*m*-HPAA (ng/g)	T (ng/g)	IAA (ng/g)
Whole brain	1.8[1]	31.8[3]	2.0[4]	10.6[5]	0.3[6]	2.3[5]	0.5[7]	11.6[8]
Hypothalamus	2.1[2]	60.1[3]	1.2[2]	4.5[5]	0.2[2]	1.2[5]	0.9[7]	23.7[8]
Striatum	4.2[2]	64.6[3]	11.1[2]	28.3[5]	3.0[2]	5.3[5]	2.9[7]	63.6[8]
Brain Stem	1.6[2]	33.1[3]	0.4[2]	8.6[5]	0.1[2]	1.8[5]	0.2[7]	11.7[8]
Cerebellum	2.1[2]	31.3[3]	0.1[2]	8.1[5]	0.01[2]	1.2[5]	0.3[7]	10.1[8]
Rest	1.4[2]	27.6[3]	0.5[2]	5.3[5]	0.2[2]	1.7[5]	0.3[7]	-

The superscripts indicate the source of reference. [1]Durden *et al*, 1973; [2]Philips unpublished, cited by Boulton, 1976; [3]Durden and Boulton, 1982; [4]Philips *et al*, 1974a; [5]Durden and Boulton, 1981; [6]Philips *et al*, 1975; [7]Philips *et al*, 1974b; [8]Warsh *et al*, 1977.

Table 3

The effect of the administration of phenylalanine, *p*-tyrosine, *m*-tyrosine and some aromatic-l-amino acid decarboxylase inhibitors on mouse striatal *p*-TA and *m*-TA

Treatment	Dose (mg/kg)	Time (hours)	*p*-TA (ng/g)	*m*-TA (ng/g)
Controls	-	-	21[1]	6.1[1]
Phenylalanine	800	2	55[1]*	11.5[1]*
p-Tyrosine	800	2	124[1]*	4.4[1]*
m-Tyrosine	2	2	17[1]*	81[1]*
NSD 1055	200	3	2.5[1]*	9.3[1]*
Controls	-	-	2.4[2]	7.1[2]
NSD 1034	2	2	10[2]*	8.8[2]
NSD 1034	400	2	1.3[2]*	1.7[2]*
Controls	-	-	2.4[3]	6.8[3]
α-Fluoromethyldopa	10	3	1.5[3]*	22[3]*
α-Fluoromethyldopa	100	3	2.0[3]*	10[3]*

The asterisk indicates that the results are statistically significantly different from controls. The superscripts indicate the source of reference. [1]Juorio and Boulton, 1982; [2]Juorio, 1983; [3]Boulton and Juorio, 1983.

doses, all increased *m*-TA concentration (Table 3). By increasing further the dose of the inhibitor, however, the *m*-TA increases were reduced (Table 3); only high doses of NSD 1034 (400 mg/kg) achieved a marked reduction in striatal *m*-TA (Table 3). An interesting and unexpected effect were the increases in mouse striatal *p*-TA and to a lesser extent *m*-TA that were induced by the administration of subanaesthetic doses of benzene (Juorio and Boulton, 1983). This effect could be explained as a result of an *in vivo* activation by benzene of aromatic-l-aminoacid decarboxylase (Juorio and Yu, 1983) which has been known to occur *in vitro* for quite some time (Lovenberg, Weissbach and Udenfriend, 1962).

Table 4

The effect of intraventricular cerebral administration of 6-hydroxydopamine (6-OHDA) (250 μg/rat) or lesion of the substantia nigra compacta on the rat striatal *p*-TA, *m*-TA and DA

Treatment	Time (days)	*p*-TA (ng/g)	*m*-TA (ng/g)	DA (ng/g)
Vehicle	10	10.6[1]	2.1[1]	8440[1]
6-OHDA	10	3.6[2]*	0.9[2]*	3150[2]
Control side	-	12.9[2]	3.5[2]	7120[2]
Operated side	2	2.8[2]*	1.5[2]*	3730[2]*
Control side	-	14.8[2]	3.5[2]	8890[2]
Operated side	11	4.1[2]*	2.2[2]*	3130[2]*
Control side	-	13.9[2]	4.2[2]	10650[2]
Operated	25	6.5[2]*	2.2[2]*	4140[2]*

The asterisks indicate that the results are statistically significantly different from controls and the supercript indicates the source of reference. [1]Boulton, Juorio, Philips and Wu, 1977; [2]Juorio and Jones, 1981.

METABOLISM

Monoamine oxidase and aldehyde reductase are the enzymes principally involved in the catabolism of PE, *p*-TA, *m*-TA and T to produce the corresponding acid metabolites that are found in the rat brain (Table 2). A more complete discussion on the effects of monoamine oxidase inhibitors on the trace amines will be found in the chapter by S.R. Philips.

NEURONAL DEGENERATION

The intravenous administration of 6-hydroxydopamine to kittens or cerebro-ventricularly to mature rats produces a degeneration of the brain catecholamine-containing fibers and a depletion of both DA and NA (Laverty *et al.*, 1965; Ungerstedt, 1968). The cerebro-ventricular administration of 6-hydroxydopamine produced a marked reduction in rat striatal *p*-TA, *m*-TA and DA (Table 4). These effects were observed 10 days after drug administration and suggest that *p*-TA and *m*-TA are stored intraneuronally.

The DA containing nerve terminals in the striatum arise from cell bodies located in the substantia nigra (pars compacta) and since the concentration of *p*-TA and *m*-TA are highest in the striatum (Table 2), it is possible that *p*-TA and *m*-TA containing neurons may also exist in this nucleus. The reductions in the concentration of these three amines observed between 2 and 25 days after the application of unilateral electrolytic lesions to the rat pars compacta of the substantia nigra (Table 4) further supports this contention, unless the *p*- or *m*-TA-containing nerves are arising in cell bodies located at a higher level. The effects of brain nuclei lesions on the concentration of PE and T have not yet been investigated.

BINDING

It has been known for quite some time that PE and T bind to rat brain homogenates and particulate fractions (Boulton *et al.*, 1972). A more recent report has claimed the existence of a high affinity and saturable binding sites for PE in the rat brain (Hauger *et al.* 1972). The presence of *p*-TA binding by rat brain homogenates (Boulton *et al.*, 1972) or crude synaptosomal preparations (Ungar *et al.*, 1977) was also found. Recent reports have shown the existence of a high affinity binding site for tritiated T in rat brain homogenate that is different from the 5-HT receptor site (Kellar and Cascio, 1982); autoradiographic studies have shown that the highest binding occurs in the striatum, hippocampus and cortex (Perry *et al.*, 1982).

Table 5

The effects of some psychotropic drugs on mouse striatal *p*-TA, *m*-TA and homovanillic acid (HVA)

Treatment	Dose (mg/kg)	Time (hours)	*p*-TA (ng/g)	*m*-TA (ng/g)	HVA (ng/g)
Control	-	-	19.1[1]	5.1[1]	920[2]
Chlorpromazine	2	2	8.3[1]*	5.6[1]	-
Fluphenazine	0.1	2	7.4[2]*	6.1[2]	2460[2]*
Haloperidol	0.2	2	6.8[1]	6.7[1]	-
Control	-	-	21.0[3]	6.3[3]	1130[2]
Amfonelic Acid	5	4	6.6[3]*	9.1[3]*	1890[3]*
d-Amphetamine	5	2	6.1[3]*	7.9[3]	-
Control	-	-	20.4[4]	6.8[4]	1060[4]
γ-Hydroxybutyrate	1000	2	10.6[4]*	18.1[4]*	3150[4]*

The asterisks indicate that the results are statistically significantly different from controls and the superscript indicates the source of reference. [1]Juorio, 1977a; [2]Juorio, 1980; [3]Juorio, 1982a; [4]Juorio, 1982b.

UPTAKE AND RELEASE

The presence of saturable high and low affinity uptake systems for *p*- and *m*-TA have been demonstrated in brain slices (Steinberg and Smith, 1971; Ross and Renyi, 1971; Dyck, 1978; Petrali *et al.*, 1979). Rat striatal slices release preloaded labelled or endogenous *p*-TA in the presence of depolarizing agents (Dyck, 1978; Dyck *et al.*, 1982). Also, a specific uptake system for tryptamine in the rat cerebral cortex has been recently proposed (Jones, 1983). For more details on uptake and release on trace monoamines see chapter by L.E. Dyck.

NEUROPHYSIOLOGICAL EFFECTS

The iontophoretic administration of small concentrations of the trace amines tend to have little effect on their own, instead they are quite potent to potentiate the changes in neuronal firing rates produced by some of the clinical transmitters (DA, NA, 5-HT) (Jones and Boulton, 1980a and b). For more details on the neurophysiological effects of trace amines see chapter by R.S.G. Jones.

EFFECTS OF PSYCHOTROPIC DRUGS

a) *p*- and *m*-TA. The concentration of striatal *p*-TA is reduced by antipsychotic (chlorpromazine, haloperidol and related compounds), psycho-stimulants (*d*-amphetamine, amfonelic acid) or γ-hydroxybutyrate while *m*-TA level is not changed or is increased (Table 5). Similar effects on *p*- and *m*-TA concentrations were observed after the increases in dopamine synthesis induced by short term lesion (within 24 hours) of the substantia nigra compacta (Juorio and Jones, 1981), administration of amfonelic acid, d-amphetamine, γ-hydroxybutyrate (Table 5) or stress (Juorio, 1979). The finding that these drugs also increase the concentration of DA acid metabolites suggested the existence of a reciprocal interrelation between *p*- and *m*-TA and DA (Table 7). This suggestion was further supported by the findings that compounds that reduce DA turnover rate (i.e. reductions in DA acid metabolite concentrations) such as dopamine receptor agonists (apomorphine, piribedil or lergotrile) or an inhibitor of DA synthesis (α-methyl-*p*-

Table 6

Drug-induced reduction of DA-turnover: effect on *p*- and *m*-TA striatal concentrations

Treatment	Dose (mg/kg)	Time (hours)	*p*-TA (ng/g)	*m*-TA (ng/g)	DOPAC (ng/g)
Controls	-	-	22.9	6.2	690
Apomorphine	20	1	37.9*	5.5	300
Piribedil	10	1	37.0*	7.3	400*
Lergotrile	10	1	40.7*	7.6	340*
Controls	-	-	19.3	5.8	-
α-Methyl-*p*-tyrosine	200	2	37.4*	3.3*	-

The asterisks indicate that the results are statistically significantly different from controls. The results are from Juorio, 1979.

tyrosine) produce an increase in *p*-TA and no change or a reduction in *m*-TA (Tables 6 and 7). One possible explanation for this reciprocal effect between the concentration of the *p*- and *m*-isomers of TA and the turnover rate of DA could be that either tyramine acts as a modulator (Juorio, 1979b). Which one of the tyramines is acting as a modulator would depend on whether tyrosine hydroxylase is activated or not; if the enzyme is activated the effect will be a shift towards the hydroxylation of *p*-tyrosine to form L-DOPA that would then be decarboxylated to form DA. The effect of this would be that the availability of *p*-tyrosine for decarboxylation will be reduced and thus the concentration of *p*-TA decreased (Table 7). *m*-TA in its turn could be formed either by hydroxylation of phenylalanine (Table 3) or by dehydroxylation of L-DOPA or DA (Boulton and Dyck, 1974) thus tending to increase its levels (Table 7). From this it follows that if tyrosine hydroxylase is activated, *m*-TA is the isomer that acts as a neuromodulator. The

Table 7

Interrelationship between changes in brain DA turnover and p-TA and m-TA concentrations

	Dopamine Turnover	p-TA Concentration	m-TA Concentration
Dopamine receptor blockers (chlorpromazine, haloperidol and related compounds) (see Table 5)	increase	reduction	tendency to increase
Dopamine releasors (d-amphetamine and amfonelic acid) (see Table 5)	increase	reduction	tendency to increase
Increase DA synthesis [short term lesions (24 h) substantia nigra pars compacta, γ-hydroxybutyrate] (Juorio and Jones, 1981 and see Table 5)	increase	reduction	increase
Stress (Juorio, 1979b)	increase	reduction	increase
Dopamine receptor agonists (lergotrile, apomorphine and piribedil), (see Table 6)	reduction	increase	reduction
Dopamine synthesis inhibition (α-methyl p-tyrosine), (see Table 6)	reduction	increase	reduction

Table 8

The effects of the administration of reserpine and tetrabenazine on the concentration of PE, _p_-TA, _m_-TA and DA on rat or mouse brain

	Dose (mg/kg)	Time (hours)	PE (ng/g)	_p_-TA (ng/g)	_m_-TA (ng/g)	DA (ng/g)
				Rat brain[1]		
Controls	-	-	2.5	8.0	2.3	11070
Reserpine	0.4	3	2.0	3.8*	1.6*	9050*
Reserpine	0.4	6	2.7	3.1*	1.1*	5900*
Reserpine	1	24	2.8	1.4*	0.4*	1260*
				Mouse brain[2]		
Controls	-	-	-	19.3	5.8	-
Reserpine	1	24	-	1.9*	1.2*	-
Tetrabenazine	20	1	-	3.1*	1.1*	-

The concentrations of PE were determined in the hypothalamus and those of _p_-TA, _m_-TA and DA in the striatum. The asterisk indicates that the results are statistically significantly different from controls. The superscripts indicate the source of reference: [1]Boulton, Juorio, Philips and Wu, 1977; [2]Juorio, 1979a.

opposite metabolic changes take place after inhibition of tyrosine hydroxylase (either by administration of α-methyl-*p*-tyrosine or dopamine receptor agonists); that is a reduction in DA turnover, increase in *p*-TA and no change or a tendency to decrease *m*-TA (Table 7). In this situation *p*-tyramine will act as a modulator.

Tyramine concentrations were reduced to less than 20% of control values one day after a reserpine injection of 1 or 10 mg/kg (Boulton et al., 1977). A single dose of reserpine (0.4 mg/kg) significantly decreased the content of

Table 9

Effect of some psychotropic drugs on the pargyline-induced accummulation of T and 5-HT in the mouse striatum

	Dose (mg/kg)	Time (hours)	T (ng/g)	5-HT (ng/g)
Pargyline control	-	-	40	1020
Chlorpromazine	2	2	81*	880*
α-Flupenthixol	2	2	112*	820*
β-Flupenthixol	2	2	42	1120
d-Amphetamine	5	0.5	30*	1090
Reserpine	5	6	59*	330*

The asterisk indicates that the results are statistically significantly different from controls. The mice were first injected with pargyline (100 mg/kg), 2 hours later one of the drugs (chlorpromazine, α- or β-flupenthixol) and they were killed at 4 hours after the beginning of the experiment. *d*-Amphetamine was injected at 3.5 hours after pargyline and killed at 0.5 h later. Reserpinized mice were injected 2 hours later with pargyline and killed at 6 hours after the beginning of the experiment. The results are from Juorio, 1982c.

both tyramines in the caudate nucleus. A single dose of reserpine (0.4 mg/kg) significantly decreased the content of both p- and m-TA in the rat striatum (Table 7). Reserpine (1 mg/kg) or tetrabenazine (20 mg/kg) produced similar reductions in mouse striatal p-TA and m-TA (Table 7).

b) PE. The hypothalamic content of PE was not affected by reserpine administration while striatal T was increased (Table 8).

c) T. The administration of chlorpromazine, haloperidol, spiperone or α-flupenthixol produced no changes in mouse striatal T (Juorio, 1982c); their administration to mice that had been pretreated with a monoamine oxidase inhibitor (pargyline), however, produced an increase in T (Table 9). d-Amphetamine on the other hand produced a reduction (Table 9).

CONCLUSIONS

It is now established beyond any doubt that the trace amines (PE, p-TA and T) are present in the mammalian nervous system. As yet there are no histochemical indications of their neuronal location. The effects of neurotoxin administration or lesions to neuronal cell bodies, subcellular fractionation and reserpine administration all indicate that at least p-TA and m-TA are present within intraneuronal granules.

The brain contains phenylalanine, p-tyrosine and tryptophan and those are the amino acid precursors of some of the trace amines. The presence of m-tyrosine has not yet been demonstrated.

The trace amines are principally inactivated by oxidative deamination by MAO and metabolization into their respective trace acids; this process is blocked by the administration of MAO inhibitors. Some of the trace amines are actively concentrated and may bind to neuronal structures.

The trace amines are compounds produced by cells within the nervous system; they possess the ability to affect the excitability of nerve cells and represent a normal link in the regulatory mechanisms that control the function of the nervous system. This role fits well with the definition of modulation proposed by Florey(1967). It is more difficult to assign a transmitter function to any of the trace amines, although it remains possible that p-TA, m-TA or T are transmitters in some restricted pathways.

ACKNOWLEDGEMENTS

I thank Dr. A.A. Boulton for his advice and encouragement, and Saskatchewan Health for financial support.

REFERENCES

Bertler A. and Rosengren E. (1959) Occurrence and distribution of catecholamines in the brain. Acta Physiol. Scand. 47, 350-361.

Bogdanski D.F., Pletscher A., Brodie B.B. and Udenfriend S. (1956) Identification and assay of serotonin in brain. J. Pharmacol. Exp. Ther. 117, 82-88.

Boulton A.A. and Juorio A.V. (1983) Cerebral decarboxylation of meta- and para-tyrosine. Exper. 39, 130-134.

Boulton A.A., Juorio A.V., Philips S.R. and Wu P.H. (1977) Effects of reserpine and 6-hydroxydopamine on the levels of some arylalkylamines in rat brain. Br. J. Pharmac. 59, 209-214.

Boulton A.A., Wu P.H. and Philips S.R. (1972) Binding of some primary aromatic amines to certain rat brainparticulate fractions. Can. J. Biochem. 50, 1210-1218.

Durden D.A. and Boulton A.A. (1981) Identification and distribution of m- and p-hydroxyphenylacetic acid in the brain of the rat. J. Neurochem. 36, 129-135.

Durden D.A. and Boulton A.A. (1982) Identification and distribution of phenylacetic acid in the brain of the rat. J. Neurochem. 38, 1532-1536.

Durden D.A., Philips S.R. and Boulton A.A. (1973) Identification and distribution of β-phenylethylamine in the rat. Can. J. Biochem. 51, 995-1002.

Dyck L.E. (1978) Uptake and release of meta-tyramine, para-tyramine and dopamine in rat striatal slices. Neurochem. Res. 3, 775-791.

Dyck L.E., Juorio A.V. and Boulton A.A. (1982) The in vitro release of endogenous m-tyramine, p-tyramine and dopamine from rat striatum. Neurochem. Res. 7, 705-716.

Fellman J.H., Roth E.S. and Fujita T.S. (1976) Decarboxylation to tyramine is not major route of tyrosine metabolism in mammals. Arch. Biochem. Biophys. 174, 562-567.
Florey E. (1967) Neurotransmitters and modulators in the animal kingdom. Fed. Proc. 26, 1164-1178.
Hauger R.L., Skolnick P. and Paul S.P. (1982) Specific [H]-phenylethylamine binding sites in rat brain. Eur. J. Pharmacol. 83, 147-148.
Jones R.S.G. (1983) A specific uptake system for tryptamine in the rat cerebral cortex. Br. J. Pharmac. 78, Proc. Suppl. 12P.
Jones R.S.G. and Boulton A.A. (1980a) Interactions between p-tyramine, m-tyramine, β -phenylethylamine and dopamine on single neurons in the cortex and caudate nucleus of the rat. Can. J. Phys. Pharmacol.58, 222-227.
Jones R.S.G. and Boulton A.A. (1980b) Tryptamine and 5-hydroxytryptamine: actions and interactions on cortical neurons in the rat. Life Sci. 27, 1849-1856.
Juorio A.V. (1976) Presence and metabolism of -phenylethylamine, p-tyramine, m-tyramine and tryptamine in the brain of the domestic fowl. Brain Res. 111, 442-445.
Juorio A.V. (1977a) Effect of chlorpromazine and other antipsychotic drugs on mouse striatal tyramines. Life Sci. 20, 1663-1668.
Juorio A.V. (1977b) Effects of d-amphetamine and antipsychotic drug administration on striatal tyramine levels in the mouse. Brain Res. 126, 181-184.
Juorio A.V. (1979a) Drug-induced changes in the formation, storage and metabolism of tyramine in the mouse. Br. J. Pharmac. 66, 377-384.
Juorio A.V. (1979b) Effect of stress and L-DOPA administration on mouse striatal tyramine and homovanillic acid levels. Brain Res. 179, 186-189.
Juorio A.V. (1980) Effects of molindone and fluphenazine on the brain concentration of some phenolic and catecholic amines in the mouse and the rat. Br. J. Pharmac. 70, 475-480.
Juorio A.V. (1982a) The effects of amfonelic acid and some other central stimulants on mouse striatal tyramine, dopamine and homovanillic acid. Br. J. Pharmac. 77, 511-515.
Juorio A.V. (1982b) The effect of γ -hydroxybutyrate on mouse striatal tyramine, dopamine and homovanillic acid. Br. J. Pharmac. 75, 447-450.

Juorio A.V. (1982c) The effects of some antipsychotic drugs, d-amphetamine and reserpine, on the concentration and rate of tryptamine and 5-hydroxytryptamine in the mouse striatum. Can. J. Phys. Pharmacol. 60, 376-380.

Juorio A.V. (1983) The effects of some decarboxylase inhibitors on mouse striatal tyramines. Neuropharmacol. 22, 71-73.

Juorio A.V. and Boulton A.A. (1982) The effects of some precursor amino acids and enzyme inhibitors on mouse striatal concentration of tyramines and homovanillic acid. J. Neurochem. 39, 859-863.

Juorio A.V. and Jones R.S.G. (1981) The effect of mesencephalic lesions on tyramine and dopamine in the caudate nucleus of the rat. J. Neurochem. 36, 1898-1903.

Juorio A.V. and Philips S.R. (1976) Arylalkylamines in octopus tissues. Neurochem. Res. 1, 501-509.

Juorio A.V. and Robertson H.A. (1977) Identification and distribution of some monoamines in tissues of the sunflower star. Pycnopodia helianthoides (Echinodermata). J. Neurochem. 28, 573-579.

Juorio A.V. and Yu P.H. (1983) Benzene-induced activation of brain aromatic-l-aminoacid decarboxylase. J. Neurochem. 41, Suppl. S75B.

Kellar K.J. and Cascio C.S. (1982) [^{3}H]-Tryptamine: high affinity binding sites in rat brain. Eur. J. Pharmacol. 78, 475-478.

Laverty R., Sharman D.F. and Vogt M. (1965) Action of 2,4,5-trihydroxyphenylethylamine on the storage and release of noradrenaline. Br. J. Pharmac. 40, 836-846.

Lovenberg W., Weissback H. and Udenfriend S. (1962) Aromatic-l-amino acid decarboxylase. J. Biol. Chem. 237, 89-93.

McQuade P.S. and Juorio A.V. (1982) The effects of the administration of β-phenylethylamine on tyramine metabolism. Eur. J. Pharmac. 83, 277-282.

Perry D.C., Manning D.R. and Snyder S.H. (1982) In vitro autoradiographic localization of [^{3}H]-tryptamine binding sites in rat brain. Soc. Neurosci. Proc. 8, 783.

Petrali E.H., Boulton A.A. and Dyck L.E. (1979) Uptake of para-tyramine and meta-tyramine into slices of the caudate nucleus and hypothalamus of the rat. Neurochem. Res. 4, 633-642.

Philips S.R., Durden D.A. and Boulton A.A. (1974a) Identification and distribution of p-tyramines in the rat. Can. J. Biochem. 52, 366-373.

Philips S.R., Durden D.A. and Boulton A.A. (1974b) Identification and distribution of tryptamine in the rat. Can. J. Biochem. 52, 447-451.
Philips S.R., Davis B.A., Durden D.A. and Boulton A.A. (1975) Identification and distribution of m-tyramine in the rat. Can. J. Biochem. 53, 65-69.
Ross S.B. and Renyi A.L. (1971) Uptake and metabolism of β-phenylethylamine and tyramine in mouse brain and heart slices. J. Pharm. Pharmacol. 23, 276-279.
Saavedra J.M. (1974) Enzymatic isotopic assay for and the presence of β-phenylethylamine in brain. J. Neurochem. 22, 211-216.
Saavedra J.M. and Axelrod J. (1973) Effect of drugs on the tryptamine content of rat tissues. J. Pharmacol. Exp. Ther. 185, 523-529.
Silkaitis R.P. and Mosnaim A.D. (1976) Pathways linking 1-phenylethylanine and 2-phenylethylamine with p-tyramine in rabbit brain. Brain Res. 114, 105-115.
Steinberg M.I. and Smith C.B. (1971) Effects of desmethylimipramine and cocaine on the uptake, retention and metabolism of H^3-tyramine in rat brain slices. J. Pharmacol. Exp. Ther. 173, 176-192.
Ungar F., Mosnaim A.D., Ungar B. and Wolf M.E. (1977) Tyramine-binding by synaptosomes from rat brain: effect of centrally active drugs. Biol. Psychiat. 12, 661-668.
Ungerstedt U. (1968) 6-Hydroxydopamine induced degeneration of central monoamine neurones. Eur. J. Pharmac. 5, 107-110.
Warsh J.J., Chan P.W., Godse D.D., Coscina D.V. and Stancer H.C. (1977) Gas chromatography-mass fragmentographic determination of indole-3-acetic acid in rat brain. J. Neurochem. 29, 955-958.

AMINERGIC CONTROL OF THE ELECTROCORTICOGRAM: A PROGRESS REPORT

C. H. Vanderwolf

Department of Psychology

University of Western Ontario, London, Ontario,

Canada

It is widely believed that the presence or absence of low voltage fast electroencephalographic activity (LVFA) in mammals can be related to a single behavioral dimension which has been variously described as sleep-waking, low arousal-high arousal, consciousness-unconsciousness, attention-inattention or excitement-relaxation. However, recent work has shown that this traditional concept is inadequate (Vanderwolf & Robinson, 1981). The neocortex in the rat appears to receive at least two diffuse activating inputs (i.e., inputs that produce LVFA) which have quite different relations to behavior. Any LVFA which occurs during anesthesia or during behavioral immobility in the waking state appears to be due to a cholinergic input from the brain stem reticular formation. Such activity can be blocked by atropinic drugs. The LVFA which occurs during such waking behaviors as face-washing, licking, chattering or gnashing of the teeth and spontaneous tremor is also largely atropine sensitive. These behaviors, together with waking immobility, can be referred to as Type 2 behavior. During the performance of a class of behaviors which includes head movement, walking and spontaneous changes in posture (Type 1 behavior) an atropine-resistant input from the reticular formation produces LVFA (Fig. 1). If Type 1 and Type 2 behaviors are performed simultaneously, e.g., a change of posture occurring during an episode of face-washing,

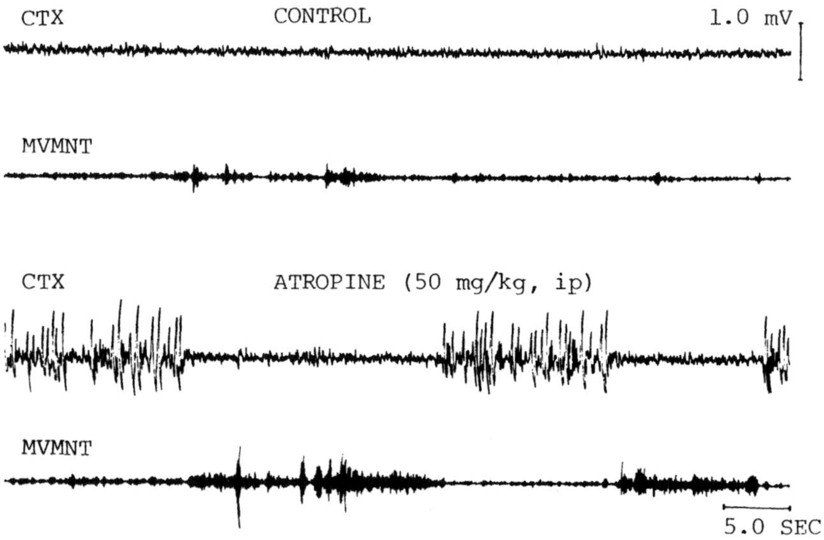

Fig. 1. Effects of atropine SO_4 on neocortical electrical activity in relation to behavior. CTX, neocortical wave activity; MVMNT, output of magnet-and-coil type of movement sensor. Upper traces; undrugged rat displays continuous low voltage fast activity (LVFA) with no relation to motor activity. Lower traces; after atropine SO_4 (i.p.) large amplitude slow waves occur during immobility but LVFA persists during spontaneous motor activity (head movements and stepping).

atropine-resistant LVFA is also present. A practical and reliable means of eliciting atropine-resistant LVFA consists of handling or pushing a rat to produce struggling or walking (push trial).

This paper summarizes a series of experiments intended to elucidate the neurochemical basis of atropine-resistant LVFA. Rats were prepared with pairs of chronically implanted stainless steel electrodes (125 μ in diameter) which permitted monopolar or bipolar surface-to-depth recording in frontal or parietal neocortex (details in Cooley & Vanderwolf, 1978; Vanderwolf & Pappas, 1980). These electrodes permitted recording of spontaneous slow waves, transcallosal evoked potentials (Leung, Harvey & Vanderwolf, 1982) and multi-unit activity (MUA). Behavior was monitored by a magnet and coil type of movement sensor and by direct visual observation. Drugs were administered by intraperitoneal injection. Doses of salts are given in terms of the weight of the salts. Colonic temperatures were maintained at 36-38°C. Non-parametric statistics were used (Siegel, 1956).

Using these methods, it was shown (Vanderwolf et al., 1978; Vanderwolf & Pappas, 1980) that reserpine abolished atropine-resistant neocortical LVFA, but did not abolish atropine-sensitive LVFA. Thus, neither reserpine nor atropine alone eliminated all LVFA, but when the two drugs were combined (reserpine, 10 mg/kg; plus atropine SO_4, 50 mg/kg; or scopolamine HBr, 5 mg/kg, which was also effective) all LVFA was eliminated, producing a rat which displayed continuous 1-6 Hz, 1-2 mV, irregular slow wave activity. Such waves continued without interruption for many hours, even during the occurrence of vigorous struggling (push trials) or strong stimulation of the reticular formation. Since the effect of reserpine on atropine-resistant LVFA was blocked by prior treatment of the rats with nialamide (50 mg/kg), it appeared that atropine-resistant LVFA is dependent on an amine which is a substrate for monoamine oxidase (Vanderwolf & Pappas, 1980).

One means of identifying this unknown amine consisted of administering various test compounds to reserpinized rats in the hope of restoring atropine-resistant LVFA. Rats were pretreated with reserpine (10 mg/kg) 14-18 hr prior to a recording session. A 5 min record of spontaneous (undisturbed) neocortical activity and behavior was made, and push trials were given. If benserazide HCl (50 mg/kg, a peripherally acting blocker of L-aromatic amino acid decarboxylase) was used, it was given at this point and a further 5 min record taken after a delay of 25 min. Then the test compound was given and a continuous record taken for 30-60 min. Finally atropine SO_4 (50 mg/kg) was given and additional records were made for 30 min. Push trials were given periodically under all conditions (10-20/condition/ rat). If benserazide was not used, the test compound was given immediately after the initial 5 min recording period.

In an alternative procedure, the atropine was given after the initial 5 min recording period at the same time as the benserazide (if it was used) and the test compound given 30 min later. The advantage of this procedure is that the absence of LVFA on push trials could be demonstrated prior to the administration of the test compound.

Use of the first of these procedures showed that many compounds can increase the level of LVFA in reserpinized rats. After treatment with reserpine (10 mg/kg) the rats became severely cataleptic. Irregular slow waves or rhythmical spindles occurred frequently and LVFA was present on only 10-25 sec/min as compared to 50-60 sec/min in normal or vehicle injected rats. However, push trials always resulted in LVFA. Benserazide produced no obvious changes but if dihydroxyphenylalanine (L-DOPA, 300 mg/kg) was administered subsequent to the benserazide, the occurrence of LVFA was increased and the rats began to walk, rear and gnaw at the experimental apparatus (Fig. 2). Closely similar effects were produced by m-tyrosine (300 mg/kg, following benserazide pretreatment; this is decarboxylated to form m-tyramine, a dopamine agonist--see Ungerstedt et al., 1973) and by d-amphetamine SO_4 (1-2 mg/kg) and apomorphine HCl (0.25-2.5 mg/kg, without benserazide pretreatment). Clonidine HCl (0.5-1.0 mg/kg) also increased the occurrence of LVFA in reserpinized rats (Vanderwolf et al., 1980). 5-Hydroxy-L-tryptophan (5-HTP, 300 mg/kg, following benserazide pretreatment) produced a rapid increase in LVFA together with tremor, head shakes, generalized muscle spasms and treading movements of the forelimbs (Fig. 2). Other amino acids, such as L-phenylalanine (400-600 mg/kg) (Fig. 2) and D-phenylalanine (400-600 mg/kg), given following benserazide pretreatment, had no obvious effect on the behavior and neocortical electrical activity of reserpinized rats.

Restoration of normal levels of spontaneous LVFA in reserpinized rats did not necessarily indicate that atropine-resistant LVFA had been restored. When atropine was administered following d-amphetamine, apomorphine, L-DOPA, clonidine, or m-tyrosine in these experiments, all LVFA rapidly disappeared and irregular slow waves were continuously present even though the rats walked spontaneously or struggled during handling (Vanderwolf et al., 1980; Table 1). If the test compounds were given following atropinization, motor activity

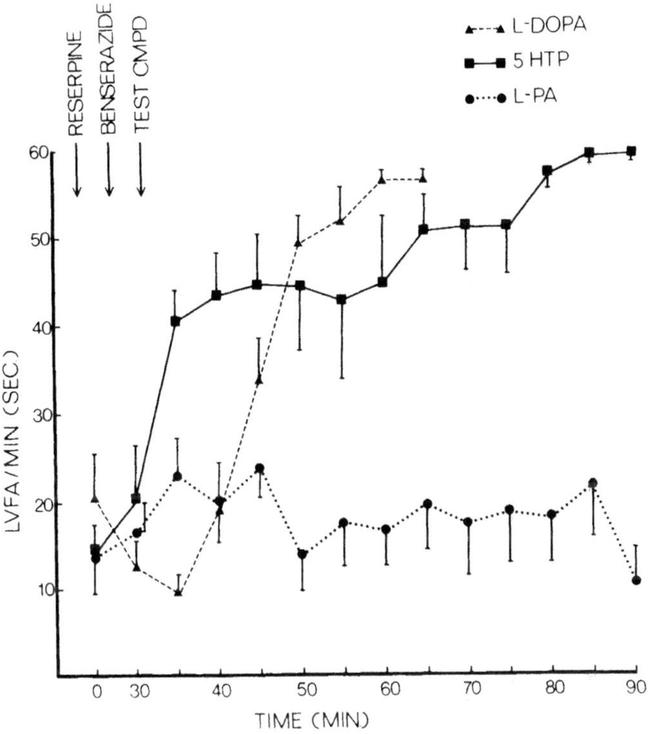

Fig. 2. Occurrence of LVFA in reserpinized rats after various treatments. Between 14 and 18 hr after reserpine (10 mg/kg) a 5 min undisturbed record of neocortical activity and behavior was taken and benserazide (50 mg/kg) was given. A further 5 min record was taken beginning 25 min later and a test compound was given. Continuous recording followed for 40-60 min. Test compounds: L-DOPA, dihydroxyphenylalanine, 300 mg/kg, N=8; 5-HTP, 5-hydroxytryptophan, 300 mg/kg, N=7; L-PA, L-phenylalanine, 400-600 mg/kg, N=6. Values are means ± standard errors of the means. Increase significant for L-DOPA ($p < .001$) and 5-HTP ($p < .01$) according to Friedman nonparametric analysis of variance. L-DOPA data are replotted from Vanderwolf et al., 1980.

TABLE 1. Percent of push trials associated with neocortical low voltage fast activity after various drug treatments.

Group	Test compound	Reserpine (10 mg/kg) + atropine (50 mg/kg)	Reserpine (10 mg/kg) + atropine (50 mg/kg) + test compound
1.	saline (N = 10)	4.6 ± 3.1	1.0 ± 1.0
2.	tranylcypromine (10 mg/kg; N = 5)	0 ± 0	67.5 ± 6.6*
3.	pargyline (5 mg/kg; N = 5)	3.0 ± 2.0	24.0 ± 19.4
4.	pargyline (50 mg/kg; N = 10)	3.0 ± 2.0	59.9 ± 11.7*
5.	pargyline (50 mg/kg) + L-tryptophan (300 mg/kg; N = 7)	2.0 ± 1.5	57.1 ± 12.0*
6.	L-tryptophan (300 mg/kg; N = 6)	0 ± 0	0 ± 0
7.	benserazide (50 mg/kg) + L-DOPA (300 mg/kg; N = 9)	---	0 ± 0
8.	benserazide (50 mg/kg) + DL-p-tyrosine (300-1000 mg/kg; N = 5)	---	10 ± 7.1
9.	benserazide (50 mg/kg) + DL-m-tyrosine (300 mg/kg; N = 6)	---	1.7 ± 1.5
10.	benserazide (50 mg/kg) + L-phenylalanine (400-600 mg/kg; N = 7)	---	0 ± 0
11.	benserazide (50 mg/kg) + D-phenylalanine (400-600 mg/kg; N = 8)	---	0 ± 0
12.	benserazide (50 mg/kg) + 5-hydroxy-L-tryptophan (300 mg/kg; N = 6)	---	77.8 ± 8.9*

* Differs from saline group, $p < .01$ (Mann-Whitney U test) Values given are means and standard errors of the means. In Groups 1-6 the sequence of drugs was reserpine, atropine, test compounds; in Groups 7-12 the sequence was reserpine, test compounds, atropine (see text).

increased but no LVFA occurred. Thus, catecholaminergic agonists exert their effects on neocortical LVFA in reserpinized rats via an atropine-sensitive (probably cholinergic) mechanism (Vanderwolf & Robinson, 1981).

However, a further series of experiments showed that β-phenylethylamine (PE) will restore atropine-resistant LVFA in reserpinized rats (Vanderwolf et al., 1980). The effect is shown in Fig. 3. As already mentioned, neocortical activity in rats 14-18 hr after treatment with reserpine (10 mg/kg) frequently consisted of 1-6 Hz irregular slow waves or of rhythmical spindle activity. These waveforms were associated with a bursting pattern of MUA. In general, positive potentials in the deep layers of the neocortex were associated with a suppression of MUA while negative potentials were associated with a burst of MUA. It is likely that the extracellular positivity and related unit silence were due to a hyperpolarization occurring simultaneously in large numbers of cells while the negativity and unit firing were due to depolarization of the same cells. When the neocortical record assumed a low voltage fast pattern (either spontaneously or as a result of push trials) MUA occurred continuously at a high rate. This pattern is similar to what was observed in normal (undrugged) waking rats. Following injection of atropine SO_4 (50 mg/kg) all LVFA disappeared in the reserpinized rats; 1-6 Hz slow wave activity and the associated bursting pattern of MUA became continuous. If push trials were given at this point, the rate of MUA discharge showed an obvious increase, but despite this, unit firing was always totally suppressed during the positive peaks of large slow waves recorded in the depths of the neocortex.

Injection of PE (80 mg/kg, N = 19) produced a behavioral syndrome including head movements, alternating treading movements of the forelimbs, and episodes of generalized muscle spasms (5-HT syndrome, Jacobs, 1976), together with some locomotion, rearing, salivation and gnawing. LVFA began to appear within 2-3 min after injection, became prominent in 5-10 min and had largely disappeared again in 20-30 min. This LVFA was associated with continuous MUA and was especially well developed during episodes of struggling or walking (spontaneous or induced by pushing). During periods of relative immobility (treading movements of the forelimbs, head

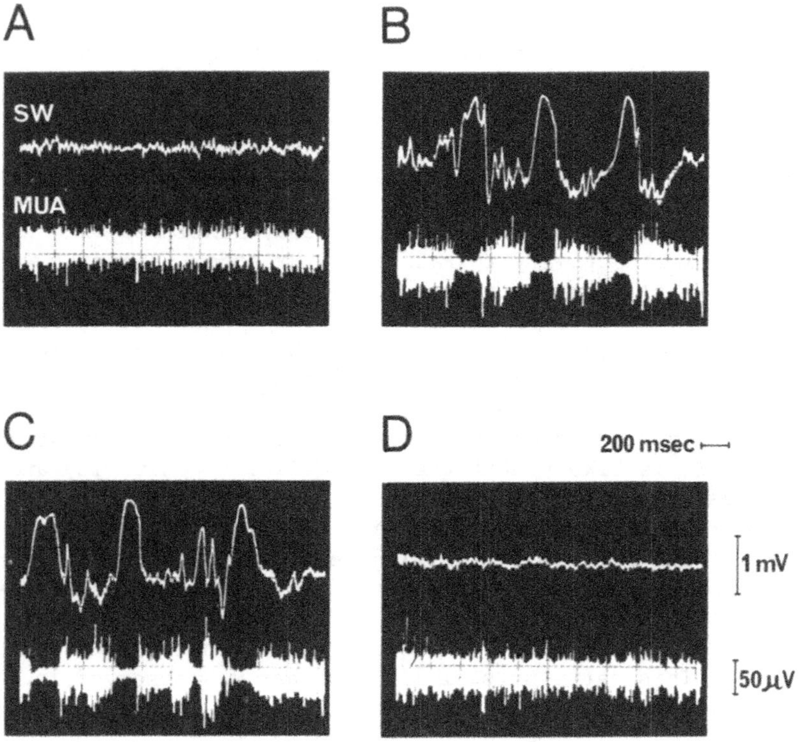

Fig. 3. Effects of atropine and β-phenylethylamine (PE) on spontaneous neocortical electrical activity in a reserpinized rat. SW, slow wave activity, band-pass, .3 Hz-3 kHz; MUA, multiunit activity, band pass, 300 Hz-10 kHz. Both records from same bipolar surface-to-depth electrode pair. Upward deflection indicates positivity in deep cortical layers. MUA derived from deep electrode. A,B. 16 hr after reserpine (10 mg/kg) neocortex may display: (A) low voltage fast activity (LVFA) associated with continuous MUA, or (B) large amplitude SW associated with bursting pattern of MUA. Units discharge during SW deep negativity and are silent during SW deep positivity. C, 30 min after atropine SO_4 (50 mg/kg) the large SW and bursting MUA are continuous; LVFA has disappeared. D, 10 min after PE (80 mg/kg) LVFA and continuous MUA have reappeared.

movements, etc. were present almost continuously during the presence of LVFA) low amplitude (<1.0 mV, 2-3 Hz) slow waves often occurred and were associated with a burst-suppression pattern of MUA activity.

The effects of PE varied with the dose given. Little or no atropine-resistant LVFA appeared in reserpinized rats (N = 4) given 20 mg/kg. Some LVFA appeared following a dose of 40 mg/kg (N = 6) but it disappeared in a few minutes.

Tryptamine HCl, given to reserpinized rats in doses of 50-100 mg/kg, sometimes produced atropine-resistant LVFA and behavioral effects resembling those of PE but the effect appeared to be less reliable. Smaller doses (10-25 mg/kg) appeared to have little effect (total N = 7).

These observations indicate that PE and tryptamine can restore atropine-resistant LVFA in rats following treatment with reserpine. Is this effect presynaptic, possibly involving the release of an endogenous amine, or is it post-synaptic, involving a direct effect on an aminergic receptor? This problem could be studied if it were possible to destroy the fibres of the reticulo-cortical pathway which produces atropine-resistant LVFA. If PE and tryptamine act presynaptically, such a procedure should abolish their effects; if PE and tryptamine act post-synaptically it should not. Previous work has shown that stimulation of the lateral or posterior hypothalamus is an effective means of eliciting atropine-resistant LVFA (Vanderwolf et al., 1975) and that hypothalamic lesions abolish atropine-resistant LVFA but may spare atropine-sensitive LVFA (Kolb & Whishaw, 1977; Whishaw & Kolb, 1979). Consequently, the hypothesized aminergic reticulocortical fibres involved in atropine-resistant LVFA may pass through the hypothalamus.

Consequently, a series of rats were prepared with large hypothalamic lesions produced by a diathermy apparatus. Subsequent histological study revealed that the lesion had destroyed most of the lateral and posterior hypothalamus, together with most of the medial thalamus (sparing only the habenula and periventricular nuclei) and the medial parts of the ventral thalamic nuclei and internal capsule, resulting in a preparation whose behavior and electrocortical activity approximated the cerveau isolé condition.

Recordings taken 20-24 hr after the placing of the lesion revealed continuous .5-6 Hz, 1-2 mV neocortical slow wave activity associated with a bursting pattern of MUA quite similar to the pattern seen, at times, in reserpinized rats. However, unlike the situation in reserpinized rats, LVFA could not be elicited by tactile or by painful stimulation. Administration of atropine (50 mg/kg) or scopolamine (5 mg/kg) produced no major changes, but the slow waves tended to increase in amplitude and become more irregular. When PE (80 mg/kg, N = 9) was administered 30 min later, the amplitude of the slow waves gradually fell by 25-50% but no LVFA appeared (Fig. 4). The MUA rate declined markedly, reaching a level of 13.6% of the initial rate 10 min after PE administration ($p < .001$; Friedman nonparametric analysis of variance). After approximately 1 hr the MUA rate and slow wave amplitude had returned to normal. These effects contrasted sharply with the effect of PE in reserpinized rats in which clear atropine-resistant LVFA appeared and the MUA discharge rate did not decline (Fig. 4).

Tryptamine (50-100 mg/kg, N = 6) administered to the diencephalon-damaged rats about 1 hr after PE, produced either no visible effect on neocortical activity or a decline in MUA and slow wave amplitude which was similar to the effect of PE.

The differing effects of PE and tryptamine in reserpinized rats as compared to rats with large diencephalic lesions suggests that these compounds may have at least two distinct actions on neocortical unit discharges and slow waves. First there may be a post-synaptic effect which results in suppression of spontaneous unit activity and correlated reduction of slow wave amplitude in the diencephalon-damaged rats. A similar reduction in the frequency of neocortical unit discharges has been described following iontophoretic application of PE in anesthetized rats (Henwood et al., 1979). Secondly, in reserpinized rats, PE and tryptamine may influence neocortical activity indirectly by releasing (or otherwise facilitating the action of) an endogenous amine which produces atropine-resistant LVFA and maintains a high rate of MUA despite the concurrent presence (presumably) of the activity suppressing effect of these compounds. PE is capable of releasing dopamine and

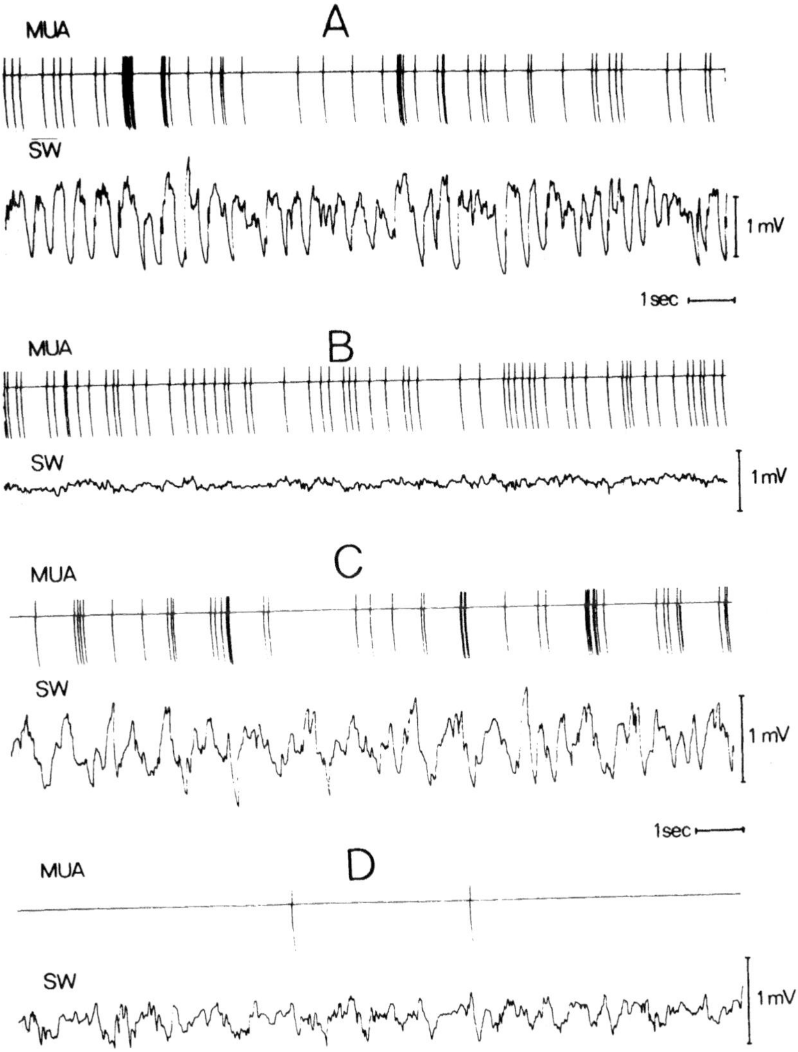

Fig. 4. Effect of PE on neocortical activity after atropinization in reserpinized or diencephalon-damaged rats. Records as in Fig. 3 except that SW polarity is reversed (deep cortical negativity up) and MUA recorded as output pulses of window discriminator (1 pulse = 4 unit discharges). A. Activity about 16 hr after reserpine (10 mg/kg) and 30 min after atropine SO_4 (50 mg/kg). B. About 10 min after addition of PE (80 mg/kg). C. Activity about 24 hr after large diencephalic lesion and 30 min after atropine SO_4 (50 mg/kg). D, about 10 min after

Fig. 4. (Cont'd)

addition of PE (80 mg/kg). In the reserpinized rat PE produces LVFA and a relatively high rate of MUA without bursting; in the diencephalon-damaged rat PE reduces SW amplitude and depresses MUA but does not produce LVFA.

norepinephrine from brain tissue (Fuxe et al., 1967). However, according to Sloviter et al (1980) it does not release 5-hydroxytryptamine (5-HT): instead, it acts as a direct agonist at 5-HT receptors. In this connection (see below) it is of interest that 5-HT may have excitatory effects in rat neocortex (Jones, 1982).

Monoamine oxidase inhibitors are also capable of restoring atropine-resistant LVFA after it has been eliminated by reserpine. This was shown in experiments in which rats were pretreated with reserpine (10 mg/kg) followed by atropine 14-18 hr later. LVFA disappeared and could no longer be elicited by inducing the rat to walk or struggle. If a control injection of saline was subsequently administered this pattern persisted unchanged (Figs. 5 & 6). Pargyline, in a dose of 5 mg/kg also had no significant effect (Fig. 6). However, when pargyline was administered in a dose of 50 mg/kg 30 min after atropinization, spontaneous LVFA appeared after a delay of 15-30 min (Figs. 5 & 6). At the same time, the rats became more active, spontaneously moving the head, face-washing, stepping or walking and, at times, making repetitive treading movements with the forepaws. The startle response to a light touch was also increased. Since many of the rats became almost continuously active, it was difficult to determine whether the restored atropine-resistant LVFA occurred in its normal relation to Type 1 behavior, i.e., there were few periods of immobility with which periods of movement could be compared. In some rats, the correlation with behavior appeared to be partially restored since 1-2 mV slow waves (1-6 Hz) tended to occur during quiescent periods while LVFA tended to occur during locomotion. However since the correlation between neocortical activity and behavior was usually less clear cut than it is in a normal rat treated with atropine alone, it appears that pargyline is largely but not entirely successful in reversing the effects of reserpine on atropine-resistant LVFA.

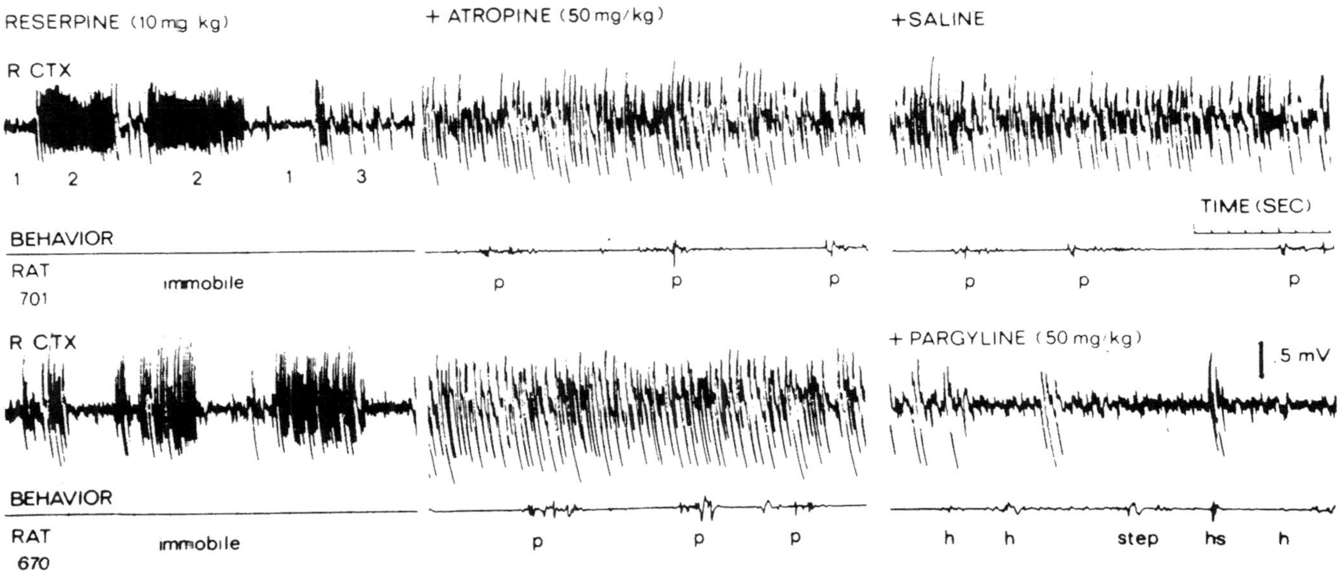

Fig. 5. Pargyline-induced restoration of atropine-resistant LVFA in a reserpinized rat. Upper traces: Control rat (#701). After reserpine, electrocorticogram includes LVFA (1), rhythmical spindles (2) and irregular SW (3) all occurring during behavioral immobility. After atropine, all LVFA is abolished, even during push trials (P). No further change after saline administration. Lower traces: Experimental rat (#670) resembles control rat after reserpine and reserpine + atropine but displays prominent LVFA during spontaneous behavior after administration of pargyline. h, head movement; step, stepping; hs, head shake.

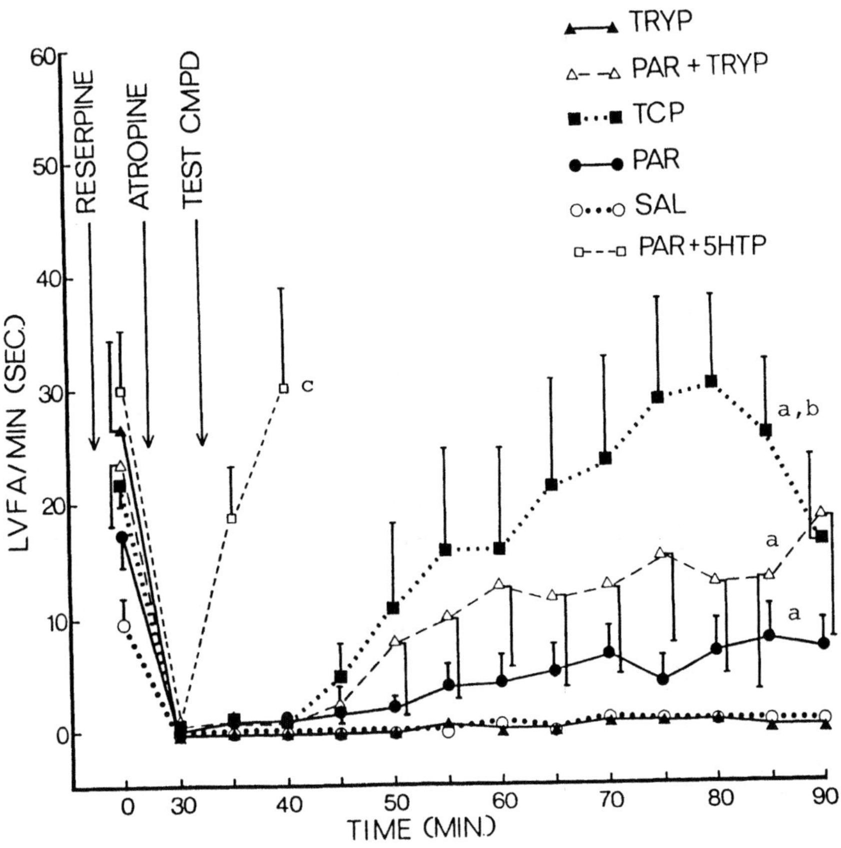

Fig. 6. Effectiveness of various compounds as restorers of LVFA in reserpinized-atropinized rats. Between 14 and 18 hr after reserpine (10 mg/kg) a 5 min undisturbed record of neocortical activity and behavior was taken and atropine SO_4 (50 mg/kg) was given. A further 5 min record was taken beginning 25 min later and test compounds were given. Continuous recording followed for 60 min. Test compounds: TRYP, L-tryptophan (300 mg/kg), N=6; PAR + TRYP, pargyline (50 mg/kg) + L-tryptophan (300 mg/kg), N=7; TCP, tranylcypromine (10 mg/kg), N=5; PAR, pargyline (50 mg/kg), N=9; SAL, saline control, N=10; PAR + 5HTP, pargyline (50 mg/kg) + 5-hydroxytryptophan (100-300 mg/kg), N=7. Note that atropine initially eliminated all LVFA in all groups. a, differs from SAL, $p < .002$, Mann Whitney U test, using total time LVFA displayed after test compound had been given; b, differs from PAR, $p < .05$; c, differs from every

Fig. 6. (Cont'd)

other group, $p < .001$, in total LVFA displayed in first 10 min after test compound.

Tranylcypromine (10 mg/kg) had behavioral and electrocortical effects similar to those of pargyline (50 mg/kg) although it tended to be more effective than pargyline in producing LVFA (Fig. 6). In addition, tranylcypromine rapidly produced a lethal hyperthermia. Within 50-60 min after receiving the drug, the rats had colonic temperatures ranging from 40.5 to 41.5°C (see Tabakoff et al., 1977). This hyperthermia appeared to be responsible for a decline in LVFA near the end of the experiment (Fig. 6).

These experiments support the hypothesis that reserpine abolishes atropine-resistant LVFA by depletion of an unknown monoamine or monoamines. Subsequent inhibition of monoamine oxidase, which increases brain monoamine levels, results in a substantial restoration of atropine-resistant LVFA together with at least a partial restoration of the normal correlation of this waveform with behavior. Can the amine responsible for these effects be identified, and is it identical with the amine involved in PE induced restoration of atropine-resistant LVFA in reserpinized rats?

As mentioned previously, a variety of other compounds were tested for their ability to restore atropine-resistant LVFA. L-Tryptophan (300 mg/kg), given following pretreatment with reserpine and atropine, failed to restore atropine-resistant LVFA. When L-tryptophan (300 mg/kg) was given simultaneously with pargyline (50 mg/kg) it did not increase significantly the effectiveness of pargyline in restoring atropine-resistant LVFA (Table 1, Fig. 6). Similarly, p-tyrosine (300-1000 mg/kg), m-tyrosine (300 mg/kg), L-phenylalanine (400-600 mg/kg), D-phenylalanine (400-600 mg/kg), and L-DOPA (300 mg/kg) all given after benserazide pretreatment, failed to restore atropine-resistant LVFA in reserpinized rats (Table 1). Octopamine HCl (100 mg/kg,no benserazide pretreatment, N = 6) produced prominent autonomic effects (piloerection, salivation, widening of the palpebral fissure), but also failed to restore atropine-resistant LVFA.

In contrast to these negative results, 5-HTP (300 mg/kg) produced a clear restoration of atropine-resistant LVFA in reserpinized rats when given after benserazide pretreatment (Table 1). When given without benserazide in a dose of 100-200 mg/kg it had no obvious effect (Vanderwolf et al., 1980). Further, when 5-HTP (100-300 mg/kg) was given simultaneously with pargyline (50 mg/kg) to rats pretreated with reserpine and atropine (Fig. 6), clear LVFA appeared in 2-4 min, accompanied by head movement, tremor, treading movements of the forepaws, and severe generalized muscle spasms. Thus, a combination of pargyline and 5-HTP is far more effective than either compound alone. However, within 10-15 min after receiving the drug combination, the rats all became cyanotic and the LVFA was replaced by a burst-suppression pattern of slow waves alternating with periods of electrical silence. Death ensued rapidly in all cases.

The foregoing experiments show that although a variety of other amino acids fail to restore atropine-resistant LVFA in reserpinized rats, 5-HTP is rapidly effective when it is given in conjunction with either benserazide or pargyline. These facts indicate that an indoleamine such as 5-HT, 5-methoxytryptamine (Green et al., 1973) or tryptamine (Saavedra & Axelrod, 1972; Boulton & Baker, 1975) is probably the actual agent which restores atropine-resistant LVFA. Of these possibilities, 5-HT is the most likely candidate since 5-HTP is its direct precursor. Further, the "vast majority" of central indoleamine-containing neurons contain 5-HT (Jonsson et al., 1974). L-Tryptophan the direct precursor of tryptamine, was relatively ineffective in restoring atropine-resistant LVFA. Evidence that PE, which also restores atropine-resistant LVFA, acts on serotonergic mechanisms (Sloviter et al., 1980) is consistent with this point of view. Similarly, the ability of pargyline and tranylcypromine to restore atropine-resistant LVFA in reserpinized rats may be attributable to a rapid increase in brain levels of 5-HT following the administration of these drugs. Additional data suggesting a role of 5-HT-containing neurons in the generation of atropine-resistant LVFA appears in the report by Robinson and Vanderwolf (1978) that electrical stimulation of the median raphé nucleus elicits atropine-resistant LVFA in the neocortex. Further,

Trulson and Jacobs (1979; 1981) showed that dorsal raphé neurons in cats increase their activity during movement in the waking state and during the muscular twitches of active sleep. Although more detailed study is needed, these relations to behavior appear similar to the relation of atropine-resistant LVFA to behavior in rats (Vanderwolf & Robinson, 1981).

However, there are difficulties with the view that serotonergic inputs to the neocortex produce atropine-resistant LVFA in non-reserpinized rats. First, several classical serotonin antagonists do not block atropine-resistant LVFA. Methysergide bimaleate (N=6, 1-30 mg/kg), cyproheptadine HCl (N=5, 1-10 mg/kg), metergoline (N=6, 1-10 mg/kg) and spiroperidol (N=5, 1-10 mg/kg) do not block the atropine-resistant LVFA occurring in association with head movement, walking or struggling. These doses are well in excess of those inhibiting 5-HTP-elicited head twitches _in vivo_ or 5-HT receptor binding _in vitro_ (Peroutka et al., 1981). A possible interpretation of this is that the classical peripheral serotonin antagonists are inactive at many serotonergic synapses in the brain (Haigler & Aghajanian, 1974; Aghajanian, 1981). Drug treatments that are known to eliminate atropine-resistant LVFA include volatile anesthetics (ethyl ether, trichloroethylene, chloroform) and large doses of morphine and reserpine. Urethane is relatively ineffective (Vanderwolf et al., 1978). Considering these facts, it is of some concern that electrophysiological studies of 5-HT agonists and antagonists are usually carried out in anesthetized animals.

Another possibility that must be considered is that 5-HT may be capable of restoring atropine-resistant LVFA in reserpinized rats but is, nonetheless, not the agent that normally produces such LVFA in non-reserpinized rats. This hypothesis might appear to be supported by the observation that DL-p-chlorophenylalanine (PCPA, a blocker of 5-HT synthesis, Koe & Weissman, 1966), in a single dose of 500 mg/kg, did not abolish atropine-resistant LVFA (Vanderwolf et al., 1978). (In fact, this lack of effect, together with the negative results of experiments with 5-HT receptor blockers, had initially lead to the conclusion that 5-HT is not involved in atropine-resistant LVFA--Vanderwolf et al, 1978). However, a reinvestigation (in progress) of this problem showed that 2 rats in a group of 7 given PCPA, 500 mg/kg, on each of 3 successive days, displayed a substantial loss of atropine-resistant LVFA. It may be that PCPA will eliminate atropine-resistant LVFA, as reserpine does, only when given in very large doses.

In conclusion, a variety of different experimental approaches indicate that 5-HT probably plays a key role in the generation of atropine-resistant LVFA in the neocortex.

ACKNOWLEDGMENTS

This research was supported by Grant AO-118 from the Natural Science and Engineering Research Council of Canada. The drugs used were purchased from Sigma Chemical Co., or were gifts from the following: Dr. G. Baker (Serpasil, Ciba); Boehringer Sohn (clonidine); Farmitalia Carlo Erba (metergoline); Health and Welfare, Canada (apomorphine); Dr. G. Mogenson (spiroperidol); Sandoz (methysergide); Smith, Kline and French (amphetamine). I thank R. K. Cooley for excellent technical assistance, L. Mitchell for her care in typing the manuscript, and D. Stewart and G. Tu who assisted in various parts of the work.

REFERENCES

Aghajanian, G. K. (1981) The modulatory role of serotonin at multiple receptors in brain, in Serotonin Neurotransmission and Behavior (Jacobs, B. L. & Gelperin, A., eds). MIT Press, Cambridge, Mass.

Boulton, A. A. and Baker, G. B. (1975) The subcellular distribution of beta-phenylethylamine, p-tyramine, and tryptamine in rat brain. J. Neurochem. 25, 477-481.

Cooley, R. K. and Vanderwolf, C. H. (1978) Construction of wire leads and electrodes for use in slow wave recording in small animals. Brain Research Bull. 3, 175-179.

Fuxe, K., Grobecker, H. and Jonsson, J. (1967) The effect of β-phenylethylamine on central and peripheral monoamine-containing neurons. Eur. J. Pharmacol. 2, 202-207.

Green, A. R., Koslow, S. H. and Costa, E. (1973) Identification and quantitation of a new indolealkylamine in rat hypothalamus. Brain Research 51, 371-374.

Haigler, H. J. and Aghajanian, G. K. (1974) Peripheral serotonin antagonists: Failure to antagonize serotonin in brain areas receiving a prominent serotonergic input. J. Neur. Transmiss. 35, 257-273.

Henwood, R. W., Boulton, A. A. and Phillis, J. W. (1979) Iontophoretic studies of some trace amines in the mammalian CNS. Brain Research 164, 347-351.

Jacobs, B. L. (1976) An animal behavior model for studying central serotonergic synapses. Life Sci. 19, 777-786.

Jones, R. S. G. (1982) Responses of cortical neurones to stimulation of the nucleus raphé medianus: A pharmacological analysis of the role of indoleamines. Neuropharmacol. 21, 511-520.

Jonsson, C., Einarsson, P., Fuxe, K. and Hallman, N. (1974) Microspectrofluorometric studies on central 5-hydroxytryptamine neurons. Adv. Biochem. Pharmacol. 10, 55-65.

Koe, B. K. and Weissman, A. (1966) p-Chlorophenylalanine: A specific depletor of brain serotonin. J. Pharmacol. exp. Ther. 154, 499-516.

Kolb, B. and Whishaw, I. Q. (1977) Effects of brain lesions and atropine on hippocampal and neocortical electroencephalograms in the rat. Exp. Neurol. 56, 1-22.

Leung, L. S., Harvey, G. C. and Vanderwolf, C. H. (1982) Combined video and computer analysis of the relation between the hemispheric response and behavior. Behav. Brain Research 6, 195-200.

Peroutka, S. J., Lebovitz, R. M. and Snyder, S. H. (1981) Two distinct central serotonin receptors with different physiological functions. Science 212, 827-829.

Robinson, T. E. and Vanderwolf, C. H. (1978) Electrical stimulation of the brain stem in freely moving rats: II. Effects on hippocampal and neocortical electrical activity, and relations to behavior. Exp. Neurol. 61, 485-515.

Saavedra, J. M. and Axelrod, J. (1972) A specific and sensitive enzymatic assay for tryptamine in tissues. J. Pharmacol. exp. Ther. 182, 363-369.

Siegel, S. (1956) Non-parametric Statistics for the Behavioral Sciences. McGraw-Hill, New York.

Sloviter, R. S., Connor, J. D. and Drust, E. G. (1980) Serotonergic properties of β-phenylethylamine in rats. Neuropharmacol. 19, 1071-1074.

Tabakoff, B., Moses, F., Phillips, S. R. and Boulton, A. A. (1977) Effects of tranylcypromine and pargyline on brain tryptamine. Experientia 33/3, 380-381.

Trulson, M. E. and Jacobs, B. L. (1979) Raphé unit activity in freely moving cats: Correlation with level of behavioral arousal. Brain Research 163, 135-150.

Trulson, M. E. and Jacobs, B. L. (1981) Activity of serotonin-containing neurons in freely moving cats, in Serotonin Neurotransmission and Behavior. (Jacobs, B. L. and Gelperin, A., eds), pp 339-365. MIT Press, Cambridge, Mass.

Ungerstedt, U., Fuxe, K., Goldstein, M., Battista, A., Ogawa, M. and Anagnoste, B. (1973) Action of m-tyrosine in experimental models: Evidence for possible anti-parkinsonian activity. Eur. J. Pharmacol. 21, 230-237.

Vanderwolf, C. H., Kramis, K., Gillespie, L. A. and Bland, B. H. (1975) Hippocampal rhythmical slow activity and neocortical low voltage fast activity: Relations to behavior, in The Hippocampus: Neurophysiology and Behavior, vol. 2. (Isaacson, R. L. and Pribram, K. H., eds), pp. 101-128. Plenum, New York.

Vanderwolf, C. H., Kramis, R. and Robinson, T. E. (1978) Hippocampal electrical activity during waking behavior and sleep: Analyses using centrally acting drugs. Functions of the septo-hippocampal system. Ciba Foundation Symposium 58 (new series). Amsterdam: Elsevier, Excerpta Medica, North-Holland, 199-226.

Vanderwolf, C. H. and Pappas, B. A. (1980) Reserpine abolishes movement-correlated atropine-resistant neocortical low voltage fast activity. Brain Research 202, 79-94.

Vanderwolf, C. H., Robinson, T. E. and Pappas, B. A. (1980) Monoamine replacement after reserpine: Catecholaminergic agonists restore motor activity but phenylethylamine restores atropine-resistant neocortical low voltage fast activity. Brain Research 202, 65-77.

Vanderwolf, C. H. and Robinson, T. E. (1981) Reticulo-cortical activity and behavior: A critique of the arousal theory and a new synthesis. Behav. Brain Sci. 4, 459-514.

Whishaw, I. Q. and Kolb, B. E. (1979) Neocortical and hippocampal EEG in rats during lateral hypothalamic lesion-induced hyperkinesia: Relations to behavior and effects of atropine. Physiol. Behav. 22, 1107-1113.

NEURONAL TRANSPORT OF TRACE AMINES: AN OVERVIEW

L. E. Dyck

Psychiatric Research Division
Cancer and Medical Research Building
University of Saskatchewan
Saskatoon, Saskatchewan S7N OWO

ABSTRACT

Recently, it has been shown that uptake of radiolabelled *meta*-tyramine (mTA) or *para*-tyramine (pTA) into slices of rat hypothalamus and striatum was very similar to that of dopamine (DA). The uptake of all three amines could be resolved into a high affinity, as well as a low affinity, uptake system. The high affinity uptakes of mTA and pTA into the striatum were sodium dependent, were decreased by lesioning of the substantia nigra or medial forebrain bundle, and by the presence of cocaine, ouabain or dinitrophenol. In addition, a depolarizing stimulus caused the Ca^{++}-dependent release of radiolabelled, as well as endogenous, mTA and pTA from rat slices. The characteristics of mTA and pTA uptake and release are those that have been associated with a neurotransmitter function; however, it appears that the release of mTA and pTA can originate from a different neuronal compartment than DA release. In keeping with this, it has been observed that two different types of drugs -- the methylphenidate-like CNS stimulants and dipropyl-substituted aminotetralins -- stimulate the release of mTA and pTA but not DA. It seems possible that alterations in the intraneuronal compartmentation of mTA and pTA may affect the quantity of catecholamine release and may also affect the nature of the post-synaptic response.

INTRODUCTION

It is well known that trace amines (such as mTA, pTA,

β-phenylethylamine (PE), β-phenylethanolamine (PEOH), *p*-octopamine (OA), tryptamine (T)) are present in nerve tissues of a number of different animal species (see Boulton, 1979; Boulton and Juorio, 1982). Their presence in such tissues and their structural similarities to catechol- and indole-amine transmitters have led to studies of their possible function as neurotransmitters. According to the existing theory of neurotransmission, an action potential generated by the cell body of a neuron travels down its axon and when it reaches the nerve terminal, its depolarization somehow causes an influx of Ca^{++}. This, in turn, causes a secretion, or release, of transmitter molecules from the storage vesicles into the synaptic cleft. The released transmitter, then, interacts with the post-synaptic receptors of another neuron. In addition to this process of transmitter release, an avid, active, specific re-uptake of (released) transmitter back into the nerve terminals can occur. One of the experimental paradigms used to study these processes (uptake and release) has been to label transmitter stores by incubating preparations of nerve tissue with tritiated or carbon-14 labelled compounds. The accumulation of the radiolabelled putative transmitter mimics the physiological re-uptake process, and the release of the previously accumulated compound by a depolarizing stimulus, such as high concentrations of KCl, mimics physiological release.

In order for a compound to be considered a neurotransmitter, its uptake and release systems must exhibit certain characteristics. First of all, its uptake must be active; i.e. require the expenditure of metabolic energy to take up the compound against a concentration gradient. Secondly, its uptake -- which probably utilizes a carrier protein to shuttle the transmitter compound across the membrane -- must have a high affinity for its transmitter(s) and a low affinity for other transmitters or non-transmitter compounds. Thirdly, the re-uptake system must be Na^{+}-dependent. Finally, stimuli which cause membrane depolarization should stimulate a Ca^{++}-dependent release of the compounds in question.

Previous studies of the transport of T, PE and PEOH have shown they are poorly taken up by a variety of tissue preparations (Ross *et al*, 1968; Ross and Renyi, 1971; Baldessarini and Vogt, 1971; Pletscher, 1976; Greenberg and Whalley, 1978; Stacey, 1961; Born *et al*, 1972; Wu and

Boulton, 1973; Costa et al, 1977; Schroder et al, 1979; Osborne, 1980; Ross and Ask, 1980). Their uptake, in general, seems to be mediated mainly by diffusion rather than an active process. Similarly, the release of these compounds appears to occur mainly by a rapid spontaneous washout and depolarizing stimuli do not increase their rates of release (Saldate and Orrego, 1977; Snodgrass and Iversen, 1974; Hery et al, 1979, 1983; and the chapter in this book by Dyck). Such transport properties are consistent with the lipophilic properties of these trace amines and seem to rule out at least a traditional type of transmitter function, since there do not appear to be specialized systems to regulate their uptake and release, and thereby regulate a neurotransmitter function.

In contrast to the above lipophilic trace amines, the more polar trace amines (mTA, pTA, OA) generally have been found to be actively accumulated by various nerve tissue preparations (see Wu et al, 1980; see Hicks, 1977 and Commarato et al 1969; Pletscher, 1976; Costa et al, 1977 Lentzen and Phillipu, 1977, 1981; Bianchi et al, 1981; Johnson et al, 1982; Dyck, 1978; Petrali et al, 1979). It has been suggested that mTA, pTA and OA could act as alternative or co-transmitters since the majority of studies have shown that both their uptake and release characteristics (Kopin et al, 1965; Hicks, 1977; Baldessarini and Vogt, 1972; Stoof et al, 1976; Saldate and Orrego, 1977; Dyck, 1978; Dyck and Boulton, 1980) are compatible with traditional or classical concepts of neurotransmission; however, more recent evidence seems to indicate that specific tyraminergic neurons exist (Petrali, 1977, 1980; Nilsson et al, 1976; Juorio and Jones, 1982).

Two questions will be addressed in the following: (1) Are the transports of mTA and pTA consistent with a transmitter function? and (2) Do separate mTA and pTA neurons exist in rat brain?

UPTAKE STUDIES

Uptake Parameters and Effect of Drugs on Uptake

The uptakes of various concentrations of trititated mTA and pTA and DA have been examined in slices of rat hypothalamus and striatum (Petrali et al, 1979). When the uptake values were analysed by a least-squares fit to a Lineweaver-Burk plot, the uptake of each amine could be resolved into both a high-affinity and a low-affinity system in both brain regions. The K_m and V_{max} values of the high-affinity uptake of all three amines into both brain areas were similar; thus, it was concluded that a common carrier system transports all three. The high affinity uptakes of mTA, pTA and DA into the striatum were inhibited by dinitrophenol, ouabain, cocaine and a decrease in the sodium concentration of the buffer (Dyck, 1978; Petrali et al, 1979); these characteristics are in accordance with a neurotransmitter function. Though both isomers of TA are often thought to be devoid of direct "catechol"-amine receptor activation, direct actions of both mTA and pTA have been reported (Luduena, 1963; Farmer, 1966; De La Lande and Waterson, 1968; Urquilla et al, 1974; Toda et al, 1978; Marin et al, 1981, 1982). Undoubtedly, mTA and pTA have indirect actions as well, and, it may well be that these effects are predominantly observed because they occur prior to or concomitantly with a direct action.

Lack of Uptake of the Non-Transmitter Amine 3-Methoxytyramine (3MT)

As can be seen in Table 1, very little ^{3}H-3MT is accumulated by rat striatal slices. Much more ^{3}H-mTA, ^{3}H-pTA and ^{14}C-DA were taken up. This lack of ^{3}H-3MT accumulation is further exemplified in Table 2. It is clear from its low T/M ratios that this amine in comparison to the others was not highly concentrated by the slices. It, therefore, appears that the uptake system in striatal slices has substrate selectivity and will not accumulate all phenylethylamines.

Effect of Lesions on Uptake

Petrali (1977) has shown that lesions of the substantia nigra decrease the high-affinity uptake of DA but not of pTA in striatal slices. From this and from the greater inhibition of DA than pTA uptake by benztropine, she concluded that pTA was taken up by a specific p-tyraminergic neuronal system. One could, however, equally conclude that pTA was taken up by other cellular elements in the slice -- such as neuronal cell bodies, capillaries, or

Table 1

Uptake of ^{14}C-DA and ^{3}H-mTA, ^{3}H-pTA or ^{3}H-3MT By Rat Striatal Slices

	Uptake (pmol/mg/15 min)	
Trititated Amine	^{3}H-Amine	^{14}C-DA
mTA	4.22 ± 0.55	4.80 ± 0.50
pTA	3.75 ± 0.38	5.71 ± 0.46
3MT	0.083± 0.001*	5.32 ± 0.5

Values are mean ± S.E.M., n = 4 in all cases. Striatal slices were incubated simultaneously for 15 min at 37°C with 10^{-7}M ^{14}C-DA and either ^{3}H-mTA, ^{3}H-pTA or ^{3}H-3MT. *Comparing uptake of ^{3}H-3MT to all other amines with Student's t-test, $p < 0.001$.

Table 2

Tissue:Medium Ratios (T/M) of ^{14}C-DA, ^{3}H-mTA, ^{3}H-pTA and ^{3}H-3MT Uptake into Striatal Slices

	T/M of	
Trititated Amine	^{3}H-Amine	^{14}C-DA
_m_-TA	52.3 ± 2.2 (15)	69.2 ± 5.3 (15)*
_p_-TA	47.3 ± 2.4 (15)	58.7 ± 4.9 (15)**
3MT	6.4 ± 0.1 (4)†	60.0 ± 5.5 (4)***

Values are mean ± S.E.M., sample number in parentheses. Experimental conditions described in Table 1. Comparing the ^{3}H-amine to ^{14}C-DA with Student's t-test, *$p < 0.01$; **$p < 0.05$; ***$p < 0.001$. Comparing 3MT to _m_-TA and _p_-TA, †$p < 0.001$.

glial cells. To preclude these latter possibilities, the effects of prior lesioning of the substantia nigra pars compacta or of the medial forebrain bundle (MFB) was investigated. Since differential effects of lesions on amine uptake could be the result of a varying degree of nerve terminal degeneration from one amine uptake experiment to another, the following studies were designed to eliminate this possibility.

Striatal slices were incubated simultaneously with ^{3}H-pTA and ^{14}C-DA. Uptake of these amines into striatal slices obtained from the lesioned hemisphere were expressed as a percentage of uptake into slices from the unlesioned side (see Figure 1). With both types of lesions, significant decreases in ^{3}H-pTA and ^{14}C-DA were observed. The MFB lesion reduced amine uptake much more effectively than the SN lesion. Essentially identical decreases in ^{3}H-mTA uptake were observed after MFB lesions (data not shown).

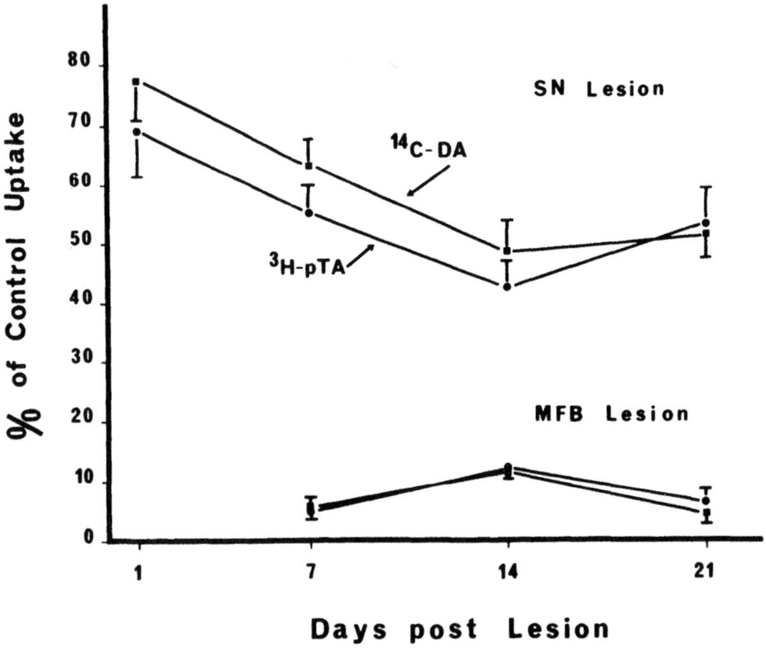

Figure 1. The effect of prior SN or MFB lesions on amine uptake. Values are mean ± S.E.M.

From these lesioning experiments, it apears most likely that the three radiolabelled amines, mTA, pTA and DA, co-exist in the same neurons. Although one cannot definitely rule out the possibility that the tyramines are present in separate cells, this would not be easily demonstrated and would require fine lesioning studies in conjunction with an assessment of endogenous DA, mTA and pTA levels for such proof. These lesion studies also demonstrate that the uptake procedure specifically labels nerve terminals, so that we can be confident that we are studying transport at these sites rather than at other sites, such as glial cells or neuronal cell bodies.

RELEASE STUDIES

K^+-Stimulated Release of Previously Accumulated Radiolabelled Amines

The experimental paradigm used to mimic physiologically-induced amine release was first to preload the striatal nerve terminals with radiolabelled amines and then to depolarize the terminals with a high concentration of KCl. Striatal slices were incubated simultaneously with ^{14}C-DA and either ^{3}H-mTA or ^{3}H-pTA (100 nM of each) for 15 min at 37°C. Then, the slices were transferred at 2 min intervals through a series of ten tubes containing Krebs buffer. The amount of radioactivity (which was virtually all unmetabolized amine) released into each tube and that left in the slices at the end of the experiment was determined. Release into each tube or fraction was expressed as a percentage of total radioactivity. As seen in Figure 2, the spontaneous wash-out (fractions 1-5) of ^{3}H-pTA and ^{3}H-mTA was faster than that of ^{14}C-DA. 27.5 mM KCl stimulated large releases of ^{3}H-pTA, ^{3}H-mTA and ^{14}C-DA into fraction 6; this releasing effect was transient such that release of all three amines had returned to basal levels by fraction 8. The lower half of figure 2 shows that the K^+-stimulated releases of all three amines was completely antagonized by the presence of 15 mM $MgCl_2$. These results indicate that the depolarization-induced releases of ^{3}H-mTA, ^{3}H-pTA and ^{14}C-DA are calcium-dependent; hence, all three amines meet the experimental criteria necessary for classification as transmitter candidates. Calcium-dependent K^+-stimulated (Stoof et al, 1976) and electrically stimulated (Saldate

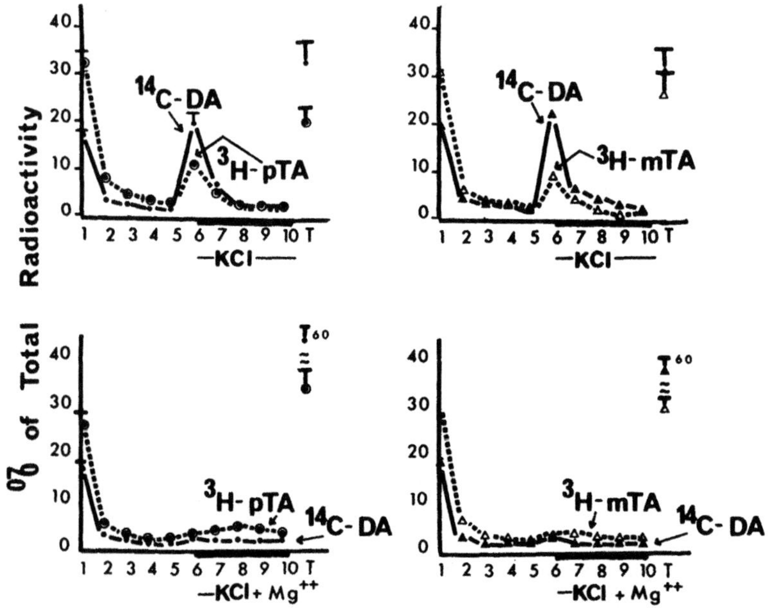

Figure 2. Calcium dependency of the K^+-stimulated releases of ^{14}C-DA and ^3H-pTA or ^3H-mTA. Values are mean ± S.E.M. 27.5 mM KCl (± 15 mM $MgCl_2$) (in a Krebs buffer) was present in fractions 6-10.

and Orrego, 1977) release of pTA has previously been reported.

To verify that depolarization induces a specific release of transmitter compounds, the release of 3MT, a nontransmitter amine was examined (see Figure 3). Striatal slices were incubated simultaneously with ^3H-3MT and ^{14}C-DA, as described above. While a large K^+-stimulated release of ^{14}C-DA was seen, no K^+-stimulated release of ^3H-3MT occurred. Furthermore, it is interesting to note that the rate of spontaneous wash-out of ^3H-3MT was much more rapid than that of ^{14}C-DA indicating a poor retention of ^3H-3MT by the tissue.

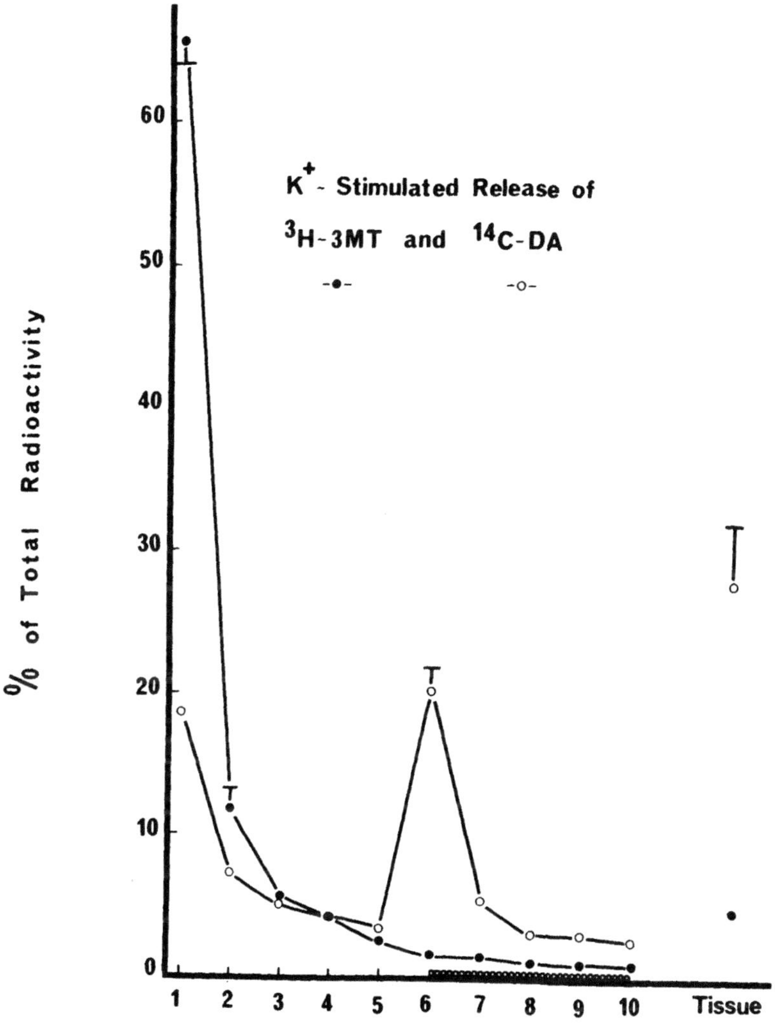

Figure 3. Failure of KCl (27.5 mM) to stimulate ^{3}H-3MT release. Values are mean ± S.E.M. KCl was present in tubes 6-10.

K^+-Stimulated Release of Endogenous Amines

Slices prepared from both striata of a rat pretreated with the MAO inhibitor phenelzine (100 mg/kg, 18 h) were transferred every 5 min through a series of five tubes containing Krebs buffer as described by Dyck et al., 1982. In one set of experiments, the amounts of DA released into the various tubes and left in the slices was determined fluorimetrically. In other experiments, the amounts of mTA and pTA released simultaneously and left in the tissue slices were determined mass spectrometrically. The spontaneous releases of pTA, mTA and DA rapidly reached a steady state so that a plateau was achieved from tubes 2 through 5. The addition of 50 mM KCl prepared in a modified Krebs buffer stimulated releases of all three amines which were decreased in the presence of 15 mM $MgCl_2$ (Dyck et al., 1982). Rats which had been pretreated with an MAO inhibitor were used so that larger amounts of mTA and pTA would be available for analysis. Although inhibition of MAO may have led to increased levels of cytoplasmic amines, rather than to increased levels of amines in the storage granules from where physiological release is believed to originate, K^+-stimulated release from the cytoplasm is not calcium-dependent (Thoa, 1975). Similarly, MAO inhibition may have increased amine levels in cellular elements other than the nerve terminals of the striatum. To confirm that the amines were accumulating primarily in nerve terminals, rats received unilateral lesions 10 days before the phenelzine treatment. This treatment,which is well known to cause degeneration of the striatal dopaminergic nerve terminals, prevented the phenelzine-induced increases in mTA and pTA levels (see Table 3). It seems, therefore, the K^+-stimulated releases of endogenous mTA and pTA originated from nigrostriatal nerve terminals. As was stated in the radiolabelled experiments, these lesioning data indicate that endogenous mTA, pTA and DA co-exist in the same neuron. This conclusion is supported by a number of other studies which show that in the brain, tyrosine hydroxylase -- which is a marker for catecholaminergic neurons -- is a key enzyme not only in DA synthesis but in mTA and pTA biosynthesis as well (see Juorio, 1979; Boulton and Juorio, 1983; Dyck et al., 1983; Juorio, this book).

Table 3

Effects of a MFB-Lesion on Striatal Amine Levels

Treatment	pTA	mTA	DA
None	1.1 ± 0.1	0.18 ± 0.01	549 ± 17
PEH	33.0 ± 2.4*	1.2 ± 0.1*	807 ± 75*
MFB lesion + PEH	2.0 ± 0.9**	0.17 ± 0.03**	---

Values, in ng, are mean ± S.E.M.; the lesioned, or unlesioned, striata from one hemisphere of one rat was pooled with that from another rat, so that this value represents ng amine in 2 striata. Using Student's t-test, *$p < 0.01$ compared to no treatment and **$p < 0.001$ compared to PEH treatment.

Veratridine-Induced Release of Previously Accumulated Radiolabelled Amines

Veratridine (VER) is an alkaloid which opens voltage-dependent sodium channels in nerve endings; thus, VER can stimulate transmitter release by causing membrane depolarization. While VER causes calcium-dependent release of catecholamine transmitters presumably from the synaptic vesicles, it causes a release of amino acid transmitters from the cytoplasm that is antagonized by Ca^{++} (Benjamin and Quastel, 1972; De Belleroche and Bradford, 1977). Thus, VER can be used to study the intraneuronal origin of released transmitter material.

Striatal slices were incubated and the release procedure performed as described above. VER was added to the buffer in tube 6. When the effects of Ca^{++}-deficiency were to be assessed, Ca^{++} was omitted from the buffer and 2 mM EGTA added (this buffer was used after the incubation step). While 10 μM VER stimulated a Ca^{++}-dependent release of ^{14}C-DA, the corresponding release of ^{3}H-pTA was not Ca^{++}-dependent (see Figure 4). In contrast, 1 μM VER stimulated a Ca^{++}-independent release of ^{14}C-DA; 1 μM VER did not stimulate ^{3}H-pTA release, but, interestingly, it did

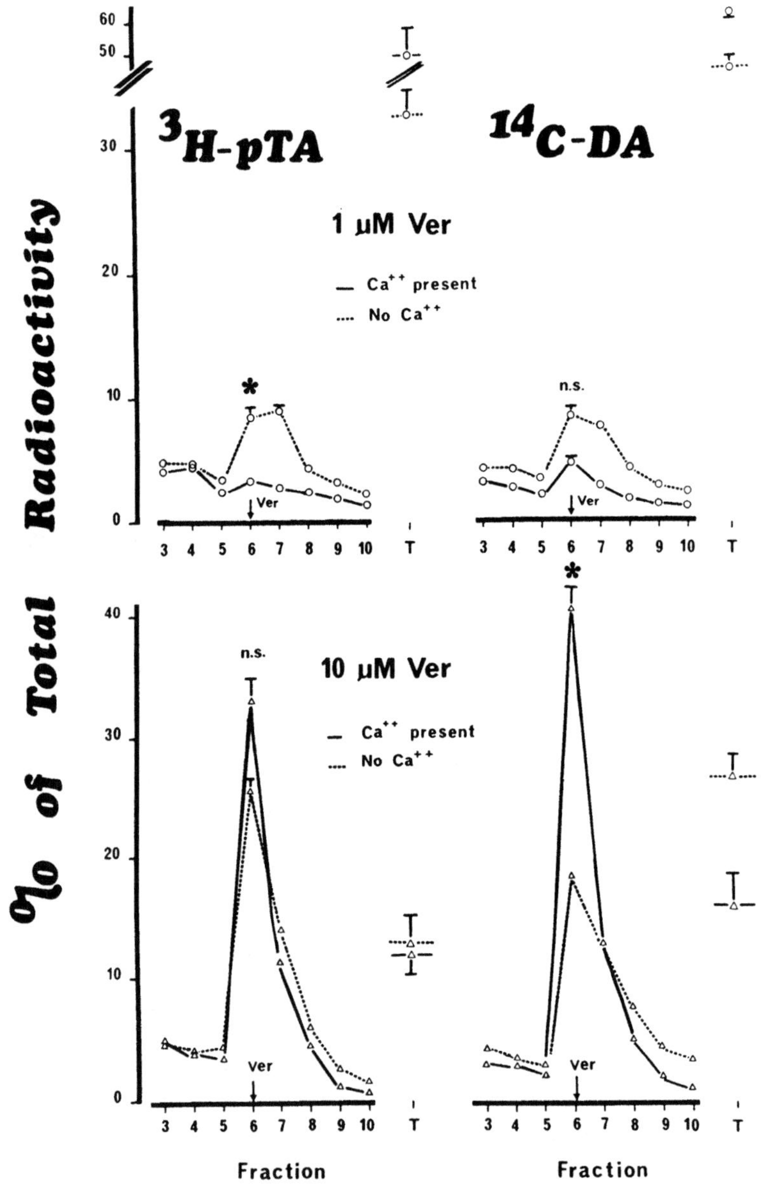

Figure 4. Veratridine-stimulated releases of ^{3}H-pTA and ^{14}C-DA. Values are mean ± S.E.M. Using Student's t-test for unequal variance, * $p < 0.05$, comparing release into fraction 6 in the presence and absence of Ca^{++}.

release ^{3}H-pTA when Ca^{++} was absent. That is, the VER-induced release of ^{3}H-pTA was antagonized by the presence of Ca^{++} as has been observed with cytoplasmic transmitters. Thus, ^{3}H-pTA release seems to originate from the cytoplasm with this stimulus, while ^{14}C-DA release still appeared to be vesicular.

Effects of CNS Stimulants on Amine Release

A comparison of the amine releasing abilities of the two types of CNS stimulants, d-amphetamine and methylphenidate, and the effect of reserpine on their releasing effects has been published (Dyck et al., 1980). Amphetamine stimulated releases of all three amines. We confirmed the original finding of Petrali (1977) that methylphenidate released very little DA but released appreciable amounts of pTA. In addition, we noted that the release of mTA was similar to pTA and, interestingly, reserpine treatment was found to enhance significantly the ability of methyl-phenidate to release DA. This latter effect was interpreted as suggestive of a cytoplasmic source of methylphenidate-induced amine release.

The methylphenidate-like stimulants -- cocaine, nomifensine and amfonelic acid -- possess several common properties. Their behavioural effects are antagonized by prior administration of reserpine; and while their mode of action is thought to occur by an inhibition of the uptake of impulse released DA (Shore, 1976; Ross, 1977; Hyttel, 1978), two of them have been found to act upon cytoplasmic stores of DA (Braestrup, 1977). This latter finding supports our suggestion that methylphenidate stimulated amine release is from a cytoplasmic source.

All four methylphenidate-like stimulants released significantly more ^{3}H-pTA and ^{3}H-mTA than DA-^{14}C (Dyck, 1981). As these drugs are not structurally similar to tyramine, it is unlikely that the mechanism of release involves a carrier-mediated exchange process. It is more likely that they interact in some way with presynaptic receptors. Such a mode of action has been suggested by Dembiec (1980), who found that prior exposure of striatal tissue to cocaine or nomifensine attenuates DA release. Perhaps the tyramines have some role in this modulation.

Effects of Aminotetralins on Amine Release

The aminotetralins (AT) (2-amino-1,2,3,4-tetrahydronaphthalenes) are semi-rigid analogs of the phenylethylamines. The mono-hydroxylated AT's -- 5-OH-AT, 7-OH-AT and 6-OH-AT -- are analogs of mTA and pTA; while the dihydroxylated AT's -- 5,6-$(OH)_2$-AT and 6,7-$(OH)_2$-AT -- are analogs of DA. Though it was originally thought that a specific release of either mTA, pTA or DA might be stimulated by the appropriate AT analog, this was not the case. What was observed, surprisingly, was that an AT with a dipropyl substituted nitrogen selectively stimulated the release of TA (*m* and *p*) but not DA regardless of the number or position of ring hydroxyl substituents (Dyck, 1981). For example, 6-OH-AT and 7-OH-AT (analogs of mTA and pTA) were shown to release both ^{3}H-pTA and ^{14}C-DA (data not shown or published), while their dipropyl derivatives released only ^{3}H-pTA or ^{3}H-mTA.

The mechanism of an AT-induced amine release probably involves an exchange process (whereby the AT is transported into the nerve terminal by a carrier, exchanges with the amine and displaces it to the outside). Davis *et al.* (1978) have shown that 6,7-$(OH)_2$-AT can be specifically concentrated by striatal nerve terminals and its carrier-mediated release can be stimulated by amphetamine. In addition, Mulder *et al.* (1980) have shown that 6,7-$(OH)_2$-AT-induced DA release appears to be carrier-mediated. It is not clear, however, whether the dipropyl substituted AT's release the tyramines by stimulating a distinct carrier or whether they cause release in some other manner. Since other uptake studies have not revealed any differences in the transport of the two TA isomers, stimulation of a distinct carrier by the dipropyl AT's seems unlikely. Perhaps these compounds, as was suggested for the methylphenidate-like stimulants, cause a preferential release of amines from cytoplasmic pools.

CONCLUSIONS

The data concerning the uptake and release of striatal mTA and pTA may perhaps best be interpreted as indicating their co-existence in DA neurons. As stated previously, studies of the synthesis of these amines by rat striatum and studies of the effects of many drugs on striatal concentrations of these amines support this conclusion. Even though these amines may co-exist in the same neurons, their preferential release, nevertheless, can be stimulated by two different classes of drugs, the methylphenidate-like stimulants and the dipropyl substituted aminotetralins. The mechanism of this preferential release is unknown.

The co-release or preferential release of either DA or the tyramines may alter post-synaptic neuronal responses as described by Jones in this book.

ACKNOWLEDGEMENTS

I wish to thank Dr. A.A. Boulton for his critical comments and encouragement throughout the many aspects of this work, Dr. R.S.G. Jones for providing the lesioned rats, and for collaboration in some of the studies, Dr. B.A. Davis for synthesizing some of the aminotetralins, Dr. A.V. Juorio for his collaboration in the endogenous release study, and Saskatchewan Health for continuing financial support.

REFERENCES

Baldessarini R.J. and Vogt M. (1971) The uptake and subcellular distribution of aromatic amines in the brain of the rat. J. Neurochem. 18, 2519-2533.

Baldessarini R.J. and Vogt M. (1972) Regional release of aromatic amines from tissues of the rat brain *in vitro*. J. Neurochem. 19, 755-761.

Benjamin A.M. and Quastel J.M. (1972) Locations of amino acids in brain slices from the rat. Biochem. J. 128, 631-646.

Bianchi L., Stella L., Dagnino G., de Gaetano G. and Rossi G. (1981) The uptake of tyramine by rat platelets. Biochem. Pharmacol. 30, 709-713.

Born G.V.R., Juengjaroen K. and Michal F. (1972) Relative activities on and uptake by human blood platelets of 5-hydroxytryptamine and several analogues. Br. J. Pharmacol. 44, 117-139.

Boulton A.A. (1979) Trace amines in the central nervous system, in, Int. Rev. Biochem., Physiol. Pharmacol. Biochem., Vol. 26 (Tipton K.F., ed) pp 179-206. University Park Press, New York.

Boulton A.A. and Juorio A.V. (1982) Brain trace amines, in, Handbook of Neurochemistry, Vol. 1 (Lajtha A., ed) pp 189-222. Plenum Publishing, New York.

Boulton A.A. and Juorio A.V. (1983) Cerebral decarboxylation of meta- and para-tyrosine. Experientia 39, 130-134.

Braestrup C. (1977) Biochemical differentiation of amphetamine vs methylphenidate and nomifensine in rats. Eur. J. Pharmacol. 29, 463-470.

Commarato M.A., Brody T.M. and McNeill J.H. (1969) The effect of various drugs on the uptake and metabolism of tyramine-H^3 in the rat heart. J. Pharmacol. Exp. Ther. 167, 151-158.

Costa J.L., Murphy D.L. and Reveille J. (1977) Evaluation of the uptake of various amines into storage vesicles of intact human platelets. Br. J. Pharmacol. 61, 223-228.

Davis A.J., Roberts P.J. and Woodruff G.N. (1978) The uptake and release of [^{3}H]-2-amino-6,7-dihydroxy-1,2,3,4-tetrahydronaphthalene (ADTN) by striatal nerve terminals. Br. J. Pharmacol. 63, 183-190.

DeBelleroche J.S. and Bradford H. (1977) On the site of origin of transmitter amino acids released by depolarization of nerve terminals in vitro. J. Neurochem. 29, 335-343.

De La Lande I.S. and Waterson J.G. (1968) The action of tyramine on the rabbit ear artery. Br. J. Pharmacol. 34, 8-18.

Dembiec D. (1980) Inhibition of potassium-stimulated release of [^{3}H]-dopamine from rat striata as a result of prior exposure to cocaine, nomifensine or mazindol. Neurochem. Res. 5, 345-349.

Dyck L.E. (1978) Uptake and release of meta-tyramine, para-tyramine and dopamine in rat striatal slices. Neurochem. Res. 3, 775-791.

Dyck L.E. (1981) Methylphenidate-like stimulants in vitro release [^{3}H]-tyramines but not [^{14}C]-dopamine. Eur. J. Pharmacol. 69, 371-374.

Dyck L.E. (1984) The uptake and release of ^{14}C tryptamine by rat brain slices, in Neurobiology of the Trace Amines Boulton A.A., Baker G.B, Dewhurst W.G. and Sandler M. (eds), Humana Press Inc., New Jersey.

Dyck L.E. and Boulton A.A. (1980) The effect of reserpine and various monoamine oxidase inhibitors on the uptake and release of tritiated meta-tyramine, para-tyramine and dopamine in rat striatal slices. Res. Commun. Psychol. Psychiat. Behav. 5, 61-78.

Dyck L.E., Juorio A.V. and Boulton A.A. (1982) The in vitro release of endogenous m-tyramine, p-tyramine and dopamine from rat striatum. Neurochem. Res. 7, 705-716.

Dyck L.E., Yang C.R. and Boulton A.A. (1983) The biosynthesis of p-tyramine, m-tyramine and β-phenylethylamine by rat striatal slices. J. Neurosci. Res. (in press).

Farmer J.B. (1966) The biphasic response of an isolated artery to tyramine. Br. J. Pharmacol. Chemother. 28, 340-347.

Greenberg R. and Whalley C.E. (1978) Studies on the uptake of phenylethylamine into the secretory vesicles of brain synaptosomes, in, Noncatecholic Phenylethylamines, Part 1. Phenylethylamine: Biological Mechanisms and Clinical Aspects (Mosnaim A.D. and Wolf M.E., eds) pp. 87-112. Marcel Dekker, New York.

Héry F., Simmonet G., Bourgoin S., Soubrie P., Artaud F., Hamon M. and Glowinski J. (1979) Effect of nerve activity on the in vivo release of [^{3}H]-serotonin continuously formed from L-[^{3}H]-tryptophan in the caudate nucleus of the cat. Brain Res. 169, 317-334.

Héry M., Faudon M. and Héry F. (1983) In vitro release of newly synthesized serotonin from superfused rat suprachiasmatic area - ionic dependency. Eur. J. Pharmacol. 89, 9-18.

Hicks T.P. (1977) The possible role of octopamine as a synaptic transmitter: a review. Can. J. Physiol. Pharmacol. 55, 137-152.

Hyttel J. (1978) Inhibition of [^{3}H]-dopamine accumulation in rat striatal synaptosomes by psychotropic drugs. Biochem. Pharmacol. 27, 1063-1068.

Johnson R.G., Carty S.E., Hayflick S. and Scarpa A. (1982) Mechanisms of accumulation of tyramine, metaraminol and isoproterenol in isolated chromaffin granules and ghosts. Biochem. Pharmacol. 31, 815-823.

Jones R.S.G. (1984) Electrophysiological studies with trace amines, in Neurobiology of the Trace Amines, Boulton A.A., Baker G.B., Dewhurst W.G. and Sandler M. (eds), Humana Press Inc., New Jersey.

Juorio A.V. (1984) The functional implications of some physiological and pharmacological interactions of brain trace amines, in Neurobiology of the Trace Amines, Boulton A.A., Baker G.B., Dewhurst W.G. and Sandler M. (eds), Humana Press Inc., New Jersey.

Juorio A.V. (1979) Drug-induced changes in the formation, storage and metabolism of tyramine in the mouse. Br. J. Pharmacol. 66, 377-384.

Juorio A.V. and Jones R.S.G. (1982) The effect of mesencephalic lesions on tyramine and dopamine in the caudate nucleus of the rat. J. Neurochem. 36, 1898-1903.

Kopin I.J., Fischer J.E., Mussachio J.M., Horst W.D. and Weise, V.K. (1965). False neurochemical transmitters and the mechanism of sympathetic blockade by monoamine oxidase inhibitors. J. Pharmacol. Exp. Therap. 147, 186-193.

Lentzen H. and Phillipu A. (1977) Uptake of tyramine into synaptic vesicles of the caudate nucleus. N.S. Arch. Pharmacol. 300, 25-30.

Lentzen H. and Phillipu A. (1981) Physico-chemical properties of phenethylamines and their uptake into synaptic vesicles of the caudate nucleus. Biochem. Pharmacol. 30, 1759-1764.

Luduena F.P. (1963) Effect of reserpine on the vasaconstrictor action of tyramine on the rabbit ear, with a discussion of the mechanisms of action of reserpine and tyramine. Acta Physiol. Lat.-Amer. 13, 221-241.

Marin J., Salaices M. and Sanchez C.F. (1982) Analysis of the effects of noradrenaline and tyramine in isolated middle cerebral and femoral arteries of cat. Gen. Pharmacol. 13, 125-132.

Marin J., Salaices M. and Sanchez C.F., Conde M.V., Gomez B. and Lluch S. (1981) Vasoconstrictor effects of tyramine and noradrenaline on goat brain vessels. Gen. Pharmacol. 12, 187-191.

Mulder A.H., Braakhuis B., DeRegt V., Dijkstra D. and Horn A.S. (1980) Effects of ADTN and various 2-aminotetralin derivatives on the efflux of ^{3}H-dopamine from rat striatal slices. Eur. J. Pharmacol. 64, 349-355.

Nilsson S. and Holmgren S. (1976) Uptake and release of catecholamines in sympathetic nerve fibres in the spleen of the cod, Gadus morhua. Eur. J. Pharmacol. 39, 41-51.

Osborne N.N. (1980) Specificity of serotonin uptake by bovine retina: comparison with tryptamine. Exp. Eye Res. 31, 31-39.

Petrali E.H. (1977) The uptake and release of some monoamines in the rat brain. Ph.D. Thesis, University of Saskatchewan.

Petrali E.H. (1980) Differential effects of benztropine and desipramine on the high affinity uptake of para-tyramine in slices of the caudate nucleus and hypothalamus.

Petrali E.H., Boulton A.A. and Dyck L.E. (1979) Uptake of para-tyramine and meta-tyramine into slices of the caudate nucleus and hypothalamus of the rat. Neurochem. Res. 4, 633-642.

Pletscher A. (1976) Transport of monoamines in membranes of adrenal chromaffin granules: physiological and pharmacological aspects. Bull. Schweiz. Akad. Med. Wiss. 32, 181-190.

Ross S.B. (1977) On the mode of action of central stimulatory agents. Acta Pharmacol. Toxicol. 41, 392-396.

Ross S.B. and Ask A.-L. (1980) Structural requirements for uptake into serotoninergic neurones. Acta Pharmacol. Toxicol. 46, 270-277.

Ross S.B. and Renyi A.L. (1971) Uptake and metabolism of β-phenylethylamine and tyramine in mouse brain and heart slices. J. Pharm. Pharmac. 23, 276-279.

Ross S.B., Renyi A.L. and Brunfelter B. (1968) Cocaine-sensitive uptake of sympathomimetic amines in nerve tissue. J. Pharm. Pharmac. 20, 283-288.

Saldate M.C. and Orrego F. (1977) Differences in the electrically induced release of ^{3}H-labeled β-phenylethylamine, tyramine, octopmaine and norepinephrine from rat neocortical slices, in, Noncatecholic Phenylethylamines, Part 1. Phenylethylamine: Biological Mecvanisms and Clnical Aspects. (Mosnaim A.D. and Wolf M.E., eds), pp 159-177. Marcel Dekker, New York.

Schroder H.M., Neuhoff V., Priggemeier E. and Osborne N.N. (1979) The influx of tryptamine into snail (Helix pomatia) ganglia: comparison with 5-hydroxytryptamine. Malacologia 18, 517-525.

Shore P.A. (1976) Actions of amfonelic acid and other non-amphetamine stimulants on the dopamine neuron. J. Pharm. Pharmac. 28, 855-857.

Snodgrass S.R. and Iversen L.L. (1974) Formation and release of ^{3}H-tryptamine from ^{3}H-tryptophan in rat spinal cord slices. Adv. Biochem. Psychopharamcol. 10, 141-150.

Stacey R.S. (1961) Uptake of 5-hydroxytryptamine byplatelets. Br. J. Pharmacol. 16, 284-295.

Stoof J.C., Liem A.L. and Mulder A.M. 1976 Release and receptor stimulating properties of p-tyramine in rat brain. Arch. Int. Phrmacodyn 220, 62-71.

Thoa N.B., Wooten G.F., Axelrod J. and Kopin I.J. (1975) On the mechanism of release of norepinephrine from sympathetic nerves induced by depolarizing agents and sympathomimetic drugs. Mol. Pharmacol. 11, 10-18.

Toda N., Hayashi S. and Hattori K. (1978) Analysis of the effect of tyramine and norepinephrine in isolated canine cerebral and mesenteric arteries. J. Pharmacol. Exp. Ther. 205, 382-391.

Urquilla P.R., Marco E.J., Balfagan G. and Lluch S. (1974) Adrenergic mechanisms in cerebral blood vessels: effect of tyramine on the isolated middle cerebral artery of the goat. Stroke 5, 447-452.

Wu P.H., Baker G.B. and Henwood R.W. (1980) Tyramine as a neurotransmitter: pros and cons, in Noncatecholic Phenylethylamines, Part 2, Phenylethanolamine, tyramines and octopamine. (Mosnaim A.D. and Wolf M.E., eds), pp. 307-339,

Wu P.H., and Boulton A.A. (1973) Distribution and metabolism of tryptamine in rat brain. Can. J. Biochem. 51, 1104-1112.

ELECTROPHYSIOLOGICAL STUDIES OF THE POSSIBLE ROLE OF TRACE AMINES IN SYNAPTIC FUNCTION.

Roland S.G. Jones

Biology Research Laboratories

Ciba-Geigy AG., Basel, Switzerland

ABSTRACT

Using electrophysiological techniques the actions of trace amines have been studied at the single cell level in the CNS. The results indicate that many of the neuronal effects of trace amines may be attributable to actions on specific receptors and as such they may possess transmitter function. In addition small amounts of these amines are able to strongly modify neuronal responses to established aminergic and non-aminergic transmitters and may also function as neuromodulators.

INTRODUCTION

Although the biochemistry of trace amines in the CNS has been extensively studied(see Boulton, 1979, 1980 for reviews) little is known about possible functional roles for these substances. It was, and still is uncertain whether trace amines could act as neurotransmitters or neuromodulators or indeed whether they possessed any active function at all. In recent years, however, a number of single cell recording studies have strongly indicated that the latter possibility, at least, is not a viable one. These studies have shown that trace amines can exert powerful influences on neuronal activity, both alone and in concert with established transmitter candidates, aminergic and non-aminergic. In the present article I will review the studies pertaining to three trace amines, p-tyramine(pTA), octopamine (OA) and tryptamine(T).and will describe firstly the effects of these amines on cell firing and secondly their ability to modify the neuronal responses to suspected transmitters.

p-TYRAMINE

Effects on Neuronal Activity

Hoffer *et al*. applied TA(presumably pTA) to Purkinje cells in the rat cerebellum by iontophoresis and observed depressant effects similar to those seen with noradrenaline(NA)(Hoffer et al, I97I). The responses to pTA were not altered by pre-treatment with 6-hydroxydopamine, indicating that release of NA was not involved. Iontophoresis of TA(again presumably pTA) into the vicinity of intracellularly recorded cat motoneurones(Engberg *et al*., I976) evoked a depolarization and decrease in membrane conductance equivalent to that seen with similar applications of NA or dopamine(DA). Again, these responses were not diminished by prior depletion of catecholamine stores.

In contrast to the above studies, other authors have concluded that the neuronal responses to pTA do involve release of catecholamines. Boakes *et al*. found that the depressant responses of caudate neurones to pTA(and mTA)in urethane anaesthetized rats were abolished by reserpine pre-treatment but could be restored by prolonged application of DA to the cell(Boakes *et al*.,I976). In the study of Bevan *et al*(I979) NA evoked both excitatory and depressant responses of cerebral cortical cells in halothane-anaesthetised rats and these responses were invariably mimicked on any given cell by (presumably) pTA. The responses to pTA were however weaker, longer in onset and offset than those evoked by NA. They were also blocked by desipramine, leading the authors to conclude that they represented the release of NA from presynaptic terminals (Bevan et al., 1979).

Henwood et al. (1980) reported that nearly 100% of neurones in the cerebral cortex and caudate nucleus were depressed to a similar extent by NA, DA, pTA, OA, phenylethylamine and phenylethanolamine in rats anaesthetised with methoxyfluorane. Finally Jones and Boulton (1980) and Jones (1981) found that in urethane-anaesthetised rats DA and NA elicited mainly depressant but occasional excitatory responses in the cortex and caudate nucleus and that these responses were mimicked by pTA. Again the effects of the latter were weaker but more persistent than those of the catecholamines.

Thus, while it is clear that pTA can alter the activity of single cells it seems likely that these effects may be indirect. However the existence of specific receptors for pTA cannot be ruled out and has been proposed previously on the basis of behavioural studies (Stoof et al., 1976). Also, the depressant actions of pTA in the cerebellum are suggested to arise from postsynaptic receptor activation (Hoffer et al., 1971) and may conceivably involve a TA receptor.

Interactions with Transmitters

The ability of pTA to modify neuronal responses to DA, NA, 5hydroxytryptamine (5HT), GABA, acetylcholine (ACh), glutamate and met-enkephalin has been tested in the cerebral cortex and caudate nucleus (Jones and Boulton, 1980; Jones, 1981; this book). When applied with weak iontophoretic currents which did not alter baseline firing rate pTA consistently and strongly potentiated both depressant and excitatory responses to NA and DA but not to 5HT (figure 1). Depressant responses to GABA were also unaffected by pTA. The excitatory effects of glutamate were not altered but

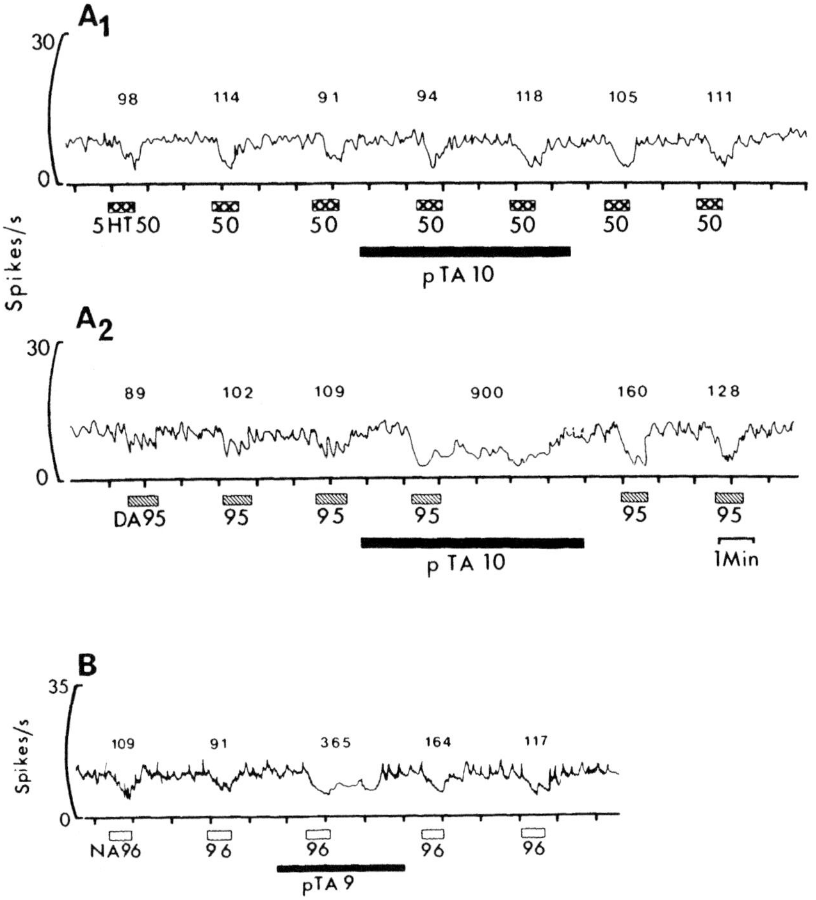

FIGURE I. Ratemeter records from two cortical neurones. The records in A are continuous from the same cell and show the failure of pTA to alter responses to 5HT on a cell on which the same application of pTA greatly potentiates responses to DA. B shows potentiation by pTA of depressant responses to NA.

occasionally excitatory responses to acetylcholine were reduced during the application of pTA (Fig.2) Finally, recent data demonstrates an ability of pTA to reduce or abolish the depressant effects of iontophoretically applied met-enkephalin on

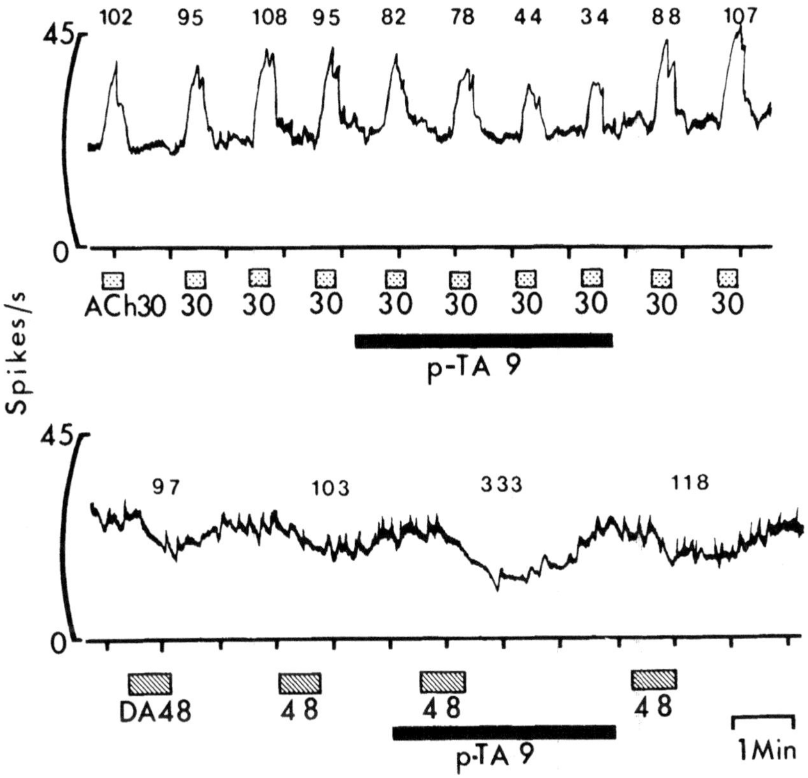

FIGURE 2. Continuous record from a single cortical cell showing reduction of ACh-excitations and potentiation of DA-induced depression.

cortical neurones (see this book).

The bases of the modifying actions of pTA are not currently known. The potentiation of responses to catecholamines could conceivably involve release and/or inhibition of re-uptake of endogenous catecholamines although this has been argued against (Jones, 1981). Unpublished data (Jones) has shown that the potentiation persists in the caudate nucleus two weeks after large electrolytic lesions in the substantia nigra, strongly suggesting a postsynaptic origin of the effect. The single cell studies clearly indicate that pTA may be a modulator of synaptic transmission mediated by several diverse substances.

This possibilty receives good support,at least in the case of DA , from the observation that various manipulations which increase striatal DA-turnover result in a simultaneous decrease in pTA levels and vice-versa(Juorio, I979; Juorio and Jones, I98I; Jones, Juorio and Boulton, I982).

OCTOPAMINE

Effects on Neuronal Firing

Iontophoretically applied OA depressed the firing of cerebellar Purkinje neurones(Kostopoulos and Yarbrough, I975). Although these effects were weaker and less consistent than those observed with NA they were not altered by prior depletion of NA. As with pTA(see above), OA was found to hyperpolarize the membrane of cat motoneurones to a similar extent as NA and DA(Engberg *et al* I976).

Several studies now indicate the existence of specific receptors for OA in the rat brain. Hicks and McLennan (I978a) demonstrated that many cells in the cortex were depressed by NA but excited by OA and vice versa. This was confirmed by Jones(I982a). Similar findings have also been reported for the dorsal horn of the spinal cord (Hicks and McLennan, I978b) and the ventral basal thalamus (Dao and Walker, I980, a,b). In addition the effects of OA are not correlated with those of DA (Hicks and McLennan, I978a,b; Dao and Walker, I980a, b).Studies with antagonists reveal that propranolol blocks responses to NA without affecting those to OA (Hicks and McLennan, I978a, b) and neuroleptics consistently block DA without altering OA (Hicks and McLennan, I978a. Dao and Walker, I980a). However, cyproheptadine, at doses which do not block responses to 5HT, appears to specifically antagonize responses to OA (Dao and Walker, I980b). These studies provide convincing evidence that OA may have important neuronal actions independant of NA and DA and indicate the likelihood of OA-mediated transmission in some central pathways.

Interactions with Transmitters

Kostopoulos and Yarbrough (I975) reported that OA was able to potentiate the depressant responses of cerebellar Purkinje cells to iontophoretically applied NA in the rat.This was confirmed in the cerebral cortex by Jones (I982a). In the latter study NA evoked both depressant and excitatory responses and both were potentiated by OA in the abscence of changes in baseline firing rate(fig 3). In contrast,responses to neither DA nor 5HT were affected by OA. While the mechanism of this interaction is unknown the results may indicate a modulatory function of OA in NA mediated transmission. In I977, Axelrod and Saavedra proposed that OA may function as a co-transmitter with NA and the electrophysiological data provide support for this suggestion.

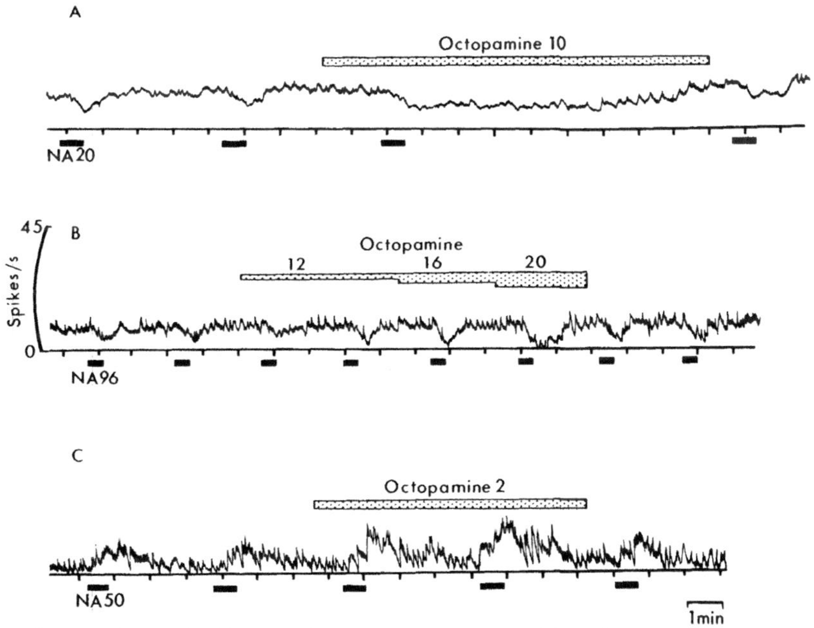

FIGURE 3. Studies from three cortical neurones showing potentiation of both excitatory and depressant responses to NA in the absence of changes in baseline firing rate.

TRYPTAMINE

Effects on Neuronal Firing

Curtis (1962) failed to observe any effect of T on spinal cord interneurones but did observe potent depression of activity in the lateral geniculate nucleus (Curtis and Davis, 1962). Other studies are unanimous in describing potent depressant actions of T on single cells. Krnjević and Phillis (1963) reported depression of cortical cells in the cat and this has recently been confirmed in the rat (Jones and Boulton, 1981; Jones, 1982b, c). Depression of activity by T in the ventral lateral geniculate, dorsal raphé and amygdala was described by Haigler and Aghajanian (1975). Several points are noteworthy. The depressant actions of tryptamine are generally more profound and prolonged than those evoked by equal applications of 5HT (Krnjević, 1965; Curtis and Davis, 1962; Haigler and Aghajanian, 1975; Jones, 1982b) and are unaffected by prior lesions of 5HT containing pathways(Jones,1982b). Many cells in cortex are depressed by T but excited by 5HT and

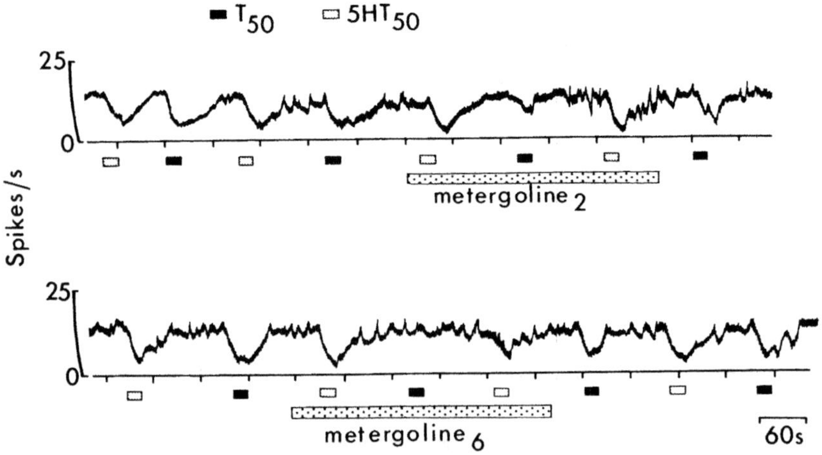

FIGURE 4. The depressant responses of this cell to T are antagonized by metergoline (2nA) but antagonism of 5HT requires a higher current(6nA).

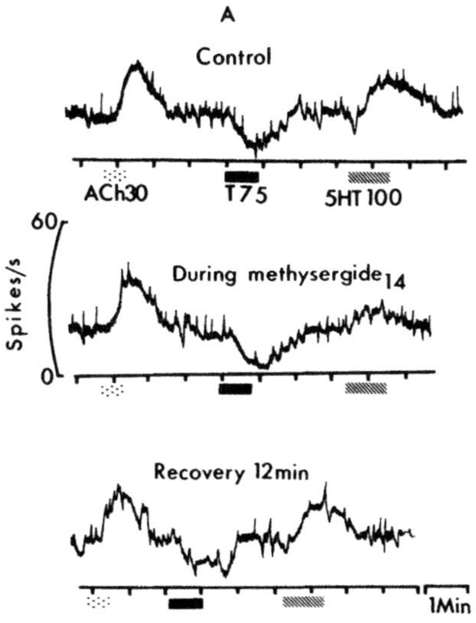

FIGURE 5. Excerpts from the ratemeter record of a study which shows the antagonism of excitatory responses to 5HT by methysergide with no concurrent alteration of the depression evoked by the application of T. The excitatory response to ACh is also unaltered during the application of methysergide.

the reverse is occassionaly true (Jones, I982b, c) Finally, dpressant responses to T may be antagonized by metergoline applied with ejecting currents which have no effect on responses to 5HT (Fig 4, Jones, I982b) and methysergide specifically blocks the excitatory effects of 5HT (Fig 5, Jones, I982c).

The above observations are indicative of the existence , particularly in the cerebral cortex, of a population of specific recognition sites for T and a role independant of its more abundant hydroxylated analogue. This possibility receives considerable support from the recent demonstration of specific binding sites for T in rat brain (Kellar and Cascio, I982) with one of the highest densities occurring in the cerebral cortex (Cascio and Kellar, I982).

Electrophysiological studies have now furnished good evidence that there is a T-containing pathway arising in the median raphé nucleus and projecting to the cerebral cortex in the rat, Stimulation of this brain nucleus evoked biphasic inhibition-excitation responses on the great majority of neurones in the fronto-parietal cortex (e.g. seeFig 6 , Jones;I982c). Pretreatment of rats

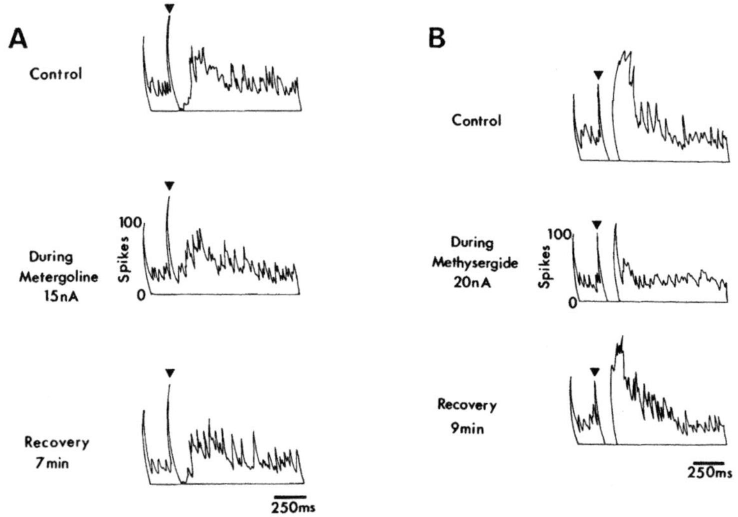

FIGURE 6. Peristimulus histograms from two cells showing biphasic responses to stimulation of the median raphe(at the arrow). Application of metergoline to the cell in A resulted in antagonism of the inhibitory phase but no change in the excitation. In B iontophoresis of methysergide considerably reduced the excitation without affecting the inhibition.

with a tryptophan hydroxylase inhibitor to increase brain T levels and decrease brain 5HT resulted in an augmentation of the inhibitory effects of raphé stimulation and a reduction in the excitatory response. These results suggested that the initial inhibition may be T-mediated and the late excitation mediated by 5HT (Jones, I982c). Other data supports this conclusion. Thus metergoline blocked the stimulus-evoked inhibition but not the excitation and methysergide had the reverse effect (cf effects of antagonists on iontophoretically applied amines above)(Fig 6, Jones, I982c).

The effects of amino acid precursors on raphé evoked responses are also consistent with the proposed roles of T and 5HT, Elevation of brain concentrations of both amines by giving L-tryptophan

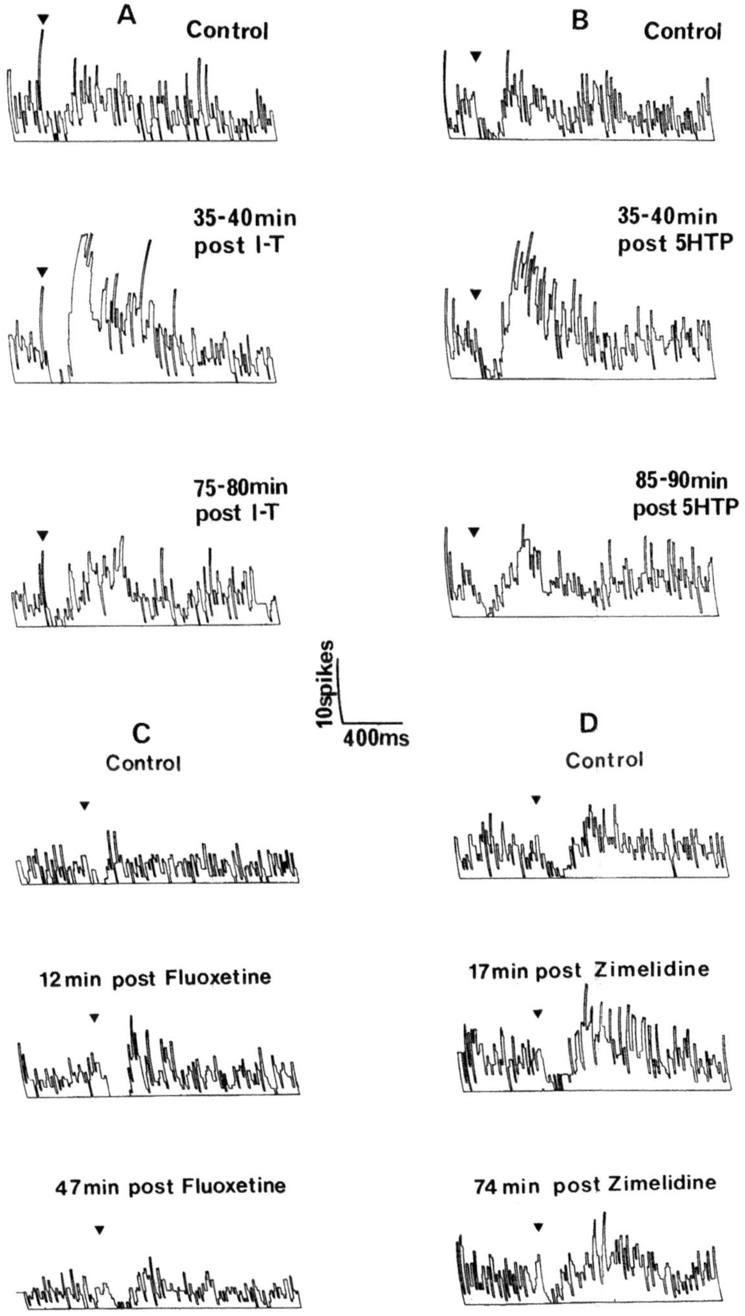
A
Control
35-40min
post I-T
75-80min
post I-T
B
Control
35-40min
post 5HTP
85-90min
post 5HTP
10spikes
400ms
C
Control
12min post Fluoxetine
47min post Fluoxetine
D
Control
17min post Zimelidine
74 min post Zimelidine

FIGURE 7 (previous page). The peri-stimulus histograms in A show that administration of L-tryptophan (ip) enhances both raphé-evoked inhibition and excitation. 5HTP, however, only increases the excitation (B). Fluoxetine (iv), which blocks both T and 5HT uptake, enhances both response phases (C), but zimelidine, which acts specifically on 5HT uptake only, potentiates the excitatory phase of the response (D).

---

caused an increase in both response phases (Fig. 7A) whereas 5hydroxytryptophan, which only elevates brain 5HT, caused only an enhancement of the excitatory phase (Fig. 7B) (Jones and Broadbent, 1982a). Raising the level of brain T alone by combining L-tryptophan with hydroxylase inhibition generally resulted in a specific increase in inhibition.

Finally, results with uptake inhibitors also were in agreement with the hypothesized role of the indoleamines. Fluoxetine, which at high enough doses inhibited both T and 5HT uptake, was able to enhance both phases of raphe-evoked responses (Fig. 7C) (Jones and Broadbent, 1982b), but zimelidine, which was specific for 5HT uptake only, augmented the excitation (Fig. 7D) (Jones and Broadbent, 1982b).

It seems very likely then that there may be a rostrally-projecting T system in the rat brain. The existence of a similar pathway with terminations in the hypothalamus has been proposed by Cox and his co-workers on the basis of experiments on body temperature (1981, 1983). In addition, work by Martin and his colleagues on spinal reflexes (see Martin et al, 1976) and more recently by Larson (1982, 1983) on nociceptive reflexes, is strongly indicative of the existence of descending T-pathways.

Interactions with Transmitters.

Jones and Boulton (1981) studied the interactions between T and 5HT on single neurones in the rat cerebral cortex. Two types of interaction were seen. On cells where 5HT evoked depressant responses, weak applications of T resulted in potentiation of these responses (Fig. 8A). In contrast, when 5HT was excitatory then weak applications of T caused an occlusion of the excitation and usually resulted in its replacement

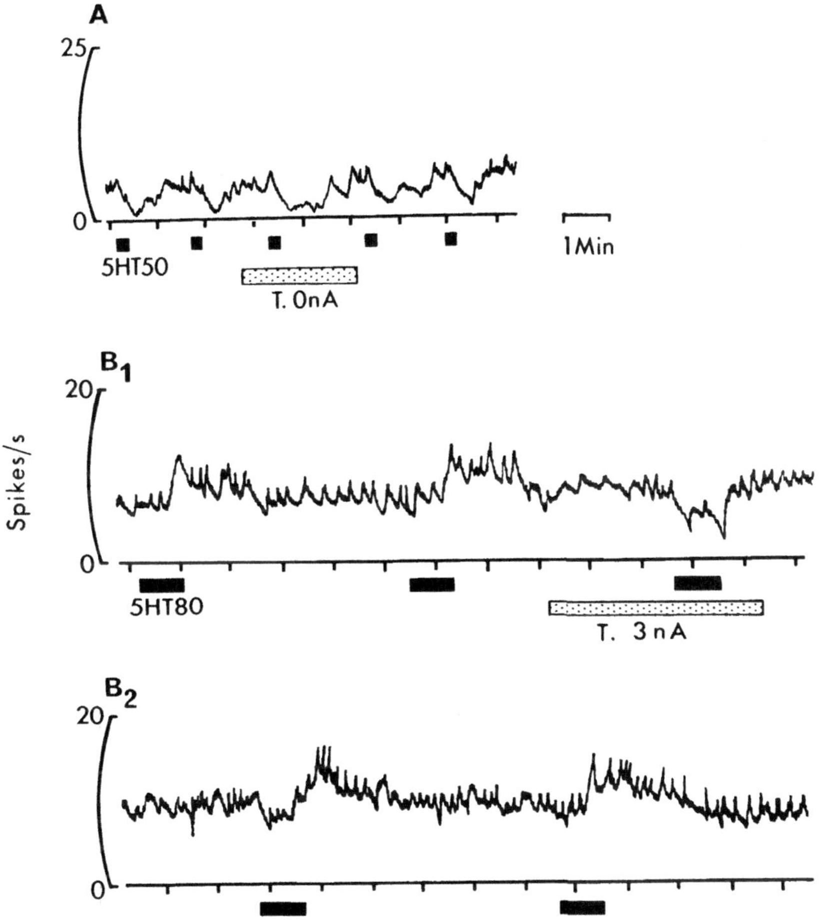

FIGURE 8. The record in A shows the ability of T to potentiate depressant responses to 5HT. B_I and B_2 are continuous and show the replacement of 5HT excitation with depression during T iontophoresis.

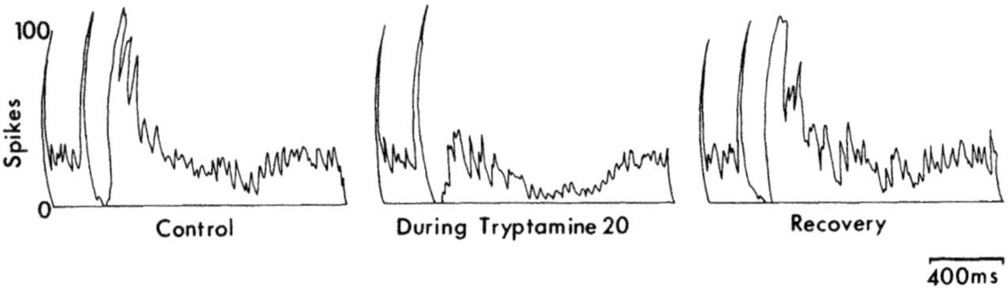

FIGURE 9. Peristimulus histograms showing the reduction of raphé evoked excitation during the application of T by iontophoresis.

with a depression (Fig 8B). It has also been shown that the excitatory consequences of median raphé stimulation which may be 5HT-mediated (see above) are reduced during the iontophoretic application of T (Jones, I982d)(Fig. 9). It is very interesting to note that the depolarization of molluscan neurones by 5HT can be effectively blocked by T (Gerschenfeld and Stefani, I966).

In contrast to its effects on 5HT-evoked excitations, T only rarely reduced the responses of cortical cells to ACh and never reversed them into depression (Jones and Boulton, I98I). However, recent studies (Jones, this book) have revealed a strong reduction of responses of cortical cells to the excitatory peptide, substance P, although reversal of responses was not seen.

Finally, neuromuscular transmission in two species appears to be modified in a biphasic, concentration-dependant manner by tryptamine. In the frog, where the transmitter is ACh, low concentrations of T increased the endplate potential and high concentrations reduced it(Friedman, I979) Similar effects were seen in the lobster, where the presumed transmitter was glutamate (Friedman, Shank and Freeman, I98I).

Although the mechanism of the modifying actions of T are not yet fully understood it should be noted that its interactions with 5HT at least, are also demonstrable at the behavioural level. Jones (I98Ib) used three models for the action of 5HT at central receptors and was able to demonstrate that T could reduce responses in two of these and potentiate them in the third. Dourish and Greenshaw have recently shown that T can potentiate the gross behavioural effects consequent on central administration of 5HT in rats(Dourish and Greenshaw, 1983).

CONCLUSIONS

Thus there are strong indications that some trace amines may be mediators or modulators of synaptic transmission in the CNS. Of course these roles are not necessarily exclusive . It is entirely conceivable that transmission of impulses in one pathway could result in simultaneous modulation of activity in another. The evidence would seem to be good to support roles of amines such as OA and T as co-transmitters with the more abundant amines exerting such dual modulator/mediator functions. The demonstration of the profound neuronal actions of trace amines both alone and in combination with other neuroactive agents is clearly suggestive of functional activities of these substances in effecting and affecting the transfer of synaptic information.

REFERENCES

Aghajanian G. K. and Haigler H. J. (1975) Hallucinogenic indoleamines: Preferential action upon presynaptic serotonin receptors. Psychopharm. Comm. **1**, 619-629.

Axelrod J. and Saavedra J. M. (1977) Octopamine. Nature **265**, 501-504.

Bevan P., Bradshaw C. M., Pun, R. Y. K., Slater T. N., and Szabadi E. (1978) Comparison of the responses of single cortical neurones to tyramine and noradrenaline: Effects of desipramine. Br. J. Pharmac. **63**, 651-657.

Boakes R. T., Dua P. R., and Baker G. B. (1976) Actions of microiontophoretically applied *p*- and *m*-tyramine on caudate neurones. Abs. Eur. Soc. Neurochem., p. 59c.

Boulton A. A. (1979) Trace amines in the central nervous system. Int. Rev. Biochem. **26,** 179-206.

Cascio C. S. and Kellar K. J. (1982) ^{3}H-Tryptamine: Characterization of binding sites in rat brain. Abs. Soc. Neurosci. **8,** 783.

Cox B., Lee T. F., and Martin D. (1981) Different hypothalamic receptors mediate 5-HT- and tryptamine-induced core temperature changes in the rat. Br. J. Pharmac. **72,** 477-482.

Cox B., Davis A., Juxon V., Lee T. F., and Martin D. (1983) A role for an indoleamine other than 5-hydroxytryptamine in the hypothalamic thermoregulatory pathways of the rat. J. Physiol. **337,** 441-450.

Curtis D. R. (1962) Action of 3-hydroxytyramine and some tryptamine derivatives on spinal neurones. Nature **164,** 292.

Curtis D. R. and Davis R. (1962) Pharmacological studies upon neurones of the lateral geniculate nucleus of the cat. Br. J. Pharmac. **18,** 217-246.

Dao W. P. C. and Walker R. J. (1980a) Octopamine responses of neurones in the rat thalamus. Br. J. Pharmac. **68,** 132P.

Dao W. P. C. and Walker R. J. (1980b) Effect of cyproheptadine on the octopamine-induced responses in the mammalian central nervous system. Experientia **36,** 584-585.

Dourish C. T. and Greenshaw A. J. (1983) Effects of intraventricular tryptamine and 5-HT on spontaneous motor activity in the rat. Res. Comm. Psychol. Psychiat. Behav. **8,** 1-9.

Engberg I., Flatman J. A., and Kadzielawa K. (1976) Lack of specificity of motoneurone responses to microiontophoretically applied phenolic amines. Acta Physiol.Scand. **96,** 137-139.

Freidman R. N. (1979) Tryptamine-induced alterations of acetylcholine release at a neuromuscular junction. Abs. Soc. Neurosoc. **5,** 480.

Freidman R. N., Shank R. P., and Freeman A. R. (1981) The effects of tryptamine on transmitter release at a lobster neuromuscular junction. Brain Res. **214,** 101-111.

Gerschenfeld H. M. and Stefani E. (1966) An electrophysiological study of 5-HT receptors of neurones in the molluscan nervous system. J. Physiol. **185**, 684-700.

Henwood R. W., Boulton A. A., and Phillis J. W. (1980) Iontophoretic studies of some trace amines in the mammalian CNS. Brain Res. **164**, 347-351.

Hicks T. P. and McLennan H. (1978a) Comparison of the actions of octopamine and catecholamines on single neurones of the rat cerebral cortex. Br. J. Pharmac. **64**, 485-491.

Hicks T. P. and McLennan H. (1978b) Actions of octopamine upon dorsal horn neurones of the spinal cord. Brain Res. **157**, 402-406.

Hoffer B. J., Siggins G. R., and Bloom F. E. (1971) Studies on norepinephrine-containing afferents to Purkinje cells of rat cerebellum. II. Sensitivity of Purkinje cells to norepinephrine and related substances administered by microiontophoresis. Brain Res. **25**, 523-534.

Jones R. S. G. (1981a) Specific enhancement of neuronal responses to catecholamines by p-tyramine. J. Neurosci. Res. **6**, 49-61.

Jones R. S. G. (1981b) In vivo pharmacological studies on the interactions between tryptamine and 5-hydroxytryptamine. Br. J. Pharmac. **73**, 485-493.

Jones R. S. G. (1982a) Noradrenaline-octopamine interactions on cortical neurones in the rat. Eur. J. Pharmac. **77**, 159-162.

Jones R. S. G. (1982b) A comparison of the responses of cortical neurones to iontophoretically applied tryptamine and 5-hydroxytryptamine in the rat. Neuropharmac. **21**, 209-214.

Jones R. S. G. (1982c) Responses of cortical neurones to stimulation of the nucleus raphe medianus: A pharmacological analysis of the role of indoleamines. Neuropharmac. **21**, 511-520.

Jones R. S. G. (1982d) Tryptamine modifies cortical neurone responses evoked by stimulation of nucelus raphe medianus. Brain Res. Bull. **8**, 435-437.

Jones R. S. G. and Boulton A. A. (1980) Interactions between p-tyramine, m-tyramine or β-phenylethylamine and dopamine on single neurons in the cortex and caudate nucleus of the cat. Can. J. Physiol. Pharmac. **58**, 222-227.

Jones R. S. G. and Boulton A. A. (1981) Tryptamine and 5-hydroxytryptamine's actions and interactions on cortical neurones in the rat. Life Sci. **27**, 1849-1856.

Jones R. S. G. and Broadbent J. (1982a) Further studies on the role of indoleamines in responses of cortical neurones evoked by stimulation of the nucleus raphe medianus: The effects of precursor loading. Neuropharmac. **21**, 1273-1278.

Jones R. S. G. and Broadbent J. (1982b) Differential effects of fluoxetine and zimelidine on the uptake of 5-hydroxytryptamine and tryptamine by cortical slices and on responses of cortical neurones to stimulation of the nucleus raphe medianus. Eur. J. Pharmac. **81**, 681-685.

Jones R. S. G., Juorio A. V., and Boulton A. A. (1983) Changes in levels of dopamine and tyramine in rat caudate nucleus following alterations of impulse flow in the nigrostriatal pathway. J. Neurochem. **40**, 396-401.

Juorio A. V. (1979) Drug-induced changes in the formation, storage and metabolism of tyramine in the mouse. Br. J. Pharmac. **66**, 377-384.

Juorio A. V. and Jones R. S. G. (1981) The effects of mesencephalic lesions on tyramine and dopamine in the caudate nucleus of the rat. J. Neurochem. **36**, 1898-1903.

Kellar K. J. and Cascio C. S. (1982) [^{3}H]-Tryptamine: High affinity binding sites in rat brain. Eur. J. Pharmac. **78**, 475-478.

Kostopoulos G. K. and Yarbrough G. G. (1975) Microiontophoretic studies of the effects of false transmitter candidates and amphetamine on cerebellar Purkinje cells. J. Pharm. Pharmac. **27**, 408-413.

Krnjevic K. (1965) Actions of drugs on single neurones in the cerebral cortex. Brit. Med. Bull. **21**, 10-14.

Krnjevic K. and Phillis J. W. (1963) Actions of certain amines on cerebral cortical neurones. Br. J. Pharmac. **20**, 471-490.

Larson A. A. (1981) Changes in pain-reflexes after intrathecal administration of tryptamine HCl in the rat. Fed. Proc. **40**, 284.

Larson A. A. (1982) Nociception is enhanced by the intrathecal injection of 5-methoxy-N,N-dimethyltryptamine in the rat. Neurosci. Lett. **33**, 323-328.

Martin W. R., Thompson J. A., and Nozaki M. (1976) Physiologic evidence for descending tryptaminergic pathways in the spinal cord. Life Sci. **19**, 1383-1386.

Stoof J. C., Liem A. L., and Mulder A. H. (1976) Release and receptor stimulating properties of p-tyramine in rat brain. Arch. Int. Pharmacodynam. **220**, 62-71.

EFFECTS OF OCTOPAMINE AND SEROTONIN ON NEURONES OF *APLYSIA CALIFORNICA*

T.P. Hicks, J.P. Edstrom and K. Lukowiak

Department of Medical Physiology,
The University of Calgary,
Calgary, Alberta. Canada T2N 4N1

ABSTRACT

Intracellular recording combined with voltage clamp methods have been used to examine the effects of octopamine (OA) and serotonin (5-HT) on voltage dependent membrane currents of various identified and unidentified neurones in *Aplysia californica*. Bath administration of drugs at known concentrations revealed that OA and 5-HT exert their effects at similar concentrations. Both amines elicit slow voltage dependent inward currents in abdominal LB cells, although concentration dependent differences in early outward currents were noted for certain voltage levels. Serotonin but not OA enhanced frequency dependent spike broadening of pleural sensory cells.

INTRODUCTION

Octopamine (OA) is one of the principal trace amines known in the nervous systems of mammalian species. It was first measured in organs innervated by sympathetic nerves in the rat (Molinoff and Axelrod, 1969) and its distribution and neurochemical properties have since been well characterized (Molinoff and Axelrod, 1972; Harmar and Horn, 1976; Robertson and Juorio, 1976; Axelrod and Saavedra, 1977; Hicks, 1977). It is nevertheless still not known what physiological role, if any, OA plays in mammalian CNS (Robertson, 1981), although theories abound (Fischer et al., 1965; Kostopoulos and Yarbrough, 1975; Boulton, 1976; Lloyd et al., 1977; Trewby et al., 1977; Baldessarini and Fischer, 1978; Diamond and Borison, 1978; Sandler et al., 1979). To date, there have not been many studies dealing with the electrophysiological effects of OA on mammalian central neurones. Kostopoulos and Yarbrough (1975), Hicks and McLennan (1978 a, b), Henwood et al., (1979) and Jones (1982) all concur that the amine elicits primarily depressant effects. Further progress in the study of OA in mammalian CNS has been hindered by the lack of a descriptive pharmacology for the amine. Even the question of whether OA reacts with receptors pharmacologically distinct from those of the catecholamine transmitters, noradrenaline and dopamine, is as yet unresolved, although there have been few studies employing a range of antagonists (Robertson, 1981).

The marine mollusc, Aplysia, has proven useful as a model system for studies of neurophysiological mechanisms (Kandel, 1976). In particular, several groups have studied OA and related amines in this invertebrate. The presence of OA in single neurones and ganglia was reported first by Saavedra et al., (1974). Subsequent reexamination by McCaman and McCaman (1978) showed high levels of OA in certain ganglia, notably the buccal, pedal and cerebral. The OA measured probably derived from the neuropil and not from cell bodies. When Aplysia abdominal ganglia are incubated with OA or 5-HT, cyclic adenosine 3', 5'-monophosphate (cAMP) levels are markedly increased; these increases are blocked by phentolamine and methysergide, respectively (Levitan, Marsden and Barondes, 1974). The amine also increased incorporation of radioactively labelled phosphate into specific protein. These and other data suggest that OA and 5-HT may produce relatively longlasting changes in a phosphoprotein in the ganglion. Since the effect was restricted to the neuropil where all the synaptic contacts are situated, the results suggest further that the process is localized to the postsynaptic cells (Levitan and Barondes, 1974).

Carpenter and Gaubatz (1974) reported that certain neurones in the cerebral, abdominal, pleural and pedal ganglia are more sensitive to OA than to phenylethanolamine, dopamine or noradrenaline. The most typical response was hyperpolarization accompanied by a conductance increase. Later, Pellmar (1981) described a slow, voltage-dependent inward current (SVDIC) evoked by OA and 5-HT which appeared only at membrane potentials above -40 mV. These currents are insensitive to changes in concentration of potassium, sodium and chloride and are blocked by the calcium antagonist, cadmium. These findings corresponded to an earlier description of the effects of 5-HT (Pellmar and Carpenter, 1980) with respect to time course, voltage dependency and ionic sensitivity. The studies concluded that the SVDIC was a regenerative inward calcium current.

In abdominal sensory neurones, 5-HT produces a slow EPSP and facilitates the release of transmitter via a cAMP-dependent process involving protein phosphorylation (Klein _et al._, 1980) which precedes a decrease in one type of potassium conductance (Hawkins _et al._, 1981; Klein and Kandel, 1980). This decrease in potassium conductance is believed to play a part in the process of sensitization by removing a restraining influence on the action potential which increases its duration. The amount of calcium entering during each action potential increases with action potential duration. The shape of the action potentials of pleural sensory neurones is susceptible to frequency dependent broadening (Edstrom and Lukowiak, 1983) (FDSB) and this spike broadening can be augmented by 5-HT (Edstrom, Hicks and Lukowiak, in preparation) in a manner suggestive of the mechanism underlying sensitization as described by Klein and Kandel (1980). The 5-HT sensitization resembles the action of 5-HT on the abdominal ganglion in that cAMP is increased. Since OA mimics the effect of 5-HT on cAMP and LB membrane currents, we wondered if it might also mimic the effect of 5-HT on FDSB in sensory cells.

Prior to speculating about the role of OA in mammals, more information about the electrophysiological actions of the amine is necessary. It will then be possible to compare this knowledge with data on synaptically-evoked effects on target neurones. Many effects of OA and 5-HT observed in molluscan model systems strongly resemble the action of 5-HT on certain central mammalian and peripheral neurones (van der Maelen and Aghajanian, 1980; Wood and Mayer, 1978). A comparison of the effects and pharmacology of OA and 5-HT on Aplysia neurones may prove useful in attempts to assess the role of the compound in other systems shown to possess OA sensitive cells - such as mammalian cerebral cortex (Hicks and McLennan, 1978 b; Henwood et al., 1979; Jones 1982), ventrobasal thalamus (Dao and Walker, 1980), caudate nucleus (Henwood et al., 1979), hypothalamus and brain stem (David and Delacour, 1980) and spinal cord (Hicks and McLennan, 1978a).

METHODS

Either pleural or abdominal ganglia taken from Aplysia californica weighing between 125 and 250 g, were pinned to the sylgard floor of a perspex perfusion chamber. The ganglion was surgically desheathed in a hypertonic sucrose/seawater solution (2 M sucrose was diluted 1:1 with normal artificial seawater (NASW), adapted from the procedure described by Connor, 1979) and the nerve trunks were cut as closely as possible to the ganglion. Following desheathing, the preparation was perfused with NASW at 15 C for between 1 and 2 hours before the recording commenced. When the perfusate was changed, fluid exhange near the ganglion began in 1 to 1.5 min and was complete after 3-4 min.

Intracellular electrodes were made from fibre-fill glass capillary tubes and contained either 1.5 M KCl or 2 M potassium citrate. These electrodes had resistances between 5 and 25 Mohms. The method of Graubard (personal communication), where the electrode is encased in an insulated sheath of conductive silver paint connected to the driven shield of the amplifier and insulated by Q-Dope from the bathing medium, was used to minimize the electrode capacitance.

Single electrode (Dagan 8100) or double electrode (Dagan 8500) voltage clamp apparata were used. In single electrode voltage clamp experiments, switching frequencies between 3 and 4 kHz were used. A MINC 11/23 computer was used to obtain current voltage relationships. It presented command voltage steps selected at random from a predetermined set of allowable values. Both the current and voltage records were digitized, stored on disk and analyzed later. Only cells depolarized by OA were selected for detailed study, although OA-induced hyperpolarizations were frequently observed and several cells showed no response to either OA or 5-HT. Presenting OA and 5-HT, either iontophoretically or by perfusion, revealed no ostensible desensitization or fatigue of effect. Perfusion times varied between 3 and 30 min.

In the spike broadening experiments the Dagan 8100 single electrode voltage clamp was used as an electrometer to record from, and pass current into, the pleural sensory cells. A stimulator (WPI 1850A), driven by a clock-pulse-generator (WPI 1830) was used to pass variable amplitude command pulses into the active bridge of the amplifier. Action potentials were evoked by passing short (3 msec), high intensity (4-8 nA) stimulus pulses into the sensory cells. Trains of stimuli were separated by 5 min intervals. Drug applications were for 15 min.

The data were recorded on a four-channel DC recorder (Gould) and on an 11/23 MINC computer. The computer was slaved to a window discriminator detecting the peak of each action potential. Up to 35 action potentials were digitized at dwell times at or above 35 microseconds. The digitized records were aligned on the peak derivative (dV/dt) of the action potential (AP).

Three substances were used in these experiments. They were $CdCl_2$ (Fisher), OA (Sigma) and 5-HT (Sigma). All compounds were dissolved in NASW without deleting any other constituent. The pH of the NASW was adjusted to 7.8 with either HCl or NaOH. The composition of NASW was NaCl, 425 mM; KCl, 10 mM; $CaCl_2$, 10 mM; $MgCl_2$, 22 mM; $MgSO_4$, 26 mM; and $NaHCO_3$, 2.5 mM. The high calcium-magnesium sea water (HCMSW) used in the LB cell experiments was the same except: NaCl, 185 mM; $CaCl_2$, 0.06 mM and $MgCl_2$, 0.132 mM.

The indifferent electrode was a calomel half cell connected with the perfusing solution through a saline bridge made from 4% agar dissolved in NASW. Before any cells were impaled, the electrode was left in the bathing medium for 15 min to 1 hour until it had equilibrated with the surrounding fluid. The data from a cell were discarded if there was a difference of more than 5 mV between the reference voltage measured before a cell was impaled and that measured following withdrawl of the electrode from the cell.

RESULTS

LB Cells and Slow Currents

It proved surprisingly difficult to determine the relative potencies of 5-HT and OA in eliciting the slow inward current and the slow depolarization of LB cells. This was due to the fact that on these cells, we encountered previously-undescribed effects of OA and 5-HT. Earlier studies were primarily concerned with the mechanisms of action of the amines. In these studies, drugs were administered iontophoretically with the cell clamped at different membrane potentials. Initially we performed some experiments of this sort to test the efficacy of various antagonist candidates. We were persuaded to alter this protocol because of uncertainty regarding relative agonist potencies and possible direct antagonist effects. Instead, we proceeded with an initial characterization of the novel drug actions and a determination of the relative potencies of the two amines.

Administration of drug via perfusion requires long-term recording stability, i.e., a steady baseline, since the drug actions were long-lasting and slowly developing. Effects of iontophoretic applications of 5-HT or OA lasted up to 5-8 min and the effect of 3 min perfusions of high (10^{-4} - 10^{-3} M) drug concentrations could last 30 min to 1 hr. Furthermore, LB cells receive periodic bursts of synaptic activity from the rhythmically active interneurone II network (Kandel 1976) and from elsewhere. In Figure 1A the considerable background synaptic activity impinging upon the LB cell is apparent and the periodic appearance of interneurone II activity (not shown) can depolarize the cell by as much as 30 mV and cause depolarization block of LB discharge. Even so, the effect of OA on the unclamped cell in NASW is apparent but cannot be measured accurately. There is no stable baseline and the level of depolarization is further obscured by frequent action potentials. The frequency of interneurone II discharge also is increased by the amines, thereby

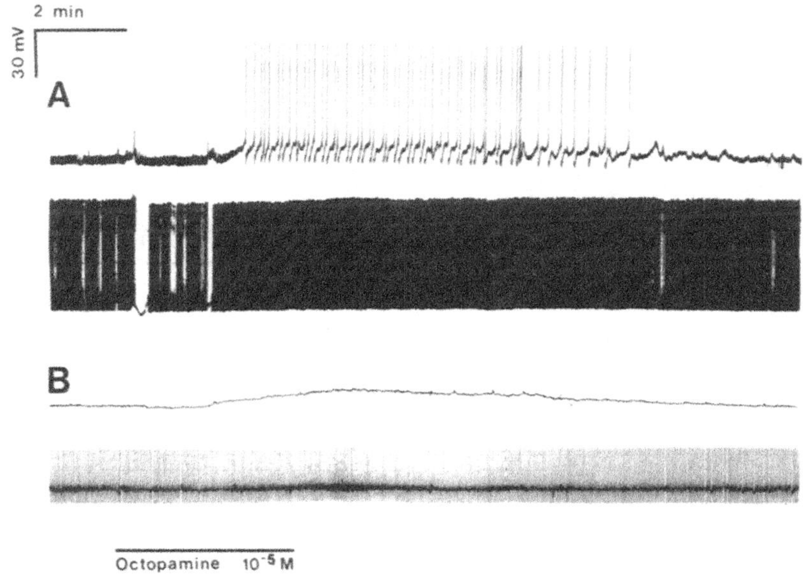

Figure 1: Recordings of membrane potential from an unclamped LB cell and from L11 showing the effect of 10^{-5} M OA in NASW (top two traces) and in HCMSW (bottom two traces). In each pair of records, the top trace is from the LB cell and the bottom is from cell L11. The peaks of the action potentials are cut off.

exacerbating the intrinsic instability. Although not obvious at this time base, the firing frequency of L11 is increased and the cell is depolarized by OA.

Changing the perfusate to one containing high concentrations of Ca^{++} and Mg^{++} and reduced Na^{+} diminishes interneurone II activity as well as most of the spontaneous synaptic input (Figure 1B). This procedure does not block monosynaptic transmission but does suppress conduction through polysynaptic circuits (Kandel, 1976). It can be seen that in the presence of HCMSW, while the resting LB membrane potential is not completely stable, it is more level and the firing rate of L11 is reduced and more constant. Under these conditions the effects of OA are still present.

Because prolonged exposure to HCMSW might alter the characteristics of the SVDIC, we determined the optimum voltage in HCMSW. This was done in two ways. The first was identical to the method of Pellmar and Carpenter (1980); the drug was administered while the membrane potential of the cell was clamped at different levels. The second was to clamp the membrane potential near the normal resting potential (-40 mV) and then step briefly to different voltages to explore the membrane currents. Testing by the first method, the maximum SVDIC was observed between 0 and -10 mV for both 5-HT and OA. At 10^{-5} M OA gave a marginally larger response than 5-HT; at 0 mV, the 5-HT induced current was approximately 90% of the OA induced current. At potentials below -40 mV, 5-HT consistently caused a larger inward current, about twice as great as that induced by the same concentration of OA. The threshold concentrations for both OA and 5-HT were between 10^{-7} and 10^{-6} M for the inward current, at both depolarized and hyperpolarized membrane potentials. OA was marginally more potent at the depolarized potentials, having a threshold concentration slightly below but within one log unit of 5-HT, although in about 25% of the cases they appeared to be equipotent. At hyperpolarized potentials, 5-HT was slightly more potent than OA. Potency analyses were complicated by the lengthy time courses of action of the amines, although this difficulty never appeared to be worsened by desensitization.

The drugs were tested on unclamped neurones in HCMSW (Figure 2). Under these conditions, the thresholds are about 10^{-6} M for OA and 10^{-7} M for 5-HT. The latency to initial effect for OA was longer than that for 5-HT. The OA response may be characterized as a long, slow depolarization while the response of the cell to 5-HT is biphasic, comprising an initial short latency depolarization followed by a slow depolarization comparable to

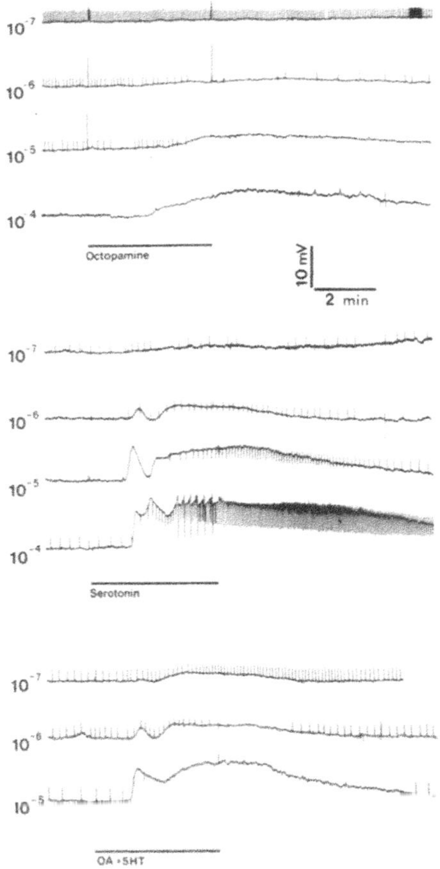

Figure 2: Unclamped membrane potential of LB cell in HCMSW with OA (1st 4 records), 5-HT (2nd 4 records) and a 1:1 mixture of OA and 5-HT (3rd set) at indicated concentrations.

that elicited by OA. Serotonin excites while OA inhibits the neurone responsible for the tonic synaptic input. The drug-induced depolarization passes through the reversal potential of part of this dual-action synaptic potential. When 5-HT and OA are administered in combination, their effects are neither mutually inhibitory nor synergistic but are nearly additive. In the combined presence of both compounds the 5-HT excitation of the interneurone is reduced.

These observations suggested that the descriptions of the action of 5-HT and OA available are incomplete and that in addition to

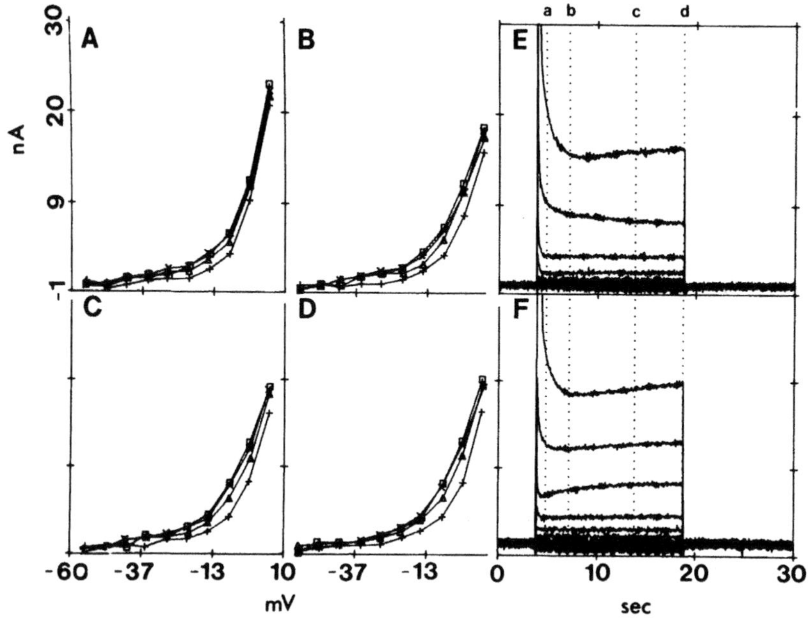

Figure 3: Voltage clamp record from an LB cell in HCMSW and various concentrations of 5-HT. Ten voltage steps between 5 and -55 mV were presented randomly. The holding potential was -40 mV. Current record samples are shown in panels E (10^{-4} M 5-HT) and F (control). The four markers, "a", "b", "c" and "d" show the time during the voltage clamp steps when the data were extracted for the four I-V curves appearing in "A", "B", "C" and "D". Symbols: open squares, control; "x", 10^{-6} M; triangles, 10^{-5} M; crosses, 10^{-4} M.

the static actions of the drugs, there are important dynamic response properties as well. Voltage clamp experiments using step changes in voltage were performed to investigate these properties and their responses to drugs.

The effects of 5-HT and OA are compared in Figures 3 to 5. Current-voltage relationships for 5-HT are shown in Figure 3, for OA in Figure 4 and a comparison of the changes elicited by both drugs is presented in Figure 5. Serotonin decreases the depolarization induced outward current (E). This effect increases with time (A through D and E). There is a voltage-dependent decrease in outward current over part of the voltage range scanned. At potentials at or near the resting potential, 5-HT has a minimal effect and this effect increases with membrane potential. In addition, 5-HT also affects the time-course of the various current pulses. At the highest step potential there is a simple decrease in outward current. At around 0 mV there is an additional time-dependent decrease in depolarization-dependent

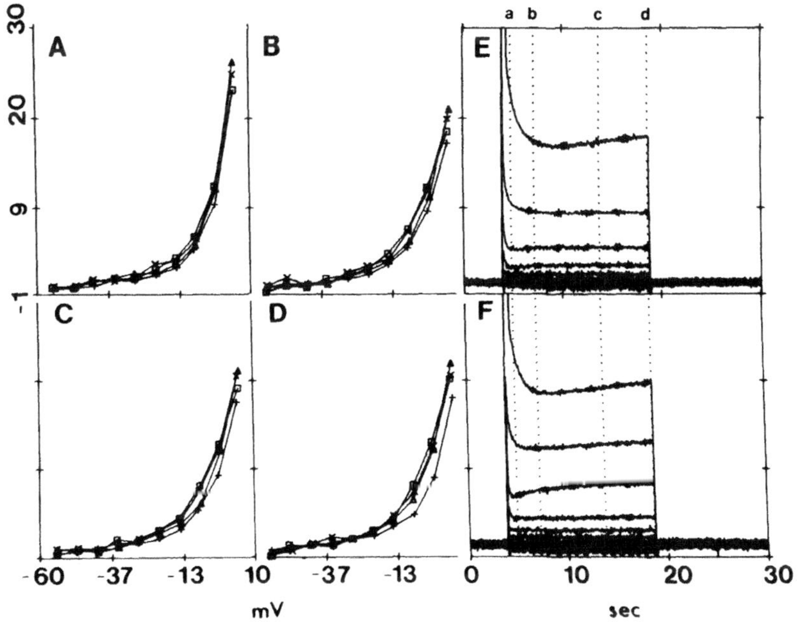

Figure 4: Same as for Figure 3, except OA was used. Data are from the same cell as Figure 3.

inward current. Panels A-D in Figure 3 include data obtained with three concentrations of 5-HT and show that the threshold is between 10^{-6} and 10^{-5} M. At all effective drug concentrations, the action of 5-HT on the outward current is qualitatively similar.

The effects of OA on the same cell are shown in Figure 4. Both amines have similar effects at the highest concentrations. At intermediate OA concentrations, however, there is an increase in early outward current at the maximum depolarization. The threshold for this effect of OA is 10^{-6} M; at higher concentrations, it converts to an effect qualitatively similar to that of 5-HT. At the scale used in these plots the differences are not easily seen. Therefore the results of pannel B in Figures 3 and 4 are plotted together in Figure 5. Here the currents at each voltage are subtracted from the currents in the control sweep at the same voltage. The result is a family of curves showing the change in current from control at different voltages and at different concentrations of drug. It can be seen that the

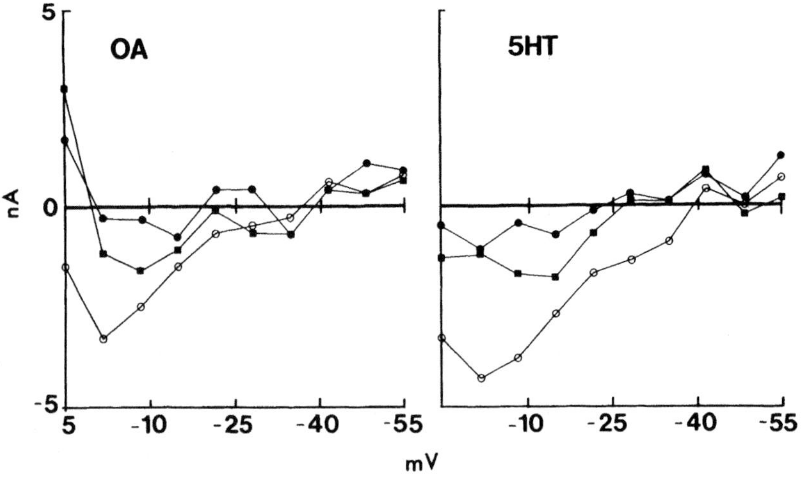

Figure 5: Data from panels "B" in Figures 3 and 4 replotted to show voltage-dependence of the inward currents. Symbols: Filled squares, 10^{-6}M; filled circles, 10^{-5} ; open circles, 10^{-4} M.

effects of the two drugs are strikingly different when they are administered at concentrations below 10^{-5} M, but they cause similar effects at or above 10^{-5} M.

Pleural Sensory Cells and FDSB

The effects of 5-HT on the FDSB of pleural sensory neurones are shown in Figure 6. At 10^{-5} M, the highest concentration tested, 5-HT caused a substantial enhancement of the FDSB. The

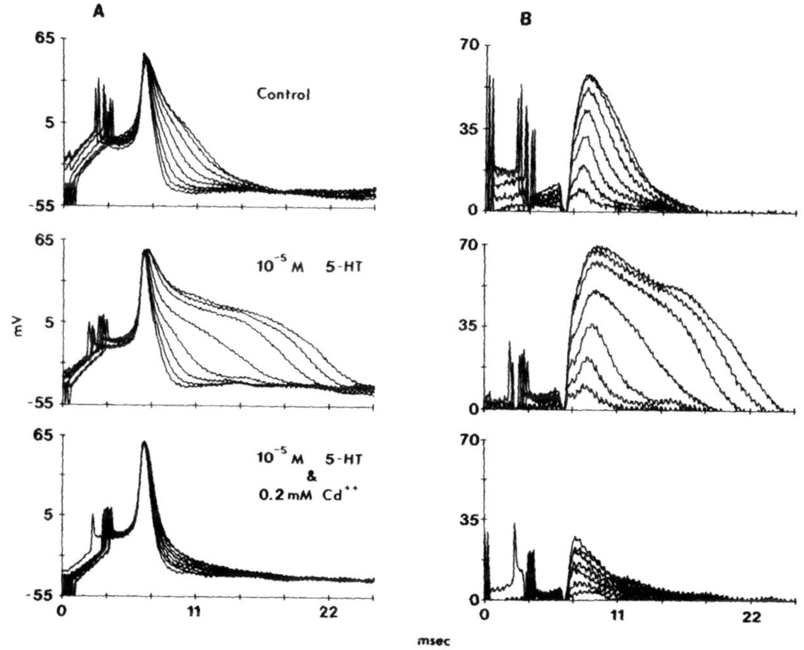

Figure 6: Effect of 5-HT and Cd^{++} on FDSB in a pleural sensory cell. Every fifth action potential (A) in a train of 35 is shown under a variety of experimental conditions. (B) Augmentation of action potentials in train; the first action potential in the train is subtracted from all subsequent action potentials and the differences plotted. The ISI was 0.14 sec.

threshold was 10^{-7} M and the maximal effect was obtained at 10^{-4} M (Figure 8). This increase in FDSB is easily blocked by the calcium antagonist, Cd^{++}. By contrast, at concentrations between 10^{-6} and 10^{-3} M, OA does not significantly affect the FDSB. The same cell is shown to be responsive to 5-HT in Figure 7 and in Figure 8. Neither drug changes the sensory cell's resting potential.

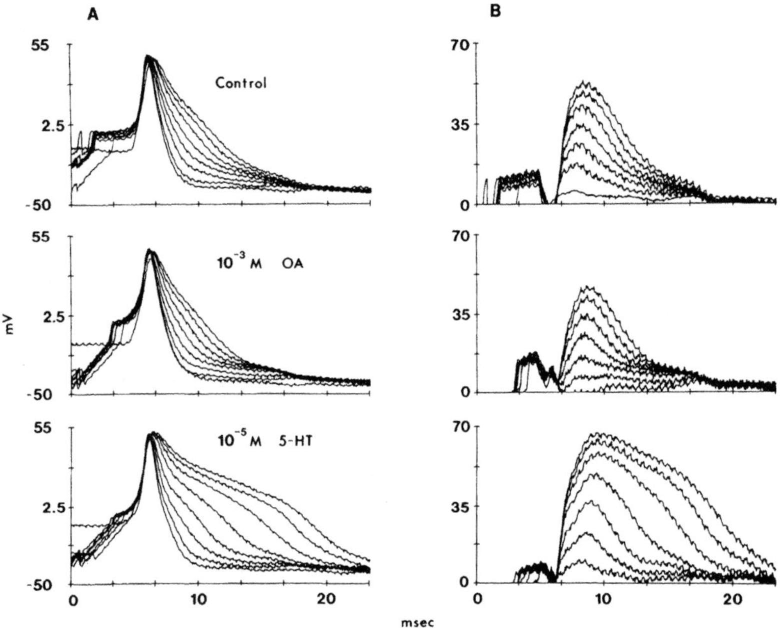

Figure 7: Same as Figure 6, different cell. The ISI was 0.15 sec.

Discussion

In the beginning stages of this study, there were two paramount questions. First, since OA and 5-HT both produce SVDIC in LB cells, presumably *via* an increase in cAMP levels, is this similarity of action due to activation of a common receptor, or are different receptors mediating a single effect? Second, because 5-HT and OA evoke a common effect in LB cells, do they also act similarly

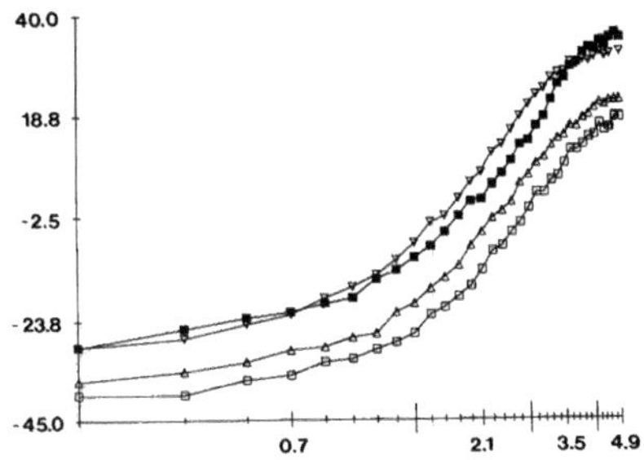

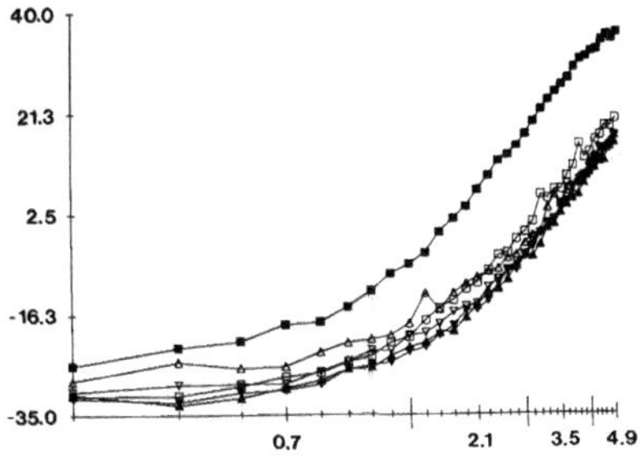

Figure 8: Activation curves for FDSB in the presence of various drugs. The figure on the left shows the effect of various 5-HT concentrations on the FDSB in the same cell as was illustrated in Figure 6. Symbols: open square, control; open triangle, 10^{-7} M; filled triangle, 10^{-6} M; filled square, 10^{-5} M. The figure on the right shows the data obtained from the same cell as was illustrated in Figure 7 and shows the effect of 10^{-6} to 10^{-3} M OA on FDSB, as well as the effect of 10^{-5} M 5-HT (filled squares) for ease of comparison. Open squares are control.

in pleural sensory cells, which are known to respond to 5-HT by an increase in cAMP? We have been able to determine that the amines possess similar pharmacological potencies and that both compounds produce similar static effects. However, the two drugs also exert dissimilar transient effects. We hoped that by comparing the threshold doses of the amines or of the responses they elicited, we might learn which was the preferred agonist. The SVDIC showed a slightly greater sensitivity to OA than for 5-HT but this difference is small and equivocal, leading us to question which parameter was appropriate as the independent variable.

Previous experiments (Carpenter and Gaubatz, 1974; Pellmar, 1981) precluded quantitative dose-response comparisons due to methodological constraints. Bath administration in NASW is inappropriate because the concentrations of amines required to elicit a response from the LB cells also have pronounced effects on their synaptic input.

With the experiments conducted in HCMSW, these difficulties were largely circumvented. However, the responses observed still did not correspond exactly to existing descriptions. Although 5-HT and OA both elicited the SVDIC as Pellmar (1981) described, we noticed further that OA also elicited an inward current at hyperpolarized potentials. This property of OA was observed at a marginally higher threshold dose than for 5-HT. Pellmar (1981) observed an inward Na^+ current with 5-HT but not OA, in NASW. In our experiments, any inward Na^+current should have been greatly diminished in HCMSW since Na^+ was reduced by 60%. Therefore the inward current we observed at hyperpolarized potentials is unlikely to be carried by Na^+. These currents do not resemble the slow Na^+ currents reported by Pellmar and Carpenter (1980) and by Pellmar (1981) inasmuch as they do the response of the A cell in Helix (Cottrell, 1979) where there is no increase in conductance until the K^+ equilibrium potential is exceeded, i.e., there is an apparent increase in the anomalous rectification of the cell. The anomalous rectifying channel is pH dependent and Meech and Thomas (1980) have shown that increasing intracellular Ca^{++} leads to decreased pH; this would enhance anomalous rectification. Either the increased extracellular Ca^{++} concentration or drug-dependent Ca^{++} movements might augment the anomalous rectification. Without extensive experimentation it is not possible to rule out the possibility that OA acted via the normal mechanism for 5-HT.

The time and voltage-dependent outward current elicited by OA at depolarizing voltages was unexpected and extended previous analyses of the mechanism of the drugs. At higher concentrations of OA and at longer clamp pulse times, the OA I-V curve resembles that of 5-HT, whereas at low concentrations and short clamp times, 5-HT effects never resembled the OA-induced response. The response to 5-HT is time dependent, since even after 15 seconds it had not developed fully. These effects were missed in previous studies because the iontophoretic administration of drugs prevented the observation of dynamic, drug-dependent conductance changes. Because the membrane potential was held at a constant value, only the **change** in holding current, and not the actual current, was recorded while the drugs were given for a brief period of time. The time course of the observed responses then, reflected the development of a steady-state conductance at a given voltage but not the time-dependent conductances at that voltage.

This same issue confuses comparisons of the drug effects on LB cells and sensory cells. On the grounds of the different time courses of the events, Pellmar (1981) suggested that the SVDIC evoked by OA and 5-HT in LB cells was a different phenomenon from the 5-HT dependent spike broadening in sensory cells demonstrated by Klein et al. (1980). This conclusion is probably correct, although it is based on unsound grounds since fast transient responses were not investigated by Pellmar, only the steady-state ones. In fact, it was probably the activation time of the response which was measured; this parameter should not be confused with the effect of the response on transient phenomena. Both our own experiments and those of Kandel (1976) on sensory cells used prolonged drug presentations which served to activate the mechanisms before the cells are tested with short-duration transient pulses.

Only tentative conclusions may be drawn regarding the effects of 5-HT and OA on the pleural sensory cells. The FDSB and its sensitivity to 5-HT is strongly reminiscent of the effects of 5-HT on unconditioned APs (Klein et al., 1980). If they are related then the drug-sensitive nature of FDSB is apt to play an important role in processes of synaptic plasticity. Although consistent with the results of Sieglebaum et al. (1982), however, this question is not immediately relevant to the present discussion and a more detailed analysis will ensue (Edstrom, Hicks and Lukowiak, in preparation). What is of particular interest is that there is such a clear differentiation between the effect of 5-HT and OA on these cells.

The relationship between the responses of sensory cells to 5-HT-induced FDSB and the LB cells' responses in the voltage clamp studies is not clear. At first appearance it may seem that the two cell types have receptors which are either more discriminative or subserve different functions. There are fundamental differences between the LB cells and the pleural sensory cell. The increase in spike duration in the sensory cell may be due to a decrease in restraining potassium current, while the increase in inward current in LB cells might reflect an increase in the current carried by Ca^{++}. The difference between the two cell types may only represent the lack of a similar mechanism for expression, rather than a fundamental difference in mechanism of drug action.

Two interesting lines of thought emerge from the foregoing discussion. The first concerns molecular aspects of the drug-elicited mechanisms. If, as has been shown previously, both amines increase cAMP in their target cells, then unless the cAMP is sequestered in "leak-proof" subcellular compartments, both drugs should exert identical or very similar effects. If this proves correct, then a situation exists wherein two presumably different inputs to a cell converge postsynaptically onto a common pathway (in molecular terms) to exert a common effect. There is another possibility however. Since LB cells respond similarly to 5-HT and high concentrations of OA, one might suppose that OA exerts a conventional effect at physiological concentrations but at higher doses, it interacts with a relatively indiscriminate 5-HT receptor to produce its complex actions. This implies that an OA-containing cell governs the nature of the post-synaptic cell's response: if a little OA is released, then rapid depolarizations will be antagonized by the outward current. High activity in the cell using OA might cause a synaptic "spill-over" of the amine that would induce the depolarization effects normally evoked by 5-HT. It is therefore important to proceed with the antagonist studies in order to see if the various components of the response of the cells to OA and 5-HT can be distinguished pharmacologically. If so, then a multiplicity of receptors exists. If not, then one is forced to assume that the cAMP intermediate step is the important common factor.

Acknowledgements: We thank C.A. Smith for technical assistance. This project was funded by grants from the Medical Research Council of Canada and the Alberta Heritage Foundation for Medical Research.

References

Axelrod J. and Saavedra J. M. (1977) Octopamine. Nature, (Lond.) 265, 501-504.

Baldessarini R. J. and Fischer J. E. (1978) Trace amines and alternative neurotransmitters in the central nervous system. Biochem. Pharmacol. 27, 621-626.

Boulton A. A. (1976) Cerebral aryl alkyl aminergic mechanisms, in Trace Amines and the Brain (Usdin E. and Sandler M., eds), pp. 21-39. Marcel Dekker Inc., New York.

Carpenter D. O. and Gaubatz G. L. (1974) Octopamine receptors on Aplysia neurones mediate hyperpolarizataion by increasing membrane conductance. Nature (Lond.) 252, 483-485.

Castellucci V. F., Kandel E. R., Schwartz J. H., Wilson F. D., Nairn A. C. and Greengard P. (1980) Intracellular injection of the catalytic subunit of cAMP dependent protein kinase stimulates facilitation of transmitter release underlying behavioral sensitization in Aplysia. Proc. Nat. Acad. Sci. USA 22, 7492-7496.

Connor J. A. (1979) Calcium current in molluscan neurones: measurement under conditions which maximize its visibility. J. Physiol. (Lond.) 286, 41-60.

Dao W. P. C. and Walker R. J. (1980) Effect of cyproheptadine on the octopamine-induced responses in the mammalian central nervous system. Experientia 36, 584-585.

David J. C. and Delacour J. (1980) Brain contents of phenylethanolamine, m-octopamine and p-octopamine in the roman strains of rats. Brain Res. 195, 231-235.

Diamond B. I. and Borison R. L. (1978) A new putative neuroregulator of the extrapyramidal system. Pharmacology 17, 210-214.

Edstrom J. and Lukowiak K. (1983) Frequency dependent action potential prolongation in Aplysia pleural sensory neurones. Submitted to: Neurosience.

Fischer J. E., Horst W. D. and Kopin I. J. (1965) β-Hydroxylated sympathomimetic amines as false neurotransmitters. Br. J. Pharmac. 24, 477-484.

Harmar A. J. and Horn A. S. (1976) Octopamine in mammalian brain: rapid post mortem increase and effects of drugs. J. Neurochem. 26, 986-993.

Henwood R. W., Boulton A. A. and Phillis J. W. (1979) Iontophoretic studies of some trace amines in the mammalian CNS. Brain Res. 164, 347-351.

Hicks T. P. (1977) The possible role of octopamine as a synaptic transmitter: a review. Can. J. Physiol. Pharmacol. 55, 137-155.

Hicks T. P. and McLennan H. (1978a) Actions of octopamine upon dorsal horn neurones of the spinal cord. Brain Res. 157, 402-406.

Hicks T. P. and McLennan H. (1978b) Comparison of the actions of octopamine and catecholamines on single neurones of the rat cerebral cortex. Br. J. Pharmac. 64, 485-491.

Jones R. S. G. (1982) Noradrenaline-octopamine interactions on cortical neurones in the rat. Eur. J. Pharmac. 77, 159-162.

Kandel E. R. (1976) Cellular Basis of Behavior, W.H. Freeman & Co., San Francisco.

Klein M., Shapiro E. and Kandel E. R. (1980) Synaptic plasticity and the modulation of the Ca++ current. J. Exp. Biol. 89, 117-157.

Kostopoulos G. K. and Yarbrough G. G. (1975) Microiontophoretic studies of the effects of false transmitter candidates and amphetamine on cerebellar Purkinje cells. J. Pharm. Pharmac. 27, 408-412.

Levitan I. B. and Barondes S. H. (1974) Octopamine- and serotonin-stimulated phosphorylation of specific protein in the abdominal ganglion of Aplysia californica. Proc. Nat. Acad. Sci. USA 71, 1145-1148

Levitan I. B., Madsen C. J. and Barondes S. H. (1974) Cyclic AMP and amine effects on phosphorylation of specific protein in abdominal ganglion of Aplysia californica; localization and kinetic analysis. J. Neurobiol. 5, 511-525.

Lloyd K. G., Davidson L., Price K., McClung H. J. and Gall D. G. (1977) Catecholamine and octopamine concentrations in brains of patients with Reye syndrome. Neurology 27, 985-988.

McCaman M. W. and McCaman R. E. (1978) Octopamine and phenylethanolamine in Aplysia ganglia and in individual neurons. Brain Res. 141, 347-352.

Meech R. W. and Thomas R. C. (1980) Effect of measured calcium chloride injections on the membrane potential and internal pH of snail neurones. J. Physiol. (Lond.) 298, 111-129.

Molinoff P. and Axelrod J. (1969) Octopamine: normal occurrence in sympathetic nerves of rats. Science 164, 428-429.

Molinoff P. B. and Axelrod J. (1972) Distribution and turnover of octopamine in tissues. J. Neurochem. 19, 157-163.

Pellmar T. C. (1981) Voltage-dependent current evoked by dopamine and octopamine in Aplysia. Brain Res. 223, 448-454.

Pellmar T. C. and Carpenter D. O. (1980) Serotonin induces a voltage-sensitive calcium current in neurons of Aplysia californica. J. Neurophysiol. 44, 423-439.

Robertson H. A. (1981) Octopamine: after a decade as a putative neuroregulator, in Essays in Neurochemistry and Neuropharmacology Vol. 5 (Youdim, M. B. H., Sharman, D. F., Lovenberg, W. and Lagnado, J. R., eds), pp. 47-73. John Wiley and Sons, Ltd., Chichester.

Robertson H. A. and Juorio A. V. (1976) Octopamine and related non-catecholic amines in invertebrate nervous systems. Int. Rev. Neurobiol. 19, 173-224.

Saavedra J. M., Brownstein M. J., Carpenter D. O. and Axelrod J. (1974) Octopamine: presence in single neurons in Aplysia suggests neurotransmitter function. Science 185, 364-

365.
Sandler M., Ruthven C. R. J., Goodwin B. L., Reynolds G. P., Rao V. A. R. and Coppen A. (1979) Deficient production of tyramine and octopamine in cases of depression. Nature (Lond.) 278, 357-358.
Siegelbaum S. A., Camardo J. S. and Kandel E. R. (1982) Serotonin and cyclic AMP close single K^+ channels in *Aplysia* neurones. Nature (Lond.) 299, 413-417.
Trewby P. N., Chase R. A., Davis M. and Williams R. (1977) The role of the false neuro-transmitter octopamine in the hypotension of fulminant hepatic failure. Clin. Sci. Mol. Med. 52, 305-310.
van der Maelen G. P. and Aghajanian G. K. (1980) Intracellular studies showing modulation of facial motoneurone excitability by serotonin. Nature (Lond.) 287, 346-347.
Wood J. D. and Mayer C. J. (1979) Serotonergic activation of tonic-type enteric neurons in guinea pig small bowel. J. Neurophysiol. 42, 582-593.

Free Communications

INDIVIDUAL HOUSING STRESS ELEVATES BRAIN AND ADRENAL TRYPTAMINE CONTENT

R.E.W. HARRISON AND S.T. CHRISTIAN

RUSH MEDICAL CENTER, CHICAGO, IL 60612 and

UNIVERSITY OF ALABAMA IN BIRMINGHAM, AL 35294

Tryptamine (T) is present in a variety of tissues, including brain and adrenal glands (Boulton and Baker, 1975; Philips and Jourio, 1978). The role of T in psychological stress has not been established, but there are several indications that tissue levels of the compound may be changed by psychological stress. As a result of stress, the amount of tryptophan available as a T precursor increases in brain (Curzon et al., 1972) and a related compound, phenylethylamine, which is also formed by the catalytic action of aromatic amino acid decarboxylase is increased in urine (Paulos and Tessel, 1982). In addition, assays of monoamine oxidase (MAO), the enzyme responsible for the degradation of T, indicate a decrease in MAO activity in animals that have been psychologically stressed (Maura and Vaccari, 1975). Since T has a rapid turnover in the CNS (Durden and Phillips, 1980), then a slight decrease in MAO activity could presumably produce large changes in T concentrations.

To further elucidate the role of T in psychological stress, we have assayed this compound in brain and adrenal glands taken from single animals which were either individually housed or group housed.

MATERIALS AND METHODS

All organic solvents were obtained from Burdick and

Jackson Laboratories, Inc., Muskegon, MI. All inorganic reagents were from J.T. Baker Chemical Company, Phillipsburg, N.J. Heptafluorobutyryl imidazole (HFBI) was purchased from Pierce Chemical Company, Rockford, IL; T was obtained from Aldrich Chemical Company, Milwaukee, WI; and α,α,β,β-tetradeutero tryptamine (DT) was purchased from Merck, Sharp and Dohme, Montreal, Canada. All aqueous solutions were prepared from glass-distilled water.

Sprague-Dawley rats (60 or 120 days old) were group housed 4 to a cage or individually housed in identical cages. All animals were sacrificed one hour into the light cycle in a separately ventilated room acoustically isolated from the housing area. Prior to sacrifice on their 60th day, this age group of animals was either group housed, or individually housed for 5.5, 8 or 28 days. Prior to sacrifice on their 120th day, this age group of animals was either group housed or individually housed for 21 or 60 days. In order to avoid post-mortem increases in brain or adrenal T, the tissues were removed from the animal as rapidly as possible and frozen in liquid nitrogen. Tissues not immediately extracted were wrapped separately and stored at -65°C until analyzed. Tissue T extraction was a modification of that used by Christian et al. (1975) for CSF. After deproteinization with 23% $HClO_4$, the tissue lipids were separated from T by 15% isooctane in hexane. The amine-containing aqueous phase was basified, and extracted 4x with $MeCl_2$. The combined $MeCl_2$ extracts were dried and the residue heated with HFBI at 85°C for 1 hour. The heptafluorobutryl derivative of T (HFB-T) was taken up in 1 ml $MeCl_2$ and stored hermetically under N_2 gas with Na_2SO_4 at -65°C until analysis. Details of this procedure are found elsewhere (Harrison, 1982).

Total mass spectra of pure HFB-derivatives of T and DT were determined on a Hewlett-Packard 5985 gas chromatograph/mass spectrometer operating in electron impact (EI) mode with a direct insertation probe. Standard and recovery curves, as well as qualitative tissue analyses were performed via the EI mode on the gas chromatographic effluent using a computer controlled selected ion monitor (SIM). The GC separation was accomplished on a Supelco 4', 2 mm internal diameter glass column containing 2% OV-17 on 100-120 mesh chromosorb W-HP operated at 150°C initially for 2 min. then 5°C/min until 170°C. Injector temperature was 150°C with a helium flow at 40 ml/min. From inspection of the total mass spectrum of HFB-T and its deuterated congener, ion characteristics suitable for quantification were chosen for use with SIM. Holding DT constant at 1000 ng, standard curves for quantification of T in SIM were generated by serial dilution of a working standard of pure T to concentrations of 0-1000 ng/ml whereas the recovery curve of T was constructed from serial spiking of 2 ml aliquots taken from an $HClO_4$ homogenate of pooled tissue samples of either brain or adrenal glands of 8 Sprague-Dawley rats. Pure standards and recovery samples were then extracted and derivatized as described. The typical injection volume was 1 μℓ. The ions monitored were m/z 326 and m/z 339 for T and m/z 328 and m/z 342 for DT. The amount of T (in ng/g wet weight) was calculated from:

$$[T] = \frac{800\left(\frac{\text{peak area proteo-ion}}{\text{peak area deutero-ion}} - N\right)}{W}$$

The peak area for the indicated ion is in arbitrary units of integrated ion current during the retention time (R_T) of HFB-T; N = 0.01 as a "worst case" noise figure and W is the wet weight of the tissue analyzed. Results were analyzed non-parametrically with Kruskal Wallis One Way Analysis of Variance by Ranks (KW) and Jonckheere-Terpstra Test for Ordered Alternatives (JT).

RESULTS

Both recovery and standard curves were linear for HFB-T corresponding to 5-1000 pg T/μℓ in the injected sample. In the assay of brain and adrenal T, the compound eluting at the R_T of HFB-T gives rise to fragments at m/z 326 and m/z 339. The ratio of the peak areas of these ions is identical to that of pure T standards and recoveries of T in brain and adrenal glands. Although T could not be determined in brains or adrenals from group housed control animals to which no T was added, the recovery curve for brain indicated a small amount of endogenous T corresponding to 200 pg. This figure is in agreement with earlier studies (Warsh et al., 1977; Artigas and Gelpi, 1979; Phillips and Boulton, 1979). Animals which had been individually housed showed

TABLE 1 - BRAIN AND ADRENAL T

Tissue	Age	Individual Housing	T Median, Range	
Brain	60 days	5.5 days	9.2, 6.9	ng/g
Brain	60	28	21.8, 150.8	ng/g
Adrenal	60	8	.27, 11.44	μg/g
Adrenal	60	28	65.7, 173.8	μg/g
Brain	120	21	54.3, 340.8	ng/g
Brain	120	60	74.0, 178.2	ng/g
Adrenal	120	21	.471, .304	μg/g
Adrenal	120	60	2.07, 1.86	μg/g

significant age- and stress-dependent increases in both brain and adrenal T ($p<.025$, KW); Table 1). Although brain T increased in both 60 and 120 day old animals, the T increase was not a function of duration of individual housing ($p<.01$, JT). The increase in adrenal T was related to the amount of time spent individual housing for both groups ($p<.006$, JT) but this increase was less for 120 day old animals. In animals sacrificed at 120 days of age, the adrenal wet weight increased from 45 ± 13 mg SD in group housed animals to 167 ± 7 mg SD in probands individually housed for 60 days. No such increase was observed in animals sacrificed at 60 days of age.

DISCUSSION

Under the stated conditions, T can be extracted in high yield (Harrison, 1982) and the HFB-derivative of T formed is stable for very long periods of time (Benington, et al., 1975). In order avoid decomposition of HFB-T during analysis, care must be taken that the injection port is free of reactive halides. This assay is sensitive to an amount of HFB-T corresponding to 5 pg of T, although greater sensitivities can be achieved if the final HFB-T solution is concentrated on ice under a N_2 gas jet and greater volumes injected.

There has been some debate as to whether individual housing constitutes a stress. In some studies adrenal weight is assumed to correlate with the amount of stress a proband receives. Most, but not all, indicate that adrenal weight is increased as a result of the stress of individual housing (Brain and Benton, 1979). We have observed age- and stress-dependent elevations in adrenal weight since this variable increases as a condition of stress in 120 but not 60 day old animals. Since previously age had not been a variable in individual housing studies, our result can explain some differences reported prior to our study.

Indolealkylamine biochemistry is affected by both age and stress. Age-related changes in brain levels of 5-HT and 5-HIAA, (Weil-Fugazza et al., 1980) and changes in activity of tryptophan hydroxylase (Sze, 1976), tryptophan pyrrolase (Green and Curzon, 1975) and MAO (Maura and Vaccari, 1975; Benedetti and Keane, 1980) indicate that aging can be observed as a progressive and interactive alteration of neuroendocrine regulatory systems. The present study lends support to such a view for T in brain and adrenals. We found T concentrations in the brain and adrenal glands in each single animal to be uncorrelated, suggesting independent regulation of sequestered T in these tissues. The large range of values observed for T may be due to individual variations in the ability to synthesize, accumulate, or metabolize T as a function of psychological stress. A large variation in brain T is not unexpected, because of the rapid turnover and degradation of the compound (Durden and Phillips, 1980). Rats stressed by individual housing of animals become hyperactive (Weinstock et al., 1978) or undergo cardiovascular status changes (Bennett and Gardiner, 1979). A metabolic role for T is not excluded for such hyperactivity (Marsden and Curzon, 1979) or cardiovascular status changes (Krstic and Djukovic, 1981; Stollak and Furchgott, 1983) provided T is available in pharmacologically active amounts, which we maintain is available endogenously as a result of stress. Since T acts on neurons (Jones and Boulton, 1980; Jones 1982) in a manner distinct from 5-HT (Kellar and Cascio, 1982) we propose that T is a neuromodulator involved in the physiological, behavioral and biochemical response to psychological stress.

REFERENCES

Artigas F. and Gelpi F. (1979) A new mass fragmentographic method for the simultaneous analysis of tryptophan, tryptamine, indole-3-acetic acid, serotonin and 5-hydroxyindole-3-acetic acid in the same sample of rat brain. Anal. Biochem. **92**, 233-242.

Benedetti M. S. and Keane P. E. (1980) Differential changes in monoamine oxidase A and B activity in the aging rat brain. J. Neurochem. **35**, 1026-1032.

Benington F., Christian S. T., and Morin R. D. (1975) Identification and separation of indolealkylamines by gas-liquid chromatographic analysis of their heptafluorobutyryl derivatives. J. Chromatog. **106**, 435-439.

Bennett T. and Gardiner S. M. (1978) Corticosteroid involvement in the changes in noradrenergic responsiveness of tissues from rats made hypertensive by short-term isolation. Br. J. Pharmacol. **64**, 129-136.

Boulton A. A. and Baker G. B. (1975) The subcellular distribution of β-phenylethylamine, *p*-tyramine and tryptamine in rat brain. J. Neurochem. **25**, 477-481.

Brain P. F. and Benton D. (1979) The interpretation of physiological correlates of differential housing in laboratory rats. Life Sci. **24**, 99-116.

Christian S. T., Benington F., Morin R. D., and Corbett L. (1975) Gas-liquid chromatographic separation and identification of biologically important indolealkylamines from human cerebrospinal fluid. Biochem. Med. **14**, 191-200.

Curzon G., Joseph M. H., and Knott P. J. (1972) Effects of immobilization and food deprivation on rat brain tryptophan metabolism. J. Neurochem. **19**, 1967-1974.

Durden D. A. and Philips S. R. (1980) Kinetic measurements of the turnover rates of phenylethylamine and tryptamine *in vivo* in rat brain. J. Neurochem. **34**, 1725-1732.

Eleftheriou B. E. and Boehlke K. W. (1967) Brain monoamine oxidase in mice after exposure to aggression and defeat. Science **155**, 1693-1694.

Green A. R. and Curzon G. (1975) Effects of hydrocortisone and immobilization on tryptophan metabolism in brain and liver of rats of different ages. Biochem. Pharmacol. **24**, 713-716.

Harrison R. E. W. (1982) Ph.D. dissertation, University of Alabama in Birmingham.

Jones R. S. G. and Boulton A. A. (1980) Tryptamine and 5-hydroxytryptamine: Actions and interactions on cortical neurones in the rat. Life Sci. **27**, 1849-1856.

Jones R. S. G. (1982) A comparison of the responses of cortical neurones to iontophoretically applied tryptamine and 5-hydroxytryptamine in the rat. Neuropharmacology **21**, 209-214.

Kellar K. S. and Cascio C. S. (1982) [^{3}H]-Tryptamine: High affinity binding sites in rat brain. Eur. J. Pharmacol. **78**, 475-478.

Krstic M. K. and Djurkovic D. (1981) Comparison of the cardiovascular responses to intracerebroventricular administration of tryptamine, 5-hydroxytryptamine, tryptophan and 5-hydroxytryptophan in rats. Arch. Int. Physiol. Biochim. **81**, 385-391.

Marsden C. A. and Curzon G. (1978) The contribution of tryptamine to the behavioural effects of L-tryptophan in tranylcypromine-treated rats. Psychopharmacol. **57**, 71-76.

Maura G. and Vaccari A. (1975) Relationships between age of submission to environmental stress and monoamine oxidase activity in rats. Experientia **31**, 191-193.

Paulos M. A. and Tessel R. E. (1982) Excretion of β-phenethylamine is elevated in humans after profound stress. Science **215**, 1127-1129.

Philips S. R. and Boulton A. A. (1979) The effect of monoamine oxidase inhibitors on some arylalkylamines in rat striatum. J. Neurochem. **33**, 159-167.

Stollak J. S. and Furchgott R. F. (1983) Use of selective antagonists for determining the types of receptors mediating the actions of 5-HT and tryptamine in the isolated rat aorta. J. Pharmacol. Exp. Ther. **224**, 215-221.

Sze P. Y. (1976) Glucocorticoid regulation of the serotonergic system of the brain. Adv. Biochem. Psychopharmacol. **15**, 251-265.

Warsh J. J., Cosina D. V., Godse D. D., and Chan P. W. (1977) Brain tryptamine in rats by a new gas chromatographic-mass fragmentographic method. Biochem. Med. **18**, 18-20.

Weil-Fugazza J., Godefroy F., and Stupfel M. (1980) Relation of sex and aging to serotonin metabolism in rats. Mech. Aging and Dev. **13**, 199-204.

Weinstock M., Speiser Z., and Ashkenazi R. (1978) Changes in brain catecholamine turnover and receptor sensitivity induced by social deprivation in rats. Psychopharmacol. **56**, 205-209.

THE UPTAKE AND RELEASE OF ^{14}C-TRYPTAMINE BY RAT BRAIN SLICES

Lillian E. Dyck

Psychiatric Research Division
Cancer and Medical Research Building
University of Saskatchewan
Saskatoon, Saskatchewan S7N OWO

INTRODUCTION

Direct tryptamine (T) effects in the rat brain and in rabbit aortic strips have been reported (Jones, 1982; Stollak and Furchgott, 1983); moreover, in the rat brain, there appear to be distinct T receptors (Kellar and Cascio, 1982; Cascio and Kellar, 1982). It is conceivable, then, that T may be a transmitter in the rat brain. If so, T should be accumulated by a high-affinity uptake system which is sodium-, temperature- and energy-dependent; in addition, depolarization induced release of T should be calcium-dependent (Iversen, 1974; Thoa *et al.*, 1975). The following study was undertaken to determine whether T transport exhibited these characteristics.

MATERIALS AND METHODS

Rat brain tissues (fronto-parietal cortex, CX; hypothalamus, HT; caudate, Cau) were weighed, chopped to form 0.2 mm thick slices and placed in a modified Krebs buffer (see Dyck, 1978) containing pargyline, in which they were pre-incubated for 15 min at 37°C; then, the tissue slices were transferred to another tube and incubated simultaneously with 500 nM ^{3}H-serotonin (^{3}H-5HT) and ^{14}C-T for various time periods. ^{3}H-(G)-5HT (500 mCi/mmol) and ^{14}C-T[side chain-2 ^{14}C] (50-55 mCi/mmol) were purchased from Amersham, Ont. and New England Nuclear, MA, respectively. Virtually all of the radioactivity was accounted for by unchanged amines. Amine uptake at 0°C (incubated in an ice-bath) was also determined and subtracted from all samples.

In order to study amine release, slices of CX, HT and Cau were incubated as described above; then, they were transferred at 2 min intervals through a series of ten tubes containing buffer at 37°C. To measure basal or spontaneous release, a normal Krebs buffer was present in all ten tubes; however, to measure K^+-stimulated release, the last five tubes contained a buffer modified to contain 50 mM KCl. Radioactivity released into the buffer and left in the slices was quantified.

RESULTS

The amounts of ^{3}H-5HT and ^{14}C-T accumulated into slices of CX, HT and Cau after 5, 10, 15 or 30 min incubations are shown in Figure 1. Clearly, much more ^{3}H-5HT than ^{14}C-T was accumulated, and marked regional differences

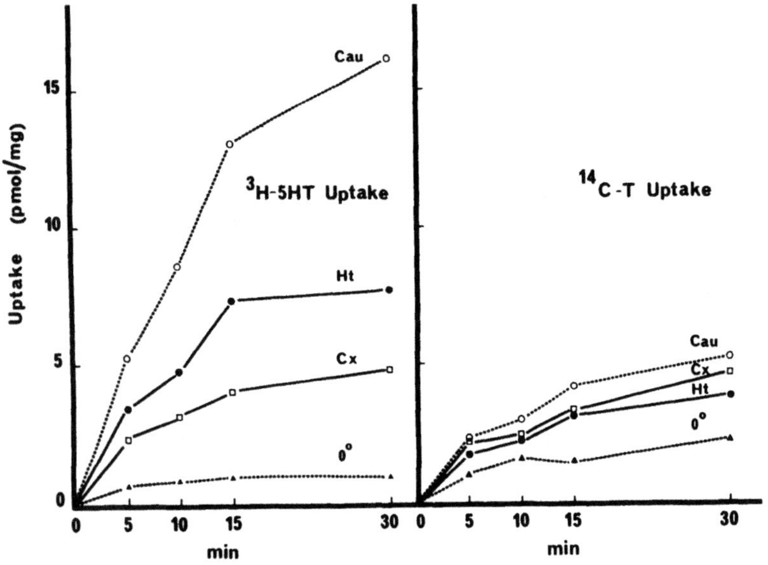

Figure 1. Effect of incubation time on uptake of ^{3}H-5HT and ^{14}C-T (substrate concentration, 500 nM).

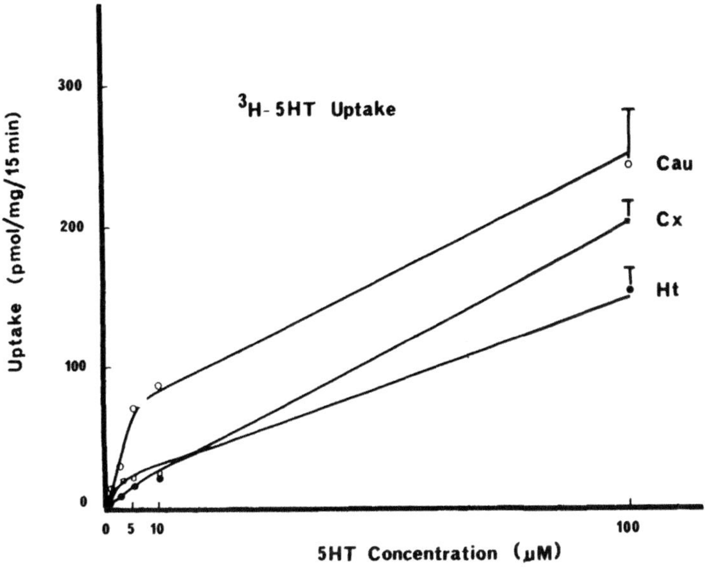

Figure 2. Effect of substrate concentrations on ^{3}H-5HT uptake (incubation time, 15 min).

in ^{3}H-5HT but not in ^{14}C-T uptake occurred. It is also important to note that the uptake of ^{14}C-T at 0°C represented a much larger proportion of total uptake than with ^{3}H-5HT.

Varying the concentration of ^{3}H-5HT from 0.5 to 100 μM revealed that ^{3}H-5HT was taken up by two processes (Figure 2). A linear diffusional process predominated from about 5 μM ^{3}H-5HT for all three brain areas. Similarly, a linear diffusional uptake of ^{14}C-T occurred (Figure 3), but in this case it seems to predominate at lower concentrations (about 2.5 μM); i.e., the saturable or active component of ^{14}C-T uptake appears to be smaller than that for ^{3}H-5HT.

Dinitrophenol (DNP), ouabain and cocaine significantly inhibited the uptake of ^{3}H-5HT and ^{14}C-T in all three brain areas (data not shown). With the exception of the effects of DNP in the CX, the uptake of ^{3}H-5HT was decreased to a greater extent than was ^{14}C-T uptake by these drugs in all three brain areas. The use of a low sodium buffer did not

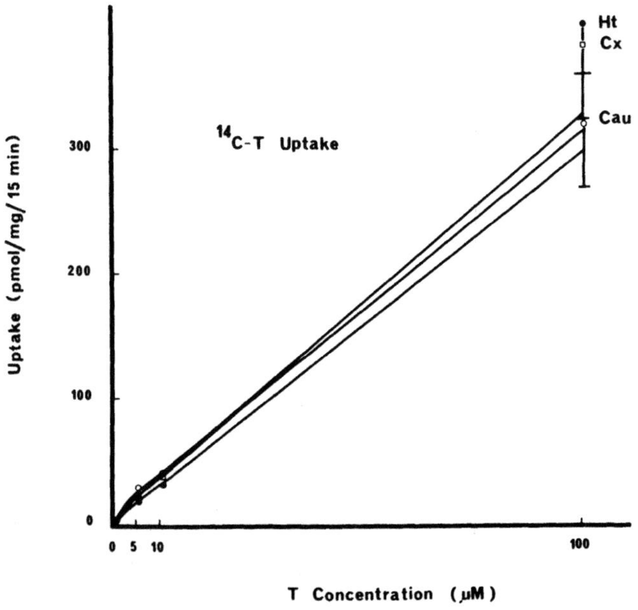

Figure 3. Effect of substrate concentration on ^{14}C-T uptake (incubation time, 15 min).

affect ^{3}H-5HT or ^{14}C-T uptake into cortical slices, but decreased uptake of both amines in the HT and Cau. Low Na^+ decreased the uptake of ^{3}H-5HT into the Cau more than the uptake of ^{14}C-T.

The concentrations of zimelidine, fluoxetine, norzimelidine and citalopram which inhibited the uptake of ^{3}H-5HT and ^{14}C-T by 50% (IC_{50}'s) were determined (Table 1). In the HT and CX, much higher concentrations of these drugs were required to inhibit ^{14}C-T uptake than ^{3}H-5HT uptake. Interestingly, this was not so in the Cau where these drugs inhibited ^{14}C-T uptake more effectively than ^{3}H-5HT uptake.

The initial basal release of ^{14}C-T into fractions 1-5 was much more rapid than that ^{3}H-5HT (data not shown). When a depolarizing concentration of KCl was introduced into tubes 6-10 (or into tubes 3-5) a marked release of

TABLE 1

IC_{50} values for the inhibition of ^{3}H-5HT and ^{14}C-T uptake into slices of HT, Cau and CX

Drug	IC_{50} (μM) ^{3}H-5HT	^{14}C-T	Ratio (T/5HT)
HT:			
Zimelidine	18.2	89.7	4.9
Fluoxetine	8.6	29.5	3.4
Norzimelidine	5.6	54.3	9.7
Citalopram	0.1	35.4	354
Cau:			
Zimelidine	50.0	28.0	0.6
Fluoxetine	20.6	9.6	0.5
Norzimelidine	27.0	12.2	0.5
Citalopram	143.2	44.3	0.3
CX:			
Zimelidine	20.8	50.9	2.4
Fluoxetine	25.1	54.9	2.2
Norzimelidine	16.6	39.3	2.4
Citalopram	2.2	77.1	35.0

IC_{50} values were determined from log probit plots using a least squares fit of the linear portion of the graphs.

^{3}H-5HT was observed, but no significant release of ^{14}C-T occurred. Similarly, veratridine (25 μM) and 100 μM concentrations of unlabelled T, 5HT, amphetamine and p-chloroamphetamine were unable to stimulate ^{14}C-T release, but did cause ^{3}H-5HT release (data not shown).

DISCUSSION

The uptake studies showed that T was taken up by slices of CX, HT and Cau predominantly by a diffusional process; however, a small active uptake was also present as indicated by Figure 3 and by the inhibitory effects of DNP, ouabain, cocaine, low Na^+ and the '5HT' uptake blockers. ^{14}C-T release occurred only spontaneously, and could not be stimulated by a variety of stimuli. Previous studies of the uptake and release of T, with the odd exception, are in excellent agreement with the present data (Stacey, 1961; Born, 1972; Wu and Boulton, 1972; Snodgrass and Iversen, 1974; Hery *et al.*, 1979, 1983; Shroder *et al.*, 1979; Osborne, 1980; Ross, 1980; Jones and Broadbent, 1982).

Though T has profound behavioural and physiological effects, it seems unlikely that it does so through a conventional transmitter action, since these data clearly show that its uptake and release properties are not transmitter-like. It may be that T has a modulatory action as first proposed by Boulton (1976).

Acknowledgements

I would like to thank Dr. A.A. Boulton for his encouragement, Mr. R.C. Mag-Atas for technical assistance and Saskatchewan Health for financial support.

REFERENCES

Born G.V.R., Juengjaosen K. and Michal F. (1972) Relative activities on and uptake by human blood platelets of 5-hydroxytryptamine and several analogues. *Br. J. Pharmacol.* *44*, 117-139.

Boulton A.A. (1976) Identification, distribution, metabolism and function of *meta*- and *para*-tyramine, phenylethylamine and tryptamine in brain. *Adv. Biochem. Psychopharmacol.* *15*, 57-67.

Cascio C.S. and Kellar K.J. (1982) Tetrahydro-β-carbolines: affinities for tryptamine and serotonergic binding sites. *Neuropharmacol.* *21*, 1219-1221.

Dyck L.E. (1978) Uptake and release of *meta*-tyramine, *para*-tyramine and dopamine in rat striatal slices. *Neurochem. Res.* *4*, 633-642.

Héry F., Simonnet G., Bourgoin S., Soubrie P., Artaud F., Hamon M. and Glowinski J. (1979) Effect of nerve activity on the *in vivo* release of [^{3}H]serotonin continuously formed from L-[^{3}H]tryptophan in the caudate nucleus of the cat. *Brain Res.* *169*, 317-334.

Hery M., Faudon M.V. and Héry F. (1983) In vitro release of newly synthesized serotonin from superfused rat suprachiasmatic area-ionic dependency. Eur. J. Pharmacol. 89, 9-18.

Iversen L.L. (1974) Uptake mechanisms for neurotransmitter amines. Biochem. Pharmacol. 23, 1927-1935.

Jones R.S.G. (1982) A comparison of the responses of cortical neurones to iontophoretically applied tryptamine and 5-HT in the rat. Neuropharmacol. 21, 209-214.

Jones R.S.G. and Broadbent J. (1982) Differential effects of fluoxetine and zimelidine on the uptake of 5-hydroxytryptamine and tryptamine by cortical slices and on responses of cortical neurones to stimulation of the nucleus raphé medianus. Eur. J. Pharmacol. 81, 681-685.

Kellar K.J. and Cascio C.S. (1982) [^{3}H]Tryptamine: high affinity binding sites in rat brain. Eur. J. Pharmacol. 18, 475-478.

Osborne N.N. (1980) Specificity of serotonin uptake by bovine retina: comparison with tryptamine. Exp. Eye Res. 31, 31-39.

Ross S.B. and Ask A.-L. (1980) Structural requirements for uptake into serotonergic neurones. Acta Pharmacol. Toxicol. 46, 270-277.

Schroder H.U., Neuhoff V., Priggemeier E. and Osborne N.N. (1979) The influx of tryptamine into snail (helix pomatia) ganglia: comparison with 5-hydroxytryptamine. Malacologia 18, 517-525.

Snodgrass S.R. and Iversen L.L. (1974) Formation and release of ^{3}H-tryptamine from ^{3}H-tryptophan in rat spinal cord slices. Adv. Biochem. Psychopharmacol. 10, 141-150.

Stacy R.S. (1961) Uptake of 5-hydroxytryptamine by platelets. Br. J. Pharmacol. 16, 284-295.

Stollak J.S. and Furchgott R.F. (1983) Use of selective antagonists for determining the types of receptors mediating the actions of 5-hydroxytryptamine and tryptamine in the isolated rabbit aorta. J. Pharmacol. Exp. Ther. 224, 215-221.

Thoa N.B., Wooten G.F., Axelrod J. and Kopin I.J. (1975) On the mechanism of release of noradrenaline from sympathetic nerves induced by depolarizing agents and sympathomimetic drugs. Mol. Pharmacol. 11, 10-18.

Wu P.H. and Boulton A.A. (1973) Distribution and metabolism of tryptamine in rat brain. Can. J. Biochem. 51, 1104-1112.

TRYPTAMINE-INDUCED CHANGES IN ENDOGENOUS 5-HYDROXYTRYPTAMINE AND $[^3H]$-5HT RELEASE FROM MOUSE HYPOTHALAMIC SLICES

C.M. Robinson and C.A. Marsden

Department of Physiology and Pharmacology

Medical School, Queen's Medical Centre
Clifton Boulevard, Nottingham NG7 2UH

INTRODUCTION

Tryptamine (T) is present in low amounts in rodent brain (Saavedra and Axelrod, 1973; Marsden and Curzon, 1974; Philips et al., 1974) and when it is administered peripherally to rats or mice pretreated with a MAO inhibitor it results in behavioural effects similar to those produced by indirect and direct acting 5HT agonists (Marsden and Curzon, 1978; Marsden, 1980; Luscombe, et al.,1982; Jones, 1982). Furthermore, interventions that disrupt 5HT functional integrity such as 5HT depletion and administration of 5HT antagonists, reduce both 5HT mediated and T-induced syndromes (Marsden and Curzon, 1978; Luscombe et al., 1982). This suggests that T acts as a 5HT agonist to induce behavioural responses, probably by releasing neuronal 5HT. Fuxe and Ungerstedt (1968) demonstrated that T increased release of extragranular 5HT using fluorescence histochemistry while Baker et al (1980) showed that T caused release of $[^3H]$-5HT from rat striatal slices. However, more recent work by Cox and Ennis (1982) and Mounsey et al (1982) has shown that T inhibits the potassium evoked release of $[^3H]$-5HT from superfused rat hypothalamic and cortical slices.

We have examined the effects of T on 5HT release in two *in vitro* systems. The results suggest that T can cause receptor mediated inhibition of 5HT release but at high concentrations displacement of neuronal 5HT occurs.

METHODS

Male CS-1 mice (20-25 g) were used in all experiments.

(a) Measurement of Endogenous 5HT Release from Hypothalamic Slices using fixed volume Conditions

Coronal hypothalamic slices were cut into 0.5 x 0.5mm cubes which were washed three times in calcium-free Krebs buffer and centrifuged at 3,000 rpm for one min. The cubes were then resuspended in 500 µl gassed (95% O_2; 5% CO_2) Krebs bicarbonate buffer pH. 7.4 containing glucose (10 mM) and pargyline (1µM), preincubated in a shaking water bath for 5 min at 37°C prior to the addition of T (10^{-6}M) and then incubated for a further 20 min. Other drugs were added immediately before the preincubation stage, e.g. paroxetine (10^{-5}M), metergoline (10^{-6}M) and atropine sulphate (10^{-6}M). At the end of the incubation period 50 µl of 0.1 M perchloric acid was added to each tube and these were centrifuged (5 min at 3,000 rpm) and the resulting supernatants frozen and stored at -20° C until assayed. 5HT was assayed by HPLC with electrochemical detection using reverse phase separation (Spherisorb 5 ODS, 25 cm x 5 mm column) and a 0.1 M acetate-citrate mobile phase buffer (pH. 4.1) containing 3% methanol. Detection used a carbon paste working electrode (Bioanalytical Systems Inc - Anachem Ltd.) maintained at a potential of +0.65 V with respect to a Ag/AgCl reference electrode.

(b) Measurement [^{3}H]-5HT Release using superfused Hypothalamic Slices

Hypothalamic cubes were prepared as in the fixed volume experiments except that 0.25 x 0.25 mm cubes were used. The superfusion was carried out as described by Cox and Ennis (1982). Slices were incubated for 20 min at 37°C in Krebs bicarbonate buffer (pH 7.4) containing 10^{-7}M tritiated 5HT ([^{3}H]-5HT) (specific activity: 13 curies/mmol, Amersham International PLC), 1 µM pargyline, 2 mM ascorbate, and glucose (10 mM), washed three times in 5 ml calcium-free Krebs buffer containing glucose, resuspended in 1.4 ml of Krebs buffer and then 140 µl aliquots added to each of the 8 superfusion chambers. Chambers were then perfused with oxygenated Krebs buffer, adjusted to a flow rate of 0.4 ml/min, and kept in a bath containing oxygenated Krebs buffer at a temperature of 37^{0}C.

After an initial standardization period of 30 min fractions of superfusate were collected every 4 min. The first three fractions were used to measure basal[^{3}H]-5HT release; two 4 min pulses of potassium (30 mM) were administered by changing the Krebs buffer reservoir during the collection of the 4th (S_1) and the 10th (S_2) fractions. Drugs were adminstered from the buffer reservoir after collection of the 4th fraction to four chambers with four chambers maintained as controls. The percentage of radioactivity released above basal by the two pulses of potassium was expressed at the ratio S_2/S_1 for both the control and drug treated slices, using the total of the three values immediately before and after each potassium pulse.

RESULTS

(a) Effect of T on 5HT release

There was a discrepancy in the effect of T (10^{-6}M) on 5HT release in the fixed volume and superfusion release experiments. In the former, T significantly increased endogenous 5HT release above untreated controls, whereas in the superfusion system inhibition of K^{+} stimulated [^{3}H]-5HT release (decrease in S_2/S_1) was observed when T was infused after S_1.

(b) Effect of Paroxetine on the T-induced Changes in 5HT Release

Paroxetine (10^{-5}M) reversed the effects of T in the fixed volume experiments so that a statistically significant decrease in 5HT release was observed rather than an increase (Fig.I). In the superfusion system addition of paroxetine did not significantly alter the S_2:S_1 ratio observed with T in the absence of paroxetine (Fig.I). Paroxetine however, did potentiate the release of label from both control and T treated slices (Fig.I). In the fixed volume experiments paroxetine also increased basal (control) release of endogenous 5HT (Fig.I). In both systems metergoline (10^{-6}M) significantly reduced the T-induced inhibition of endogenous 5HT and [^{3}H]-5HT release. The decreased 5HT release induced

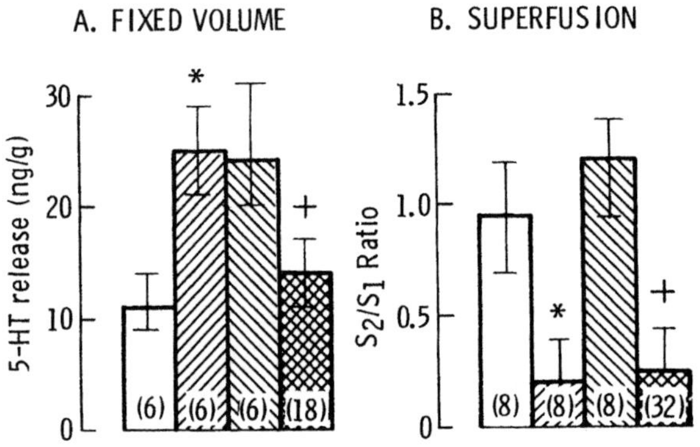

Figure I Effect of T (10^{-6}M) on endogenous 5HT and K^+ (30 mM) stimulated [^{3}H]-5HT release from mouse hypothalamic slices with and without the 5HT uptake inhibitor paroxetine (10^{-5}M). □ = control ▨ + T ▧ + paroxetine ▩ + paroxetine and T. Results are given as means ± range of values. *P < 0.05 V control + P < 0.05 V paroxetine

by T in the presence of paroxetine under fixed volume or superfusion conditions was also reversed by the muscarinic receptor antagonist atropine (10^{-6}M).

DISCUSSION

T inhibited potassium evoked [^{3}H]-5HT release (Cox and Ennis 1982; Mounsey et al 1982) In contrast, under fixed volume conditions T increased release of endogenous 5HT release. This effect was blocked by inhibition of 5HT uptake suggesting the increased release of 5HT in the presence of T was due to the displacement of presynaptic 5HT. T administration to rodents results in behavioural effects which are dependent on the presence of 5HT and may involve release of 5HT (Marsden and Curzon, 1978; Luscombe et al., 1982). The most likely explanation of the difference between the effects of T under fixed volume and superfusion conditions is that with the former T is added to the slice preparation at the beginning of incubation, accumulates in presynaptic neurones and causes release of 5HT. In contrast with the superfusion system the added T is constantly flushed through the chambers preventing its uptake and accumulation be neurones.

The T-induced inhibition of 5HT release was antagonised by metergoline, demonstrating the involvement of "tryptamine" receptors.

Cox and Ennis (1982) reported that the T-induced inhibition of [^{3}H]-5HT release in the rat hypothalamus was attenuated by atropine suggesting the involvement of muscarinic receptors and a second neurotransmitter system. The present results in the mouse support this view as atropine (10^{-6}M) reversed the T-induced decrease in [^{3}H]-5HT release.

In summary, the 5HT releasing action of T may be a pharmacological effect caused by the accumulation of high levels of T in the presynaptic 5HT neurone and this may be responsible for the observed behavioural syndrome *in vivo*. However, the physiological effect of hypothalamic T is more likely to be inhibition of 5HT release but it remains to be determined whether this effect on 5HT release is mediated via specific T receptors (Jones 1982; Kellar and Cascio, 1982) or whether T acts on one of the 5HT receptor subtypes.

REFERENCES

Baker G B, Hiob L E, Martin I L, Mitchell P R and Dewhurst W G (1980) Interactions of tryptamine analogues with 5-hydroxytryptamine and dopamine in rat striatum *in vitro*. Proc.West.Pharmcol.Soc.23: 167-170.
Cox B and Ennis C (1982) The effects of tryptamine on serotonin release from hypothalamic slices is mediated by a cholinergic interneurone. Psychopharmacology 78: 85-88.
Fuxe K and Ungerstedt U (1968) Histochemical studies on the effect of (+)-amphetamine, drugs of the imipramine group and tryptamine on central catecholamine and 5-hydroxytryptamine neurones after intraventricular injection of catecholamines and 5-hydroxytryptamine. Eur J Pharmacol 4: 134-144.

Jones R S G (1982b) Tryptamine: a neuromodulator or neurotransmitter in mammalian brain? Progress in Neurobiology 19: 117-139.

Kellar K J and Cascio C S (1982) [^{3}H] -tryptamine: high affinity binding sites in rat brain. Eur J Pharmacol 78: 475-478.

Luscombe G, Jenner P and Marsden C D (1982) Tryptamine-induced myoclonus in guinea-pigs pretreated with monoamine oxidase inhibitor indicates pre and post-synaptic actions of tryptamine upon central indoleamine systems. Neuropharmacol 21: 819-831.

Marsden C A (1980) The involvement of 5-hydroxytryptamine and dopamine in the behavioural effects of α-methyltryptamine . Neuropharmacol 19: 691-698.

Marsden C A and Curzon G (1974) Effects of lesions and drugs on brain tryptamine. J Neurochem 23: 1171-1176.

Marsden C A and Curzon G (1978) The contribution of tryptamine to the behavioural effects of L-tryptophan in tranylcypromine treated rats. Psychoparmacol 57: 71-76.

Mounsey I, Brady K A, Carroll J, Fisher R and Middlemiss D.N. (1982) K^+- Evoked [^{3}H]-5-HT release from rat frontal cortex slices: The effect of 5-HT agonists and antagonists. Biochem. Pharmacol. 31 49-53

Philips S R, Durden D R, Boulton A A (1974) Identification and distribution of tryptamine in the rat. Can J Biochem 52: 447-451

Saavedra J M and Axelrod J (1973) Effect of drugs on the tryptamine content of rat tissues. J Pharmacol exp Ther 185: 523-529

We thank the Wellcome Trust for financial support and Beechams PLC (paroxetine) and Carlo Erba Ltd. (metergoline) for generous gifts of drugs.

EFFECT OF CHRONIC HALOPERIDOL ON THE LEVELS OF BLOOD AND URINARY PHENYLETHYLAMINE AND PHENYLACETIC ACID IN RATS

Philip A. Shea, Stephen E. Wade, Stephen D. Dunlop, and Hugh C. Hendrie

Institute of Psychiatric Research, Department of Psychiatry, Indiana University School of Medicine, Indianapolis, IN 46223

INTRODUCTION

Phenylethylamine (PE) and/or its major metabolite phenylacetic acid (PAA) have been suggested to be involved in the pathophysiology of schizophrenia and the major affective disorders. In the majority of past studies on PE and PAA levels in urine of psychiatric patients, it has not been clear whether subjects were still receiving or recently withdrawn from antipsychotic medication. It is possible that such medication may contribute to changes in PE dynamics either during treatment or following withdrawal, though this possibility has not been thoroughly tested. Therefore, in a preliminary experiment we have directly tested the effects of a chronic neuroleptic schedule on the levels of PE in both blood and urine, and the levels of PAA in urine, of rats. Urinary excretion of these compounds was studied prior to drug treatment, during chronic treatment, and following withdrawal, while blood PE levels were observed immediately after and five days after the completion of a chronic treatment schedule.

METHODS AND MATERIALS

Male Wistar rats (Harlan Industries, Indianapolis) were used throughout. Rats were housed individually in wire-bottomed cages, under a 12/12 light/dark schedule with lights on at 1900h, and received food and water *ad lib*.

After one week of adaptation to housing conditions, daily injections were given to each animal between 0800 and 0900h; animals were allocated at random to one of two groups according to injection regime: (1) saline, i.p. for 21 days (SAL group), or (2) saline, i.p. for 2 days, followed by haloperidol (1 mg/kg) i.p. for 19 days (HAL group). Animals were weighed daily; mean weight on day 1 was 296 g. Injections were stopped after day 21. On day 22, half the animals in each injection group were chosen at random for sacrifice as described below; the rest were maintained for 5 days without injections before being similarly sacrificed. The combination of 2 injection regimes and two times of sacrifice created four treatment groups: saline injected, killed on day 22(SAL 22); saline injected, killed on day 26 (SAL 26); haldol injected, killed on day 22 (HAL 22), and haldol injected, killed on day 26 (HAL 26).

Urine Collection: For 15 of the 26 days of their experimental treatment (days 2,3,4,5,6,7,8,9,12 16,19,23,24,25,26) the animals of the HAL 26 treatment group were transferred from their home cages to standard metabolic cages within the same room. While in the metabolic cages, rats received food and water ad lib., and urine voided during the preceeding 24h period was collected between 0800 and 0900h of the designated day. Urine collections were made without preservative and frozen immediately upon completion; they were stored at -20°C until analysis. The series of urine collections for this treatment group thus represents urines voided (1) during the period of saline injection, prior to haloperidol treatment (days 2,3), (2) during the period of haloperidol treatment (days 4,5,6,7,8,9,12,16,19), and (3) during the period following haloperidol treatment, in which no injections were made (days 23,24,25,26).

Sacrifice procedure and blood collection. Each rat was thoroughly anaesthetized by exposure to ether within 2 min after removal from the home cage. The thoracic cavity was surgically opened, and the left ventricle of the heart penetrated by a 20 ga needle attached to a 20 ml syringe containing 1 ml 4 mM EDTA.

The rat was then rapidly exsanguinated. Clotting of blood was prevented by rinsing the needle with EDTA as well as by providing EDTA solution within the syringe. Pargyline (final concentration $5 \cdot 10^{-5}$M) was added to whole blood immediately upon collection; the samples were then frozen and stored at -20°C until analysed for PE.

Chemical analyses. Unconjugated PE was measured in whole blood and in urine and total PAA in urine by selected ion montoring GC/MS. Haldol in urine was measured by the radioreceptor method of Creese and Snyder (1977)

Data analysis. Group differences in blood PE content were evaluated by t-test. Changes in PE excretion of the HAL 26 group during the experiment were examined by analysis of variance.

RESULTS AND DISCUSSION

Rats treated with 1 mg/kg haloperidol for 19 days showed higher blood PE levels than rats similarly injected with saline when blood was taken 1 day after the last injection, but not when blood was taken five days after the last injection. Blood PE levels were significantly higher in a group of haloperidol-treated rats killed one day after the last injection than in a similar group killed 5 days after the last injection (Table 1).

Urinary excretion of PE also responded to haldol treatment. PE excretion averaged 130% of the baseline rate in the HAL 26 rats during the period of drug injection, and fell again to pre-treatment levels by the end of the experiment (Table 2). This PE normalization after drug withdrawal was not immediate, however; urine excreted during the period 24-48h after the last drug injection (day 22) contained PE in amounts similar to those seen during haloperidol injection.

TABLE 1. Whole Blood PE Levels in Saline-Injected and Haloperidol-Injected Rats, 1 Day and 5 Days After Completion of 19 Day Injections

	Whole Blood PE, ng/ml mean (std dev)	
	1 day after final injection (Day 22)	5 days after final injection (Day 26)
TREATMENT		
Haldol (1 mg/kg body wt)	4.56 (1.005)* n = 10	1.99 (0.98) n = 5
Saline	3.30 (0.660) n = 8	2.49 (0.81) n = 5

*This value is significantly greater than each of the others. (HAL 22 vs SAL 22, $t = 3.058$ with 16 d.f., p .01; HAL 22 vs HAL 26, $t = 4.75$ with 13 d.f., p .001.)

TABLE 2. Urinary PE Output Before, During and After 19 Days Chronic Haloperidol Treatment

	PE, ug/24 hr (mean, n = 10 animals throughout)
Before Haloperidol (Days 2,3)	1.087
During Haloperidol (Days 4,5,6,7,8,9,12,16,19)	1.405
After Haloperidol (Days 24,25,26)	1.040

Analysis of variance shows $F_{2,18} = 9.466$, $p = .002$. Mean PE excretion is higher during haloperidol treatment than before or after.

The blood and urine data support the concept of a reversible elevation of blood PE levels and urinary PE excretion as a consequence of haloperidol treatment. A plausible mechanism for this effect would be reversible inhibition of monamine oxidase type B as a result of haloperidol treatment. Such inhibition might be expected to result in lowered production of PAA from PE. However, comparison of urinary PAA output of rats before and during drug treatment showed no reliable decrease in PAA as a result of treatment (paired t with 7 d.f. - .594, no significant difference).

PE excretion has been reported to differ between paranoid schizophrenic patients and control subjects, as well as between paranoid schizophrenic patients and other schizophrenic patients (Karoum et. al. 1980; Potkin et. al., 1979). Although the authors of these reports recognized the possibility that PE blood levels and excretion rate might be influenced by drug treatment, they argued that differences in drug consumption were unlikely to be the cause of the between-group differences. This argument was supported by the observation that phenothiazines did not consistently alter urinary PE concentrations in patients studied with and without medication (Potkin et. al., 1979).

The results reported here show a clear effect of the neuroleptic drug haloperidol in elevated PE excretion and also in elevated PE blood levels. The effect appears to be reversible upon withdrawal of haloperidol.

Whether a similar neuroleptic induced elevation of PE levels may have contributed to the reported differences between clinical and control samples or between paranoid and nonparanoid schizophrenic (Potkin et. al., 1979; Karoum et. al., 1980), is unknown. The time course of development and reversal of such effects would be critically important to their detection in a clinical population, and no information is available on these points in humans. Also, the phenothiazines and the butyrophenones may differ in this regard. However, the results presented here suggest that investigation of neuroleptic drug effects upon indices of trace-amine production and metabolism should be part of a thorough program to explore diagnostic usefulness or etiological significance of the trace amines in psychiatry.

REFERENCES

Creese I, Snyder SH (1977) A simple and sensitive radioreceptor assay for antischizophrenic drugs in blood. Nature 270:180-182.

Karoum F, Potkin SG, Murphy DL, & Wyatt RJ (1980). Quantitation and metabolism of phenylethylamine and tyramine's three isomers in humans. In: Noncatecholic Phenylethylamines, part 2, ed. Mosnaim AD & Wolf ME. pp 177-200.

Potkin SG, Karoum F, Chuang LW, Cannon-Spoor HE Phillips I, & Wyatt RJ (1979). Phenylethylamine in paranoid chronic schizophrenia. Science 206:470-471.

A COMPARISON OF THE EFFECTS OF ACUTE AND CHRONIC ADMINISTRATION OF PHENELZINE AND TRANYLCYPROMINE ON BRAIN CONCENTRATIONS OF 2-PHENYLETHYLAMINE, *p*-TYRAMINE AND TRYPTAMINE IN THE RAT

Glen B. Baker, Donald F. LeGatt and Ronald T. Coutts
Neurochemical Research Unit, Department of Psychiatry and Faculty of Pharmacy and Pharmaceutical Sciences, University of Alberta, Edmonton, Alberta, Canada T6G 2G3

INTRODUCTION

The "trace" amines 2-phenylethylamine (PE), *p*-tyramine (*p*-TA) and tryptamine (T) have been implicated in the etiology of a number of psychiatric disorders, including depression (Dewhurst, 1968; Boulton and Milward, 1971; Sabelli and Mosnaim, 1974; Sandler *et al.*, 1979). It is now well known that monoamine oxidase (MAO) inhibitors, drugs that are used in treatment of psychiatric disorders, cause rather dramatic increases in brain concentrations of these trace amines (Boulton *et al.*, 1973; Axelrod and Saavedra, 1975; Philips *et al.*, 1980; Philips and Boulton, 1979; McKim *et al.*, 1980). However, these reports on MAO inhibitors are acute studies, and there is a paucity of information available on the effects of chronic administration of MAO inhibitors on brain levels of these compounds. In recent years, we have conducted experiments on the effects of chronic injection of two clinically efficacious MAO inhibitors, tranylcypromine (TCP) and phenelzine (PLZ) on brain levels of PE (Baker *et al.*, 1982), *p*-TA (Baker *et al.*, 1983) and T in the rat. The findings from these experiments are reviewed here.

METHODS

Male Sprague-Dawley rats (120-140 g) were injected intraperitoneally once daily with either tranylcypromine (10 mg/kg) or phenelzine (15 mg/kg). Various groups were killed 6 h after injection on days 1, 2, 8 or 19. Brains were frozen immediately in isopentane on solid carbon dioxide and stored at -60° C until the time of analysis. The brains were analyzed using gas chromatography with electron-capture detection (Baker *et al.*, 1981). MAO activity was measured using the procedure of Wurtman and Axelrod (1963), with ^{14}C-5-HT as substrate for MAO-A and ^{14}C-PE as substrate for MAO-B.

RESULTS AND DISCUSSION

The results of these experiments are shown in Table 1 and indicate that there were some interesting differences between the effects of the two drugs.

(a) Tryptamine

Brain concentrations of T were higher in the TCP-treated groups than in the corresponding PLZ groups at all time intervals. This marked difference between the two drugs cannot be explained on the basis of MAO inhibition, as both drugs inhibited MAO by greater than 90% by day 2. In this regard, it is interesting that administration of TCP to rodents has been reported to cause an increase in brain concentrations of tryptophan, the precursor of T (Grahame-Smith, 1971; Tabakoff *et al.*, 1977). The reason for this increase in brain levels of tryptophan is unclear, but it may be that TCP facilitates tryptophan transport, since other MAO inhibitors are known to influence transport of this amino acid (Rafaelson, 1976). In addition, it has been reported (Robinson *et al.*, 1979) that chronic administration of TCP is associated with an increase in activity of aromatic amino acid decarboxylase activity; presumably, this would result in increased formation of T from its precursor amino acid, tryptophan.

(b) 2-Phenylethylamine

Concentrations of PE were significantly higher ($p < 0.001$-0.05, Student's *t*-test) in the brains of TCP-treated rats than in the PLZ-treated animals on days 1, 2 and 8. But by day 19, PEA levels in the PLZ group surpassed those in the corresponding TCP group. It is

TABLE 1. Levels (ng/g) of PE, T and p-TA in rat brain after acute and chronic administration of TCP or PLZ.

TCP	Saline Controls	Day 1	Day 2	Day 8	Day 19
PE	2.2 ± 0.6	72.9 ± 12.3	38.9 ± 7.3	42.8 ± 4.5	60.1 ± 2.6
T	< 1	17.0 ± 1.6	15.4 ± 1.8	19.9 ± 4.3	33.3 ± 5.2
p-TA	2.4 ± 0.3	7.8 ± 0.8	34.8 ± 1.2	17.5 ± 1.5	14.0 ± 0.3
PLZ					
PE	2.2 ± 0.6	2.0 ± 0.3	2.1 ± 0.4	6.1 ± 0.6	177 ± 111
T	< 1	< 1	2.1 ± 0.2	4.2 ± 0.5	9.1 ± 1.4
p-TA	2.4 ± 0.3	10.2 ± 1.6	23.1 ± 1.2	33.3 ± 4.9	42.4 ± 1.8

Results represent means ± S.E.M. (n = 6). Doses of drugs: TCP, 10 mg/kg i.p. and PLZ, 15 mg/kg i.p. Rats were sacrificed 6 h after drug administration on the day indicated.

conceivable that a considerable portion of this PEA may arise directly from PLZ since it has been reported that extensive conversion of PLZ to PEA can occur in the body (Dyck *et al.*, 1981; Baker *et al.*, 1982).

(c) *p*-Tyramine

Concentrations of *p*-TA were similar with the two drugs on day 1, but on day 2 *p*-TA levels were significantly higher ($p < .001$) with TCP than with PLZ. Thereafter, brain concentrations of *p*-TA in the PLZ-treated rats were significantly greater ($p < .02$ on day 8 and $p < .001$ on day 19) than in the TCP-treated rats. This phenomenon may be related to the progressively increasing PE brain levels on days 8 and 19 in the PLZ-treated rats, since metabolic conversion of PEA to *p*-TA is reported to occur (Boulton, 1976).

These studies have demonstrated that interesting changes in brain concentrations of trace amines can occur during chronic administration. These changes are of particular interest since it is known that MAO inhibitors generally take 10 days to 3 weeks to be effective in treatment of depression. It is also worth noting that there are a number of differences between TCP and PLZ with regard to their actions on trace amine levels; it is conceivable that these differences may be related to differences in pharmacological profiles and incidence of side effects between these two MAO inhibitors.

Although there was a large mean increase in brain levels of PEA by day 19, there was also considerable variation in PE concentrations at this time. These experiments were repeated using another group of rats, and similar results were found. These findings suggest that some important changes may be occurring at this time, and that time studies at longer intervals should be conducted.

The doses of drugs used in these experiments were ones that are reported in animal experiments in the literature. Although the ratio of doses (PLZ/TCP = 1.5) was chosen because this is the approximate ratio of doses used clinically, the absolute doses are much higher, on a mg/kg basis, than those employed in humans. To be more meaningful, these experiments should also be performed at a dose similar to that used clinically; such studies are now underway in our laboratory.

ACKNOWLEDGEMENTS

The authors gratefully acknowledge funding from the Medical Research Council of Canada, the Alberta Heritage Foundation for Medical Research, and the Alberta Mental Health Research Fund.

REFERENCES

Axelrod J. and Saavedra J. M. (1974) Octopamine, phenylethanolamine, phenylethylamine and tryptamine in the brain, in Aromatic Amino Acids in the Brain, Ciba Foundation Symp. 22 (new series), pp. 51-59. Elsevier, Amsterdam.

Baker G. B., Coutts R. T., and LeGatt D. F. (1983) Levels of p-tyramine in rat brain after chronic administration of MAO-inhibiting antidepressants. Progr. Neuro-Psychopharmacol. Biol. Psychiat. (in press)

Baker G. B., Coutts R. T., and Martin I. L. (1981) Measurement of amines in the central nervous system using gas chromatography with electron-capture detection. Progr. Neurobiol. **17**, 1-21.

Baker G. B., Le Gatt D. F., and Coutts R. T. (1982) Effects of acute and chronic administration of phenelzine on 2-phenylethylamine levels in rat brain. Proc. West Pharmacol. Soc. **25**, 417-420.

Boulton A. A. (1976) Cerebral aryl alkyl aminergic mechanisms, in Trace Amines and the Brain (Usdin E. and Sandler M., eds), pp. 21-40. Marcel Dekker, New York.

Boulton A. A. and Milward L. (1971) Separation, detection and quantitative analysis of urinary β-phenylethylamine. J. Chromatogr. **57**, 287-296.

Boulton A. A., Philips S. R., and Durden D. A. (1973) The analysis of certain amines in tissues and body fluids as their dansyl derivatives. J. Chromatogr. **82**, 137-142.

Dewhurst W. G. (1968) New theory of cererbral function and its clinical application. Nature **218**, 1130.

Dyck L. E., Durden D. A., and Boulton A. A. (1981). 1,1,2,2-Tetradeutero-2-phenylethylhydrazine (D_4PEH) is a more potent drug than 2-phenylethylhydrazine (PEH). Proc. 8th Meet. Int. Soc. Neurochem., pp. 249.

Grahame-Smith D. G. (1971) Studies in vivo on the relationship between brain tryptophan, brain 5-HT synthesis, and hyperactivity in rats treated with a monoamine oxidase inhibitor and L-tryptophan. J. Neurochem. **18**, 1053.

McKim H. R., Calverley D. G., Philips S. R., Baker G. B. and Dewhurst W. G. (1980) The effects of tranylcypromine on the levels of some cerebral amines in rat diencephalon, in Recent Advances in Canadian Neuropsychopharmacology (Grof P. and Saxena B., eds), pp. 7-13. Karger, Basel.

Philips S. R., Baker G. B., and McKim H. R. (1980) Effects of tranylcypromine on the concentrations of some trace amines in the diencephalon and hippocampus of the rat. Experientia **36**, 241-242.

Philips S. R. and Boulton A. A. (1979) The effect of monoamine oxidase inhibitors on some arylalkylamines in rat striatum. J. Neurochem. **33**, 259-267.

Rafaelson O. J. (1976), in Monoamine Oxidase and its Inhibition, Ciba Foundation Symp., **39** (new series), pp. 294. Elsevier, Amsterdam.

Robinson D. S., Campbell I. C., Walker M., Statham N. H., Lovenberg W., and Murphy D. L. (1979) Effects of chronic monoamine oxidase inhibitor treatment on biogenic amine metabolism in rat brain. Neuropharmacology **18**, 771-776.

Sabelli H. C. and Mosnaim A. D. (1974) Phenylethylamine hypothesis of affective behavior. Am. J. Psychiat. **131**, 695-699.

Sandler M., Ruthven C. R. J., Goodwin B. L., and Reynolds G. P. (1979) Deficient production of tyramine and octopamine in depression. Nature **278**, 357.

Tabakoff B., Moses F., Philips S. R., and Boulton A. A. (1977). Effects of tranylcypromine and pargyline on brain tryptamine. Experientia **33**, 380-381.

Wurtman R. J. and Axelrod J. (1963) A sensitive and specific assay for the estimation of monoamine oxidase. Biochem. Pharmac. **12**, 1439-1440.

PHENYLETHYLAMINE DEAMINATION IN THE NORADRENERGIC NEUROTRANSMITTER SYSTEM

Nancy A. Garrick and Dennis L. Murphy

Clinical Neuropharmacology Branch
National Institute of Mental Health
Bethesda, Maryland 20205 U.S.A.

Phenylethylamine's locomotor stimulating properties have generally been attributed to indirect, amine-releasing effects dependent upon functioning brain noradrenergic and dopaminergic systems, as well as to contributions from direct effects on dopaminergic and serotonergic receptors, and possibly specific brain receptors for phenylethylamine itself (ANTELMAN *et al.*, 1977; HAUGER *et al.*, 1982; JACKSON, 1978; SLOVITER *et al.*, 1980).

In the classic studies of amine release, phenylethylamine administered to rats in doses (100 mg/kg) which produce amphetamine-like stimulation were found to be accompanied by a marked depletion of norepinephrine in whole brain and heart, while brain dopamine and serotonin concentrations were unchanged (JONSSON *et al.*, 1966). A series of subsequent studies of phenylethylamine administration in rats, mice and guinea pigs similarly demonstrated consistent reductions of vesicular norepinephrine in brain, heart and other peripheral tissues using both biochemical and fluorescence histochemical techniques (see review by Jackson, 1978). In contrast, brain dopamine concentrations were reported to be increased following single phenylethylamine injections but decreased following multiple doses (FUXE *et al.*, 1967; JACKSON, 1971) or decreased in specific brain areas such as the caudate but not in other areas (JACKSON & SMYTHE, 1973). Serotonin concentrations in whole brain were unaltered in all of these studies (JACKSON, 1978).

Both pharmacologically administered and endogenously synthesized phenylethylamine is inactivated principally via oxidative deamintion. Rapid metabolism of phenylethylamine by monoamine oxidase (MAO) is reflected in phenylethylamine's much shorter half-life (0.4 min) in rodent brain compared to that for norepinephrine (205 min) (DURDEN & PHILIPS, 1980). Treatment with MAO inhibitors leads to elevations in brain phenylethylamine in rodents and increased urinary excretion of phenylethylamine in patients receiving MAO inhibiting antidepressants such as pargyline (PHILIPS & BOULTON, 1979; MURPHY *et al.*, 1981).

On the basis of increasing knowledge regarding substrate-selective subtypes of MAO (MURPHY, GARRICK & COHEN, 1983), we wondered whether some of the differential releasing effects of phenylethylamine on norepinephrine in the brain and peripheral noradrenergic neurons compared to dopamine or serotonin might be related to recently observed differences in the localization of the two MAO subtypes in specific neuronal systems in rodents and primates (GARRICK & MURPHY, 1980; GARRICK & MURPHY, 1982).

Phenylethylamine was shown several years ago to be preferentially deaminated by MAO type B (Yang & Neff, 1973). However, almost all studies of phenylethylamine deamination have been done in tissues containing both MAO-B and MAO-A. In the present study, we examined some tissues containing exclusively or predominantly MAO-B or MAO-A, and compared the deamination of phenylethylamine, norepinephrine and in some cases dopamine and serotonin in these tissues with other tissues containing both MAO-A and MAO-B, including whole brain and cortex from rodents and primates, to evaluate whether MAO subtype localization might be related to phenylethylamine's reported neurotransmitter-selective and brain region-selective amine releasing actions.

RESULTS

In the tissues listed in Table 1, the deamination of ^{14}C-phenylethylamine (20 μM) was compared to that of ^{14}C-ℓ-norepinephrine (1.7 mM) using previously reported techniques for tissue preparation and the assay of MAO activity (GARRICK & MURPHY, 1982). In the other studies, ^{14}C-dopamine (0.5 mM) and ^{14}C-serotonin (1.0 mM) were used (GARRICK & MURPHY, 1980). Tissue homogenates were assayed at pH 7.2 using saturating monoamine concentrations

Table 1. Phenylethylamine and Norepinephrine Deamination by Monoamine Oxidase in Rodent and Primate Tissues

Tissue	Ratio of Phenylethylamine to Norepinephrine Deamination	MAO Specific Activity (nmol/mg protein/h)	
		Phenylethylamine	Norepinephrine
RODENT:			
MAO-A containing noradrenergic tissues:			
Mouse neuroblastoma N1E-115 cells	0.03	0.4	11
Rat superior cervical ganglia	0.24	29	121
Rat spleen	0.36	4	11
Mixed neuronal tissues:			
Mouse whole brain	1.50	21	14
Rat whole brain	0.70	30	43
Rat cortex	1.18	33	28
PRIMATE:			
MAO-A containing tissue:			
Human placenta	0.09	33	373
Serotonergic tissue:			
Human platelet	1.50	12	8
Mixed neuronal tissues:			
Human cortex	2.57	36	14
Vervet cortex	2.05	43	21
Rhesus cortex	3.04	70	23

approximately five times the previously determined K_m's for each substrate. Deaminated products were obtained using cation exchange column chromatography.

When phenylethylamine deamination was expressed relative to norepinephrine deamination in these tissues, an association between tissue type and the phenylethylamine/norepinephrine deamination ratio could be discerned, as indicated in Table 1. Most striking was that phenylethylamine deamination was extremely low in proportion to norepinephrine deamination in noradrenergic tissues (e.g. mouse N1E-115 neuroblastoma cells, rat spleen and rat superior cervical ganglia) compared to mouse brain, rat cortex and whole rat brain (Table 1). The N1E-115 neuroblastoma cell line grown in tissue culture consists of norepinephrine-synthesizing cells with axons, and has previously been found to contain essentially exclusively MAO-A activity (DONNELLY et al., 1976).

In primate tissues, the MAO-A rich human placenta also showed a very low capacity to deaminate phenylethylamine compared to human cortex, vervet monkey cortex or rhesus monkey cortex. One serotonin-storing cell studied, the human platelet, had a phenylethylamine/norepinephrine deamination ratio of 1.50 and a phenylethylamine/serotonin ratio of 3.08.

In contrast, when phenylethylamine/dopamine deamination ratios were examined in the dopamine rich caudate nucleus compared to the cerebral cortex from rat, human and nonhuman primate brain (Table 2), only small differences in these ratios were observed, indicating that phenylethylamine was inactivated as readily as dopamine in all of these areas.

INTERPRETATIONS AND CONCLUSIONS

Our data indicate that the capacity of some representative noradrenergic tissues to deaminate phenylethylamine relative to norepinephrine is low in comparison to tissues containing many different neuronal elements such as whole brain or cerebral cortex, or to one tissue specialized for the uptake and storage of serotonin, the human platelet. These data indicate that the noradrenergic neurotransmitter system may have a limited capacity to inactive phenylethylamine, and along with some other tissues like human placenta, may be especially vulnerable to the amine releasing or synaptic neuromodulating effects of phenylethylamine.

Table 2. Phenylethylamine and Dopamine Deamination by Monoamine Oxidase in Rodent and Primate Tissues

Tissue	Ratio of Phenylethylamine to Dopamine Deamination	MAO Specific Activity (nmol/mg protein/h	
		Phenylethylamine	Dopamine
RODENT			
Caudate	1.03	34	33
Cortex	0.67	32	48
HUMAN			
Caudate	0.93	53	57
Cortex	0.74	36	49
VERVET MONKEY			
Caudate	2.35	80	34
Cortex	1.13	43	38

Although comparisons of the amine releasing effects of phenylethylamine in all of these tissues have not yet been done, our data are consistent with several studies reviewed above demonstrating that phenylethylamine administration to rodents leads to large reductions in norepinephrine in brain, heart and other peripheral tissues, while dopamine concentrations are usually unchanged or are reduced in some regions but not others, and brain serotonin concentrations remain unaltered. We hypothesize that phenylethylamine may be less rapidly inactivated in noradrenergic neurons which contain mostly MAO-A and little, if any, MAO-B; hence, phenylethylamine may accumulate intracellularly in higher concentrations and for longer durations and produce proportionately greater norepinephrine release, or act during a more prolonged time period to modulate the synaptic effects of norepinephrine.

While the higher concentrations of phenylethylamine and its major metabolite phenylacetic acid found endogenously in the hypothalamus (DURDEN & BOULTON, 1982) might be contributed to by relatively lower phenylethylamine deaminating capacity in noradrenergic neurons in this brain area, such a mechanism cannot explain the even higher endogenous phenylethylamine and phenylacetic acid concentrations found in the corpus striatum compared to the rest of brain, as slightly higher rather than lower phenylethylamine/dopamine deamination ratios were found in the caudate nucleus from both rodents and primates. Other data, for example the evidence that phenylethylamine concentrations are reduced following destruction of catecholamine neurons by 6-hydroxydopamine treatment, suggest that the higher density of catecholaminergic neurons (and the related availability of dopa decarboxylase) may be responsible for a greater rate of production of phenylethylamine in this area, and that production rates rather than MAO-related degradation rates may govern phenylethylamine steady-state concentrations in some brain areas.

As human and nonhuman primate brain contain more MAO-B than rodent brain, further explorations of primate-rodent species differences in phenylethylamine deamination in brain regions and in peripheral tissues may be of importance in understanding possible functional roles for phenylethylamine in the sympathetic nervous system, in mammalian brain function, and possibly in human psychobiology.

REFERENCES

ANTELMAN S.M., EDWARDS D.J. & LIN M. (1977) Phenylethylamine: evidence for a direct, postsynaptic dopamine-receptor stimulating action. Brain Res. 127, 317-322.

DONNELLY C.H., RICHELSON E. & MURPHY D.L. (1976) Properties of monoamine oxidase in mouse neuroblastoma N1E-115 cells. Biochem. Pharmacol. 25, 1639-1643.

DURDEN D.A. & BOULTON A.A. (1982) Identification and distribution of phenylacetic acid in the brain of the rat. J. Neurochem. 38, 1532-1536.

DURDEN D.A. & PHILIPS S.R. (1980) Kinetic measurements of the turnover rates of phenylethylamine and tryptamine in vivo in the rat brain. J. Neurochem. 34, 1725-1732.

FUXE K., GROBECKER, H. & JONSSON J. (1967) The effect of β-phenylethylamine on central and peripheral monoamine-containing neurons. Eur. J. Pharmacol. 2, 202-207.

GARRICK N.A. & MURPHY D.L. (1982) Monoamine oxidase type A: Differences in selectivity towards ℓ-norepinephrine compared to serotonin. Biochem. Pharmacol. 31, 4061-4066.

GARRICK N.A. & MURPHY D.L. (1980) Species differences in the deamination of dopamine and other substrates for monoamine oxidase in brain. Psychopharmacology 72, 27-33.

HAUGER R.L., SKOLNICK P.H. & PAUL S.M. (1982) Specific [^{3}H] β-phenylethylamine binding sites in rat brain. Eur. J. Pharmacol. 83, 147-148.

JACKSON D.M. (1971) The effect of β-phenylethylamine on noradrenaline concentrations in guinea-pig brain. J. Pharm. Pharmacol. 23, 623-624.

JACKSON D.M. & SMYTHE D.B. (1973) The distribution of β-phenylethylamine in discrete regions of the rat brain and its effect on brain noradrenaline, dopamine and 5-hydroxytryptamine levels. Neuropharmacology 12, 663-668.

JACKSON D.M. (1978) β-Phenylethylamine: Studies on the mechanism of its stimulant effects, in Noncatecholic Phenylethylamines (Mosnaim A.D. & Wolf M.E., eds.) pp. 289-313. Marcel Dekker, Inc., New York.

JONSSON J., GROBECKER H. & HOLTZ P. (1966) Effect of β-phenylethylamine on content and subcellular distribution of norepinephrine in rat heart and brain. Life Sci. 5, 2235-2246.

MURPHY D.L., Pickar D., Jimerson D., Cohen R.M., Garrick N.A., Karoum F. & Wyatt R.J. (1981) Biochemical indices of the effects of selective MAO inhibitors (clorgyline, pargyline and deprenyl) in man, in Clinical Pharmacology in Psychiatry (Usdin E., Dahl S., Gram L.F. & Lingjaerde O., eds.) pp. 307-316. MacMillan Press, London.

MURPHY D.L., GARRICK N.A. & COHEN R.M. (1983) Monoamine oxidase inhibitors and monoamine oxidase: Biochemical and physiological aspects relevant to human psychopharmacology (Burrows J.D., Norman T.R. & Davies E., eds.) pp. 209-227. Elsevier/North Holland Biomedical Press, Amsterdam.

PHILIPS S.R. & BOULTON A.A. (1979) The effect of monoamine oxidase inhibitors on some arylalkylammes in rat striatum. J. Neurochem. 33, 159-167.

SLOVITER R.S., CONNOR J.D. & DRUST E.G. (1980) Serotonergic properties of β-phenylethylamine in rats. Neuropharmacology 19, 1071-1074.

YANG H.-Y.T. & Neff N.H. (1973) β-Phenylethylamine: A specific substrate for type B monoamine oxidase of brain. J. Pharmac. exp. Ther. 187, 365-371.

EFFECT OF LIGNOCAINE ON TYRAMINE AND SEROTONIN OXIDATION IN BRAIN

S. J. Haque and M. K. Poddar

Department of Biochemistry
University of Calcutta
Calcutta-700019, India

Lignocaine (2-[diethylamino]-N-[2-6-dimethylphenyl] acetamide), a lipophilic compound, is widely used as local anesthetic as well as antiarrhythmic drug. Local anesthetics are known to alter various physico-chemical properties of biological membranes (Roth and Seeman, 1971; Lee, 1976; Bojta and Antoni, 1979; Haque et al., 1983). A balanced coordination between the activities of sympathetic and parasympathetic neurons in relation to the functional involvement of various neurotransmitters is believed to play an important role in the regulation of the cardiovascular functions. One of the important mitochondrial membrane-bound enzymes, monoamine oxidase (MAO), is primarily responsible for the oxidative deamination of monoamine transmitters in brain and other organs of man and animals. MAO, as based on its substrate specificity and inhibitor sensitivity, is assumed to exist in two distinct forms, MAO-A and MAO-B (Johnston, 1968; Achee et al., 1977; Murphy, 1978). Several studies have shown that changes in the membrane lipid microenvironment modulate MAO activity (Houslay and Tipton, 1973; Yu, 1979). Although it has been reported that lignocaine inhibits mitochondrial MAO (Haque et al., 1983; Fowler et al., 1980), possibly by increasing the lipid fluidity of mitochondrial membrane (Haque et al., 1983), nothing is known about lignocaine's effect on the oxidation of monoamines in relation to the changes in the kinetic properties of brain MAO. In the present study, the effect of lignocaine on tyramine (TA) and serotonin (5-HT) oxidation in rat brain is reported under *in vivo* and *in vitro* conditions.

EXPERIMENTAL PROCEDURES

Treatment with Lignocaine: (a) *In vivo* - A single dosage (100-150 mg/kg) of lignocaine was administered (i.p.) to adult male albino rats (Charles Foster strain; 125-175 g) and animals were sacrifieced by cervical dislocation at different time intervals (10, 30, 60, 120 and 240 min.). For long term treatment, rats were injected (i.p.) with lignocaine (50 mg/kg/day) for 15 and 30 consecutive days and were sacrificed at 1 hr after the last injection. Rats of control groups corresponding to acute and long term treatments received an equal volume of vehicle (0.9% saline) through the same route and were sacrificed by the same way as described above. (b) *In vitro* - Rat brain mitochondria were preincubated with varying concentrations (2-25 mM) of lignocaine at 37° C for 20 min., unless otherwise stated, prior to the addition of amine substrate in the incubation medium.

Preparation of Brain MAO-A and MAO-B: Rat brain MAO of either type by *in vivo* treatment with specific inhibitors was prepared according to Mitra and Guha (1980). Rats were injected (i.p.) with either clorgyline (2 mg/kg/day for 1 day) or pargyline (2 mg/kg/day for 3 consecutive days) to inhibit selectively MAO-A or MAO-B type and the corresponding control animals were injected with the same volume of normal saline. The animals were sacrificed by decapitation 24 hr after the last injection.

Preparation of Brain Mitochondria: Rat brain mitochondria were prepared according to the sucrose density gradient method of Gray and Whittaker (1962), as modified by Bradford et al. (1973).

Assay of MAO Activity: MAO activity, using TA or 5-HT as substrate, was determined by the method of Green and Haughton (1961). All incubations were carried out at 37° C using air as gas phase. The protein content of enzyme preparations was measured by the method of Lowry et al. (1951). K_m and V_{max} of MAO were determined by Lineweaver-Burk plots using varying substrate concentrations (0.1-1.0 mM).

Statistical Analysis: The statistical significance was tested by two-tailed student *t*-test.

RESULTS AND DISCUSSION

It appears from Fig. 1 that lignocaine (2-20 mM) produces a concentration-dependent inhibition (*in vitro*) of TA and 5-HT oxidation in rat brain mitochondria. But at higher concentrations (> 20 mM) the inhibition does not progress further. The degrees of inhibition of 5-HT oxidation, irrespective of lignocaine concentrations, are greater than that of TA oxidation. The IC_{50} values of lignocaine are found to be 2.6 mM and 5.0 mM for 5-HT and TA oxidation respectively. It is known that 5-HT and NA are chiefly oxidized by MAO-A and PE and benzylamine by MAO-B, which are sensitive to lower concentrations of clorgyline (10^{-11}-10^{-8} M) and deprenyl (10^{-9}-10^{-6} M) respectively; TA and DA are common substrates (Achee et al., 1977; Murphy, 1978). Thus it may be suggested that lignocaine preferentially inhibits MAO-A activity in rat brain.

Fowler et al. (1980) have also shown that lignocaine inhibits MAO activity in rat liver. This may further be supported by the finding (Fig. 2) that lignocaine produces significant and concentration-dependent *in vitro* inhibition of 5-HT (30-79%) and TA (24-70%) oxidation in rat brain MAO-A preparation. No significant effect on 5-HT or TA oxidation is noted in MAO-B preparation under

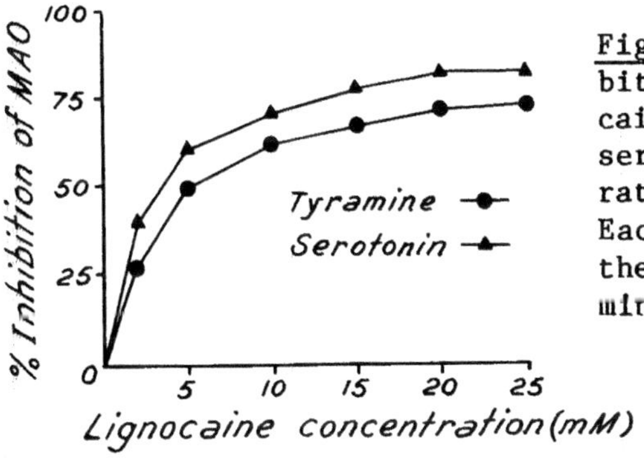

Fig. 1 *In vitro* inhibitory effect of lignocaine on tyramine and serotonin oxidation in rat brain mitochondria. Each point represents the mean of four determinations.

similar conditions of lignocaine treatment. The purity of the selective MAO-type preparation of rat brain mitochondria is checked by the in vitro clorgyline dose-response curve of TA oxidation (Fig. 3). It is found that TA oxidation is completely inhibited by 10^{-8} M clorgyline in the MAO-A preparation, while in the MAO-B preparation TA oxidation is quite insensitive to the same concentration of clorgyline. Complete inhibition of TA oxidation in the MAO-B preparation is achieved at 10^{-4} M clorgyline, which is a characteristic of rat brain MAO-B (Johnston, 1968; Mitra and Guha, 1980). The reversibility of lignocaine-induced in vitro inhibition of TA and 5-HT oxidation in rat brain mitochondria is tested by Ackermann-Potter plots (Ackermann and Potter, 1949), and repeated washing experiments (Haque and Poddar, 1983).

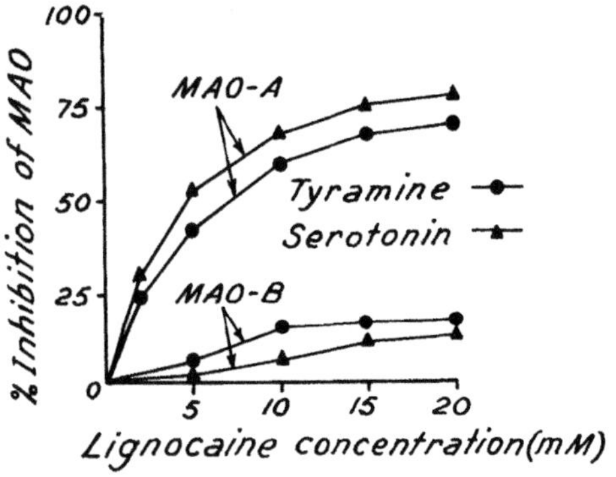

Fig. 2 In vitro inhibitory effect of lignocaine on tyramine and serotonin oxidation in rat brain MAO-A and MAO-B preparations. Each point represents the mean of four separate determinations.

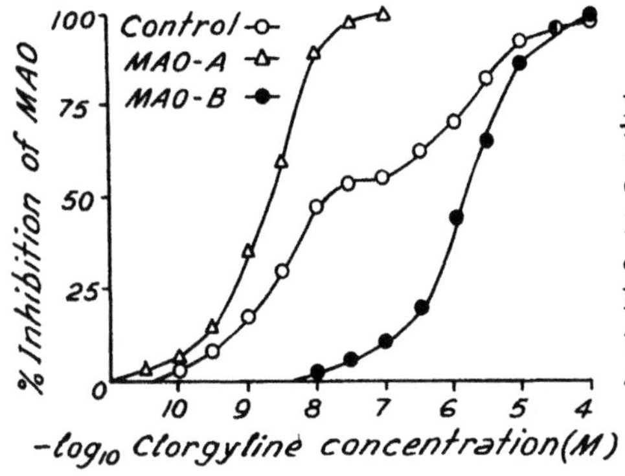

Fig. 3 In vitro inhibition of tyramine oxidation by clorgyline in rat brain mitochondrial preparation. Each point represents the mean of five separate determinations.

TABLE 1. In vitro effects of lignocaine on K_m and V_{max} of rat brain mitochondrial MAO

Lignocaine (mM)	Tyramine K_m (mM)	Tyramine V_{max} (O.D/hr/ mg Pr.)	Serotonin K_m (mM)	Serotonin V_{max} (O.D/hr/ mg Pr.)
0.0	151.0± 4.3	0.53±0.02	124.0±3.7	0.32±0.011
2.0	225.0± 5.8*	0.33±0.01*	132.0±2.9	0.20±0.009*
5.0	317.0± 6.9*	0.27±0.01*	227.0±6.4*	0.15±0.004*
10.0	427.0±11.8*	0.22±0.01*	308.0±7.7*	0.13±0.005*

Each value represents the mean ± SEM of four separate determinations.
Significantly different from control, *P < 0.001.

The results of Lineweaver-Burk plots (Table 1) show that lignocaine (2-10 mM) produces a concentrations-dependent increase in the K_m of MAO for TA (49-182%) and 5-HT (6-148%) in rat brain mitochondria. The V_{max} values of TA and 5-HT oxidation are decreased by 38-58% and 35-58% respectively with the increase of drug concentration (2-10 mM). The lignocaine-induced decrease in substrate affinity (K_m^{-1}) and catalytic property (V_{max}) of MAO towards exogenous monoamine substrates reveals that lignocaine produces a mixed type inhibition of TA and 5-HT oxidation in rat brain mitochondria.

Single dosage of lignocaine in vivo (100-150 mg/kg i.p.) produces no significant change in TA or 5-HT oxidation in rat brain mitochondria (results not shown).

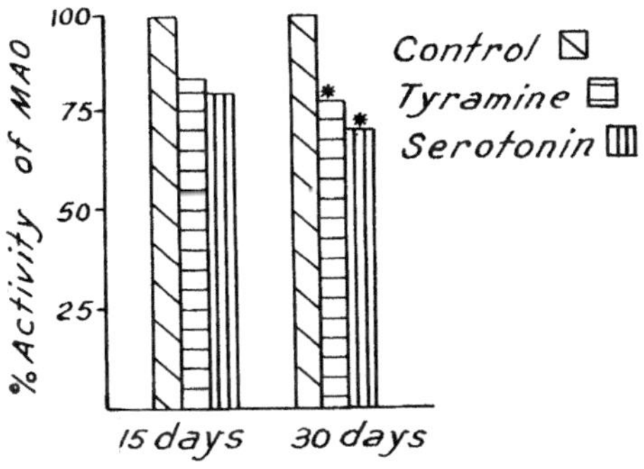

Fig. 4 In vivo effect of long term lignocaine administration (50 mg/kg/day, i.p.) on tyramine and serotonin oxidation in rat brain. Each bar represents the mean of 4 separate determinations.
*P < 0.05

Long term treatment with lignocaine (50 mg/kg/day) for 15 and 30 consecutive days, on the other hand, inhibits appreciably the oxidation of TA and 5-HT (Fig. 4). The lack of effect of lignocaine on TA and 5-HT oxidation in rat brain mitochondria under acute conditions may be explained by the fact that lignocaine has high hepatic extraction and metabolic rate (Opie, 1980; Burke, 1981). The slight but appreciable inhibitory effect of lignocaine depends upon the period of treatment; under chronic conditions, this may possibly be due to its slower metabolism (unpublished observation, Haque and Poddar, 1983).

Recently we have found that lignocaine increases the lipid fluidity of mitochondrial membrane (Haque et al., 1983). Thus the present results, the lignocaine-induced reversible inhibition of MAO activity and the changes of its kinetic properties, suggest that in vitro inhibition of TA and 5-HT oxidation catalysed by mitochondrial MAO is associated with the modulation of membrane microenvironment.

Acknowledgements: This work was supported by the Indian Council of Medical Research, New Delhi, India.

REFERENCES

Achee F. M., Gabay S. and Tipton K. F. (1977) Some aspects of monoamine oxidase in brain. Prog. Neurobiol. **8**, 325-348.

Ackermann W. W. and Potter V. R. (1949) Enzyme inhibition in relation to chemotherapy. Proc. Soc. Exp. Biol. Med. **72**, 1-9.

Bojta J. and Antoni F. (1974) Effect of lignocaine on the membrane of human lymphocytes. Biochem. Pharmacol. **28**, 2441-2444.

Bradford H. F., Bennett G. W. and Thomas A. J. (1973) Depolarising stimuli and the release of physiologically active amino acids from suspensions of mammalian synaptosomes. J. Neurochem. **21**, 495-505.

Burke M. D. (1981) Cytochrome P-450: a pharmacological necessity or a biochemical curiosity? Biochem. Pharmacol. **30**, 181-187.

Fowler C. J., Callingham B. A., Mantle T. J. and Tipton K. F. (1980) The effect of lipophilic compounds upon the activity of rat liver mitochondrial MAO-A and MAO-B. Biochem. Pharmacol. **29**, 1177-1183.

Gray E. G. and Whittaker V. P. (1962) Isolation of nerve endings from brain: an electron microscopic study of cell fragments derived by homogenization and centrifugation. J. Anat. **96**, 79-88.

Green A. L. and Haughton T. M. (1961) A colorometric method for the estimation of monoamine oxidase. Biochem. J. **78**, 172-175.

Haque S. J. and Poddar M. K. (1983) Effect of lignocaine on monoamine oxidase activity of brain and liver. Method. Find. Exp. Clin. Pharmacol. (in press)

Haque S. J., Roychoudhury B. and Poddar M. K. (1983) Effect of lignocaine on MAO activity and membrane fluidity of rat brain and liver mitochondria. Ind. J. Biochem. Biophys. (in press)

Houslay M. D. and Tipton K. F. (1973) The nature of electrophoretically separable multiple forms of rat liver MAO. Biochem. J. **135**, 173-186.

Johnston J. P. (1968) Some observations upon a new inhibitor of MAO in brain tissue. Biochem. Pharmacol. **17**, 1285-1297.

Lee A. G. (1976) Interaction between anesthetics and lipid mixture amines. Biochem. Biophys. Acta **448**, 34-44.

Lowry O. H., Rosebrough N. J., Farr A. L. and Randall R. J. (1951) Protein measurement with the folin phenol reagent. J. Biol. Chem. **193**, 265-275.

Mitra C. and Guha R. (1980) Serotonin oxidation by type B MAO of rat brain. Biochem. Pharmacol. **29**, 1213-1216.

Murphy D. L. (1978) Substrate selective MAO-inhibition: tissue, species and functional differences. Biochem. Pharmacol. **27**, 1889-1893.

Opie L. H. (1980) Drugs and the heart: IV. Antiarrhythmic agents. Lancet **1**, 861-868.

Roth S. and Seeman P. (1971) All lipid soluble anesthetics protect red cell. Nature, New Biol. **231**, 284-285.

Yu P. H. (1979) Effect of lipid depletion of type A and type B MAO from rat heart and bovine liver. In: Monoamine Oxidase: Structure, Function and Altered Function (T. P. Singer, R. W. Vonkorff and D. L. Murphy, eds), pp. 233-244. Academic Press, New York.

DEGRADATION KINETICS BY MAO OF PEA DERIVATIVES. A MODEL FOR THE MOLECULAR BASIS OF THEIR ANALGESIC AND BEHAVIORAL EFFECTS?

A.D.Mosnaim, M.E. Wolf, & E.A. Zeller

Chicago Med.School & Shriners Hospital

N. Chicago, Illinois 60064

INTRODUCTION

The classical endogenous noncatecholic phenylethylamines, phenylethylamine (PEA), phenylethanolamine, *p*-tyramine and *p*-octopamine appear to be involved in peripheral and central nervous system synaptic transmission mechanisms (1,2). The physiological and pharmacological actions of PEA appear of particular importance as alterations in the metabolism of this amine has been postulated to contribute to the pathophysiology of a number of neuropsychiatric disorders (3-7), migraine (8,9) and diabetes (10). In order to better understand the molecular basis of PEA actions we have examined the relationship between the chemical structure, the analgesic activity and kinetic parameters of degradation by monoamine oxidase (MAO) of a series of monosubstituted derivatives of PEA.

MATERIALS AND METHODS

Phenylethylamine and PEA derivatives (as bases) were obtained from Aldrich Chem. Corp. WI and crystallized as HCl salts. Other chemicals were of the highest quality commercially available and used without further purification. Substances were injected in normal saline (i.p., 0.25 ml/mouse). The hot-plate ($55 \pm 0.5^{\circ}C$) method was used to study anti-nociceptive response. Mice (male, Swiss albino 25-30 g each) were conditioned to the hot-plate by 3 exposures; animals with a latency time (jumping time) of over 4 sec. (placebo reactors) were omitted from the study. After drug injections, mice (groups of 10) were tested at several intervals (5', 20', 40', 60', and 180') (Table 1). Table 2 represents a protocol, which was carried out for every amine, to determine the effects of naloxone on the observed analgesic effects of the PEA derivatives studied. Intact mitochondria, isolated from 0.25 M sucrose suspension, was used to test mice and rat liver MAO activity (11,12) (peroxidation of leuco-2',7'-dichlorofluorescin by H_2O_2, produced by MAO action, and quinone measurement at 502 nm). MAO types A and B were evaluated, respectively after mitochondria incubation (30 min. prior to substrate addition) with either deprenyl (5×10^{-7} M) or clorgyline (5×10^{-7} M). Five to 6 substrate concentrations were used to determine K_m- and V_{max}-values.

RESULTS

PEA and its derivatives monomethylated in the ortho-, para-, β-, and N-methyl position produce analgesia in mice as determined by the hot-plate method (13). The intensity and duration of this effect varies for each one of the compounds, appears to be generally dose-dependent and in every case is enhanced by pretreatment with the MAO inhibitor pargyline (Tables 1 and 2). Administration of naloxone did not significantly block the analgesic effects of PEA or PEA derivatives with the exception of N-methyl PEA (for comparison see Table 2; other substances act similarly to PEA). The compounds studied, with the exception of α-Me-PEA (amphetamine), which also failed to produce analgesia, are good substrates for rat and mice liver mitochondrial MAO, preferentially MAO type B, with the exception of N-methyl PEA (MAO-A and MAO-B) (Table 3).

DISCUSSION

PEA and monomethylated PEA derivatives produce analgesia in mice. This effect has a rapid onset of action, is of relatively short duration and its intensity

TABLE 1. [1]ANALGESIA PRODUCED BY MONOMETHYLATED DERIVATIVES OF PHENYLETHYLAMINE (PEA)[2]

	[4]5 min		[4]20 min	
	Saline	[3]P	Saline	[3]P
Saline	2.4±0.4	3.4±0.6	1.9±0.3	2.8±0.4
p-Me PEA (15mg/kg)	34.2±2.3	45.0	28.9±3.4	45.0
p-Me PEA(100mg/kg)	45.0	45.0	26.9±3.4	45.0
PEA (15mg/kg)	5.4±1.3	14.0±1.6	3.1±0.4	6.0±3.2
PEA (100mg/kg)	45.0	45.0	9.1±0.8	45.0
N-Me PEA (15mg/kg)	9.1±1.1	12.5±0.9	4.0±0.4	5.1±0.6
N-Me PEA(100mg/kg)	22.3±2.1	33.1±3.4	5.4±0.8	5.3±0.6
o-Me PEA (15mg/kg)	2.8±0.6	8.3±1.1	5.1±0.8	2.7±0.4
o-Me PEA(100mg/kg)	17.0±1.5	21.9±2.1	7.0±0.6	6.1±0.7
β-Me PEA (15mg/kg)	3.8±0.8	7.3±1.6	2.4±0.5	2.5±0.7
β-Me PEA(100mg/kg)	9.5±1.1	27.3±2.8	3.1±0.8	40.1±1.3
d-α-Me PEA (5mg/kg)	3.7±0.9	—	3.3±0.4	—
d-α-Me PEA(10mg/kg)	6.6±1.1	—	2.4±0.3	—

[1] Results are expressed as the mean of the medians (jumping time in sec.) ± S.E.M.; 10 mice per group at least 4 groups per drug treatment. Cut off time 45 sec.

[2] Drugs were dissolved and injected in saline; i.p., 0.25 ml/mouse

[3] Pargyline (P) (15mg/kg, i.p.) was given 90 min. prior to PEA's injection

[4] Testing time after injection of PEA derivative.

depends on the drug and dose used (Table 1). It should be noted that all the substances studied also produce behavioral effects similar to PEA e.g., increased exploratory activity, stereotypy (1), the onset of action and intensity of which generally paralleled with their analgesic properties; behavioral manifestations however, last

TABLE 2. [1]ANALGESIA PRODUCED BY MONOMETHYLATED DERIVATIVES OF PHENYLETHYLAMINE (PEA)[2]: EFFECTS OF [3]NALOXONE

		[1]ANALGESIA (MICE, HOT-PLATE METHOD)	
		[4]5 min	[4]20 min
Saline		2.4±0.4	1.9±0.3
Pargyline	(15mg/kg)	3.4±0.6	2.8±0.4
Naloxone	(15mg/kg)	1.9±0.5	2.7±0.3
Pargyline+Naloxone		2.8±0.4	3.4±0.7
PEA	(15mg/kg)	5.4±1.3	3.1±0.4
NL+PEA	(15mg/kg)	4.8±0.7	4.3±0.3
P+PEA	(15mg/kg)	14.0±1.6	6.0±0.2
P+NL+PEA	(15mg/kg)	11.3±0.6	8.4±0.4
PEA	(100mg/kg)	45.0	9.1±0.8
NL+PEA	(100mg/kg)	45.0	7.8±0.4
P+PEA	(100mg/kg)	45.0	45.0
P+NL+PEA	(100mg/kg)	45.0	45.0
N-Me PEA	(15mg/kg)	9.1±1.1	4.0±0.4
P+N-Me PEA	(15mg/kg)	12.5±0.9	5.1±0.6
P+NL+N-Me PEA	(15mg/kg)	1.9±0.4	2.1±0.3
N-Me PEA	(100mg/kg)	22.3±2.1	5.4±0.8
NL+N-Me PEA	(100mg/kg)	4.0±0.6	2.1±0.2
P+N-Me PEA	(100mg/kg)	33.1±3.4	5.3±0.6
P+NL+N-Me PEA	(100mg/kg)	6.1±0.8	4.7±0.4

[1]Results are expressed as the mean of the medians (jumping time in sec.) $\pm$ S.E.M.: 10 mice per group, at least 4 groups per drug treatment. Cut off time 45 sec.
[2]Drugs were dissolved and injected in saline; i.p., 0.25 ml/mouse
[3]Pargyline (P) and Naloxone (NL) were injected i.p. 90 or 5 min. prior to PEA or N-MePEA injection, respectively
[4]Testing time after injection of PEA derivative

considerably longer (a more complete report will be published elsewhere). As expected from the enzymatic data, showing their rapid metabolism by MAO, pretreatment with pargyline significantly increased the PEAs' analgesic actions. These results would also indicate that the

TABLE 3. MONOMETHYL DERIVATIVES OF PHENYLETHYLAMINE. DEGRADATION BY MONOAMINE OXIDASE FROM [1]RAT AND [2]MOUSE LIVER MITOCHONDRIA

	[1]V_{max} nmoles/mg protein/min			[2]$\tilde{K}m$ (n molar)		Km (n molar x 10^3)	
Compound	A+B	A	B	A	B	A	B
[1]*p*-Me PEA	5.40	0.42	4.48	28	3	11.4	1.2
[1]PEA	5.23	1.41	4.15	105	21	71.4	3.3
[2]PEA	4.62	0.25	4.04	38	8.6	9.3	2.13
[1]N-Me PEA	4.98	2.82	3.15	69	8	41.7	4.3
[2]N-Me PEA	2.85	0.51	2.49	553	16.1	197	5.74
[1]β-Me PEA	1.25	0.08	1.01	260	100	52.6	20.2
[1]*o*-Me PEA	6.31	1.15	5.23	290	115	15.9	4.6

[1]V_{max} is given for untreated mitochondria (type A+B) and for deprenyl ($5x10^{-7}M$, type A) or clorgyline ($5x10^{-7}M$, type B) treated mitochondria (30 min. incub.).
[2]$\tilde{K}m$ values represent concentration of non-protonated amines.

observed anti-nociceptive effects are produced by the amines and not by the corresponding aldehydes, acids or other products of MAO degradation (11). Our kinetic studies (Table 3) (Km and Vmax) indicate that in general the pharmacologically more active compounds are also relatively better substrates for MAO, suggesting perhaps that in this case substrate-enzyme interactions could be used as a model for better understanding the mechanism of action of these substances at the molecular level (11). One could suggest a similarity between the MAO active site and PEA receptor (14,15), and speculate that the strong actions of *p*-Me PEA would indicate that at the PEA receptor, at the *locus* of the para-position, there is space for a methyl group which enhances hydrophobicity (11). We have also found a similar relationship between the kinetics of degradation by MAO and the behavioral effects elicited by *p*-monohalogenated (F, Cl, Br and I) PEA derivatives; however, due to their relatively high toxicity, analgesic studies with these substances are very difficult to interpret (16). An important exception to these findings was found in d-amphetamine, which produces behavioral effects similar in many respects to those observed after PEA, but which is not a MAO substrate [in fact displays for the rat enzymes a K_i of 1×10^{-6} M(MAO-B) and 4×10^{-8} M(MAO-A)] and fails to produce analgesia in mice.

Similarly to the behavioral studies, several mechanism of action could be suggested to account for the analgesic properties of the PEA derivatives. These would include involvement of monoaminergic (1,2,17-22) and opiate (23) systems. Although the analgesic effects of these compounds (Table 3, including monomethylated and monohalogenated compounds) was not blocked by naloxone pretreatment (with the exception of N-Me PEA), the possible involvement of the opiod system should not be discarded yet as our preliminary results show that the administration of PEA (i.p., 100 mg/kg) causes a small, rapid (<2 min) and short-lasting (<15 min) methionine-enkephalin increase in selected mouse brain areas.

ACKNOWLEDGEMENTS

We thank Dr. K. Arora and Mr. S. Myles for excellent technical assistance. This work was supported in part by the National Migraine Foundation, the State of Illinois Department of Mental Health and Development Disabilities, and Shriners Hospital for Crippled Children.

REFERENCES

1. Mosnaim A. D. and Wolf M. E. (eds.) (1978) Non-catecholic Phenylethylamines, Part 1: Phenylethylamine: Biological Mechanisms and Clinical Aspects. Marcel Dekker, New York.
2. Mosnaim A. D. and Wolf M. E. (eds.) (1980) Non-catecholic Phenylethylamines, Part 2: Phenylethanolamine, Tyramines and Octopamine. Marcel Dekker, New York.

3. Mosnaim A. D., Inwang E. E., Sugerman J. H., DeMartini W., and Sabelli H. C. (1973) Ultraviolet spectrophotometric determination of 2-phenylethylamine in biological samples and its possible correlation with depression. Biol. Psychiat. **6**, 235-257.
4. Sabelli H. C. and Mosnaim A. D. (1974) Phenylethylamine hypothesis of affective behavior. Am. J. Psychiat. **131**, 695-699.
5. Sandler M. and Reynolds G. P. (1976) Does phenylethylamine cause schizophrenia? Lancet **1**, 70-71.
6. Karoum F., Linniola M., Potter W. Z., Chuang L.-W., Goodwin F. K., and Wyatt R. J. (1982) Fluctuating high urinary phenylethylamine excretion rates in some bipolar affective disorder patients. Psychiat. Res. **6**, 215-222.
7. Wolf M. E. and Mosnaim A. D. (1983) Phenylethylamine in neuropsychiatric disorders. Gen. Pharmacol. **14**, 385-390.
8. Diamond S., Dalessio D., Graham J., and Medina J. L. (eds.) (1975) Vasoactive Substances Relevant to Migraine. Charles C. Thomas, New York.
9. Mosnaim A. D., Wine R., Karoum F., Diamond S. and Wolf M. E. (1982) Methionine enkephalin, phenylethylamine, phenylacetic acid, platelet monoamine oxidase, and prolactin in plasma of migraine patients. Clinical Pharmac. and Therap. **31**, 251-252.
10. Mosnaim A. D., Karoum F., Zeller A., Callaghan O. H., Singh S. P., and Wolf M. E. (1982) Platelet monoamine oxidase activity and plasma levels of noncatecholic phenylethylamines in insulin-dependent diabetic subjects. Clin. Chim. Acta **126**, 237-242.
11. Zeller E. A., Mosnaim A. D., Borison R. L., and Huprikar S. V. (1976) Phenylethylamine: Studies on the mechanism of its physiological action, in Advances in Biochemical Psychopharmacology, Vol. 15: First and Second Messengers - New Vistas, pp. 75-86. Raven Press, New York.
12. Kochli H. and von Wartburg J. P. (1978) A sensitive photometric assay for monoamine oxidase. Anal. Biochem. **84**, 127-135.
13. Ehrenpreis S., Balagot R. C., Comaty J. E., and Myles S. B. (1979) Advances in Pain Research and Therapy, Vol. 3, pp. 479-488. Raven Press, New York.

14. Ungar F., Mosnaim A. D., Ungar B., and Wolf M. E. (1978) Preliminary studies of the sodium borohydride stabilizable binding of phenylethylamine and tyramine to brain preparation. Res. Commun. Chem. Pathol. Pharmacol. **19**, 427-434.
15. Hansen T. R., Greenberg J., and Mosnaim A. D. (1980) Direct effect of phenylethylamine upon isolated rat aortic strip. Eur. J. Pharmacol. **63**, 95-101.
16. Mosnaim A. D., Vazquez A., and Nair V. (1982) First World Congress on Toxicology and Environmental Health, Washington, D. C. (Abst.).
17. Jackson D. M. and Smythe D. B. (1973) The distribution of β-phenylethylamine in discrete regions of the rat brain and its effect on brain noradrenaline, dopamine and 5-hydroxytryptamine levels. Neuropharmacology **12**, 663-668.
18. Sabelli H. C., Vazquez A. J., Mosnaim A. D., and Madrid-Pedemonte L. (1974) 2-Phenylethylamine as a possible mediator for Δ^9-tetrahydrocannabinol-induced stimulation. Nature **248**, 144-145.
19. Borison R. L., Havdala H. S., and Diamond B. I. (1977) Chronic phenylethylamine-induced stereotypy in rats: A new animal model for schizophrenia? Life Sci. **21**, 117-122.
20. Dourish C. T. (1982) A pharmacological analysis of the hyperactivity syndrome induced by β-phenylethylamine in the mouse. Br. J. Pharmac. **77**, 129-139.
21. Sloviter R. S., Connor J. D., and Drust E. G. (1980) Serotonergic properties of β-phenylethylamine in rats. Neuropharmacology **19**, 1071-1074.
22. Goudie A. J. and Buckland C. (1982) Serotonin receptor blockade potentiates the behavioral effects of β-phenylethylamine. Neuropharmacology **21**, 1267-1272.
23. Garzon J., Moratalla R., and Del Rio J. (1980) Potentiation of the analgesia induced in rats by morphine or [D-Ala^2]-Metenkephalinamide after inhibition of brain type B monoamine oxidase: The role of phenylethylamine. Neuropharmacology **19**, 723-729.

THE KINETICS OF HYDROXYLATION OF PHENYLETHYLAMINE, AMPHETAMINE AND PHENYLALANINE IN RODENT TISSUES

O. Callaghan, A. Mosnaim, J. Chevesich, and M. Wolf

The Chicago Medical School
N. Chicago, Illinois

INTRODUCTION

Phenylethylamine (PEA) and amphetamine (α-MePEA) are closely related structurally and pharmacologically. Both amines elicit a number of similar clinical and behavioral (animal) responses, appear to use similar transport systems and to act via some common neurochemical mechanisms (1, 2). These observations have led several investigators to suggest that at least some of the α-MePEA actions are mediated via PEA, an endogenous amine (3) which has been conceptualized as a naturally occurring α-MePEA (4). We now report for the first time that both amines are p-hydroxylated, in vitro, by a common rat liver microsomal enzyme. It seems likely that PEA and α-MePEA might compete for the same catalytic site.

MATERIALS AND METHODS

The purity of radioactive substrates and standards [$1-^{14}C$-PEA HCl (NEN), D-($7-^{14}C$)-α-MePEA sulfate (Res. Prod. Int. Corp.), $7-^{14}C$-Tyramine HCl (NEN) and L-$1-^{14}C$-Phenylalanine (PhAl) from ICN Corp.; sp. act. 48.3mCi/mmol, 49mCi/mmol, 57.9mCi/mmol and 40mCi/mmol, respectively] was checked by thin-layer chromatography (TLC) separation followed by radiochromatogram-scanning (over 95% pure) (for details see below and Ref. 5). Other chemicals were of the highest purity commercially available and used without further purification.

Male Sprague-Dawley rats (250-300g) were killed by decapitation and brains and livers were removed separately (pooling of 2 or 3 animals per experiment) and homogenized in ice-cold buffer (100mM potassium phosphate pH 7.4, 10mM $MgCl_2$, 0.01mM $MnCl_2$) so as to produce a 40% (w/v) homogenate of each tissue. After separation of the 10,000xg pellet, the supernatant was centrifuged again (100,000xg, 1 hr at 2°C) and the new supernatant (9-10mg protein/ml) was used for studies of PhAl hydroxylation while the 100,000xg pellet was resuspended in an equal volume of buffer (2-3mg protein/ml) and used as enzyme source for PEA and α-MePEA hydroxylation. Incubation mixtures (final vol. 0.9ml) included buffer (one-half dilution), 10^{-5}M pargyline, labeled substrate (either 0.031mM 1-^{14}C-PEA, 0.046mM D-7-^{14}C- α-MePEA or 0.050mM 1-^{14}C-PhAl), and tissue preparation (0.2ml). According to the protocol some tubes also contained NADPH (10^{-3}M) alone or in combination with 10^{-3}M 6,7 dimethyl-5,6,7,8 tetrahydropterine ($DMPH_4$) or 10^{-5}M p-chlorophenylalanine methyl ester (p-Chloro PhAl).

For determination of kinetic parameters (Km and Vmax), incubations were carried out in the presence of NADPH and a range of substrate concentration of 0.0025 to 0.025mM, 0.011 to 0.11mM, and 0.225 to 4.5mM, for PEA, α-MePEA and PhAl, respectively (Fig. 2 and 3; other data to be published elsewhere). The effect of α-MePEA and of PEA on each other's hydroxylation reaction was studied by adding cold amine to the incubation mixture (range $1x10^{-6}$M - $1.7x10^{-4}$M). Each point represents the mean of three experiments. Tubes were incubated (37°C, 20 min.) and the reaction stopped by ice-cold acetone (0.9ml). After centrifugation (2,500xg, 2 min.), products in the supernatant were separated by TLC (50 µl aliquots) (Fig. 1) and autoradiographed (2 weeks) (for TLC details see Ref. 5). Developed films, and the simultaneous use of appropriate radioactive and cold standards (visualized with ninhydrin), allowed us to locate and identify raction products, which were then scraped off the plate and counted.

RESULTS

Rat liver microsomal fraction converts PEA and α-MePEA, but not PhAl, to their corresponding p-hydroxy-

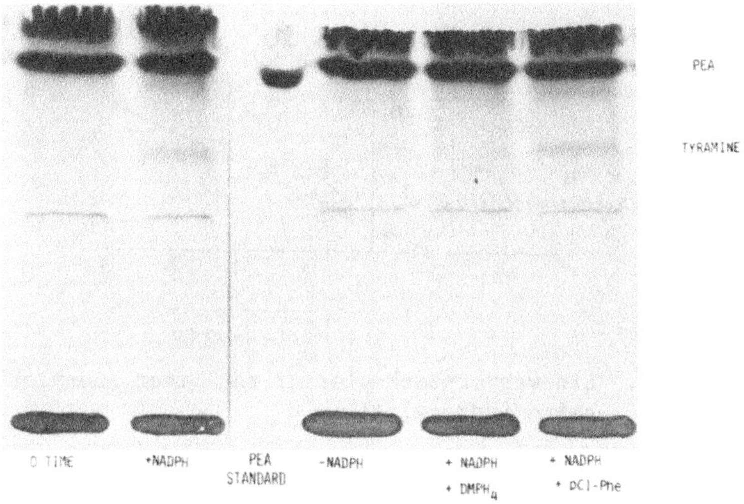

Figure 1.

lated metabolites *p*-tyramine and *p*-hydroxyamphetamine, respectively. Both reactions require NADPH and are inhibited by $DMPH_4$ but not by *p*-chloro PhAl (Fig. 1; other data to be published elsewhere). Hydroxylation of PhAl to tyrosine is accomplished by the 100,000xg supernatant fraction. At the substrate concentration used, a dependence on NADPH could not be demonstrated and the reaction does not appear to be strongly inhibited by $DMPH_4$. Under similar conditions, we failed to detect hydroxylation by rat brain preparation of either amine or L-PhAl.

Figures 2 and 3 show that PEA and α-MePEA competitively inhibit each other's hydroxylation. It should also be noted that the Vmax and Km for PEA are lower than that of α-MePEA and PhAl (Vmax 0.25, 6.25 and 35.7nmoles/mg protein/20 min. and Km 1.25×10^{-5}M, 4.2×10^{-5}M and 1.67×10^{-2}M, respectively). In contrast to these findings, Jonsson (Ref. 6 and 7) reported that the Km for PEA was higher than that of α-MePEA.

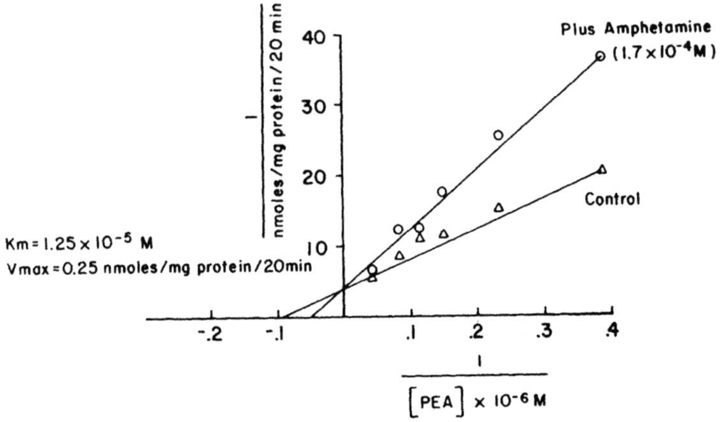

Figure 2. Lineweaver-Burk plot of rat liver phenylethylamine hydroxylase.

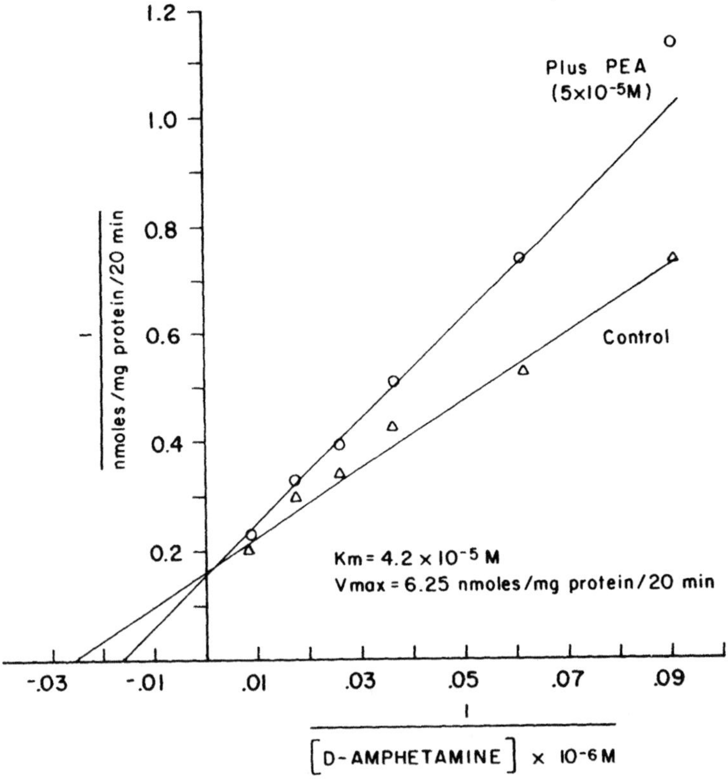

Figure 3. Lineweaver-Burk plot of rat liver D-amphetamine hydroxylase.

DISCUSSION

PEA is an endogenous neuroamine mostly derived from L-phenylalanine decarboxylation. Its conversion to tyramine, a reaction occurring in mammalian brain (*in vivo*) (8) and in peripheral tissues (9), is normally a minor metabolic pathway (10); in fact, we have yet to detect it in our *in vitro* rat brain studies. However, PEA hydroxylation to tyramine could prove of importance in conditions characterized by abnormal dopaminergic function (e.g. Parkinsonism, tardive dyskinesia) as this metabolic reaction is one of the steps of a novel biochemical pathway linking L-phenylalanine to dopamine bypassing the formation of tyrosine and L-dopa (1, 10, 11).

We have now characterized a "PEA-hydroxylating" enzymatic system in the rat liver microsomal fraction (100,00xg pellet) which requires NADPH and is inhibited by $DMPH_4$ but not by *p*-Chloro PhAl. This system also converts α-MePEA to *p*-OH-α-MePEA but does not hydroxylate L-phenylalanine nor L-tyrosine. Competitive inhibition of PEA hydroxylation by α-MePEA is obtained at concentrations (Fig. 2) in the same order of magnitude reached by this substance *in vivo* (rat brain, unpublished observation) after the administration of behaviorally active doses of α-MePEA. These results offer a new mechanism of action for α-MePEA and further support previous suggestions that some of its actions may be mediated by PEA.

ACKNOWLEDGEMENTS

This work was supported, in part, by the National Migraine Foundation and the Department of Mental Health and Developmental Disabilities of the State of Illinois.

REFERENCES

1. Mosnaim A. D. and Wolf M. E. (eds.) (1978) *Noncatecholic Phenylethylamines, Part 1: Phenylethylamine: Biological Mechanisms and Clinical Aspects.* Marcel Dekker, New York.
2. Mason M., Wolf M. E., and Mosnaim A. D. (1983) Studies in the mechanism of phenylethylamine uptake by rabbit erythrocytes. *Comp. Biochem. and Physiol.* (in press).

3. Inwang E. E., Mosnaim A. D., and Sabelli H. C. (1973) Isolation and characterization of phenylethylamine and phenylethanolamine from human brain. J. Neurochem. **20**, 1469-1473.
4. Borison R. L., Mosnaim A. D., and Sabelli H. C. (1975) Brain 2-phenylethylamine as a mediator for the central actions of amphetamine and methylphenidate. Life Sci. **17**, 1331-1343.
5. Mosnaim A. D., Callaghan O. H., and Wolf M. E. (1981) Determination of noncatecholic phenylethylamines and monomethylated derivatives of phenylethylamine. J. Chromatogr. **224**, 481-487.
6. Jonsson J., Lindeke B., and Cho A. K. (1975) Oxidation of phenylethylamine yielding tyramine by rat liver microsomes. Acta Pharmacol. et Toxicol. **37** 352-360.
7. Jonsson J. (1974) Hydroxylation of amphetamine to para-hydroxyamphetamine by rat liver microsomes. Biochem. Pharmacol. **23**, 3191-3197.
8. Mosnaim A. D., Silkaitis R., and Wolf M. E. (1980) Rabbit brain metabolism of phenylethylamine and tyramine: Drug effects. Life Sci. **27**, 557-566.
9. Boulton A. A., Dyck L. E., and Durden D. A. (1975) Hydroxylation of β-phenylethylamine in the rat. Life Sci. **15**, 1673-1683.
10. Mosnaim A. D. and Wolf M. E. (eds.) (1980) Noncatecholic Phenylethylamines, Part 2: Phenylethanolamine, Tyramines and Octopamine. Marcel Dekker, New York.
11. Silkaitis R. P. and Mosnaim A. D. (1976) Pathways linking L-phenylalanine and 2-phenylethylamine with p-tyramine in rabbit brain. Brain Res. **114**, 105-115.

CHRONOTROPIC AND INOTROPIC SELECTIVITY OF SUBSTITUTED PHENYLETHYLAMINES IN THE ISOLATED PERFUSED RABBIT HEART

D.M. Ferguson and A.J. Vazquez

Dept. Pharmacol., UHS/Chicago Medical School

North Chicago, Illinois 60064, U.S.A.

INTRODUCTION AND OBJECTIVES

The cardiovascular actions of a number of endogenous amines, eg. catecholamines as well as the noncatecholic phenylethylamines, were first studied systematically by Barger and Dale in 1910. Subsequently, many authors have differentiated the direct and indirect actions of these amines in vascular beds. The present study examines the effect of intracoronary artery injection of these amines on heart rate and contractility in the isolated, perfused rabbit heart, with particular attention towards establishing mechanism of action (direct and/or indirect) and detecting any selectivity for chronotropic or inotropic activity. These results could lead to a better understanding of the cardiac consequences of pathologic or iatrogenic conditions which produce increased levels of these amines. It is hoped that they will also lead to a more rational design of cardiac drugs.

METHODS

Hearts from male, New Zealand white rabbits were perfused at 30 degrees C in a Langendorff apparatus in the manner of Anderson and Craver (1948). A force transducer was connected to the apex of the heart by means of a hook and thread. A measure of contractility, dF/dt, was generated by differentiating the force signal. Heart rate was measured by using a tachometer triggered from dF/dt. Bolus injections of amines (see Table 1) were made through a side arm of the perfusion cannula; cardiac effluent was not recirculated. Antagonists were introduced into the reservoir and allowed to recirculate for at least one hour before resuming measurements.

Initially, a series of increasing doses of the amines (dissolved in ascorbic acid-saline) were tested for cardiac activity and selectivity for either chronotropic or inotropic augmentation of cardiac function. Repeated tests in the presence of cocaine, propranolol, or butoxamine defined those amines acting directly on beta-1 receptors.

Responses to norepinephrine (NE) were used as an internal standard to determine the maximum chronotropic and inotropic responsiveness of each heart. Experimental data are expressed as percent of maximum response and plotted against the negative logarithm of the dose (in moles) to yield dose-response curves for the heart rate and dF/dt responses. In order to quantify any separation in the chronotropic and inotropic responses to individual amines, the difference in the normalized chronotropic and inotropic responses at a particular dose was used as a selectivity index. The individual data for each drug series were pooled to yield mean selectivity indices. If the index is greater than zero, the drug shows a selectivity for chronotropic stimulation; if the index is less than zero, the drug shows inotropic preference. Mean indices were tested for their difference from zero using a paired-t test. The responses before and after antagonists were also compared by paired-t tests.

RESULTS

All of the amines tested had cardiac stimulating activity. PEA, pTRM, pOA, mOA, DA, and ALB showed significant ($P < 0.05$) chronotropic selectivity, whereas PE showed a significant preference for inotropic activity (Table 1). None of the other amines studied exhibited any significant selectivity. Experiments using cocaine, propranolol, and butoxamine showed that only pOA and PE were selective, direct-acting agonists, and that their selective actions involved interacting with only beta-1 receptors. Table 2 summarizes results obtained with the entire series of amines. Figures 1 and 2 show chronotropic (HR) and inotropic (dF/dt) dose-response curves for pOA and PE, respectively. Figure 3 shows a plot of the selectivity indices for NE, pOA, and PE.

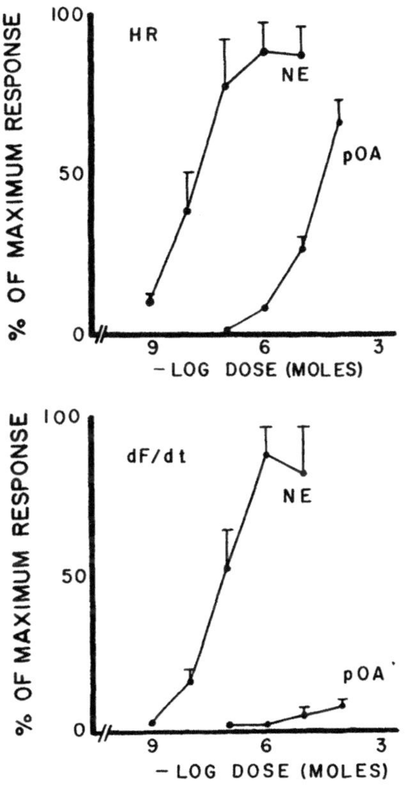

Figure 1. Chronotropic and inotropic responses to para-octopamine. The hearts' responses to para-octopamine and norepinephrine are normalized with respect to the maximum responses obtained with norepinephrine. Data points are means; plungers are standard errors. N=5.

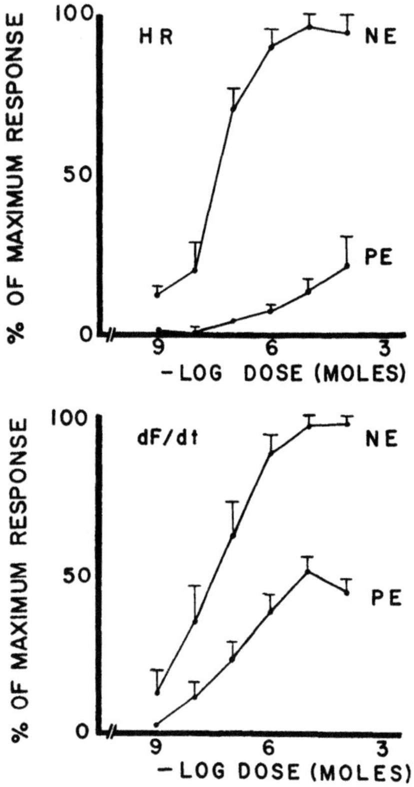

Figure 2. Chronotropic and inotropic responses to phenylephrine. The hearts' responses to phenylephrine and norepinephrine are normalized with respect to the maximum responses obtained with norepinephrine. Data points are means; plungers are standard errors. N=5-8.

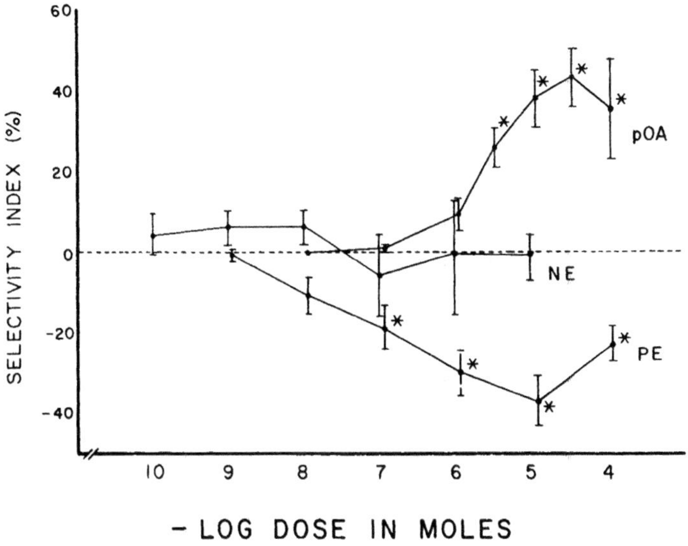

Figure 3. Selectivity of norepinephrine, para-octopamine, and phenylephrine. The selectivity index is the normalized heart rate response minus the normalized dF/dt response. Positive values indicate chronotropic selectivity, and negative values indicate inotropic selectivity. Data points are means; plungers are standard errors. *Indicates mean differs significantly from zero ($P < .05$).

Table 1

Mean Selectivity Index of Substituted Phenylethylamines

-Logarithm of Doses in Moles

Drug	12	11	10	9	8	7	6	5.5	5	4.5	4
PEA (1)					-0.7	4.7	20.7		74.3*		
pTRM (2)					-5.2	-7.2	10.0		46.8*		75.5*
mTRM (3)					-1.9	-0.2	13.1		52.7*		87.7*
PET (4)					5.1	7.7	9.4		41.9		72.9
pOA (5)					0.0	1.8	9.1	26.4*	37.9*	43.2*	35.6*
mOA (6)					9.2*	-8.0	-13.1		6.5		12.0*
SYN (7)					0.0	-1.0	0.8		0.1		
PE (8)				-1.0	-11.2	-19.1*	-30.2*		-37.3*		-22.9*
DA (9)						0.0	9.4		25.3*		23.5*
NE (10)			4.7	6.6	6.5	-6.1	-1.1		-1.3		
E (11)	-4.2	-3.5	-2.6	-20.5	-32.0*	-11.2	15.9		-11.7		
ISP (12)	-1.3	1.3	3.7	-2.3	0.8	-2.4	0.0				
ALB (13)					3.5	12.3	19.3*		25.9*		28.3

*Selectivity index mean differs significantly from zero at the .05 probability level.
1. phenylethylamine; 2. *p*-tyramine; 3. *m*-tyramine; 4. phenylethanolamine; 5. *p*-octopamine; 6. *m*-octopamine; 7. synephrine; 8. phenylephrine; 9. dopamine; 10. norepinephrine; 11. epinephrine; 12. isoproterenol; 13. albuterol.

Table 2

Structure and Action of Substituted Phenylethylamines[1]

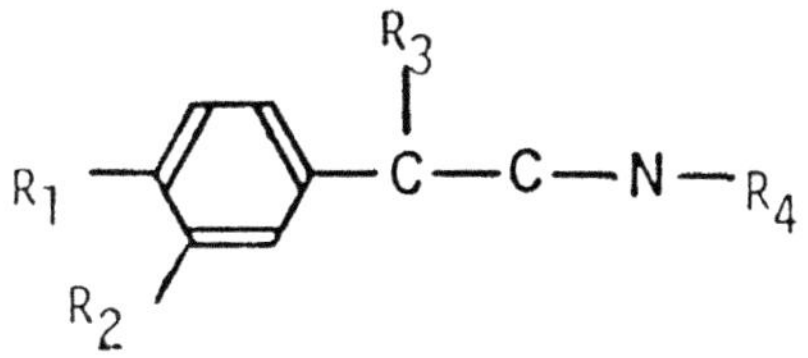

Drug	R_1	R_2	R_3	R_4	Selective	Type	Direct	Beta-1
PEA	H	H	H	H	Yes	HR	No	---
pTRM	OH	H	H	H	Yes	HR	No	---
mTRM	H	OH	H	H	Yes	HR	No	---
PET	H	H	OH	H	No		Y/N	---
pOA	OH	H	OH	H	Yes	HR	Yes	Yes
mOA	H	OH	OH	H	No		---	---
SYN	OH	H	OH	CH_3	No		Y/N	---
PE	H	OH	OH	CH_3	Yes	dF/dt	Yes	Yes
DA	OH	OH	H	H	Yes	HR	Y/N	---
NE	OH	OH	OH	H	No		Yes	Yes
E	OH	OH	OH	CH_3	No		Yes	Yes
ISP	OH	OH	OH	$CH(CH_3)_2$	No		Yes	Yes
ALB	OH	CH_2OH	OH	$C(CH_3)_3$	Yes	HR	No	---

[1]Dashed lines indicate no experimental data from this study.

DISCUSSION AND CONCLUSIONS

The chronotropic selectivity observed for the indirect-acting amines (Table 2) is most likely due to the fact that the density of sympathetic nerve endings is much greater in the atria where the heart beat originates than in the ventricles (Euler, 1954). The contractility measured in this study essentially consists of venticular contractility, and therefore amines which induce release of endogenous NE would be expected to have chronotropic selectivity. The data from this study generally support this hypothesis, but it should be noted that SYN showed no tendency for selectivity but did act indirectly. The reason for this discrepancy has not been explored; it may lie in a balance of direct and indirect actions. Similarly, DA is promoted as an inotropic selective agent.

The direct-acting selectivity of pOA and inotropic selectivity of PE at cardiac beta-1 receptors suggest that in rabbit these receptors may be differentiated into chronotropic and inotropic subtypes.

These amines might play a physiological role modulating cardiac responsiveness to sympathetic activity. Alterations in their concentration ratio in serum (or in the presynaptic pool) as a consequence of hormonal, dietary, or pharmacological influences, might determine whether the heart will respond with tachycardia or increase in force to central noradrenergic discharges. If such a mechanism is proven operative in man, there could be important pathologic consequences and therapeutic opportunities. Development of selective antagonists of the cardiac beta-1 receptor subtypes could provide amelioration of the symptomatic expression of alterations in the pool of trace amines.

REFERENCES

Anderson F.F. and Craver B.N. (1948) A Pyrex apparatus for the perfusion of the coronary circulation of mammalian hearts. J. Pharmacol. Exp. Ther. 93:135-141.

Barger G. and Dale H.H. (1910) Chemical structure and sympathomimetic action of amines. J. Physiol. (Lond.) 41:19-59.

Euler U.S. von. (1954) Adrenaline and noradrenaline. Distribution and action. Pharmacol. Rev. 6:15-22.

TRACE AMINE-PEPTIDE INTERACTIONS ON CENTRAL NEURONES: I. TRYPTAMINE AND SUBSTANCE P.

Roland S. G. Jones

Biology Research Laboratories

Ciba-Geigy AG., Basel,Switzerland

Much attention is being paid to the recent observations of co-existence of amines and peptides in neurones in several areas of the CNS and possible implications of functional interactions of two transmitters released from the same neurone The existence of a descending spinal pathway containing 5-hydroxytryptamine(5HT) and substance P (SP) is now established(Hokfelt et al.,I978) and it has been demonstrated recently that 5HT is occlusive to the effects of SP on dorsal horn interneurones(Davies and Roberts,I98I). Although it is not known whether 5HT and SP co-exist in ascending pathways this appears possible(Savard et al.,I982) and the present study has investigated the interactions of 5HT with SP on single neurones in the anterior cingulate cortex. In view of the probable existence of a raphe-cortical pathway containing tryptamine(T)(see Jones,I982) the interactions between this amine and SP were also investigated.

METHODS

Conventional electrophysiological techniques were used to record from and apply substances by iontophoresis to single spontaneously active neurones in the anterior cingulate cortex.The rats were anaesthetized with chloral hydrate(400 mg/kg IP). When a cell was located control responses to SP were established and then weak applications of 5HT or T were made without interrupting the cycle of peptide applications. On some cells acetylcholine was used as a control agonist.

RESULTS

5HT

The effects of 5HT on responses of I5 cells excited by SP were tested. On 3 cells the SP-responses were potentiated during the application of 5HT. A further 7 cells showed a reduction of excitatory responses during the period of 5HT ejection and the remainder were not affected within the current range tested(0-80 nA). In general responses to SP were only altered when there was a concurrent change in baseline firing rate. Acetylcholine responses were usually altered at the same time and in the same direction as the SP responses(5 cells). It should be noted however that on several cells there was little change in the response to SP even when the baseline was substantially affected. Some examples of interaction studies are shown in figure I.

T

Nine cells which were excited by SP were tested with weak applications of T(0-30 nA). on all but one of these neurones there was a strong reduction or an abolition of the SP responses during the ejection of T. In contrast to 5HT this reduction of responses occurred in the abscence of any detectable change in baseline firing rate excepton one cell. In addition the reduction of SP responses could be elicited by ejecting currents of T which did not alter the responses of the same neurone to alternate applications of Acetylcholine. Some examples of SP-T interactions are shown in figure 2

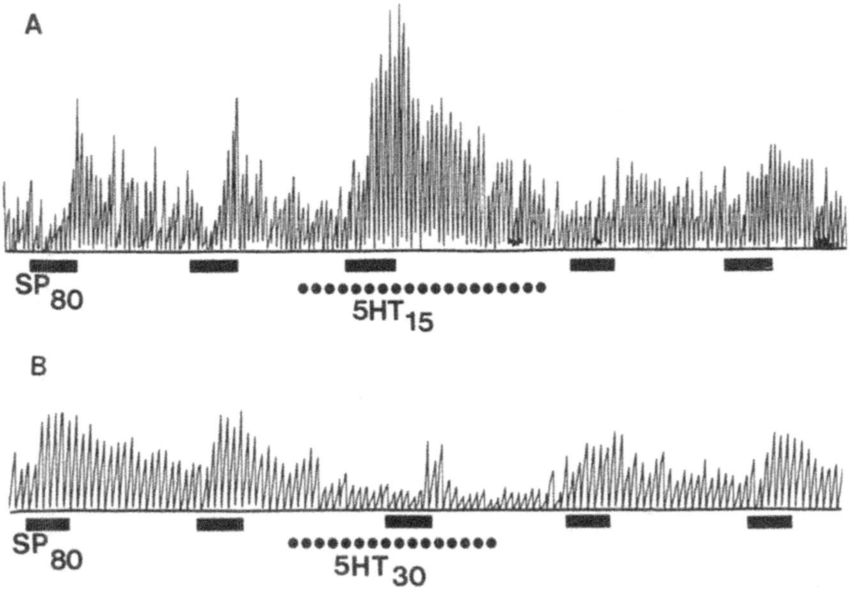

FIGURE I. Ratemeter records from two different cells illustrating SP/5HT interactions. In (A) the excitatory response to SP was potentiated during the application of 5HT. The amine itself did cause a slight increase in firing rate. The cell in (B) shows a reduction of SP responses in the prescence of 5HT but only concurrent with a considerable decrease in baseline firing rate.

DISCUSSION

In their study of neurones in the dorsal horn Davies and Roberts (I98I) reported both depression and potentiation of SP-responses by 5HT. Whereas the latter effect was always accompanied by an increase in baseline firing rate the former usually occurred in the abscence of baseline changes. In the present study of neurones in the cingulate cortex both potentiation and reduction of SP-responses were concurrent with corresponding alterations of baseline firing rate. Also, in the present experiments responses to acetylcholine were affected concurrently with SP-responses but

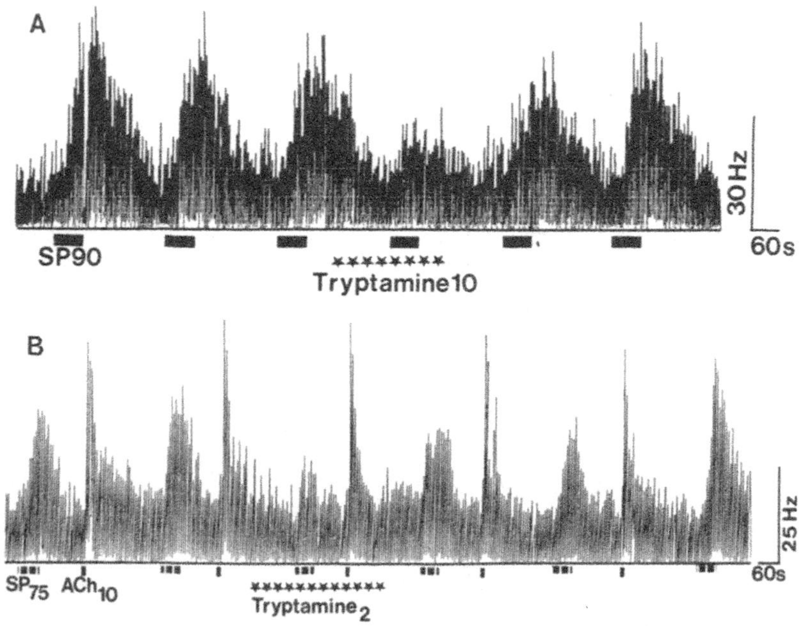

FIGURE 2. T/SP interaction studies on two cortical neurones. The cell in (A) showed a strong reduction in responses to the peptide although T did not noticeably alter the baseline firing rate. A similar effect is seen in (B) although responses to acetylcholine are not concurrently altered.

Davies and Roberts used glutamate as a control agonist and found no such correlation. It seems likely then, that whereas in the spinal cord the depression of SP responses by 5HT may reflect a specific 5HT/SP interaction, in the cortex the action of 5HT may be more related to non-specific changes in membrane excitability.

The latter conclusion would not seem applicable in the case of T/SP interactions in the cingulate cortex, however. T consistently reduced the SP-evoked excitations without altering the cell firing rate and without concurrent changes in responses to acetylcholine. The mechanism of this action is unknown at present but it would not appear to involve a non-specific decrease in membrane excitability. However, this can only be confirmed or denied by the use of intracellular recording techniques. It may be speculated that the interaction involves a direct effect of T at the receptor for SP or an action on the receptor-excitation coupling mechanism.

In view of the recent evidence implicating T as a transmitter in a raphe-cortical pathway (see Jones, 1982 and also this book) it is very tempting to speculate that the T/SP interaction seen here may reflect a functional one between amine and peptide neuronal systems in this brain area.

REFERENCES

Hokfelt, T., Ljungdahl, A., Steinbush, H., Verhofstad, A., Nilsson, G., Brodin, E., Pernow, B and Goldstein, M.(1978) Immunohistochemical evidence of SP-like immunoreactivity in some 5HT-containing neurones in the rat CNS. Neuroscience. 3, 517-538.

Davies, J.E. and Roberts, M.H.T.(1981) 5HT reduces SP responses on dorsal horn interneurones: a possible interaction of neurotransmitters. Brain Res., 217, 399-404.

Jones, R.S.G. (1982) Tryptamine: a possible neurotransmitter or neuromodulatorin the mammalian CNS. Prog. Neurobiol. 19, 117-139.

Savard, P., Merand, Y., Bedard, P., Dussault,J.H. and Dupont, A. (1982) Interrelationship between 5HT and SP system in rat brain. Abs. Soc. Neurosoc. 8, 118.

TRACE AMINE-PEPTIDE INTERACTIONS.II. PHENYLETHYLAMINE AND ENKEPHALIN, P-TYRAMINE AND ENKEPHALIN.

Roland S. G. Jones

Biology Research Laboratories

Ciba-Geigy AG., Basel, Switzerland

A study by Garzon et al.(1980) has shown that the analgesia induced by intraventricular injection of methionine-enkephalin(mENK) in rats was potentiated by parenteral administration of phenylethylamine(PEA). The present study determined whether this peptide-amine interaction was demonstrable at the cellular level by applying PEA and mENK iontophoretically to single cells in the rat cerebral cortex. In addition, similar studies investigated the interactions between mENK and p-tyramine(pTA). The latter amine has been shown to modify responses of central neurones to amine transmitters(Jones, 1981) and it was of interest to determine its effects on responses to a non-amine transmitter candidate.

METHODS

Experiments were performed on male albino rats anaesthetized with either urethane or chloral hydrate. Conventional techniques were used to record from,and apply substances by iontophoresis to,single spontaneously active cells in the fronto parietal cortex.When a cell was located and control responses to mENK established a weak amine application was initiated. The latter was adjusted so that no change in baseline firing rate occurred and the mENK applications continued uninterrupted.

RESULTS

PEA

The effects of PEA on depressant responses to mENK were studied on 8 cells. On 7 of these PEA potentiated the responses without detectable alteration of baseline firing rate. (figure I). On no occasion was there a reduction in mENK-response and the amine did not affect responses to GABA.

TA

In contrast to the effects of PEA, TA consistently reduced or abolished the depressant responses of cortical neurones to mENK. Examples of this effect are shown in figure 2 on three

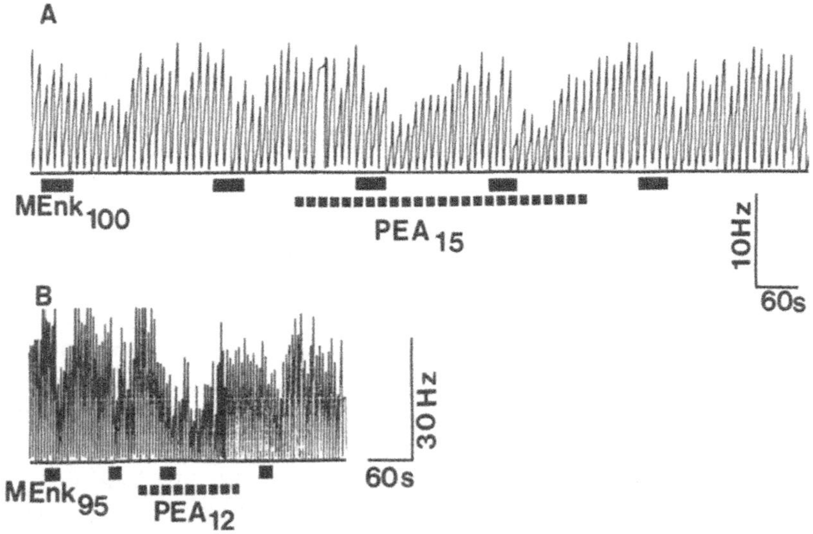

Figure I Ratemeter records from two cells showing potentiation of responses to mENK by PEA.

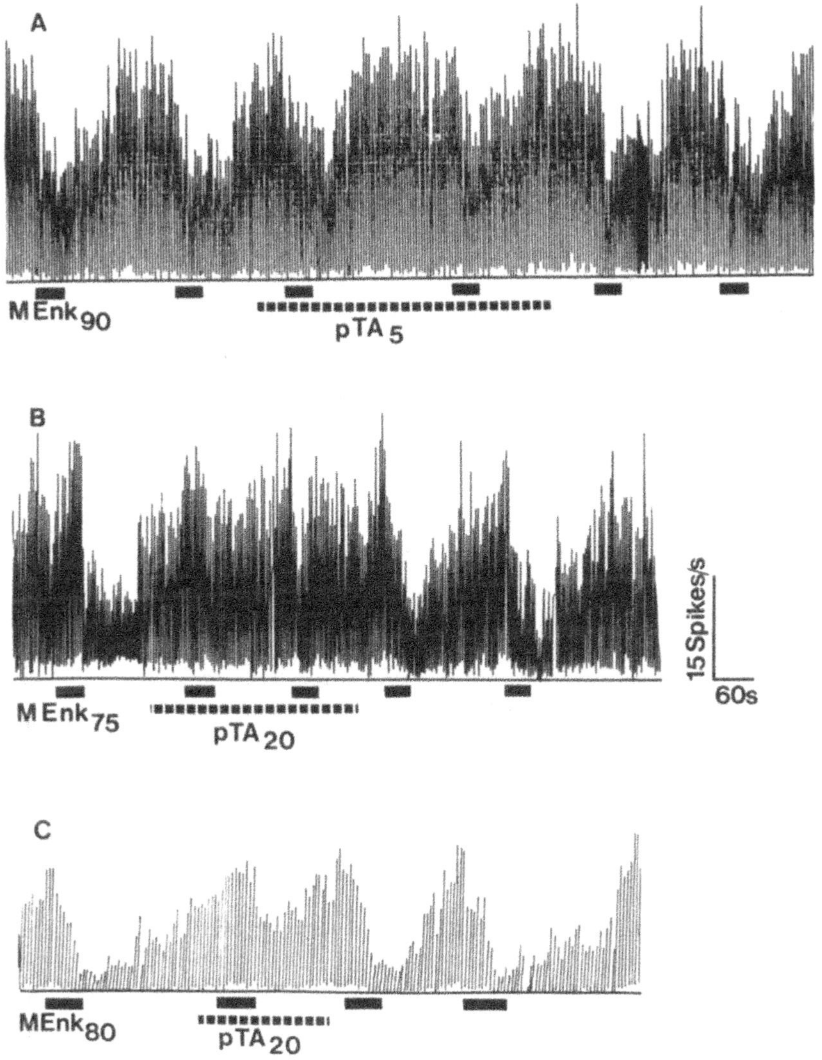

Figure 2. Ratemeter records from three cells showing reduction of responses to mENK by TA.

cells. Similar effects of TA were seen on a further three cells. On one other neurone TA application resulted in a weak potentiation of the response to mENK and on one cell there was no detectable alteration in response. TA did not alter responses to GABA.

DISCUSSION

Thus the present results on single cells agree with the data of Garzon et al(I980) showing that PEA can potentiate the analgesia induced by central administration of mENK. The mechanism of this effect is not known at present. One possibilty which must be considered is that the potentiation of mENK is secondary to an action of PEA on catecholamine systems . PEA potentiates responses to catecholamines in the cortex (Jones and Boulton, I980) so if the responses to mENK comprised a catecholamine component as has been suggested(Palmer and Hoffer, I980) then PEA could potentiate the peptide responses through an action on the catecholamine systems. It should be noted however that TA also potentiates catecholamine responses(Jones, I98I) but had the opposite effect on responses to mENK.

The reduction of responses by TA was an unexpected observation.There are no indications in the literature to suggest any interaction of TA with mENK. It is noteworthy, however, that TA appears to be a competitive antagonist of striated muscle contractions induced by the neuropeptide, proctolin, in the cockroach(Brown, I975). This peptide has the structure H-Arg-Tyr-Leu-Pro-Thr-OH and it is suggested that the antagonistic action of TA may be directed towards the interaction of the tyrosine residue with its receptor(Brown, I975) Since the biological activity of mENK resides in part with the tyrosine residue then the antagonism of mENK responses seen here could be similarly explained.

Whatever the explanations of the present results it is clear that the trace amines so far tested can markedly alter the neuronal effects of mENK. It is entirely possible that the trace amines could have a similar modifying action in the endogenous situation and consideration should be given to a role of these substances as modulators of the analgesic and/or non-analgesic actions of opioid peptides.

REFERENCES

Brown, B.E. (I975) Proctolin: a peptide transmitter candidate in insects. Life Sci. I7, I24I-52

Garzon, J., Moratalla, R. and Del Rio, J. (I980) Potentiation of the analgesia induced in rats by morphine or mENK after inhibition of brain type B MAO: the role of PEA. Neuropharm. I9, 723-729.

Jones, R.S.G. (I98I) Specific enhancement of neuronal responses to catecholamines by p-tyramine J. Neurosci. Res., 6, 49-6I

Jones R.S.G. and Boulton,A.A. (I980)Interactions between pTA, mTA, or PEA and dopamine on single neurones in the cortex and caudate nucleus of the rat. Can. J. Phys. Pharm. 58, 222-227.

Palmer, M.R. and Hoffer, B.J. (I980) Catecholamine modulation of enkephalin-induced electrophysiological responses in cerebral cortex. J. Pharm. Exp. Ther., 2I3, 205-2I5.

D. BEHAVIOR

BETA-PHENYLETHYLAMINE: SOME PRELIMINARY CHRONIC STUDIES.

David M. Jackson and Owen F. Jenkins

Department of Pharmacology, University of Sydney, N.S.W., 2006, Australia.

Abstract: Beta-phenylethylamine (PE) (75 mg/kg once daily) was administered to mice for 1, 4 or 9 days, and behavioural and biochemical experiments performed 1 day after the last treatment dose. There was no quantitative change in sensitivity of PE treated mice to the post synaptic dopamine (DA) receptor stimulant effects of apomorphine, nor was there any change in striatal [^{3}H]-spiperone binding after 4 or 9 days of PE treatment. After 9 days of treatment (but not after 4) there was a marked trend for PE treated animals to be subsensitive to the locomotor depressant effects of low doses of apomorphine. The effect of PE treatment on the stimulant effects of amphetamine was complex, with both subsensitive and supersensitive components being evident. This was dependent upon duration of treatment and habituation to the environment. PE treatment did not affect tyrosine hydroxylase activity in the striatum (as assessed by DOPA accumulation after inhibition of DOPA decarboxylase) and did not affect the rate of striatal DA disappearance after tyrosine hydroxylase inhibition.

Many of the acute effects of PE in laboratory animals resemble those of amphetamine. Like amphetamine, PE produces locomotor excitation in mice (Mantegazza and Riva, 1963; Jackson, 1975a) and stereotyped behaviour in rats (Braestrup and Randrup, 1978). The similarities between the pharmacological effects of these two amines are probably due, at least in part, to some shared biochemical actions, with both PE and amphetamine exerting some of their central stimulant effects via the release of (newly synthesised) DA (and noradrenaline (NA))

(Carlsson et al., 1966; Fuxe et al., 1967). However, while it seems likely that virtually all the central stimulant effects of amphetamine can be explained by such indirect actions, there is considerable evidence that a major component of PE induced stimulation depends upon direct DA receptor stimulation (Jackson, 1972; Jackson, 1975a; Antelman et al., 1977), perhaps by a deaminated metabolite of PE.

While some of the acute pharmacological effects of PE are now understood, it is becoming increasingly clear from studies with drugs such as amphetamine that the chronic effects of psychoactive drugs are often different from, or even opposite to, their acute effects. Thus neuroleptics produce DA receptor supersensitivity, tricyclic antidepressants down-regulate central and peripheral alpha 2-adrenergic receptors and amphetamine shows either an increased or decreased response depending on the behaviour being measured, etc. While there is a copious literature on the chronic effects of amphetamine, reports on the effects of repeated PE administration are scarce. Such studies are of interest because if PE is a neurotransmitter or plays some type of central modulatory role in neurotransmission (see Sabelli et al., 1978 for discussion), then it is important to elucidate it's long-term effects on known neurotransmitter function. In addition, long-term studies with agents such as PE provide an important contribution to our understanding of synaptic homeostasis under conditions where a neuronal system is pushed to the limits by repeated agonist or antagonist treatments.

In the present experiments we have looked at some behavioural and biochemical effects of short term chronic treatments with PE.

MATERIALS AND METHODS

The subjects were male albino mice (20 - 35 g, University of Sydney Animal House) kept under constant lighting (6 a.m. - 6 p.m. light) and temperature (22 +/- 2°C) conditions, with free access to food and water. Treatment consisted of once daily (morning) injections with PE (75 mg/kg as the hydrochloride) for 1, 4 or 9 days, and all experiments were conducted 20-28 h after the last PE dose. This dose of PE was chosen as the chronic treatment because it is the lowest that produces both the indirect and direct behavioural effects of this amine (Jackson, 1972). The indirect effects appear within 2 to 3 minutes and peak 5 minutes after injection, and the direct effects are evident between 20 and 60 minutes after injection (early and late phases respectively). The locomotor activity of groups of 3 mice was measured in 4 matched rectangular photocell cages as previously described (Bailey and Jackson, 1978), beginning immediately after the injection of various challenge drugs (except in one case).

For catechol determinations, groups of 2 to 4 mice were decapitated and the striata excised into 0.4 M perchloric acid solution containing 0.05 % sodium metabisulphite and 0.2 % disodium EDTA. Dihydroxyphenylalanine (DOPA) and DA were extracted according to the cation exchange chromatographic method of Atack and Magnusson (1978) and assayed by the methods of Kehr et al. (1972) and Atack (1973) respectively.

Spiperone binding assays were conducted as described previously (Jackson et al., 1981). Briefly, the striata of 2 mice were excised into 0.32 M sucrose. After homogenization and centrifugation, a washed P2 fraction was used for binding in assay tubes containing [^{3}H]-spiperone (2 nM) and tissue (0.3 - 0.4 mg protein) (final assay volume 2 ml). Tubes were incubated at 27°C for 20 minutes and membranes collected on GF/F filters, washed twice with 5 ml 0.05 M TRIS buffer at 0°C (pH 7.4 at 25°C) and bound radioactivity determined by liquid scintillation spectroscopy at an efficiency of 20 - 30 %. Specific binding was defined as the difference between radioactivity bound in the presence and absence of 10uM spiperone. Protein was estimated according to the method of Lowry et al. (1951).

Drugs used were d-amphetamine sulphate (Smith, Kline and French, Australia); alpha-methyl-p-tyrosine methyl ester hydrochloride (AMPT), DA hydrochloride, bovine serum albumin, L-DOPA, reserpine, m-hydroxybenzylhydrazine dihydrochloride (NSD1015) (Sigma, U.S.A.); apomorphine hydrochloride (Sandoz, Switzerland); spiperone (Janssen, Belgium); 1-phenyl-4-[^{3}H]-spiperone (23 Ci/mmol) (New England Nuclear, U.S.A.); PE hydrochloride (prepared from the base by Dr. R. Allan, Pharmacology Department, University of Sydney). Apomorphine was dissolved in 0.05 % ascorbic acid and stored in the dark at 0°C. Spiperone and reserpine were dissolved in a minimum volume of acetic acid and diluted with buffer and water respectively. d-Amphetamine, AMPT, NSD1015 and PE were dissolved in saline. All drugs were injected I.P. in a dose volume of 10 ml/kg, and doses are expressed in terms of the salts.

Statistical analysis was performed by multifactor analyses of variance with, in some cases, repeated measures on one factor (time) (Winer, 1971). Further comparisons were made using t tests.

RESULTS

When injected into mice, PE produces a complex behavioural syndrome characterised by locomotor activity alternating with periods of immobility, sniffing, headbobbing, rearing and other stereotyped behaviours (Braestrup and Randrup, 1978; Dourish, 1982). This repertoire of behaviour makes quantification of PE induced behaviour in activity meters difficult, and suggests that open field observation and/or assessment of stereotypy may be more useful. Nevertheless, using photocell activity meters we have previously reported that PE produces a biphasic stimulatory action consisting of an early phase (seen with doses equal to or less than 75mg/kg) which is dependent upon the release of newly synthesised catecholamines by PE, and a later phase (seen with doses equal to or greater than 75mg/kg) which is dependent upon direct postsynaptic DA receptor stimulation by PE. In preliminary experiments, the biphasic nature of PE induced

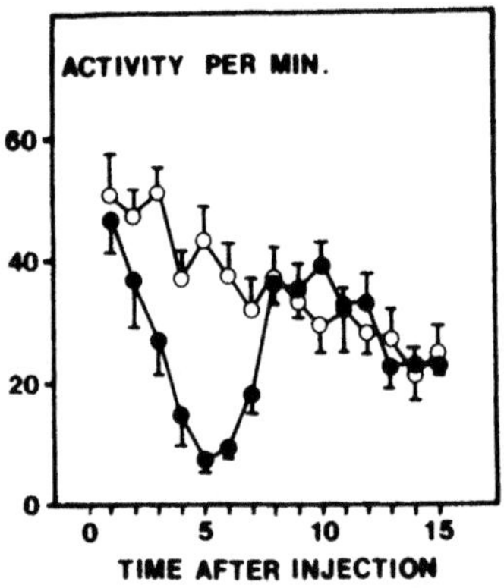

FIG. 1. The effect of PE, 25 mg/kg (●) or saline (○) on the locomotor activity of groups of 3 mice (counts per minute, mean +/- s.e.m., n = 8 in each case.

excitation was confirmed, but in addition it was noted that a low dose (25 mg/kg) produced marked locomotor depression in the first 10 minutes (Fig. 1) rather than the excitation seen with 50 or 75 mg/kg. The sedation was characterised by the absence of normal exploratory behaviours such as locomotion, grooming and rearing and may be due to the DA released by PE (Fuxe et al., 1967) stimulating DA autoreceptors, thus producing a reduction of further DA synthesis and release (Carlsson, 1975), and, as a consequence, locomotor depression (Strombom, 1976).

In the present study we treated mice for 1 or 4 days with PE as described, and 20 - 28 h after the last dose, challenged them with various doses of PE. The data for the 4 day treatment experiment are shown in Fig. 2. In vehicle treated animals, challenge with 75 mg/kg of PE (Fig. 2a) produced a biphasic stimulatory response, with an early phase being evident 3 to 7 minutes after injection (inset) and a late phase occurring 15 to 30 minutes after injection. In PE treated animals, however, although the late phase of stimulation was essentially unchanged, the early phase activity was absent. Essentially similar data (not shown) were obtained after the 1 day PE treatment indicating that, in both cases, PE treatment prevented the early phase stimulation (dependent upon catecholamine

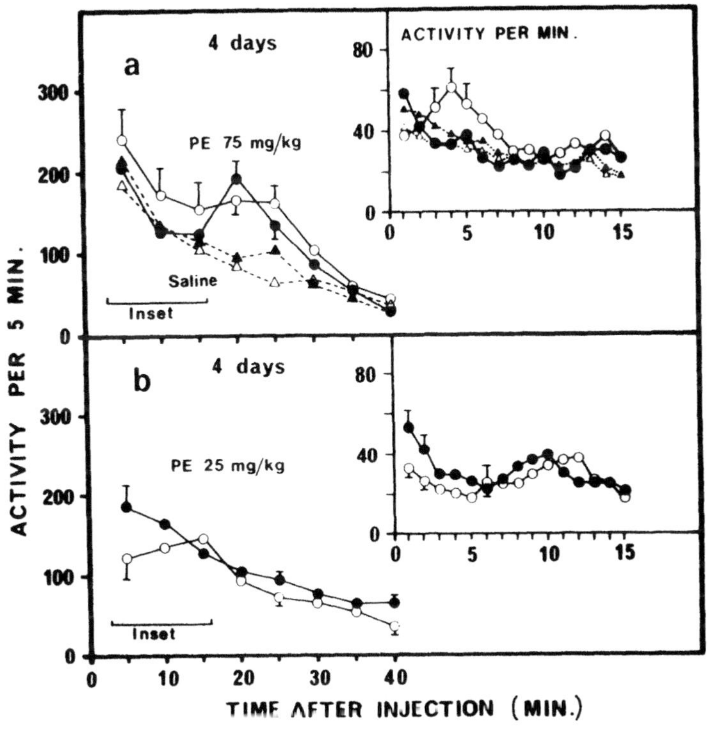

FIG. 2. The effect of a 4 day treatment with PE (filled symbols) or vehicle (open symbols) on the response of mice to a challenge with saline (▲) or PE (25 or 75 mg/kg) (●). Activity was monitored every minute for 15 minutes (inset) and thereafter each 5 minutes. The data are the mean locomotor activities of 10 replicates +/- s.e.m.´s.

release) but did not affect the late phase stimulation. After a challenge with 25 mg/kg no significant differences in response were seen (Fig. 2b).

We also examined the response of the DA autoreceptors to depressant doses of apomorphine and found no change in sensitivity after 4 days of PE treatment (Fig 3a, $F_{1/55} = 0.014$, $P > 0.1$). However, after 9 days of PE treatment, there was a strong trend for the PE treated animals to be less sensitive than vehicle treated animals to the depressant effect of apomorphine (Fig 3b, $F_{1/47} = 3.015$, $P = 0.089$). This marginal difference was due primarily to the higher dose of apomorphine and consequently the time course data (counts per 5 minutes for the half hour) were analysed further. This analysis indicated that there was a

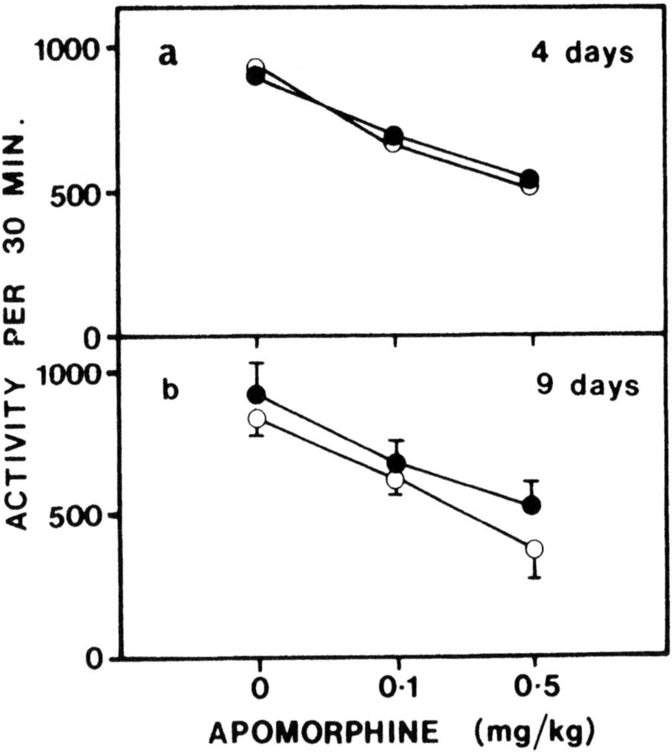

FIG. 3. The effect of a 4 (a) or 9 (b) day treatment with PE (●) or vehicle (○) on the response of mice to low doses of apomorphine. Locomotor activity was monitored each 5 minutes for 30 minutes beginning immediately after injection, and the data represent the mean locomotor activities of 7 to 10 replicates +/- s.e.m.'s.

significant difference in the shape of the two time course curves ($F_{4/42}$ = 21.82, P < 0.001) with the difference being due to less apomorphine induced depression in the PE treated animals 5 and 10 minutes after injection (data not shown).

To examine the sensitivity of postsynaptic DA receptors after PE treatment, mice were premedicated with reserpine plus AMPT to eliminate catecholamine storage and synthesis, and challenged with apomorphine. The locomotor

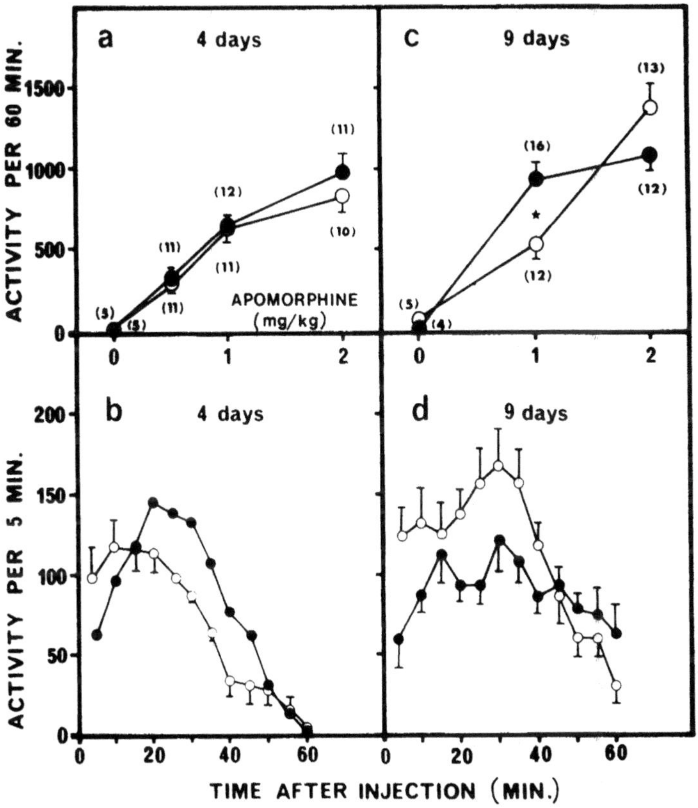

FIG. 4. The locomotor activity of PE (●) or vehicle (○) treated mice after 4 (a,b) or 9 (c,d) days of treatment, in response to challenge with apomorphine. Four hours prior to the challenge mice were premedicated with reserpine (10 mg/kg) plus AMPT (250 mg/kg). The upper diagrams show the total 1 h activity (mean +/- s.e.m., (n)), while the lower diagrams show the time curves for animals challenged with the high dose of apomorphine.
★ P < 0.05, see text for complete analysis.

stimulation induced by a range of doses of apomorphine lasted up to 60 minutes after injection, and the total activity during this period was unaffected by either 4 (Fig. 4a) or 9 (Fig. 4c) days of PE treatment (effects of treatment respectively, $F_{1,68} = 0.87$, $P > 0.05$ and $F_{1,56} = 0.038$, $P > 0.05$). However a marginally significant interaction term ($F_{2,56} = 3.2$, $P = 0.048$) was obtained from the 9 day data due to the crossing of the two dose-response curves (Fig. 4c). The data for both experiments were further analysed after drawing up the time course curves (activity per 5 minutes). After challenge with 2 mg/kg of apomorphine (but not with 0.5 or 1.0 mg/kg) there were significant changes in the shape of the time curves after both 4 (Fig.4b) and 9 (Fig. 4d) days of PE treatment (interaction terms respectively, $F_{11,198} = 3.03$, $P < 0.001$ and $F_{11,253} = 2.88$, $P < 0.005$). Thus, although the total activity was unaltered, PE treated animals responded in a qualitatively changed manner to the high dose of apomorphine. Overall these data indicate that neither 4 nor 9 days of PE treatment affected the sensitivity of the DA receptors mediating the apomorphine induced locomotor stimulation, with the proviso that the qualitative (rather than the quantitative) response to the high dose was affected in both cases.

The response of mice to the stimulant effect of d-amphetamine was dependent upon two variables: the duration of the PE treatment, and the exposure of mice to the activity cages for a 1 h habituation period prior to d-amphetamine challenge. The response of PE treated mice not habituated to the environment (Fig. 5, left panels) was unchanged after 1 day of PE treatment (Fig. 5a, effect of treatment, $F_{1,37} = 1.02$, $P > 0.05$). However, animals treated for 4 days with PE were subsensitive to the stimulant effect of d-amphetamine (Fig. 5b, effect of treatment, $F_{1,89} = 5.69$, $P < 0.05$), while those treated for 9 days were supersensitive (Fig. 5c, effect of treatment, $F_{1,31} = 9.79$, $P < 0.005$). The results obtained in animals adapted to the environment were somewhat different (Fig. 5, right panels). These animals were unchanged in their sensitivity after 1 day (Fig. 5d, effect of treatment, $F_{1,62} = 0.416$, $P > 0.05$) and 9 days (Fig. 5f, effect of treatment, $F_{1,44} = 2.23$, $P > 0.05$) but were supersensitive to d-amphetamine after 4 days (Fig. 5e, effect of treatment, $F_{1,63} = 6.67$, $P < 0.05$). At no time, with or without environmental habituation, were there any significant differences in response to saline

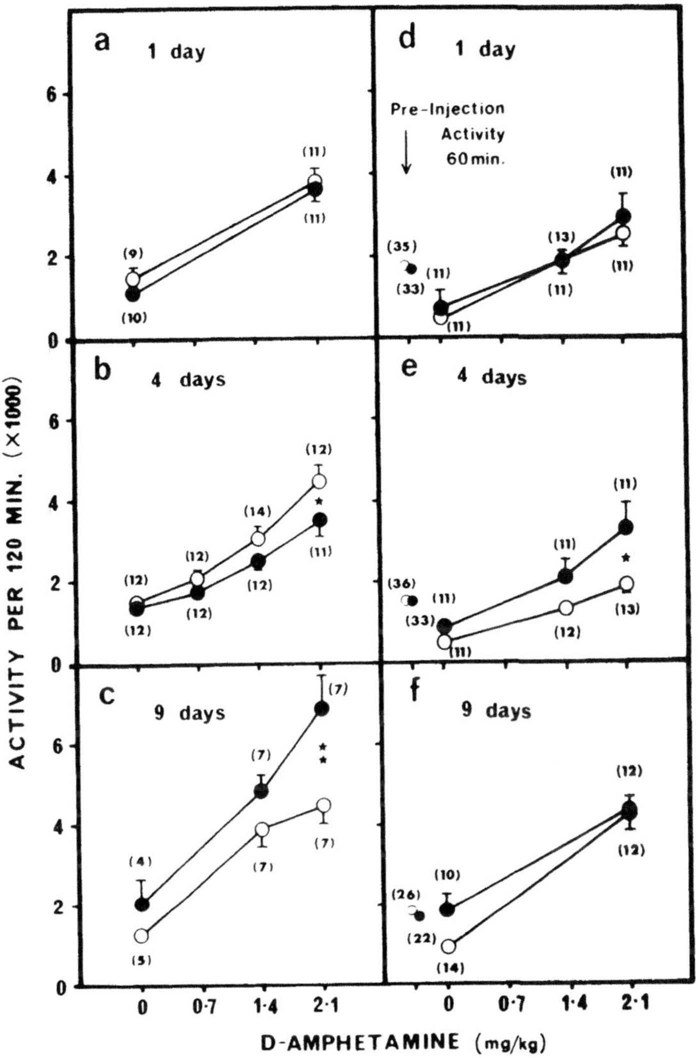

FIG. 5. The effect of 1 (a,d), 4 (b,e) or 9 (c,f) days of treatment with PE on the locomotor response of groups of 3 mice to challenge with d-amphetamine. Activity was monitored for 2 hours and the data represent the means (+/- s.e.m.) of (n) replicates. The left panels (a,b,c) represent animals naive to the activity cages, while the right panels (d,e,f) represent data from animals placed in the activity cages for 1 h prior to d-amphetamine challenge (recorded as pre-injection activity).
★ $P < 0.05$, ★★ $P < 0.01$.

challenge, although there was a strong trend in one case (Fig. 5f). It is noteworthy that in the cases where significant sensitivity changes occurred (Fig. 5,b,c,e), it was most marked with the highest dose of d-amphetamine used.

[^{3}H]-Spiperone binding was unchanged by PE treatment. The total binding in the 4 day PE and saline, or 9 day PE and saline treated animals, was 168 +/- 14, 162 +/- 10, 156 +/- 19 and 172 +/- 23 fmoles bound/mg of protein, respectively (mean +/- s.e.m., n=5 in each case). PE treatment altered neither the accumulation of DOPA after inhibition of DOPA decarboxylase nor the disappearance of DA after tyrosine hydroxylase inhibition (Table).

TABLE. The effect of PE or vehicle treatment on the amount of DOPA accumulated 30 minutes after inhibition of DOPA decarboxylase with NSD1015 (100 mg/kg) and on the levels of DA 60 minutes after the injection of saline or AMPT (250 mg/kg) (ng/ml protein, mean +/- s.e.m., n = 5-8).

Treatment		DOPA Accumulation	DA level after saline	DA level after AMPT
4 days	vehicle	10.16 +/- 0.43	123 +/- 8	89 +/- 10
	PE	9.48 +/- 0.38	139 +/- 13	114 +/- 14
9 days	vehicle	9.96 +/- 0.65	139 +/- 7	104 +/- 7
	PE	9.42 +/- 0.34	135 +/- 7	119 +/- 10

DISCUSSION

The present experiments show that PE treatment did not change the sensitivity of postsynaptic DA receptors, as assessed by the stimulant effects of apomorphine (a direct acting DA receptor agonist) and by [^{3}H]-spiperone binding to striatal membranes. Although there was no sign after 4 days of PE treatment of any changed response to the depressant effects of low, presumably DA autoreceptor selective doses of apomorphine, there was a marked, though non-significant trend towards subsensitivity in PE treated animals after 9 days of treatment. This was particularly marked with the higher dose of apomorphine and was statistically significant when the time course curves were analysed. We believe this subsensitivity might reflect a moderate down-regulation of the DA autoreceptors, in the absence of any change in the sensitivity of postsynaptic receptors. A differential lability of the two populations of DA receptor to sensitivity alteration would be consistent with findings obtained with other DA agonists. Thus, treatment of mice and rats with d-amphetamine for short periods of time, while producing behavioural supersensitivity to d-amphetamine challenge (Bailey and Jackson, 1978; Rebec and Segal, 1979), does not cause any change in post synaptic DA receptor sensitivity (Jackson et al., 1981; Muller and Seeman, 1979) but does apparently cause DA autoreceptor subsensitivity (Muller and Seeman, 1979). Similarly, the novel DA autoreceptor agonist 3-PPP (Hjorth et al., 1980) and the DA precursor L-DOPA can produce apparent DA autoreceptor subsensitivity in the absence of postsynaptic DA receptor changes (Jackson et al., 1982a; 1982b). These changes (or lack of changes) are, however, dependent upon many factors. For example, chronic treatment of rats with high doses of d-amphetamine (or methamphetamine) can causes changes in DA uptake sites and down-regulation of postsynaptic DA receptors due apparently to a direct neurotoxic action (Wagner et al., 1980; Robertson, 1982). In another vein, duration of chronic L-DOPA treatment has been shown to be an important determinant of receptor sensitivity changes (Pycock et al., 1982). All these (and other) factors are probably of similar importance in determining the chronic effects of PE. In any event, bearing in mind such considerations, it is apparent that the PE treatment schedule utilised in the present studies produced no clear cut change in postsynaptic DA receptor sensitivity.

The response to d-amphetamine challenge was unexpectedly complex with both subsensitive and supersensitive responses being obtained. The direction of the changes were dependent upon one of two variables: the duration of the PE treatment and the habituation of the mice to the activity cages prior to the challenge. After 4 days of PE treatment, a subsensitive response to d-amphetamine was obtained in non-habituated PE treated mice, but a supersensitive response in habituated PE treated mice. After 9 days of treatment, non-habituated mice were supersensitive but habituated mice were unchanged in their sensitivity to d-amphetamine challenge. Note that there was no change in response after 1 day of PE treatment. The aetiology of these changes are unknown, but they are obviously not dependent upon DA auto or postsynaptic receptor sensitivity changes. This is particularly interesting because DA autoreceptor subsensitivity has been invoked by several authors as underlying, at least in some cases, supersensitivity to d-amphetamine challenge (Muller and Seeman, 1979). Nor are they apparently due to any change in tyrosine hydroxylase activity in the striatum since DOPA accumulation in this region after inhibition of DOPA decarboxylase was unchanged by PE treatment. Nor was there any change in the disappearance of DA after inhibition of tyrosine hydroxylase in this brain region. Perhaps the factor most worthy of consideration is the possibility of an interaction between exploration (and the effects it would have on DA and NA metabolism) and d-amphetamine (with its well documented effects on DA and NA release). In non-habituated mice, the stimulant effects of d-amphetamine (caused by release of newly synthesised DA) are being superimposed on high levels of basal activity. In habituated mice, the basal level is much lower. In view of the complex nature of the d-amphetamine response and the obvious difficulties in it´s interpretation, we suggest that this compound should not be utilised as a tool to investigate receptor sensitivity changes after chronic drug treatments.

Although not investigated in detail, the behavioural response to PE itself was altered after PE treatment. This change was restricted to the early phase stimulation (see above). Biochemically we could not explain this finding since there was no change after 4 or 9 days of PE treatment in the synthesis or disappearance rate of DA in the striatum at the time when behavioural tests were conducted. Since the stimulant effects of PE are to some extent dependent upon NA and 5-hydroxytryptamine, it would be worthwhile investigating in detail the effects of chronic PE treatment on these two amines. The only relevant study of which we are aware (Karoum et al., 1982) reported no sustained change in NA (or DA) metabolism in various brain regions of rats withdrawn for 24 hours from a chronic treatment schedule of PE.

The preliminary studies reported here indicate that the chronic effects of PE are in some ways different to it's acute effects and resemble some of the changes produced by chronic treatment with a number of other DA agonists. These findings warrant further investigation. However, future studies should utilise a technique of administration which will maintain constant brain levels of PE.

ACKNOWLEDGEMENTS

The authors are grateful to Smith, Kline and French, Australia, Sandoz Australia and Janssen Pharmaceutica, Belgium for generous donations of drugs and to Peter Lafferty for conducting the binding experiments.

REFERENCES

Antelman S.M., Edwards D.J., and Lin M. (1977) Phenylethylamine: evidence for a direct postsynaptic dopamine-receptor stimulating action. Brain Research 127, 317-322.

Atack C.V. (1973) The determination of dopamine by a modification of the dihydroxyindole fluorimetric assay. Br. J. Pharmacol. 48, 699-714.

Atack C.V., and Magnusson T. (1978) A procedure for the isolation of noradrenaline (together with adrenaline), dopamine, 5-hydroxytryptamine and histamine from the same tissue sample using a single column of strongly acidic cation exchange resin. Acta Pharmacol. Toxicol. (Copenh). 42, 35-57.

Bailey R.C. and Jackson D.M. (1978) A pharmacological study of changes in central nervous system receptor responsiveness after long-term dexamphetamine and apomorphine administration. Psychopharmacol. 56, 317-326.

Braestrup C., and Randrup A. (1978) Stereotyped behaviour in rats induced by phenylethylamine, dependence on dopamine and noradrenaline, and possible relation to psychosis?, in Noncatecholic Phenylethylamines, Part I. Phenylethylamine: Biological Mechanisms and Clinical Aspects (Mosnaim A. and Wolf M.E., eds), pp. 245-269, Marcel Dekker Inc., New York.

Carlsson A. (1975) Receptor-mediated control of dopamine metabolism, in Pre- and Postsynaptic Receptors (Usdin E., and Bunney W.E., eds), Marcel Dekker Inc., New York.

Carlsson A.K., Fuxe K., Hamberger B., and Lindqvist M. (1966) Biochemical and histochemical studies on the effects of imipramine like drugs and amphetamine on central and peripheral catecholamine neurons. Acta Physiol. Scand. 67, 481-497.

Dourish C.T. (1982) An observational analysis of the behavioural effects of beta-phenylethylamine in isolated and grouped mice. Prog. Neuropsychopharmacol. Biol. Psychiat., 6, 143.

Fuxe K., Grobecker H., and Jonsson J. (1967) The effect of beta-phenylethylamine on central and peripheral monoamine-containing neurons. Europ. J. Pharmacol. 2, 202-207.

Hjorth S., Carlsson A., Lindberg P., Sanchez D., Wiksrom H., Arvidsson L.E., Hacksell U., Nilsson J.L.G., and Svensson U. (1980) A new centrally acting DA receptor agonist with selectivity for autoreceptors. Psychopharmacol. Bull. 16, 85-90.

Jackson D.M. (1972) The effect of beta-phenethylamine upon spontaneous motor activity in mice: a dual effect on locomotor activity. J. Pharm. Pharmacol. 24, 383-389.

Jackson D.M. (1975a) Beta-phenylethylamine and locomotor activity in mice. Interaction with catecholaminergic neurones and receptors. Arzneimittel-Forschung 25, 622-626.

Jackson D.M., Bailey R.C., Christie M.J., Crisp E.A., and Skerritt J.H. (1981) Long-term d-amphetamine in rats: lack of change in post-synaptic dopamine receptor sensitivity. Psychopharmacol. 73, 276-280.

Jackson D.M., Carlsson A., Hjorth S., and Lindberg P. (1982a) A behavioural study of the changes in the central nervous system of mice after subchronic treatment with the selective dopamine autoreceptor agonist 3-PPP (dl-3-[3 hydroxyphenyl]-N-n-propylpiperidine). J. Neural Transm. 53, 233-245.

Jackson D.M., Walters J.R., and Miller L.P. (1982b) Chronic L-DOPA-pretreatment of rats: an electrophysiological and biochemical study in the basal ganglia. Brain Research 250, 271-182.

Karoum F., Speciale S.G., Chuang L-W., and Wyatt R.J. (1982) Selective effects of phenylethylamine on central catecholamines: a comparative study with amphetamine. J. Pharmacol. Exp. Ther., 223, 432-439.

Kehr W., Carlsson A., and Lindqvist M.(1972) A method for the determination of 3,4-dihydroxyphenylalanine (DOPA) in brain. Naunyn-Schmiedeberg's Arch. Pharmacol. 274, 273-280.

Lowry O.H., Rosebrough N.J., Farr A.L., and Randall R.J.

(1951) Protein measurement with the Folin phenol reagent. J. Biol. Chem. 193, 265-275.

Mantegazza P., and Riva M. (1963) Amphetamine-like activity of beta-phenylethylamine after a monoamine oxidase inhibitor in vivo. J. Pharm. Pharmacol. 15, 472-478.

Muller P., and Seeman P. (1979) Presynaptic subsensitivity as a possible basis for sensitisation by long-term dopamine mimetics. Europ. J. Pharmacol. 55, 149-158 (1979).

Pycock C., Dawbarn D., and O´Shaughnessy C. (1982) Behavioural and biochemical changes following chronic administration of L-DOPA to rats. Europ. J. Pharmacol. 79, 201-215.

Rebec G.V., and Segal D.S. (1979) Enhanced responsiveness to intraventricular infusion of amphetamine following its repeated systemic administration. Psychopharmacol. 62, 101-102.

Robertson H.A. (1982) Chronic phencyclidine, like amphetamine, produces a decrease in [^{3}H]-spiroperidol binding in rat striatum. Europ. J. Pharmacol. 78, 363-365.

Sabelli H.C., Borison R.L., Diamond B.I., May J., and Havdala H.S. (1978) 2-Phenylethylamine as a neuromodulator of wakefulness, affect, and extrapyramidal function: recent advances, in Noncatecholic Phenylethylamines, Part I. Phenylethylamine: Biological Mechanisns and Clinical Aspects (Mosnaim A., and Wolfe M.E., eds), pp. 345-376. Marcel Dekker Inc., New York.

Strombom U. (1976) Catecholamine receptor agonists: effects on motor activity and rate of tyrosine hydroxylation in mouse brain. Naunyn-Schmiedeberg´s Arch. Pharmacol. 292, 167-176.

Wagner G.C., Ricaurte G.A., Seiden L.S., Schuster C.R., Miller R.J., and Westley J. (1980) Long-lasting depletions of striatal dopamine-uptake sites following repeated administration of methamphetamine. Brain Research 181, 151-160.

Winer B.J. (1971) Statistical Principles in Experimental Design. McGraw-Hill Kogakusha, Ltd., Tokyo.

β-PHENYLETHYLAMINE: A FUNCTIONAL ROLE AT THE BEHAVIOURAL LEVEL?

A.J. Greenshaw

Psychiatric Research Division,
Cancer and Medical Research Building,
University of Saskatchewan,
Saskatoon, Saskatchewan S7N OWO

ABSTRACT

It has been proposed that β-phenylethylamine (PE) may be equivalent to, or may mediate, the effects of d-amphetamine. This hypothesis has been assessed in terms of a comparison of the general stimulus profiles of these two compounds. From the literature it is apparent that PE and d-amphetamine share similar discriminative stimulus properties. This has been demonstrated for PE with and without MAO inhibition. An analysis of effects of PE and d-amphetamine on self-stimulation indicates that their effects are equivalent in this paradigm when PE is administered in the presence of a MAO inhibitor but not in its absence. Self-administration experiments have shown that both compounds may act as rewards or reinforcers in different species. PE is, however, ineffective as a stimulus for the induction of taste aversions, in marked contrast to d-amphetamine. The implications of these data, and the results of some neurochemical analyses, for a possible functional role of PE are discussed.

INTRODUCTION

It is now well established that β-phenylethylamine (PE) is a natural constituent of mammalian brain tissue (Durden *et al.*, 1973). Despite the low concentrations of this amine under normal conditions, its high turnover rate

(Durden and Philips, 1980); heterogeneous distribution (Durden *et al.*, 1973); and subcellular localisation (Boulton and Baker, 1975) point to a probable physiological role in the regulation of neural activity. The proposed role for PE as a neuromodulator (Boulton, 1979) has received further support from recent electrophysiological studies (Jones and Boulton, 1980). Furthermore, preliminary reports suggest the existence of specific PE receptors within the rat central nervous system (Hauger *et al.*, 1982).

In reviewing studies of structure-activity relationships with respect to amphetamine and analogues thereof, Biel (1970) has pointed out that the PE moiety of amphetamine is crucial to most of its pharmacological and biochemical properties. This observation is of course consistent with the early studies of Barger and Dale (1910) which indicated that PE formed the basic carbon structure of amines which have sympathomimetic effects. In view of these considerations PE has been characterised as an "endogenous amphetamine" (Sandler and Reynolds, 1976) which may possibly act as a mediator of the effects of amphetamine and related compounds (Chuang *et al.*, 1982).

At the level of effects on behaviour the actions of PE and of amphetamine are in many respects similar (Stein, 1964; Huang and Ho, 1974; Dourish and Boulton, 1981), although notable differences have also been reported (Howard *et al.*, 1976; Jackson, 1972). One fairly prominent factor which may at least partially explain the differences between the behavioural effects of the two compounds is their relative pharmacodynamic profiles. Whereas PE is rapidly destroyed by monoamine oxidase (MAO) amphetamine has a long duration of action. Thus, in the absence of a MAO inhibitor, fairly large doses of PE are necessary to induce significant changes in most behavioural measures (Howard *et al.*, 1976). In contrast to this, when given in the presence of a MAO inhibitor the effects of PE appear to be largely equivalent to those of amphetamine, both qualitatively and in terms of the effective range of doses (Stein, 1964). This point notwithstanding, the question of whether PE and amphetamine are functionally equivalent remains both important and controversial. In relation to the "endogenous amphetamine" hypothesis and the proposed role for PE in neural regulation: a corollary of this issue is that changes in brain PE activity should be functionally related to changes in some aspects of behaviour.

The paper is generally directed to an analysis of these two issues. Other papers in this volume concerning behavioural pharmacology largely address changes in spontaneous motor activity in response to administration of PE. The approach adopted here deals with conditioned behaviour. This approach is mainly directed to an analysis of the general stimulus properties of PE. The procedures adopted here, therefore, tend to focus on the effects of this compound on patterns of behaviour which reflect the animals' choice in response to variable but clearly defined response options. Stimulus profiles of drugs can conveniently be subdivided into three major categories:

Discriminative stimulus properties. Within this paradigm (Barry, 1974) animals are trained to emit a specific response, dependent on the presence or absence of a specific drug-state (i.e. presumably reflecting the animals' subjective response to drug effects).

Rewarding stimulus properties. This component of the stimulus properties of drugs is typically defined by the effects of drugs on measures of brain stimulation reward (i.e. behaviour maintained by electrical stimulation of local brain areas (Liebman, 1983), or in terms of the amount and pattern of voluntary drug intake by laboratory animals (Schuster and Thompson, 1969).

Aversive stimulus properties. This aspect of the stimulus profiles of drugs has been most clearly demonstrated in experiments involving gustatory avoidance or conditioned taste aversion (CTA) learning (Goudie, 1979). In this paradigm aversion is defined as reduced intake of a flavoured liquid, the ingestion of which has been followed by the administration of drugs or certain other stimuli.

DISCRIMINATIVE STIMULUS CONTROL AND β-PHENYLETHYLAMINE

In terms of discriminative stimulus control it is well established that d-amphetamine is a very effective cue (D'Mello and Stolerman, 1977). An early study by Huang and Ho (1974) demonstrated that animals trained to discriminate between amphetamine and saline exhibited generalisation to a low dose of PE in the presence of a MAO inhibitor (iproniazid). These authors suggested that the drug states elicited by amphetamine and by PE were equivalent.

In a more recent study Goudie (1982) assessed the effectiveness of relatively high doses of PE alone as a discriminative stimulus. In this study only 6 out of 12 subjects successfully learned the discrimination between this compound and saline. The acquisition and post-criterion maintenance data for the successful discriminators are displayed in Table 1. In view of the large numbers of sessions required for acquisition of PE/saline discrimination and the relatively low accuracy of performance (e.g. in comparison to that reported for d-amphetamine), these results indicate that PE is not a very effective cue in drug discrimination learning. In subsequent generalisation tests half of the "discriminators" exhibited generalisation to both cocaine and amphetamine. The remaining "discriminators" did not show generalisation to either cocaine or amphetamine.

The animals which treated the PE, amphetamine and cocaine cues as equivalent also generalised to deprenyl, the others did not. As deprenyl is preferentially a type B MAO inhibitor, and PE is preferentially a substrate for MAO B, this latter result is surprising. However, it is consistent with a previous report of generalisation of the deprenyl state to cocaine discrimination learning (Colpaert et al., 1980). It seems most likely that the partial generalisation observed by Goudie may possibly be attributed to a multi-dimensional cue being elicited by PE (Goudie, 1982). The failure of the deprenyl state to generalise to PE is, however, unlikely to be of great significance. The deprenyl dose used in this study to which some subjects exhibited generalisation was 10 mg/kg (IP 15 min prior to testing). This dose does not preferentially inhibit MAO B (the specificity of deprenyl is dose-dependent) and, as Goudie has pointed out, it is important to note that deprenyl is converted to amphetamine in vivo (Philips, 1981). These observations may possibly detract somewhat from the hypothesis that endogenous PE may mediate the stimulus properties of certain stimulants in view of the doses of MAO inhibitors on which this proposal was based (Colpaert et al., 1980).

It is apparent from these studies that PE and d-amphetamine may elicit an equivalent discriminative state in the presence of a MAO inhibitor and that some animals may acquire an amphetamine-like PE cue in the absence of MAO inhibition.

Table 1

Acquisition and post-criterion maintenance of PE discrimination

Subject	Sessions to criterion	Post-criterion per cent correct lever selections	Response level under 30 mg/kg PE (Mean±S.E.)	Response level under 40 mg/kg PE (Mean±S.E.)
5	62	82%	105.2 ± 4.3	28.1 ± 3.6
6	84	95%	80.0 ± 3.9	38.2 ± 2.3
7	62	96%	37.4 ± 7.7	27.3 ± 2.2
8	52	87%	85.4 ± 6.4	74.8 ± 3.1
9	68	91%	53.7 ± 5.3	28.0 ± 3.0
10	45	82%	83.0 ± 15.8	49.5 ± 4.4
Mean (±S.E.)	62.2±5.5	88.8±2.5%	74.1%	41.0%

From Goudie, 1982 with permission.

β-PHENYLETHYLAMINE AND REWARD

Behaviour Maintained by Electrical Hypothalamic Stimulation

PE in the absence of MAO inhibition. A fairly early study demonstrated that low doses of PE in the presence of a MAO inhibitor (iproniazid) may induce an amphetamine-like facilitation of lever pressing maintained by electrical stimulation of the lateral hypothalamus and act to reduce reward thresholds (Stein, 1964). In a subsequent study Howard *et al.* (1976) reported opposite effects on response rate when high doses of PE were employed. This latter study has been cited as an example of a clear difference between the effects of PE and of d-amphetamine: notably because both compounds increased measures of spontaneous motor activity. It is apparent, however, that equivalent increases in motor activity measured by photo-beams may actually reflect different behavioural response patterns (see Dourish this volume). In our experience, the activity induced by amphetamine is markedly different to that elicited by PE alone. The most important difference being that effective doses of the latter compound tend to elicit stereotyped behaviour without any prolonged increase in coordinated locomotor activity - a prominent feature of low dose amphetamine effects and of effects of low PE doses in the presence of a MAO inhibitor.

In the Howard *et al.* (1976) study, rate of lever pressing was the only measure of self-stimulation that was employed. It is evident that the use of a number of measures of operant performance may facilitate the analysis of drug action particularly in view of the problems inherent to the interpretation of effects of drugs on response rate per se (Sanger, 1981). In order to compare further the effects of PE with those of d-amphetamine the effects of these two compounds on behaviour maintained by a fixed-interval schedule of electrical lateral hypothalamic stimulation were investigated (Greenshaw *et al.*, submitted). The schedule and training procedure were as described by Greenshaw *et al.* (1981). Each animal was allowed to regulate the duration of lateral hypothalamic stimulation on a single lever. The stimulation was available according to a fixed-interval schedule whereby stimulation was only available if a fixed period of time had elapsed since the offset of the previous train of stimulation. The rate of lever pressing, duration of lateral hypothalamic stimulation and the duration of responses occurring during the fixed-interval served as behavioural measures. With this schedule the duration of stimulation under the animal's control is inversely related to current intensity (Greenshaw et al., 1981, 1983). The duration of responses occurring during the interval and overall response rate are, however, not related to the intensity of stimulation (Greenshaw et al., 1983). Thus the duration of stimulation is a good correlate of reward magnitude whereas the other measures provide an index of effects of drugs on the animals' ability to perform the required response.

The effects of a range of doses of d-amphetamine (0.25 to 4.0 mg/kg IP) and of PE (1.25 to 50.0 mg/kg IP) were assessed. Each compound was injected immediately prior to

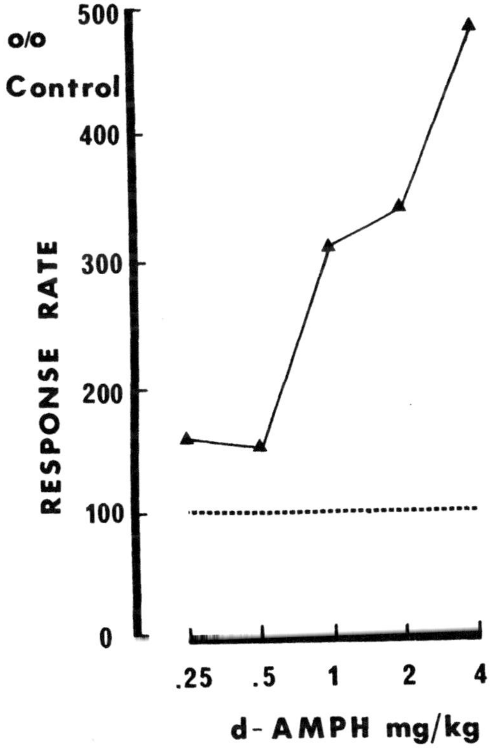

Fig. 1. The facilitatory effects of d-amphetamine on fixed-interval responding maintained by lateral hypothalamic stimulation.

the test sessions which were of 30 minutes duration. Dose-response curves were assessed within animals. At least three control days intervened between drug days. Equivalent injections of 0.9% saline were given on all non-drug days.

Administration of d-amphetamine to four animals responding on a fixed-interval 60 seconds schedule resulted in a dose-dependent increase in rates of lever pressing. These effects are displayed in Fig. 1.

The increases were significant at 1.0, 2.0 and 4.0 mg/kg (α = 0.05). Interestingly, and in contrast to the effects of this drug on other schedules of self-stimulation (Zacharko and Wishart, 1979; Zacharko and Kokkinidis, 1982), d-amphetamine did not affect the duration of lateral hypothalamic stimulation. The duration of non-reinforced responses (occurring during the fixed interval) was also unaffected. These results appear to indicate that, despite the stimulant effects of d-amphetamine, the rats were able to maintain their 'normally optimal' duration of hypothalamic stimulation.

With a further group of four animals, responding on a similar fixed-interval 30 second schedule, the effects of PE and of a single (1.25 mg/kg) dose of d-amphetamine were assessed. Administration of PE resulted in a marked decrease in overall rates of lever pressing. This effect, which was significant at 50 mg/kg, is consistent with the previous finding of Howard et al. (1976). With the same animals 1.25 mg/kg of d-amphetamine significantly increased response rate. These effects are illustrated by the data displayed in Figure 2.

The duration of hypothalamic stimulation was increased slightly by PE, however, this effect was not statistically significant. Non-reinforced response duration was significantly increased by this compound. These effects were significant at 25.0 and 50.0 mg/kg. As with the animals used for the d-amphetamine dose-response curve, the present group did not exhibit any changes in response duration or in the duration of stimulation after administration of 1.25 mg/kg of d-amphetamine. These response duration data are presented in Figure 3.

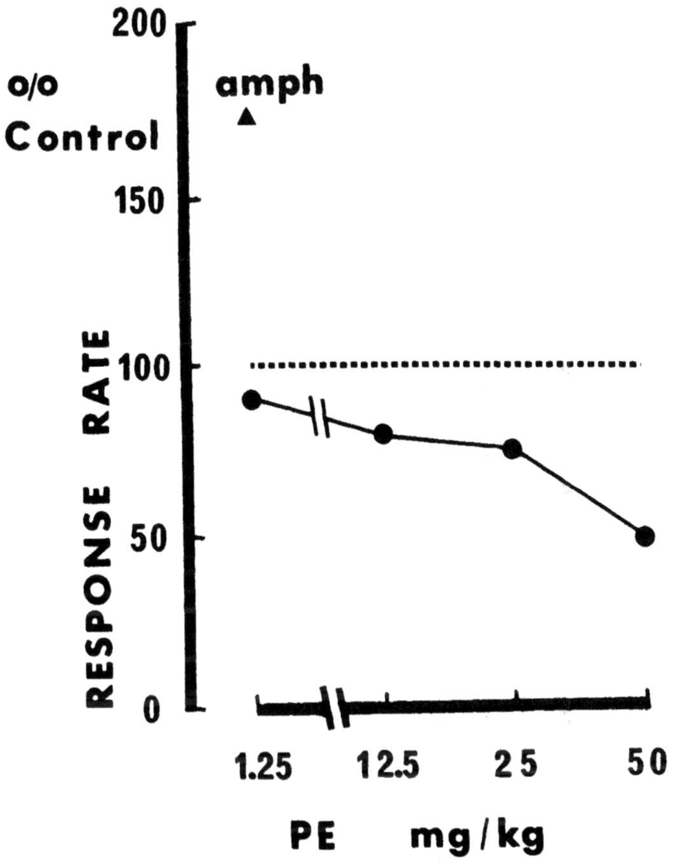

Fig. 2. The depressant effects of PE on fixed-interval responding maintained by lateral hypothalamic stimulation, in contrast to the effects of d-amphetamine.

The results described above are generally consistent with Howard et al.'s (1976) data. However, the effects of PE on response duration tend to illustrate that the animals' ability to perform the required response was impaired by the high doses of PE. Although no analysis of local response rates was conducted in our study, it was apparent that the effects of PE on responding paralleled the syndrome of stereotyped behaviour induced by this compound. Thus, responding was markedly impaired for approximately the first 15 to 20 minutes of the session, after

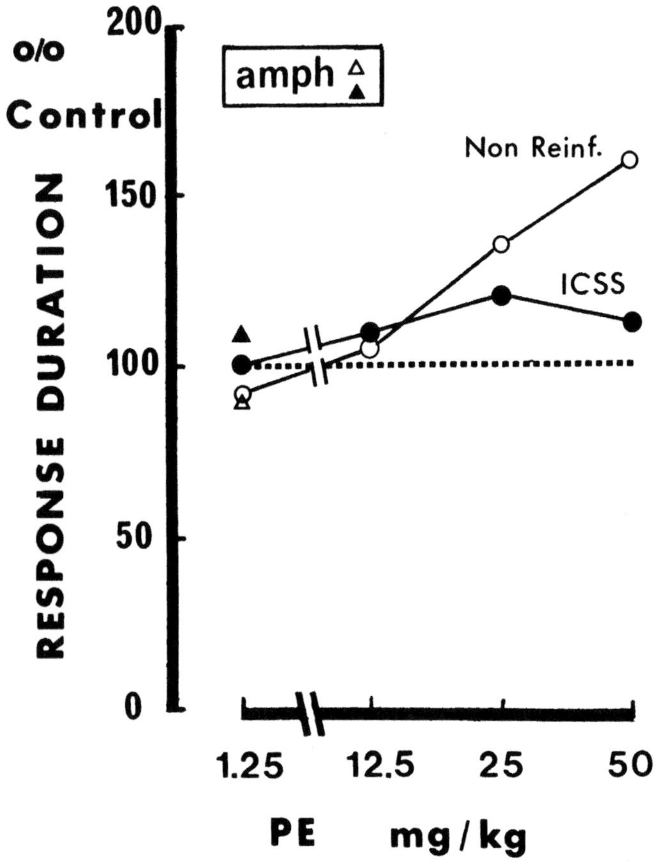

Fig. 3. The effects of PE and d-amphetamine on the duration of stimulation (ICSS) and of non-reinforced responses on a fixed-interval schedule.

which near normal responding was observed. This time-course was very similar to that observed with baselines of water intake (Broadbent et al., 1984) and with lever pressing maintained by food delivery (Goudie, personal communication).

MAO inhibition induced by l-deprenyl. No systematic investigations of the effects of low doses of PE (in the presence of a MAO inhibitor) have been reported in this context. In order to reevaluate and to extend the work reported by Stein (1964), therefore, we assessed the effects of PE and of l-deprenyl alone and in combination (Greenshaw et al., submitted). In this study l-deprenyl was chosen as a MAO inhibitor in view of its preferential action on MAO-B (Philips and Boulton, 1979). We assessed the effects of various doses of l-deprenyl (1 to 16 mg/kg s.c.), injected 3.5 h prior to the 30 min tests. PE was tested at various doses (0.5 to 4.0 mg/kg IP), injected immediately prior to testing in combination with the 4.0 mg/kg l-deprenyl treatment. The behavioural baseline against which the effects of these treatments were assessed was as described earlier. As l-deprenyl is an irreversible MAO inhibitor each animal was tested only once with a single drug treatment. In view of the demanding nature of such experimental designs, a fixed-interval three seconds schedule was used to maintain behaviour. This schedule allowed rapid acquisition of responding by the rats and necessitated a training period of only three to five 30-minute sessions.

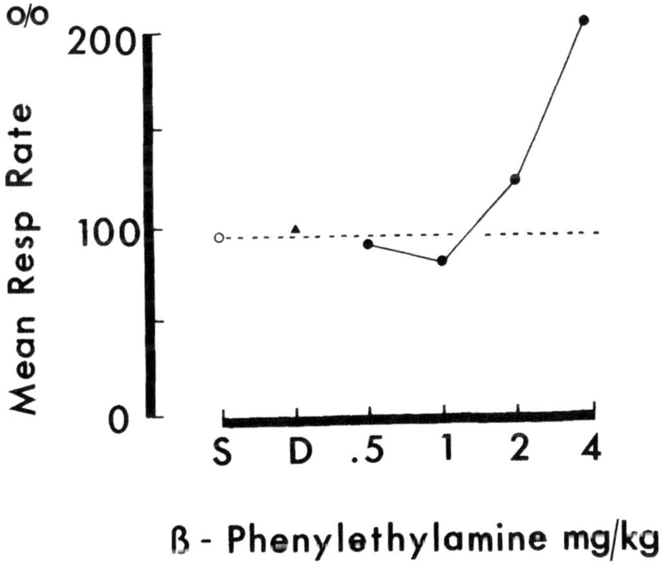

Fig. 4. The facilitatory effects of PE in the presence of l-deprenyl in relation to deprenyl (D) and saline (S). Responding was maintained by a fixed-interval schedule of lateral hypothalamic stimulation.

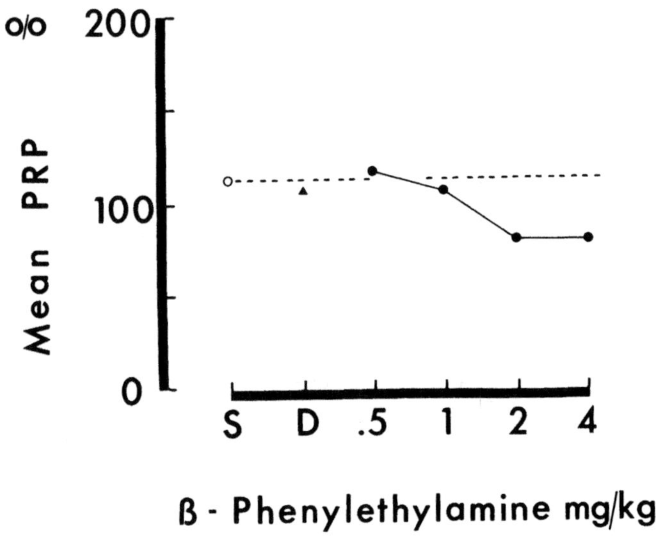

Fig. 5. The effects of PE in the presence of deprenyl on post-reinforcement pause duration. See Fig. 4 for details.

Following completion of behavioural testing (i.e. 4 h post-deprenyl and 0.5 h post-PE administration respectively) each animal was killed and striatal and hypothalamic PE levels were measured mass spectrometrically as described by Durden (1984). The neurochemical analysis was carried out in an attempt to determine the levels of PE that may be associated with changes in behaviour.

Although 1-deprenyl has been reported to induce decreases in fixed-ratio responding (Colpaert et al., 1980), no significant behavioural effects of this compound were observed in our study. This lack of effect of 1-deprenyl may be related to the time-course of drug administration. Administration of PE resulted in a significant increase in overall response rates, this effect being significant at 4.0 mg/kg. Although this compound tended to increase this measure at 2.0 mg/kg, this effect was not statistically significant. The average duration of post-reinforcement pauses (i.e., the average latency to the first response after the offset of hypothalamic stimulation) was monitored as a further variable in this study. This measure was significantly decreased by PE at 2.0 and 4.0 mg/kg. As with the effects of d-amphetamine described earlier, the present combination of PE and 1-deprenyl had no significant effect on the response duration measures. The effects of these treatments on response rate and post-reinforcement pause duration are displayed in Figures 4 and 5 respectively.

Neurochemical Measurement of PE in the Behavioural Context

The failure of 1-deprenyl treatment to induce changes in positively reinforced behaviour is inconsistent with previous reports (Colpaert *et al.*, 1980). Nevertheless the effects of a wide range of doses were assessed and these treatments did significantly elevate levels of endogenous PE. This is evident from the data presented in Table 2. It is interesting to note that, in the absence of drug treatment, PE levels were unaffected by hypothalamic stimulation. This finding is consistent with a previous report (Danielson *et al.*, 1976).

Nevertheless, this observation does not preclude an effect of stimulation on endogenous PE, as the turnover of this amine is so rapid. Thus, a relatively small effect may be revealed by an analysis of levels of metabolites (e.g. phenylacetic acid). We have not yet conducted such an analysis. With the present data, analysis of variance revealed that, despite higher basal levels of PE in the hypothalamus (in accord with previous reports, Durden *et al.*, 1973), there was a preferential increase in striatal levels of PE.

When PE was administered to rats pretreated with 1-deprenyl there was a marked increase in striatal PE levels. Due to sample loss, insufficient data were available for an analysis of hypothalamic levels. These data (presented in Table 2) reveal that increases in PE levels to upwards of approximately 75 ng/g are correlated with significant behavioural changes. This result is in striking contrast to the levels of this amine that are achieved after a 50 mg/kg dose of PE (25.7 μg/g and 6.0 μg/g after 15 and 30 minutes respectively, from Dyck, 1983).

Table 2

Effects of l-deprenyl alone or in combination with PE on striatal and hypothalamic PE levels in self-stimulators

l-Deprenyl (4 h prior to sacrifice)	Dose mg/kg					
	0	1.0	2.0	4.0	8.0	16.0
Hypothalamus	4.2	5.9	8.3	7.8	7.8	16.2
Striatum	1.3	9.9	11.0	11.5	13.7	14.6

PE (0.5 h prior to sacrifice + 4 mg/kg l-deprenyl as above)	Dose mg/kg				
	0	0.5	1.0	2.0	4.0
Striatum	11.5	35.6	59.6	73.6	322.2

PE levels are ng/g derived from pooled pairs of brain samples. Each determination is a mean of 4-6 pairs of samples.

The significance of the differential increase in endogenous striatal vs hypothalamic PE after l-deprenyl treatment is unclear. It is possible, however, that this result may reflect increased PE activity in the striatum as a consequence of repetitive motor activity (vis a vis self-stimulation). This is, of course, highly speculative. Nevertheless, in view of the proposed role of PE in the modulation of dopaminergic activity (Jones and Boulton, 1980) and the intimate relationship between striatal dopamine and motor activity (Iversen, 1977), it is a tenable hypothesis. Clearly, further work focussing on this issue would be of value in assessing the functional significance of PE activity.

With the present behavioural schedule it is not possible to attribute drug effects on response rate to altered efficacy of brain stimulation reward (Greenshaw *et al.*, 1981). In relation to this, the failure of either amphetamine or PE to alter the duration of stimulation seem to preclude a simple hypothesis relating the effects of these compounds to reward processes (Stein, 1964). Indeed, the apparent schedule-dependent nature of effects of amphetamine on self-regulated duration of hypothalamic stimulation (Greenshaw et al., submitted) are a clear indication of the complexity of measuring effects of drugs on "brain stimulation reward." Nevertheless, the behavioural data described above do indicate that, in the presence of MAO inhibition, the effects of d-amphetamine and of PE on self-stimulation are functionally equivalent even when a number of dependent variables is measured. This observation supports the original work of Stein (1964). The rate-decreasing effects of PE (in the absence of MAO inhibition) described above are consistent with the work of Howard et al. (1976). However, the significance of high dose effects of PE is doubtful in the present context (a) because of the apparent incompatibility of PE-induced stereotypy with co-ordinated responding; and (b) because of the tremendous increases in brain PE concentrations after such treatment (Dyck, 1983). It seems reasonable to accept, therefore, that the effects of PE and of d-amphetamine on self-stimulation are functionally equivalent.

In view of the lack of effect of these compounds on the duration of hypothalamic stimulation, with the present approach, further experiments are in progress in our laboratory which serve to investigate the rewarding properties of PE.

Self-administration of PE

There have, to date, been two reports of self-administration of PE (Risner and Jones, 1977; Shannon and De Gregorio, 1982). These two studies both demonstrated that intravenous infusions of PE may serve to maintain behaviour in the dog.

No studies have been reported which are directed to an assessment of the rewarding properties of PE in the rat. In our laboratory we have, therefore, initiated a study of the possible rewarding properties of intracerebroventricular (ICV) infusions of PE in this species. Our preliminary data indicate that, as with d-amphetamine (Gustafson and

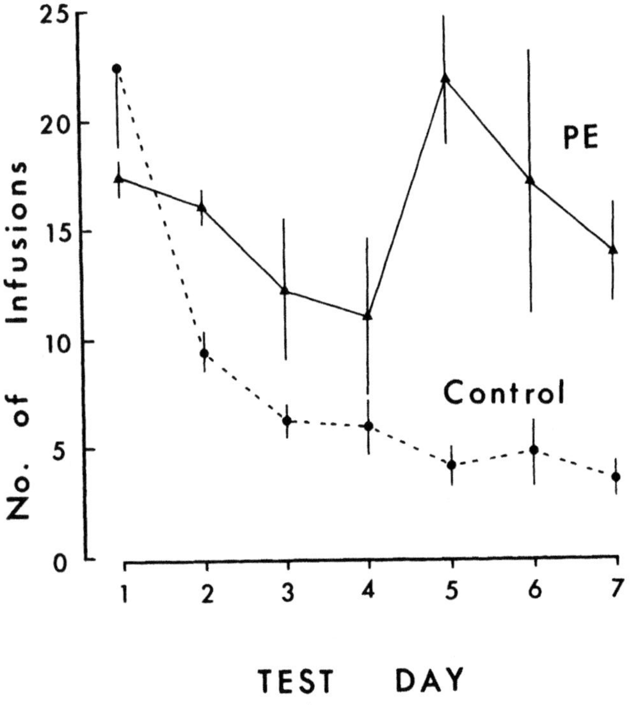

Fig. 6. The relative number of infusions in the PE vs saline self-administration groups after fixed-interval training. Each infusion (60 s, 1 μl) delivered PE at 1.67 μg/s or an equimolar quantity of saline (control).

Pickens, 1975), it is possible to maintain lever pressing with infusions of PE directly into the cerebral ventricles. Our approach has so far been to train animals to respond on a fixed-interval schedule of electrical hypothalamic stimulation. Following the initial training period the animals are given access to ICV PE on a schedule whereby a lever press results in a 60 second (1 μl) infusion of this compound into the third ventricle at a rate of 1.67 μg/s. Lever-presses during ongoing infusions have no programmed consequences. The effects of availability of this relatively high concentration of PE in contrast to an equimolar solution of saline are displayed in Figure 6 (independent groups). It can be seen from these data that, over a period of seven test sessions (60 minutes or a maximum of

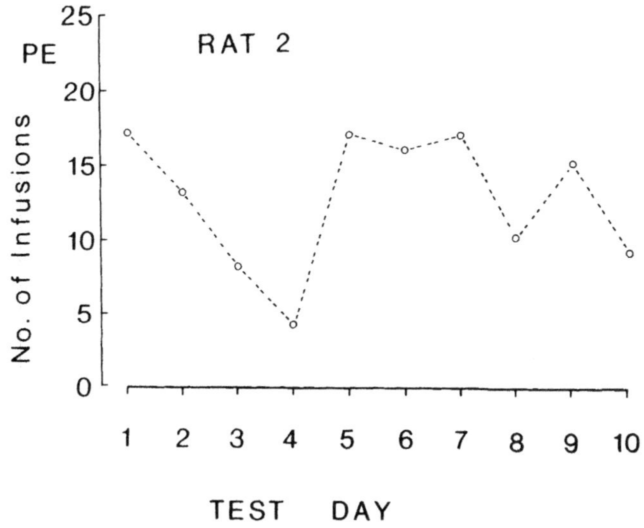

Fig. 7. The acquisition of PE self-administration for a single animal, rat 2. Details as for Fig. 6.

40 infusions per session), significantly more infusions were obtained by the PE group ($p < 0.01$).

The data for the best performer in the PE group over a period of ten test sessions are displayed in Figure 7.

In an attempt to control for possible drug-induced lever-pressing as an interpretation of these data the effects of an equivalent of ten PE infusions (over 600 seconds) were assessed on fixed-interval responding maintained by lateral hypothalamic stimulation. PE at this infusion rate markedly inhibited responding to an extent which was greater than that of the equimolar saline control infusion (saline: 63.6 ± 20.1%; PE 23.9 ± 7.1%. Mean ± S.D. relative to baseline, $p < 0.01$ within-group analysis). Although this is a very high total dose of PE, pilot data indicate that half of this amount delivered in an equivalent manner has very little effect on responding. This depressant effect of PE is consistent with previous results and tends to rule out drug-induced facilitation of responding as a contributing factor for the between-group difference in self-administration. Further work is in progress which is directed to a dose-response analysis of ICV PE self-administration in rats.

These data together with the previous reports of PE self-administration in the dog indicate that this compound may act as a reinforcer in different species. Whether this pharmacological phenomenon is of any physiological significance (i.e., in terms of neural substrates of reward or reinforcement) remains an open question which should merit further study.

AVERSIVE STIMULUS PROPERTIES OF β-PHENYLETHYLAMINE

Amphetamine is an extremely potent elicitor of CTA learning. Surprisingly, recent studies in our laboratory have clearly demonstrated that PE is one of the weakest pharmacological stimuli for the induction of CTA learning. This compound elicits only the most transient of taste aversions at doses as high as 41% of the LD50 (estimated from Heller and Lumbreras, 1976) in contrast to the marked potency of d-amphetamine at $\leq$ 4% of the LD50 for this compound (Greenshaw and Dourish, 1983). Repeated conditioning trials failed to enhance CTA learning with PE, as did prolonging its duration of action with deuterium substitution (Dourish *et al*., 1983) or MAO inhibition (Greenshaw and Dourish, 1984).

The differential potency of PE and of d-amphetamine in this context is particularly surprising in view of the general similarities of effects of these compounds described earlier. It is possible that this may be due to an important difference in the neuropharmacological profiles of d-amphetamine and PE. This might conceivably be due to a differential effect on dopamine and 5-HT systems. However, the observation that the substituted amphetamine fenfluramine (which has marked effects on 5-HT activity) is a potent elicitor of CTA learning (Booth *et al*., 1977) would tend to preclude such an interpretation. The differential CTA eliciting potency of d-amphetamine and PE is perhaps the one observation that clearly sets the behavioural effects of these compounds apart, particularly as the difference is not dependent on the rapid breakdown of PE by MAO.

The general significance of this difference is presently unclear, however, it remains that these two compounds are not functionally equivalent in this context.

CONCLUSION

In assessing stimulus profiles of PE and of d-amphetamine respectively it is apparent that there is a general similarity between these compounds when PE is protected from breakdown by MAO.

Both PE and d-amphetamine may act as discriminative stimuli, and to some extent exhibit cross-generalisation in this paradigm. Whether PE would be a more effective compound (in terms of the acquisition of discriminative control) in the presence of a MAO inhibitor is not known. The lack of such data is undoubtedly due to the lack of short-acting reversible MAO inhibitors.

In terms of behaviour maintained by electrical hypothalamic stimulation the effects of the two compounds on a number of measures seem to be equivalent. This equivalence is clearly dependent, however, on the pharmacodynamic profile of PE. When PE is administered alone at high doses, depressant effects on self-stimulation are evident. In marked contrast to this PE induces amphetamine-like effects when given at low doses in combination with a MAO inhibitor. In terms of reward or reinforcement per se, our preliminary results suggest that PE is an effective reinforcer when infused intraventricularly in the rat. This result is consistent with earlier studies of PE self-administration in dogs. It is noteworthy that in these self-administration studies no MAO inhibitors were used.

The results summarised above seem to be in good agreement with the proposal that PE and d-amphetamine may have equivalent behavioural effects. Nevertheless, in the area of CTA learning it is apparent that PE is virtually devoid of aversive stimulus properties. This clear difference between the effects of PE and d-amphetamine is intriguing. Most importantly the differential potency of PE and d-amphetamine in CTA learning is not dependent on the rapid breakdown of PE by MAO.

The analysis of PE levels in self-stimulation revealed that, with MAO inhibition, behavioural effects of PE are apparent at upwards of 75 ng/g in the rat striatum. Lower elevations of PE induced by l-deprenyl had no apparent behavioural effects despite a significant increase in both hypothalamic and striatal PE levels. It is of particular interest that with the present analysis a differential increase in striatal as opposed to hypothalamic PE was observed after l-deprenyl treatment. This observation may have important implications for a role of endogenous striatal PE in the regulation or modulation of dopaminergic neurotransmission.

The available data demonstrate clearly that PE has some interesting properties in terms of effects on conditioned behaviour. The similarities of effects of PE and d-amphetamine are striking. Nevertheless the CTA data indicate that the actions of these compounds are not simply equivalent.

It is apparent that PE activity may be physiologically important, possibly in terms of behavioural processes. Further work, however, involving an analysis of PE and its metabolites in relation to conditioned behaviour is necessary to determine whether this trace amine may play a functional role at the behavioural level.

ACKNOWLEDGEMENT

I thank Dr. A.A. Boulton for his advice and encouragement, the Saskatchewan Health Research Board for the provision of a research fellowship and the Medical Research Council of Canada and Saskatchewan Health for continuing financial support.

REFERENCES

Barger G. and Dale H.H. (1910) Chemical structure and sympathomimetic action of amines. J. Physiol. (Lond.) 41, 19-59.

Barry H. III (1974) Classification of drugs according to their discriminable effects in rats. Fed. Proc. 33, 1814-1824.

Biel J.H. (1970) Structure-activity relationships of amphetamine and derivatives, in Proc. International Symposium on Amphetamines and Related Compounds (E. Costa and S. Garattini, eds), pp 1-19, Raven Press, New York.

Booth D.A., Pilcher C.W.T., D'Mello G.D. and Stolerman I.P. (1977) Comparative potencies of amphetamine, fenfluramine and related compounds in taste aversion experiments in rats. Br. J. Pharmacol. 61, 664-677.

Boulton A.A. (1979) The trace amines: neurohumors (cytosolic, pre- and/or post-synaptic, secondary, indirect?). Behav. and Brain Sci. 2, 418.

Boulton A.A. and Baker G.B. (1975) The subcellular distribution of β-phenylethylamine, p-tyramine and tryptamine in rat brain. J. Neurochem. 25, 477-481.

Broadbent J., Greenshaw A.J. and Boulton A.A. (1984) Hypodipsic effects of β-phenylethylamine, phenylethanolamine, N-methyl phenylethylamine and d-amphetamine: a temporal analysis, in this book.

Chuang L.W., Karoum F. and Wyatt R.J. (1982) Different effects of behaviourally equipotent doses of amphetamine and methoxyamphetamine on brain biogenic amines: specific increase of phenylethylamine by amphetamine. Eur. J. Pharmacol. 81, 385-392.

Colpaert F.C., Niemegeers C.J.E. and Janssen P.A.J. (1980) Evidence that a preferred substrate for type B monoamine oxidase mediates stimulus properties of MAO inhibitors: a possible role for β-phenylethylamine in the cocaine cue. Pharmacol. Biochem. Behav. 13, 513-517.

Danielson J., Wishart T.B. and Boulton A.A. (1976) Effect of acute and chronic injections of amphetamine on intracranial self-stimulation and some aryl alkyl amines in the rat brain. Life Sci. 18, 1237-1243.

D'Mello G.D. and Stolerman I.P. (1977) Comparison of the discriminative stimulus properties of cocaine and amphetamine in rats. Br. J. Pharmacol. 61, 415-422.

Dourish C.T. and Boulton A.A. (1981) The effects of acute and chronic administration of β-phenylethylamine on food intake and body weight in rats. Prog. Neuro-Psychopharmacol. 5, 411-414.

Dourish C.T., Greenshaw A.J. and Boulton A.A. (1983) Deuterium substitution selectively enhances the effects of β-phenylethylamine on spontaneous motor activity in the rat. Pharmacol. Biochem. Behav. (in press).

Durden D.A. (1984) Quantification of trace amines and their metabolites by high resolution or metastable analysis using double focussing mass spectrometry, in this book.

Durden D.A., Philips S.R. and Boulton A.A. (1973) Identification and distribution of β-phenylethylamine in the rat. Can. J. Biochem. 51, 995-1002.

Durden D.A. and Philips S.R. (1980) Kinetic measurements of the turnover rates of phenylethylamine and tryptamine *in vivo* in the rat brain. J. Neurochem. 34, 1725-1732.

Dyck L.E. (1983) Release of monoamines from striatal slices by phenelzine and β-phenylethylamine. Prog. Neuropsychopharmacol. (in press).

Goudie A.J. (1979) Aversive stimulus properties of drugs. Neuropharmacol. 19, 971-979.

Goudie A.J. (1982) Discriminative properties in an operant task of β-phenylethylamine, in Drug Discrimination: Applications in CNS Pharmacology (F.C. Colpaert and J.L. Slangen, eds), pp 165-180, Janssen Research Foundation, Elsevier Biomedical, Amsterdam.

Greenshaw A.J. and Dourish C.T. (1983) Differential aversive stimulus properties of β-phenylethylamine and of d-amphetamine. Psychopharmacology (in press).

Greenshaw A.J. and Dourish C.T. (1984) β-phenylethylamine and d-amphetamine: differential potency in the conditioned taste aversion paradigm, in this book.

Greenshaw A.J., Juorio A.V. and Boulton A.A. (1984) Behavioural and neurochemical effects of 1-deprenyl and β-phenylethylamine in the context of self-stimulation. (submitted)

Greenshaw A.J., Sanger D.J. and Blackman D.E. (1981) The effect of drugs on fixed-interval responding maintained by electrical brain stimulation in rats, in Quantification of Steady State Operant Behaviour (C.M. Bradshaw, E Szabadi and C.F. Lowe, eds) pp 457-460, Elsevier Biomedical, Amsterdam.

Greenshaw A.J., Sanger D.J. and Blackman D.E. (1983) Effects of chlordiazepoxide on self-regulated duration of lateral hypothalamic stimulation. Psychopharmacology (in press).

Greenshaw A.J., Sanger D.J., Blackman D.E. and Boulton A.A. (1984) Effects of d-amphetamine and β-phenylethylamine on fixed-internal responding maintained by self-regulated electrical hypothalamic stimulation in rats. (submitted)

Gustafson L.K. and Pickens R. (1975) Intraventricular amphetamine self-administration. Fed. Proc. 34, 780.

Hauger R.L., Skolnick P. and Paul S.M. (1982) Specific [^{3}H]-β-phenylethylamine binding sites in rat brain. Eur. J. Pharmacol. 83, 147-148.

Heller B. and Lumbreras N. (1976) Studies on the role of phenylethylamine in methylamphetamine mechanisms. Experientia 32, 210-211.

Howard J.L., Pollard E.T., Rohrbach K.W. and Hartone N.E. Effect of B-phenylethylamine and d-amphetamine on electrical self-stimulation of brain. Pharmacol. Biochem. Behav. 5, 661-664.

Huang J.T. and Ho B.T. (1974) The effect of pretreatment with iproniozid on the behavioural activities of β-phenylethylamine in rats. Psychopharmacol. 35, 77-81.

Iversen S.D. (1977) Brain dopamine systems and behaviour, in Handbook of Psychopharmacology Vol. 8 L.L. Iversen, S.D. Iversen and S.H. Snyder (eds) pp 333-384, Plenum Press, New York.

Jackson D.M. (1972) The effect of β-phenylethylamine upon spontaneous motor activity in mice: a dual effect on locomotor activity. J. Pharm. Pharmacol. 24, 383-389.

Jones R.S.G. and Boulton A.A. (1980) Interactions between p-tyramine, m-tyramine or β-phenylethylamine and dopamine on single neurons in the cortex and caudate nucleus of the rat. Can. J. Physiol. Pharmacol. 58, 222-227.

Liebman J.M. (1983) Discriminating between reward and performance: a critical review of intracranial self-stimulation methodology. Neurosci Biobehav. Rev. 7, 45-72.

Philips S.R. (1981) Amphetamine, p-hydroxyamphetamine and β-phenylethylamine in mouse brain and urine after (-) and (+) deprenyl administration. J. Pharm. Pharmacol. 33, 739-741.

Philips S.R. and Boulton A.A. (1979) The effect of monoamine oxidase inhibitors on some arylalkylamines in rat striatum. J. Neurochem. 33, 159-167.

Risner M.E. and Jones B.E. (1977) Characteristics of β-phenylethylamine self-administration by dog. Pharmacol. Biochem. Behav. 6, 689-696.

Sandler M. and Reynolds G.P. (1976) Does β-phenylethylamine cause schizophrenia? Lancet 1, 70-71.

Sanger D.J. (1981) The role of operant behaviour in the pharmacology laboratory, in Quantification of Steady State Operant Behaviour (C.M. Bradshaw, E. Szabadi and C.F. Lowe, eds), Elsevier Biomedical, Amsterdam.

Schuster C.R. and Thompson T. (1969) Self-administration of and behavioral dependence on drugs. Ann. Rev. Pharmacol. 9, 483-502.

Shannon H.E. and DeGregorio C.M. (1982) Self-administration of the endogenous trace amines β-phenylethylamine, N-methyl phenylethylamine and phenylethanolamine in dogs. J. Pharmacol. Exper. Therap. 227, 52-60.

Stein L. (1964) Self-stimulation of the brain and the central stimulant action of amphetamine. Fed. Proc. 23, 836-841.

Zacharko R.M. and Kokkinidis L. (1982) Preference for fixed vs self-selected durations of brain stimulation in rats administered d-amphetamine sulphate. Life Sci. 30, 1029-1034.

Zacharko R.M. and Wishart T.B. (1979) Facilitation of self-stimulation with high doses of amphetamine in the rat. Psychopharmacology 64, 247-248.

DIFFERENTIATION OF PHENYLETHYLAMINE (PE)- AND AMPHETAMINE (AMPH)-INDUCED BEHAVIORS

Bruce I. Diamond, Ana Hitri, Chandresh Shah and Richard L. Borison

Department of Psychiatry, Medical College of Georgia and Augusta V.A. Medical Center, Augusta, Georgia 30912

ABSTRACT

Stereotyped behavior elicited by the chronic administration of the endogenous trace amine, PE, is similar to the behavior induced by AMPH in animals. Both PE and AMPH stereotypies have been proposed as animal paradigms of schizophrenia. In all studies male Sprague-Dawley rats (200-250 grams) were used. PE (50 mg/kg;i.p.) and AMPH (3.75 mg/kg;i.p.) were administered for a minimum of five weeks and animals were rated by a four point stereotypy scale. Pharmacological pretreatments with dopamine (DA) blockers, cholinergic blockers and serotonin (5-HT) blockers were studied on these behaviors. We found that all DA receptor blockers except metaclopramide, antagonized PE stereotypy, whereas clozapine, sulpiride, molindone, and fluperlapine failed to block AMPH behavior. Moreover, 5-HT blockers differentially affected both PE and AMPH stereotypies. In other studies, the direct injection of DA into the nucleus accumbens enhanced stereotypies induced by PE more than those induced by AMPH. The opposite was true when DA was applied to the caudate nucleus. Finally, electrophysiological recordings showed fast frequency spike activity in the caudate nucleus of AMPH treated rats. This same activity was present in the nucleus accumbens of PE treated animals. These results, along with other evidence, suggest that PE and AMPH stereotypy are mediated by different brain areas and mechanisms.

INTRODUCTION

The administration of AMPH or the endogenous trace amine PE to animals produces stereotyped behavior consisting of repetitive sniffing, head and limb movements (Randrup and Munkvad, 1970; Borison et al, 1977b; Borison and Diamond, 1978). Antipsychotic drugs, which are DA blockers, antagonize stereotypies induced by AMPH and PE (Angrist et al, 1974; Borison and Diamond, 1978; Moja et al, 1978). Both AMPH and PE stereotypy have been proposed as pharmacologic models for schizophrenia (Snyder, 1973; Borison et al, 1977b). Moreover, both AMPH and PE behavior appear to be mediated via brain 5-HT, noradrenaline and DA (Dourish, 1981, Braestrup and Randrup, 1978; Perlow et al, 1980).

The studies performed within this work were aimed at demonstrating the role of PE in behavior, its differences from AMPH behavior and its superiority as a predictor of antipsychotic drug activity.

METHODS

Animals used in all studies were albino male Sprague-Dawley rats (200-250 grams). They were housed one per cage under constant conditions controlled for temperature and humidity with 12-hours light-dark cycles and access to food and water ad libertum. Behavioral observations were conducted between 1000 and 1400 hours with animals in wire mesh cages (60 x 40 x 30 cm). Ratings for behavior were made by two independent observers who were blind to treatment using a modified four-point rating scale (Costall and Naylor, 1974) for stereotypy. The higher the score, the more intense the stereotypy, which goes from sniffing to head bobbing. Injection of all drugs was by an intraperitoneal route with drugs dissolved in saline and injected in a volume of 0.1 ml/100 gm. The effects upon stereotypy elicited by five weeks of PE (50 mg/kg) or AMPH (3.75 mg/kg) were studied after acute pretreatment with non-behaviorally active doses of various DA receptor blockers, cholinergic and 5-HT agents.

In other behavioral studies the effects of PE or AMPH were compared to DA when locally applied to either the nucleus accumbens (Nacc) or caudate nucleus (CN) of rats.

Animals were anesthetized with pentobarbital (35 mg/kg) and chronically implanted with bilateral 24 gauge stainless steel cannula using a David Korp stereotaxic apparatus. Coordinates for the Nacc and CN were obtained from the atlas of Pellegrino and Cushman (1967). Dental acrylic and jewelers screws were implanted in the skull to hold the cannula in place. Injections with 30 gauge steel tubing attached to a Hamilton syringe were made in freely moving rats. Animals were administered PE, AMPH or saline intraperitoneally for five weeks and DA was applied via the cannula located in the Nacc or CN. In other studies PE, AMPH and DA were applied to the Nacc or CN of naive rats. Cresyl violet staining for confirmation of cannula placement was performed for most animals.

In order to measure the electrical activity of the Nacc and CN of rats treated for five weeks with PE or AMPH, bipolar electrodes (Silver-silver chloride wire) with a tip separation of 0.5 mm and insulated with teflon, except 0.5 mm from the tip, were implanted stereotaxically into the CN and NAcc. These electrodes were connected to an amphenol plug which was fixed to the skull with dental acrylic. This connector was attached via a cable to a Beckman (Offner/Type R) dynograph which enabled the animals to move freely while electrocencephalographic recordings and behavioral observations were made.

RESULTS

The administration of PE or AMPH to rats produced little stereotypy on day 1 which progressed to intense head bobbing and sniffing on day 21 and remained at this level with continued drug administration for at least nine months. Neither the AMPH or PE rats displayed gnawing or chewing behavior; however, some of the PE treated animals did show limbic seizures (rearing with clonic forepaw and mouthing movements) after six months of treatment. PE stereotypy lasted for 30 minutes and AMPH stereotypy for three hours from week 3 to week 36. Beside the duration of stereotypy, other differences between PE and AMPH were the piloerection and salivation observed in PE treated animals. When the MAO inhibitor pargyline (20 mg/kg) was administered two hours prior to PE or AMPH treatment, the effects of PE and AMPH on stereotypy were not altered except that the duration of PE behavior was longer (four hours). This inhibition of MAO lasted for 72 hours as judged from subsequent PE daily injections.

The pharmacology of acute (one day) and chronic (five week) PE and AMPH behaviors were different. In figure 1, the effects of DA blockers on PE and AMPH behaviors are shown. In nonsedating doses haloperidol and pimozide completely antagonized the acute and chronic effects of PE and AMPH whereas thioridazine and clozapine, DA blockers with high anticholinergic activity, antagonized chronic PE and, to a lesser extent, AMPH stereotypy. However, thioridazine and clozapine markedly potentiated the acute effects of AMPH producing intense gnawing whereas clozapine only potentiated the acute effects of PE while thioridazine was without significant effects on acute PE behavior.

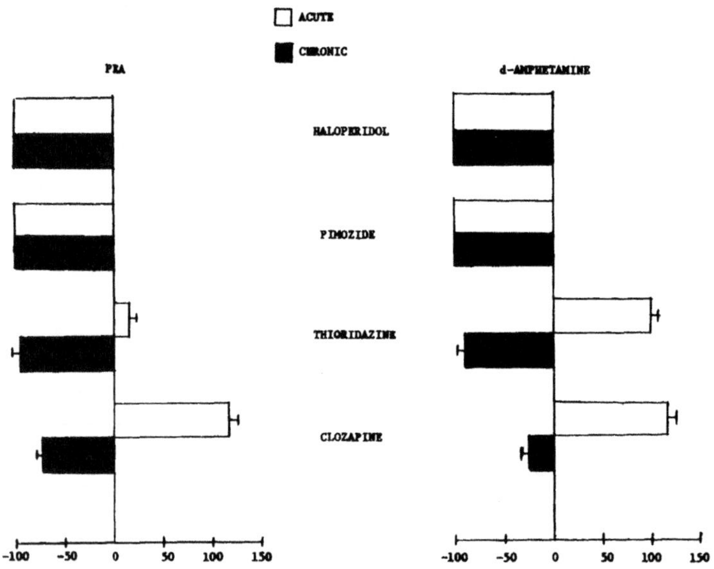

Figure 1 - Percent inhibition (-) or potentiation (+) of acute and chronic PE (50 mg/kg) or AMPH (3.75 mg/kg) stereotypy in rats by DA blockers (haloperidol 0.5 mg/kg, 30 min; pimozide 1 mg/kg, 1 hr; thioridazine 30 mg/kg, 1 hr; clozapine 30 mg/kg, 30 min).

The effects of cholinergic and 5-HT drugs on PE and AMPH behavior are shown in figure 2. Physostigmine, a cholinesterase inhibitor, slightly antagonized both acute and chronic PE and AMPH stereotypy. The cholinergic blocker, trihexyphenidyl, markedly potentiated the acute behavioral effects of AMPH and PE, and in AMPH treated animals produced gnawing behavior. This anticholinergic effect was absent in chronically treated AMPH animals and produced slight blockade of stereotypy in PE animals. The 5-HT precursor, 5-hydroxytryptophan, potentiated the acute behavioral effects of AMPH while antagonizing acute PE effects. It slightly antagonized both chronic AMPH and PE stereotypy. The 5-HT blocker, methysergide, markedly potentiated the acute behavioral effects of AMPH and PE while significantly antagonizing only chronic PE stereotypy.

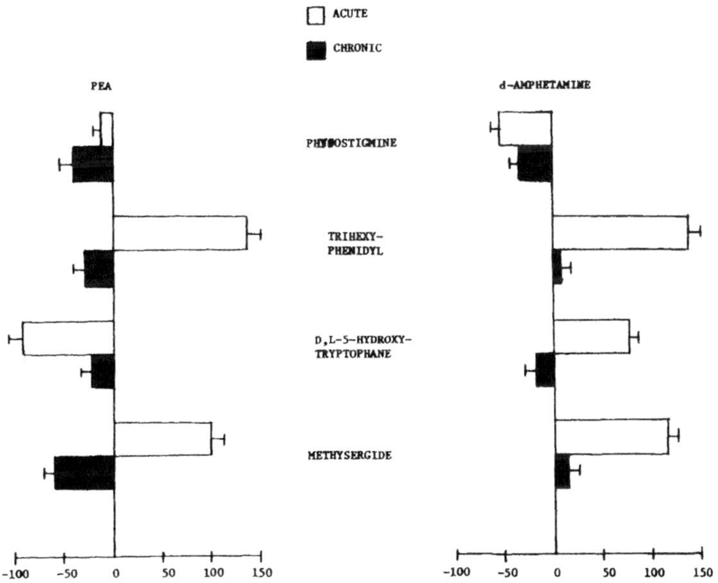

Figure 2 - Percent inhibition (-) or potentiation (+) of acute and chronic PE (50 mg/kg) or AMPH (3.75 mg/kg) stereotypy in rats by cholinergic and serotonergic agents (Physostigmine 0.5 mg/kg, 15 min; trihexphenidyl 1.0 mg/kg, 1 hr; D,L-5-hydroxytryptophan 250 mg/kg, 1 hr; methysergide 30 mg/kg, 15 min).

The pharmacology of five week PE and AMPH rats does not differ with other classical antipsychotic DA blockers. Acute pretreatment with fluphenazine (0.7 mg/kg; 1 hr), chlorpromazine (10 mg/kg; 30 min), or 7-hydroxychlorpromazine (10 mg/kg; 30 min) antagonized both PE and AMPH stereotypy. In contrast, the atypical dopamine blocker, sulpiride (10-30 mg/kg; 1 hr), selectively antagonized PE (80% inhibition) and not AMPH stereotypy. Moreover, fluperlapine (a derivative of clozapine), a DA blocker with few extrapyramidal effects in patients, as well as molindone, block selectively PE and not AMPH stereotypy at relevant doses (figure 3). In contrast, the DA blocker without antipsychotic activity, metoclopramide (1-5 mg/kg; 1 hr), failed to alter PE stereotypy while antagonizing selectively AMPH stereotypy.

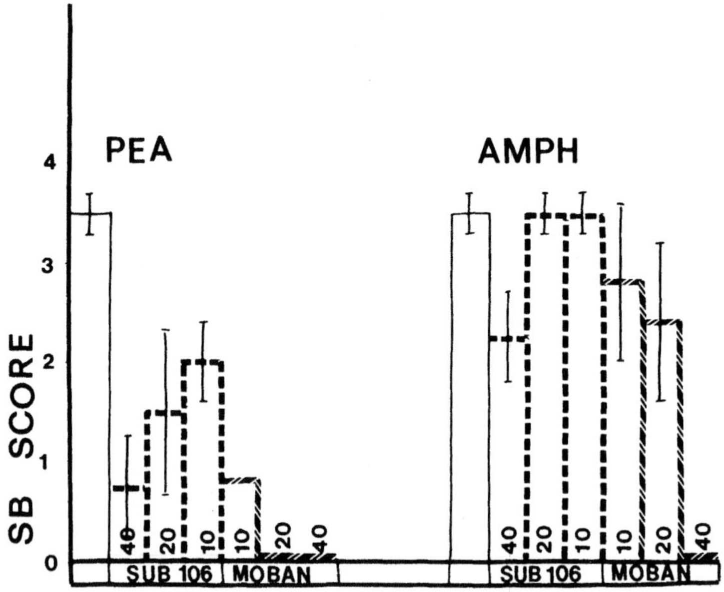

Figure 3 - Stereotypy behavior score (SB) of chronic PE and AMPH treated rats and the effects of various doses of fluperlapine (SUB 106) and molindone (MOBAN) on their stereotypy.

Blockers of 5-HT, namely methysergide and mianserin, did not significantly alter five-week stereotypy induced by AMPH although higher doses of methysergide did antagonize stereotypy by 20% (figure 4). Mianserin did not affect chronic PE stereotypy, whereas methysergide, at a low dose, potentiated and at a higher dose produced a slight but non-significant antagonism of PE stereotypy (figure 4).

In figure 5, the results of the single administration of depot fluphenazine enanthate (0.7 mg/Kg; s.c.) on chronic PE and AMPH stereotypy are shown. Three days after depot fluphenazine, PE stereotypy was inhibited by 72% and AMPH stereotypy by 43%. Although diminishing over time, this inhibition of PE and AMPH stereotypy lasted for 14 days and by day 17 was only 5%. PE stereotypy was inhibited more than AMPH stereotypy during the 14 day period.

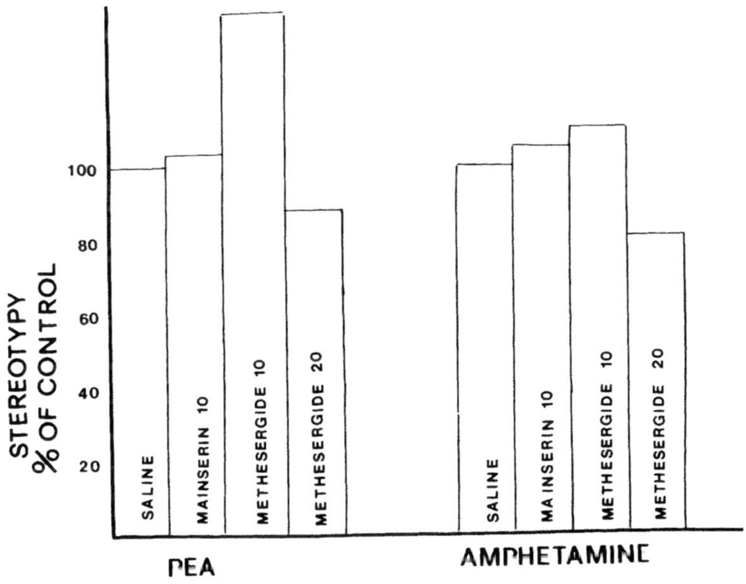

Figure 4 - Effect of two 5-HT antagonists in different doses (10 and 20 mg/kg, 15 min prior) on PE and AMPH stereotypy in rats. Data is expressed as percent of stereotypy of chronic PE and AMPH treated animals.

Serum neuroleptic levels were also monitored in these animals and there was a linear negative correlation (r - 0.92) between serum neruoleptic activity and stereotyped behavior induced by PE and AMPH. Thus, there was an inverse relationship between PE and AMPH stereotypy and serum neuroleptic levels. On day 3 there was 120 ng/ml while on day 17 there was only 2 ng/ml chlorpromazine equivalents in the serum of rats. On day 14, which was the last effective day of depot fluphenazine inhibition of PE and AMPH stereotypy, 35 ng/ml chlorpromazine equivalents were detected in the serum of rats.

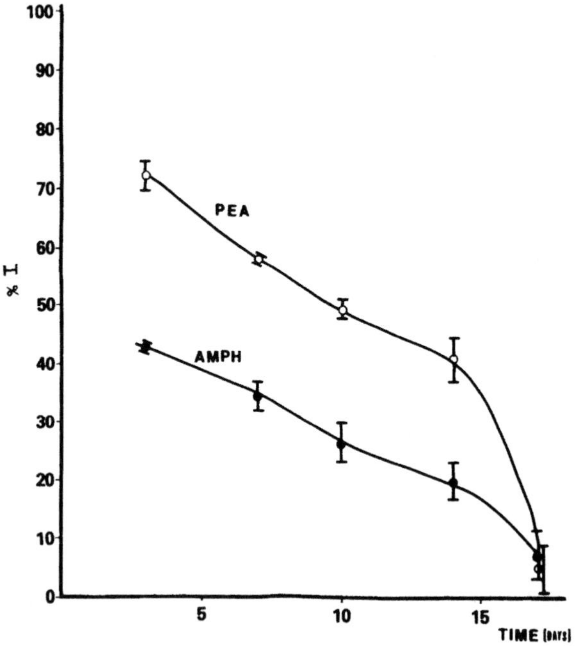

Figure 5 Effect of a single injection of fluphenazine enanthate (0.7 mg/kg, s.c.) on chronic PE and AMPH stereotypy for 17 days in rats. Data is expressed as percent inhibition of stereotypy.

The local application of 100 μg of PE to the Nacc of naive rats produced a stereotypy score of 3.3 which was equivalent to the effects of 200 μg of DA applied there. AMPH (100 μg) induced a stereotypy score of 2.6 when applied to the Nacc which was less than the effects of 100 μg of DA. In contrast, when the same dose of PE or AMPH was injected into the CN, AMPH induced intense stereotyped behavior (3.9) which was equivalent to the effects of 200 μg of DA, whereas PE produced only slight effects (1.8) on stereotyped behavior. When DA (5 μg) was applied to the Nacc of five week treated PE rats, the stereotypy elicited by this dose of DA was intense (3.5) and equivalent to 100 μg of DA injected into naive animals. AMPH treated animals did not show this response when DA was injected into the Nacc.

Electrophysiological recordings in freely moving animals failed to show acute and subacute AMPH and PE effects on electrical activity from striatal or limbic areas of the brain. However, by the third week of AMPH treatment, there was a significant increase in fast, high voltage activity in the right CN and the right Nacc of PE treated animals. At five weeks, there was a bilateral high voltage, fast activity in the Nacc of PE treated animals and the CN of AMPH animals. Both PE and AMPH animals at this time had stereotypy scores of 3.5.

DISCUSSION

PE and AMPH produce similar stereotyped behavior when administered chronically to rats. The increase in intensity of stereotypy over the three week period is a "reverse tolerance" and has been noted before (Borison et al, 1977b). The fact that subthreshold initial doses for inducing stereotypy eventually produce full-blown head and limb repetitions mirrors the kindling process suggested for electrical stimuli and its production of seizures (Goddard and McIntyre, 1969). The duration of PE stereotypy was increased from 30 minutes to 4 hours when the MAO inhibitor pargyline was used. This is not surprising since PE is an endogenous amine known to be metabolized by MAO type B (Yang and Neff, 1973). Moreover, others have found that MAO inhibitors potentiate PE stereotypy (Moja et al, 1976; Braestrup et al, 1974).

The ability of classical DA blockers to antagonize PE and AMPH stereotypy in animals supports the notion that these behaviors are mediated via brain DA. The atypical DA blockers clozapine, sulpiride, fluperlapine (a derivative of clozapine), thioridazine and molindone antagonized selectively PE stereotypy. Some of these drugs are antipsychotic agents with few extrapyramidal side effects (Gerlach and Simmelsgaard, 1978), are entirely missed when screening for antipsychotic activity (Kornetsky and Markowitz, 1975), and do not increase serum prolactin levels (Sachar, 1977). These agents also bind preferentially to limbic rather than to striatal DA receptors in animal and human brain (Borison et al, 1983; Hitri et al, 1983). Moreover, the ability of metoclopramide to antagonize AMPH and not PE stereotypy supports the different site of action of AMPH since metoclopramide is a DA blocker known to produce extrapyramidal side effects and yet is devoid of antipsychotic actions (Hassan et al, 1983). Also, AMPH stereotypy has been proposed to be mediated via the striatum (Randrup and Munkvad, 1970).

The acute effects of the atypical DA blockers thioridazine and clozapine on acute PE and AMPH stereotypy are opposite to their effects on chronic PE and AMPH behavior. In fact, the potentiation by thioridazine and clozapine of acute PE and AMPH stereotypy as well as the similar effect observed with trihexyphenidyl on these behaviors supports the notion that this potentiation is due to anticholinergic actions of these neuroleptics (Snyder and Yamamura, 1977). However, chronic PE treatment did not respond to these anticholinergic treatments with potentiation. One would thus argue that chronic PE alters DA receptors that are not modified by cholinergic blockade and these receptors are located outside the extrapyramidal system. This is based upon the well-known DA-acetylcholine balance in this system and the data demonstrating an increase in striatal DA receptors after AMPH and not PE treatment (Borison et al, 1979).

The results of mianserin and methysergide, two 5-HT antagonists, downplay a major role for 5-HT in PE and AMPH stereotypy as has been suggested (Dourish, 1981-1982). Although high doses of methysergide produce antagonism to stereotypy, these are not significant and yet higher doses, probably non-specific in action, were used in other studies (Dourish, 1981). Moreover, the effect of mianserin in potentiating chronic PE behavior and not AMPH behavior provides evidence for an inhibitory role of 5-HT in chronic PE behavior as well as differentiating PE from AMPH actions.

Both PE and AMPH stereotypy did not demonstrate tolerance to fluphenazine enanthate, a depot neuroleptic with long action (Hersch et al, 1973). The correlation of serum neuroleptic levels with antagonism of PE and AMPH stereotypy was significant and demonstrated a good correlation. PE stereotypy was antagonized more than AMPH stereotypy at similar serum levels, indicating a preferential blockade of different receptors for PE and AMPH actions. This is substantiated by the enhanced effects of DA when applied to the Nacc of chronically treated PE and not AMPH animals and the stereotypy induced by AMPH when applied to the CN and not Nacc of naive animals. Notwithstanding, the increased fast, high voltage activity observed in the Nacc of chronic PE and CN of chronic AMPH animals further differentiates neuroanatomically the actions of PE and AMPH.

Chronic PE stereotypy is antagonized by antipsychotic agents, including atypical agents such as clozapine, sulpiride and fluperlapine, which fail to antagonize AMPH stereotypy. Thus, chronic PE stereotypy has been proposed as a model to predict antipsychotic drug action (Borison et al, 1977b; Borison and Diamond, 1978; Borison et al, 1978). Antipsychotic agents appear to exert their therapeutic action via the blockade of limbic DA receptors (Lloyd, 1978), yet AMPH stereotypy is not blocked by all antipsychotic agents, is associated with DA receptor number increases in the CN (Borison et al, 1979), is associated with intense electrical activity in the CN and is not enhanced by DA application to the Nacc while PE stereotypy is blocked by all antipsychotic drugs, is associated with intense electrical activity in the Nacc, limbic seizures in animals, and enhanced stereotypy when DA is applied to the Nacc. Thus, PE stereotypy appears to be mediated via the DA mesolimbic system and AMPH stereotypy via the striatal DA system. (Supported in part by BRSG #S-07RR05365-21)

ACKNOWLEDGEMENT

We are grateful to Peggy Beasley for typing the manuscript.

REFERENCES

Angrist, B., Sathananthan, G., Wilk, S., et al (1974) Amphetamine psychosis: Behavioral and biochemical aspects. J. Psychiatr. Res. 11:13-23.

Borison, R.L. and Diamond, B.I. (1978) A new animal model for schizophrenia: Interactions with adrenergic mechanisms. Biol. Psychiatry 13:217-225.

Borison, R.L., Lemus, F., Reyes, M., Havdala, H.S. and Diamond, B.I. (1977a) Regional localization of 2-phenylethylamine in human brain. Research Communications in Psychology, Psychiatry and Behavior 2:193-201.

Borison, R.L., Havdala, H.S. and Diamond, B.I. (1977b) Chronic phenylethylamine induced stereotypy in rats: A new animal model for schizophrenia? Life Sciences 21:117-121.

Borison, R.L., Hitri, A., Klawans, H.L. and Diamond, B.I. (1979) A new animal model for schizophrenia: Behavioral and receptor binding studies, in Catecholamines: Basic and Clinical Frontiers (Usdin, E.,ed) pp. 719-721. Pergamon Press, New York.

Borison, R.L., Hitri, A., Blowers, A.J. and Diamond, B.I. (1983) Antipsychotic drug action: Clinical, biochemical and pharmacological evidence for site specificity of action. Clinical Neuropharmacology 6:137-150.

Braestrup, C. and Randrup, A. (1978) Stereotyped behavior in rats induced by phenylethylamine, dependence on dopamine and noradrenaline and possible relations to psychoses?, in Noncatecholic Phenylethylamines (Mosnaim, A.D. and Wolf, M.E., eds) pp. 245-269.

Costall, B. and Naylor, R.J. (1974) Detection of the neuroleptic properties of clozapine, sulpiride and thioridazine. Psychopharm. 43:69-73.

Dourish, C.T. (1981) Behavioral effects of acute and chronic B-phenylethylamine administration in the rat: Evidence for the involvement of 5-hydroxytryptamine. Neuropharmacology 20:1067-1072.

Dourish, C.T. (1982) An observational analysis of the behavioral effects of phenylethylamine in isolated and grouped mice. Brit. J. Pharmacol. 77:129-139.

Gerlach, J. and Simmelsgaard, H. (1978) Tardive dyskinesia during and following treatment with haloperidol, haloperidol + biperiden, thioridazine and clozapine. Psychopharmacologia 59:105-112.

Goddard, G.V., McIntyre, D.C. and Leech, C.K. (1969) A permanent change in brain function rsulting from daily electrical stimulation. Exp. Neurol. 25:295-330.

Hassan, M.H., Reches, A., Kuhn, C.M., Lopez, J. and Fahn, S. (1983) Metoclopramide blocks dopamine receptors. Neurology 33:90.

Hersch, S.R., Gaind, R., Rohde, P.D., et al (1973) Outpatient maintenance of chronic schizophrenia patients with long-acting fluphenazine: double-blind placebo trial. Br. Med. J. 1:633-637.

Hitri, A., Borison, R.L. and Diamond, B.I. (1983) Neuroleptic binding to human brain tissue. Transactions of the American Psychiatric Association 136:263.

Lloyd, K.G. (1978) The biochemical pharmacology of the limbic system: Neuroleptic drugs, in Limbic Mechanisms, The Continuing Evolution of the Limbic System Concept (Livingston, K.E. and Hornykiewicz, O., eds), pp. 263-305. Plenum Press, New York.

Moja, E.A., Stoff, D.M., Gillian, C.J. and Wyatt, R.J. (1976) Dose-response effects of β-phenylethylamine on stereotyped behavior in pargyline-pretreated rats. Biol. Psychiatry 11:731-742.

Pellegrino, L.J. and Cushman, A.J. (1967) A Stereotaxic Atlas of the Rat Brain. Appleton-Century Crofts, New York.

Perlow, M.J., Chiugh, C.C., Lake, C.R. and Wyatt, R.J. (1980) Increased dopamine and norepinephrine concentrations in primate CSF following amphetamine and phenylethylamine administration. Brain Res. 186:469-473.

Randrup, A., and Munkvad, I. (1970) Biochemical, anatomical and psychological investigations of stereotyped behavior induced by amphetamines, in Amphetamines and related compounds (Costa, E. and Garattini, S., eds), pp. 695-713. Raven Press, New York.

Sachar, E.J., Gruen, P.H., Altma, N., Langer, G., Halpen, F.S. and Crefer, M. (1977) Neuroregulators and psychiatric disease, pp. 242-246. Oxford University Press, New York.

Snyder, S.H., Taylor, R.M., Coyle, J.T. and Meyerhoff, J.L. (1970) d- and l-amphetamine effects in animals. Amer. J. Psychiat. 127:199.

Snyder, S.H. and Yamamura, H.I. (1977) Antidepressants and the muscarinic acetylcholine receptor. Arch. Gen. Psychiatry 34:236-239.

Yang, H.Y. and Neff, N.H. (1973) A specific substrate for type B monoamine oxidase of the brain. J. Pharmacol. Exp. Ther. 187:365-371.

STUDIES ON THE MECHANISM OF ACTION OF β-PHENYLETHYLAMINE STEREOTYPY IN RODENTS: IMPLICATIONS FOR A β-PHENYLETHYLAMINE ANIMAL MODEL OF SCHIZOPHRENIA

Colin T. Dourish

Psychiatric Research Division
Cancer and Medical Research Building
University of Saskatchewan
Saskatoon, Saskatchewan S7N OWO

ABSTRACT

The utility and limitations of using β-phenylethylamine (PE) stereotypy in rodents as an animal model of psychosis are examined. Stereotypy elicited by acute or chronic PE in rats and mice consists of a number of distinctive components which have separate neurochemical substrates. The dominant component of PE stereotypy in mice is abortive grooming which may be an hallucinogenic response corresponding to parasitosis in humans. The atypical neuroleptic clozapine which possesses both 5-HT and DA antagonist properties is the most potent blocker of PE stereotypy in rodents. It is proposed that stereotypy elicited by acute and chronic PE in mice may be a useful tool in prediction of the clinical usefulness of pharmacological agents in the treatment of a defined subgroup of psychiatric patients characterised by abnormally high and variable excretion rates of PE.

INTRODUCTION

Interest in the stimulant and stereotypic effects of PE arises both from the structural and pharmacological similarities between PE and amphetamine (AMPH, α-methyl PE) and from the possible role of PE in schizophrenia and affective illness suggested by clinical studies (see Davis *et al.*, 1982; Murphy *et al.*, 1983; Reynolds, 1983). It is well established that chronic AMPH abuse in non-schizophrenic individuals produces a syndrome which closely resembles paranoid schizophrenia (Connell, 1958) and exacerbates symptoms when given acutely to schizophrenics (Janowsky and Davis, 1974). The stereotyped behaviour elicited by large doses of AMPH in rodents is very similar to the compulsive behaviour observed in AMPH addicts who consume large amounts of the drug (Snyder *et al.*, 1974). Consequently, AMPH stereotypy in rodents has been proposed as a useful animal model of schizophrenia (Snyder, 1973). AMPH is thought to produce its behavioural stimulant effect by an indirect action (i.e. stimulation of release and blockade of reuptake) on the brain catecholamines dopamine (DA) and noradrenaline (NA) (Moore, 1978). Evidence in favour of the AMPH model of schizophrenia is that most clinically effective neuroleptic agents are DA receptor antagonists which potently antagonize AMPH stereotypy (Janssen *et al.*, 1966). However, it should be noted that both AMPH and most neuroleptics also have pharmacological actions on non-catecholaminergic neurotransmitter mechanisms (Sloviter *et al.*, 1978; Niemegeers and Janssen, 1979) and the importance of this factor will be discussed at length below.

Acute or chronic administration of PE in rodents produces stereotypy which is similar in many respects to AMPH stereotypy (Randrup and Munkvad, 1966; Braestrup and Randrup, 1978). One recent suggestion is that chronic PE stereotypy may be a more useful animal model of schizophrenia than AMPH stereotypy since PE stereotypy is selectively antagonised by the atypical neuroleptics clozapine and thioridazine (Borison *et al.*, 1977). These compounds have been claimed to be selective antagonists of mesolimbic DA receptors, the DA receptor population which is thought to be principally involved in the pathophysiology of schizophrenia (Burki *et al.*, 1975; Borison *et al.*, 1977). The specificity of these compounds for mesolimbic DA receptors has recently been questioned however (e.g. Reynolds *et al.*, 1982) and in addition the experimental data outlined in this chapter casts considerable doubt on the hypothesis of preferential antagonism of PE stereotypy by selective mesolimbic DA receptor blockade. The present chapter reviews recent studies on the behavioural and neurochemical mechanisms involved in PE stereotypy and considers the utility and limitations of PE stereotypy as an animal model of psychosis.

METHODS

A Methodological Critique

The majority of previous studies on the stimulant and stereotypic effects of PE utilized either automatic activity recording devices (e.g. photocell cages, Mantegazza and Riva, 1963; Jackson, 1972) or stereotypy rating scales (e.g. Borison *et al.*, 1977), or in some cases a combination of both methods (e.g. Moja *et al.*, 1976) as the measure of behavioural response. Recently, it has become increasingly evident that the usefulness of both of these methods is very limited.

Stereotypy may be defined as the purposeless repetition of a particular behaviour or group of behaviours from the animals normal behavioural repertoire to the exclusion of all other normal behaviours of that repertoire. For example, a rat injected with a large dose of PE will continually sniff and make head and paw movements but will not eat, drink or groom during the period of drug action. Stereotypy rating scales (as recently pointed out by Fray *et al.*, 1980) tend to confound the measurement of stereotypy by requiring that an observer not only judge whether a particular response has occurred but also assess its stereotypic nature. This type of procedure may lead to confusion and loss of information about the particular behavioural response since in many scales a single category may include different responses with similar degrees of stereotypy. For example in the Borison *et al.* (1977) scale +2 represents occasional head-bobbing, licking or gnawing at cage floor, easily distracted by auditory stimuli. A further, and perhaps more important, difficulty is that intensification of stereotypy is often inferred by the transition from one type of response to another. To take the example of the Borison *et al.* (1977) scale once again, +1 represents increased exploration and discontinuous sniffing whereas +4 represents continuous headbobbing. Often, however, stereotypy does not follow such a monotonic dose-response curve and different behavioural components may have different neurochemical substrates and display different temporal parameters.

Limitations in the use of photocell cages as the sole measure of behavioural response have also been pointed out (Krsiak *et al.*, 1970; Robbins, 1977). Photocell counts tend to exaggerate drug effects but more importantly such apparatus used in isolation cannot distinguish between increased counts produced by co-ordinated locomotion and increased counts produced by stereotyped movements performed in one location. In the studies reviewed in this chapter an attempt was made to overcome these various limitations by utilizing a detailed observational analysis of various components of stereotyped behaviour in combination with a novel photocell activity measure.

Apparatus and Test Procedures

Rat test apparatus. Rats were tested in individual plexiglass cages (40 cm square, 23 cm high) positioned in photobeam-based activity recording devices (Opto-Varimex Minor, Columbus Instruments, Columbus, Ohio). A logic circuit in these devices enabled a distinction to be made between whole-body ambulatory movements and photobeam interruptions due to other movements made by the animals. Total horizontal activity (including grooming, scratching, stereotypy, head swaying, tail movements, etc.) was determined by the interruption of any one of 12x12 infrared photobeams (3 cm apart), in any order. Ambulatory activity (co-ordinated locomotion), on the other hand, was determined by the interruption of consecutive photobeams. Vertical activity (rearing and jumping) was recorded by units equipped with a series of 12 photobeams (3 cm apart) (Opto Vertimex, Columbus Instruments) which were suspended (15 cm above the cage base) from the walls of the cage. Interruption of any photobeam produced a 1 ms pulse which was counted by a microprocessor (built by J.A. Koehler, Dept. of Physics, University of Saskatchewan, Canada) linked to the activity devices by standard 0.6 cm telephone jacks. The microprocessor operated as a completely self-contained data-collecting sub system according to standard ASCII single-letter commands, which it received through a serial RS-232 compatible interface from an Apple II (plus) microcomputer (Apple Computer, Cupertino, CA). The Apple II was a completely standard unit that made use of the Apple communications interface card to communicate with the microprocessor and to display the data collected either continuously or at preselected intervals in the form of tables or histograms.

Table 1

Rating scale used in the observational analysis of components of spontaneous activity

Score	Frequency
0	Absent
1	Mild intensity or present 1-2 times during time sample.
2	Moderate intensity or present 3-4 times during time sample.
3	High intensity or present 5 or more times during time sample.
4	Severe or present for prolonged periods during time sample.

Details of rat observational analysis. Rats were observed either continuously or during predetermined time sample periods (at intervals of 5-10 min) and their behaviour rated on a 5 point scale according to the intensity of responses observed (see Table 1). The behavioural responses rated were determined by a comprehensive series of preliminary studies and are most descriptive of PE's behavioural effects at high doses (see Table 2).

Mouse test apparatus and observational analysis. Mice were observed in circular plexiglass cages of 22 cm diameter and 25 cm high and their behavioural responses rated at intervals on the scale described (Table 1). The cages were not equipped with photocells, however, and the observational analysis included additional categories which were not used in the rat studies. The additional categories were: freezing (inactive in one location), forward walking (coordinated locomotion), licking (compulsive licking or biting of cage walls or floor), jumping and vocalization. All drugs were administered IP in a volume of 1 ml/kg (rats) or 10 ml/kg (mice).

Table 2

Definition of behavioural components scored in the observational analysis of rat behaviour

Behavioural Component	Description
Sniffing	Distinctive pattern of air inspiration accompanied by vibrissae movements.
Headweaving	Repetitive side-to-side movements of the head.
Headbobbing	Repetitive up and down movements of the head.
Splayed hindlimbs	A dramatic extension of the hindlimbs causing a flattening of the body posture.
Hyperreactivity	Startle response to a pencil tap on the cage top.
Grooming	Purposeful licking and cleaning of the body.
Forepaw padding	Repetitive placing movements of the forepaws - "piano playing".
Rearing	Both front paws off the ground.
Straub tail	Tail stiff and erect.
Piloerection	Hair standing on end.
Tremor	Shaking of head, limbs or whole body.
Rigidity	Assessed both by grasping the rat around the torso and by passively extending and flexing the hindlimbs.
Backward walking	Backward walking or circling.

EFFECTS OF β-PHENYLETHYLAMINE ON
SPONTANEOUS MOTOR ACTIVITY IN THE RAT

The dose response effects of PE on certain components of rat behavior are illustrated in Figure 1. Doses of 3.125, 6.25 and 12.5 mg/kg PE do not produce stereotypy. At a dose of 25 mg/kg PE, however, repetitive sniffing, head movements and hyperreactivity are evident. A "full blown" stereotypy is elicited by 50 mg/kg PE and animals given this drug dose display sniffing, head movements, forepaw padding, splayed hindlimbs, hyperreactivity, Straub tail, rigidity and piloerection; grooming is abolished by a high dose of PE.

The automatic photocell activity data (Figure 2) illustrate that 12.5 mg/kg PE stimulates horizontal and vertical activity without producing stereotypy. The dose-

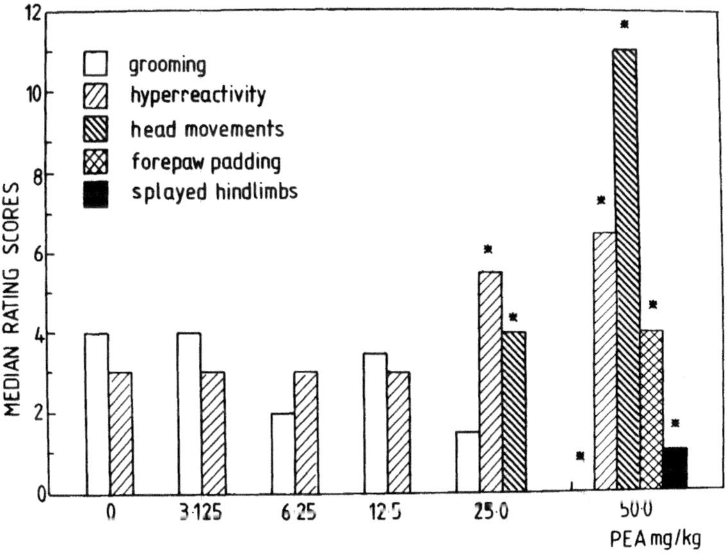

Fig. 1. Dose response relationships of β-phenylethylamine-induced stereotypy in rats. Significant differences refer to 2-tailed Mann Whitney U test comparisons between drug and control groups: * $p<0.05$. From Cooper and Dourish (1983).

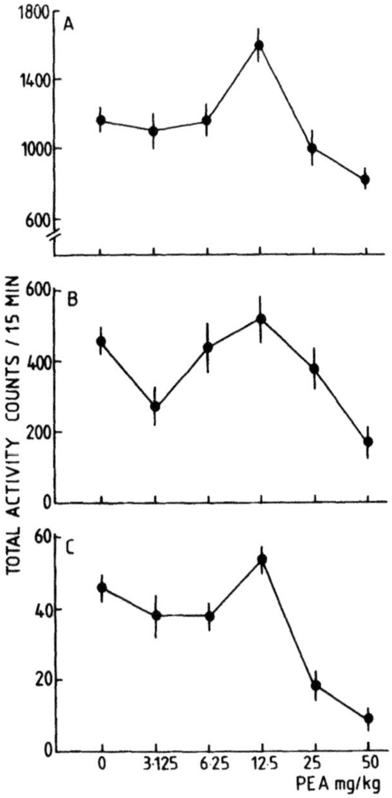

Fig. 2. Effects of β-phenylethylamine on automated measures of spontaneous motor activity in rats. A, Total Horizontal Activity; B, Ambulation; C, Vertical Activity (see Methods for details). Each point represents the mean for 8 animals during a 15 min test. From Cooper and Dourish (1983).

response curve resembles an inverted U shape since activity counts tend to decrease at 25 and 50 mg/kg PE due to the appearance of stereotypy. Both the stimulant and stereotypic effects of PE are brief at these doses with a duration of drug action of approximately 20 min (for further details see Dourish et al., 1983). The end of the PE stereotypy syndrome is marked by the appearance of "suppressed behaviours," in particular grooming and exploration. The stereotypic effects of PE in rats had previously been ascribed to an action of the compound on brain catecholamines, particularly DA (Braestrup and Randrup, 1978; Borison et al., 1977). However, many of the components of behaviour which are produced or increased by high dose PE treatment (i.e. headweaving, forepaw padding, splayed hindlimbs, rigidity, hyperreactivity) have been described as the "serotonin (5-HT) syndrome" and are associated with an increase in central 5-HT neurotransmission (see review by Jacobs, 1976). Further studies in our laboratory (Dourish, 1981; Dourish and Dewar, 1982) and by others (Sloviter et al., 1980; Widelitz, 1979) have confirmed that high doses of PE do indeed elicit 5-HT related behavioural responses. Acute administration of 50 mg/kg PE or chronic administration of 25 mg/kg PE produces the 5-HT syndrome in Sprague-Dawley, Wistar and Long-Evans rats (Dourish, 1981; Dourish and Dewar, 1982).

Rats given daily injections of 25 or 50 mg/kg PE show sensitization to the compound's behavioural effects (Table 3) and after three weeks treatment at these doses all rats show the 5-HT syndrome. This sensitization phenomenon is reflected in a decreased latency to onset of the syndrome and an increased duration of the response (see Dourish, 1981). Sensitization of the stereotypy induced by 50 mg/kg PE had previously been described by Borison et al. (1977) who claimed that the response was mediated by mesolimbic DA receptors, since it was selectively antagonized by clozapine and thioridazine (see Introduction). This interpretation now appears dubious in light of the following evidence. First, the group of behavioural responses induced by 50 mg/kg PE are those termed the "5-HT syndrome" and which are elicited by 5-HT agonists such as quipazine or precursors such as L-tryptophan when given with a monoamine oxidase inhibitor (Jacobs, 1976; Sloviter et al., 1978). Second, the putative 5-HT mediated responses are selectively antagonized by the 5-HT antagonists mianserin and methysergide in addition to clozapine (Table 4). Mianserin and methysergide prevented forepaw padding, splayed hindlimbs, rigidity and headweaving but potentiated hyperreactivity and elicited hyperactivity (Table 4). In contrast, haloperidol prevented hyperreactivity and attenuated headweaving but had no effect on the other components of the syndrome. Clozapine prevented all the components of the PE syndrome.

Table 3

Dose-response effects of acute and chronic administration of β-phenylethylamine in evoking the 5-HT behavioural syndrome in rats

Number of days of β-phenylethylamine treatment	Drug dose (mg/kg)	No. of rats tested	% of rats displaying the 5-HT syndrome
1 (Acute injection)	0	10	0
	12.5	5	0
	25	15	0
	50	20	60
7	0	10	0
	12.5	5	0
	25	15	60
	50	20	100
14	0	10	0
	12.5	5	0
	25	10	80
	50	10	100
21-42	0	10	0
	12.5	5	0
	25	10	100
	50	10	100

From Dourish (1981).

It had previously been suggested (Jacobs, 1976; Deakin and Dashwood, 1981) that hyperactivity and hyperreactivity, unlike the other components of the 5-HT syndrome may not be a reflection of serotonergic activation but instead are mediated by DA. The data in Table 4 are consistent with this proposal and therefore it seems that the most parsimonious explanation for the potency of clozapine against PE stereotypy is that this neuroleptic, which has a relatively high affinity for 5-HT receptors compared to DA receptors (Leysen et al., 1978) suppresses both the 5-HT- and DA-mediated responses which are associated with the PE

Table 4

Effects of pretreatment with 5-HT or dopamine antagonists on the behavioural syndrome induced by chronic injection of β-phenylethylamine (25 or 50 mg/kg)

Antagonist	Dose (mg/kg)	No. of rats tested	Effects on β-phenylethylamine behavioural syndrome
Mianserin	20 or 30	5	Inhibition of all components except hyperreactivity in all rats.
Methysergide	20 or 30	10	Inhibition of all components except hyperreactivity in all rats.
Clozapine	20 or 30	10	Sedation and inhibition of all components in all rats.
Haloperidol	0.2	10	Inhibition of hyperreactivity and disappearance of headweaving in 4 out of 10 rats.

From Dourish (1981).

syndrome. The recent report that PE inhibits 5-HT binding in human brain tissue is consistent with this proposal (Reynolds, 1983).

EFFECTS OF β-PHENYLETHYLAMINE ON SPONTANEOUS MOTOR ACTIVITY IN THE MOUSE

The principal behavioural effects elicited by PE in mice contrast with those observed in rats. A 25 mg/kg dose of PE induces sedation reflected in a significant increase in freezing scores (Table 5). Mice treated with 25 mg/kg PE tend to remain in one location of the cage and display decreased rearing and grooming accompanied by increased headweaving, hyperreactivity and sniffing (Table 5).

Table 5

Dose response effects of β-phenylethylamine in mice

Behavioural	Dose of β-phenylethylamine (mg/kg)					
Category	0	25	50	75	100	150
Freezing	3.13	4.00*	1.80	1.00	0.40*	0.33*
Forward Walking	1.96	1.50	4.00*	3.05	2.70	2.91
Rearing	1.50	0.50	0.80	0.41	0.10	0.99
Grooming	1.55	0*	2.00	3.12*	3.70***	3.33***
Sniffing	1.91	3.75*	4.00*	4.00**	4.00*	4.00**
Backward Walking	0	0	0	0	0.70**	1.33***
Headweaving	0	1.50**	4.00***	3.37***	3.30***	3.88***
Padding	0	0	0	2.37**	2.20***	2.55***
Hyperreactivity	0	2.25***	2.25***	3.00***	2.10**	1.40**
Tremor	0	0	0	0	0	1.30***
Jumping	0.5	0	0	0	0	0
Licking	0	0	0.25	2.12***	0.80	0.99**
Vocalization	0	0	0	0	0	0

Each score represents a median value of 8 animals averaged across time samples on a 0–4 scale. Statistical comparisons (Mann-Whitney U Test) are between drug and control values: * $p<.05$; ** $p<.02$; *** $p<.01$. N.B. Mice were tested individually since group testing modulates the behavioural effects of PE (see text). Data from Dourish (1982a).

Since it has been claimed that PE has a direct agonist action at DA receptors (see Jackson, 1978; Antelman et al., 1977), Jackson and Jenkins (1983) have speculated that the depression of activity produced by 25 mg/kg PE may be equivalent to the sedation elicited by low doses of DA agonists (e.g. apomorphine, piribedil) via the activation of DA autoreceptors (Carlsson, 1975; Dourish and Cooper, 1981). However, this interpretation appears doubtful since PE (25 mg/kg) treated animals exhibit repetitive headweaving, sniffing and hyperreactivity whereas animals given low doses of piribedil or apomorphine display repetitive yawning, stretching and penile erection but not headweaving, sniffing or hyperreactivity (Mogilnicka and Klimek, 1977; Cooper et al., 1983). Further studies including an examination of low dose PE effects on DA synthesis and release are required to shed more light on the neurochemical and pharmacological basis of PE hypomotility. In contrast to the effects described above, PE at doses of 50-150 mg/kg elicits a hyperactivity syndrome which includes headweaving, forepaw padding, hyperreactivity, sniffing, splayed hindlimbs, rearing, grooming, tremor, backward walking, licking and gnawing (Table 5). A 50 mg/kg dose of PE produces a brief stimulant effect which has a duration of approximately 20 min. Doses of 75-150 mg/kg PE, however, produce a syndrome of three distinct phases. The brief initial phase (0-5 min after injection) which consists of forward walking, sniffing and headweaving is succeeded by a locomotor depressant phase (5-20 min after injection, referred to as "early phase stereotypy") which consists of abortive grooming (the dominant response), headweaving, splayed hindlimbs, forepaw padding, sniffing and hyperreactivity (see Fig. 3) and a late locomotor stimulant phase (20-45 min after injection, referred to as "late phase stereotypy") which is characterised by forward walking, sniffing, hyperreactivity, rearing and licking.

A biphasic stimulant effect of PE on activity was first reported to occur by Jackson (1972) in grouped mice (n=5) but not in mice tested individually. However, these data were generated by photobeam interruptions, a method which can produce misleading results (as outlined in the Methodological Critique). Observational analysis has revealed that group testing produces an increase in the hyperreactive responses of running, jumping and vocalization (at the expense of stereotyped responses) which tends to increase photocell counts (Dourish 1982a). When PE

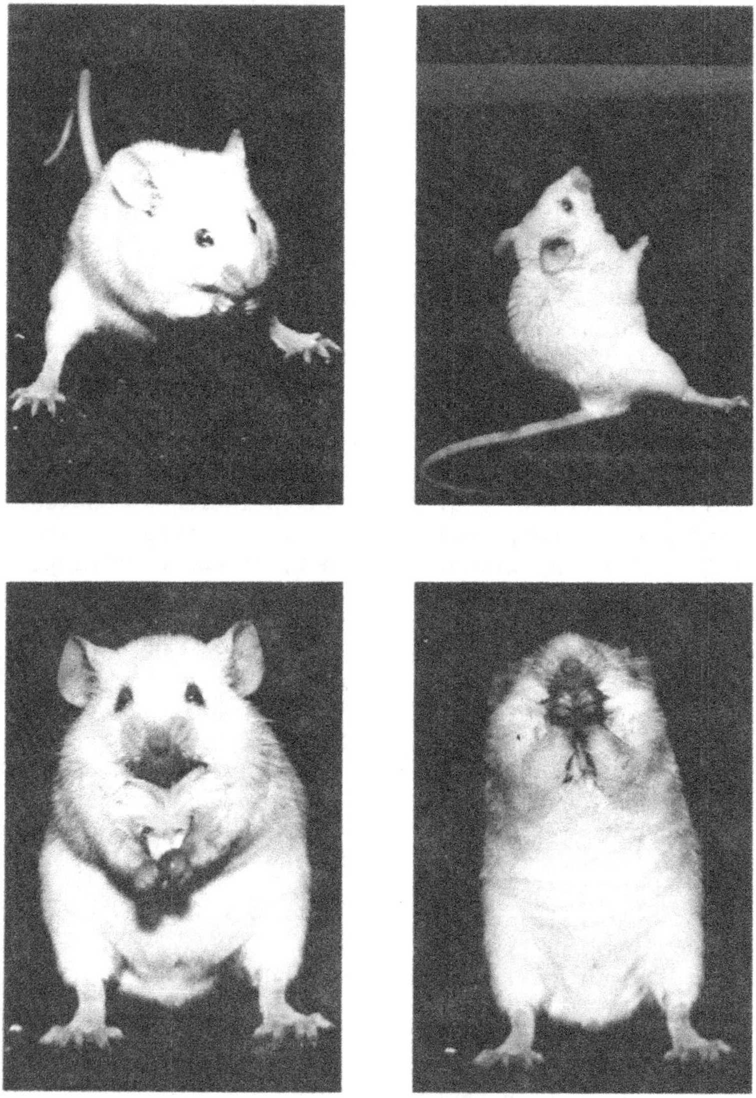

Fig. 3. Stereotypy induced by high doses (50-75 mg/kg) of β-phenylethylamine in the mouse. Top: splayed hindlimbs, loss of balance, Straub tail. Bottom: abortive grooming, salivation.

treated mice are tested individually the stereotyped components of abortive grooming, headweaving, forepaw padding, rearing and licking are dominant, and, therefore, photocell counts are lower (Dourish, 1982a). Notwithstanding this criticism, data generated by observational analysis are in good agreement with the original suggestion by Jackson (1972) of at least two distinct phases of PE stimulation in mice. Jackson (1978) proposed that early phase stimulation (0-5 min post injection) involves both DA and NA but that late phase stimulation (20-45 min post injection) is due to a direct agonist action of PE on postsynaptic DA receptors. Observational data from this laboratory are in general agreement with Jackson's (1978) proposal. However, an examination of the effects of compounds with predominantly NA antagonist, DA antagonist or 5-HT antagonist properties on individual components of stereotypy elicited by 75 mg/kg PE in mice has revealed that indoleaminergic mechanisms are of considerable importance in the mediation of early phase stereotypy (Dourish, 1982b).

The major effects of various putative neurotransmitter antagonists on PE stereotypy are summarized in Table 6. Headweaving and splayed hindlimbs are antagonized by mianserin, methysergide, metergoline, clozapine and propranolol. This suggests that these responses are 5-HT-mediated since since all of these drugs are reported to block central 5-HT receptors (Leysen et al., 1978; Weinstock et al., 1977; Niemegeers et al., 1983). 5-HT mechanisms also appear to be involved in the mediation of forepaw padding since this response was blocked by mianserin, methysergide, metergoline and clozapine. However, DA is also implicated in this response since haloperidol and clozapine attenuated forepaw padding. In addition, padding was strongly potentiated by adrenoreceptor blockade which suggests that stimulation of NA mechanisms may inhibit the expression of this behaviour. Rearing and licking were prevented by DA blockade and by metergoline but potentiated by 5-HT or adrenoreceptor blockade. These data suggest that activation of NA or 5-HT mechanisms has an inhibitory effect on rearing and licking which are DA mediated. This is consistent with the finding that NA release inhibits the expression of PE-induced gnawing and licking in rats (Mogilnicka and Braestrup, 1976; Braestrup, 1977). The blockade of licking by metergoline may appear inconsistent with the drug's established 5-HT antagonist effects. However, metergoline is reported to act as a DA antagonist in the striatum

(Spano et al., 1978) which may explain the observed suppression of licking. Hyperactivity was prevented by the DA antagonists haloperidol and thioridazine but potentiated by the 5-HT blockers metergoline and methysergide. These data indicate that PE hyperactivity in mice may be mediated by DA and inhibited by 5-HT, which is consistent with a similar proposal regarding the hyperactivity elicited by tranylcypromine plus L-tryptophan in rats (Green et al., 1981). Backward walking was rarely observed in the antagonist interaction studies since 75 mg/kg PE is below the threshold dose required to elicit this response (Dourish, 1982a). However, mice given phenoxybenzamine prior to 75 mg/kg PE consistently displayed backward walking. Therefore, it appears likely that backward walking, like rearing and licking, may be suppressed by PE-induced release of NA.

Table 6

Summary of the effects of putative antagonist drugs on β-phenylethylamine stereotypy in mice

Behavioural Response	Inhibited by	Potentiated by	Interpretation
Headweaving Splayed hindlimbs	metergoline mianserin methysergide clozapine propranolol	-	5-HT mediated.
Forepaw padding	metergoline mianserin methysergide haloperidol clozapine thioridazine	phenoxybenzamine propranolol	5-HT mediated. Possible DA involvement and inhibitory NA influence.
Rearing Licking	haloperidol thioridazine clozapine metergoline	phenoxybenzamine propranolol mianserin methysergide phentolamine	DA mediated. NA and 5-HT inhibitory influence.

Table 6 (continued)

Behavioural Response	Inhibited by	Potentiated by	Interpretation
Hyperre-activity	clozapine thioridazine haloperidol metergoline mianserin propranolol phenoxybenzamine	-	DA mediated.
Hyper-activity	haloperidol thioridazine	metergoline methysergide	DA mediated with 5-HT inhibition.
Sniffing	phenoxybenzamine	-	NA mediated.
Backward walking	not tested	phenoxybenz-amine	Mediated by DA and 5-HT with NA inhibition.
Abortive grooming	haloperidol clozapine thioridazine mianserin methysergide metergoline phentolamine phenoxybenzamine propranolol	-	May involve DA, NA and 5-HT but further studies using lower doses required.

Sniffing appears to be mediated by NA receptors since it was prevented only by phenoxybenzamine. In contrast, abortive grooming was prevented by all the antagonists tested, suggesting that both catecholamines and 5-HT are involved in the mediation of this response.

CONCLUSIONS AND CLINICAL CONSIDERATIONS

It is apparent that various components of PE stereotypy in rodents have separate neurochemical substrates. The behavioural responses which have been examined to date involve an interaction with brain DA, NA and 5-HT mechanisms. In mice, PE stereotypy is dominated by abortive grooming (see Figure 3), a behavioural response which shows sensitization following chronic PE administration (Dourish, unpublished data). PE-induced abortive grooming in mice appears to bear a close resemblance to the abnormal grooming seen in monkeys following around-the-clock AMPH and which has been suggested to correspond to the "parasitosis" commonly seen in human AMPH psychosis and in certain types of paranoid psychopathology (Munro, 1978; Nielsen et al., 1983). Since progressive behavioural alterations are often observed in the course of schizophrenia (Docherty et al., 1978; Silberman and Post, 1980) the further examination of alterations in the abortive grooming response during chronic PE would appear to be of immediate clinical relevance.

It is well established that schizophrenia and depression are heterogeneous disease states which encompass numerous sub-groups (e.g. unipolar depressive, bipolar depressive, phobic-anxious depressive, paranoid schizophrenic, non-paranoid schizophrenic). Recently, a distinctive subgroup of patients with major affective disorders has been described who are characterized by extremely high and fluctuating excretion rates of PE. These patients may excrete daily up to one thousand times more PE than other patients with apparently similar disorders or controls (Murphy et al., 1983). Similarly, fluctuating high PE excretion rates have been reported in certain bipolar depressive patients correlated with the onset and offset of manic-psychotic episodes (Karoum et al., 1982). There are also reports of increased PE in the urine of paranoid schizophrenic patients compared to non-paranoid schizophrenics or controls (Potkin et al., 1979; Jeste et al., 1980). Clearly a PE animal model of psychosis should be directly relevant to such clinical findings. Therefore, it appears possible that PE stereotypy and chronic PE intoxication in rodents may be of particular value in the prediction of response to pharmacological agents of the "high PE excretion patient". A relevant starting point may be the further investigation of clozapine which is the most potent antagonist of all components of PE stereotypy in rodents.

ACKNOWLEDGEMENTS

I thank Dr. A.A. Boulton for his encouragement and the provision of facilities, C. Yang for photography (Fig. 3) and Saskatchewan Health for continuing financial support.

REFERENCES

Antelman S.M., Edwards D.J., Lin M. (1977) Phenylethylamine: evidence for a direct, postsynaptic dopamine-receptor stimulating action. Brain Res. 127, 317-322.

Borison R.L., Havdala H.S. and Diamond B.I. (1977) Chronic phenylethylamine stereotypy in rats: a new animal model for schizophrenia? Life Sci. 21, 117-122.

Braestrup C. (1977) Changes in drug-induced stereotyped behaviour after 6-OHDA lesions in noradrenaline neurones. Psychopharmacol. 51, 199-204.

Braestrup C. and Randrup A. (1978) Stereotyped behaviour in rats induced by phenylethylamine, dependence on dopamine and noradrenaline and possible relation to psychoses?, in Noncatecholic Phenylethylamines Part I. Phenylethylamine: Biological Mechanisms and Clinical Aspects (Mosnaim A. and M. Wolf, eds), pp. 245-269. Marcel Dekker, New York.

Burki H.R., Ruch W. and Asper H. (1975) Effects of clozapine, thioridazine, perlapine and haloperidol on the metabolism of the biogenic amines in the brain of the rat. Psychopharmacol. 41, 27-33.

Carlsson A. (1975) Receptor mediated control of dopamine metabolism, in Pre- and Post-synaptic Receptors (Usdin E. and W.E. Bunney, eds), pp. 50-63, Marcel Dekker, New York.

Connell P.H. (1958) Amphetamine psychosis. Maudsley Monographs, No. 5, Oxford University Press.

Cooper S.J. and Dourish C.T. (1983) Hypodipsia, stereotypy and hyperactivity induced by β-phenylethylamine in the water-deprived rat. Pharmac. Biochem. Behav. (in press).

Cooper S.J., Dourish C.T. and Philips S.R. (1983) Central sites of action of piribedil in the induction of yawning, stretching and sexual excitement in the male rat. Br. J. Pharmac. 79, 392P.

Davis B.A., Yu P.H., Carlson K., O'Sullivan K and Boulton A.A. (1982) Plasma levels of phenylacetic acid, *m*- and *p*-hydroxyphenylacetic acid and platelet monoamine oxidase activity in schizophrenic and other patients. Psychiat. Res. 6, 97-105.

Deakin J.F.W. and Dashwood M.R. (1981) The differential neurochemical bases of the behaviours elicited by serotonergic agents and by the combination of a monoamine oxidase inhibitor and L-Dopa. Neuropharmacol. 20, 123-130.

Docherty J.P., van Kammen D.P., Siris S.G. and Marder S.R. (1978) Stages of onset of schizophrenic psychosis. Am. J. Psychiat. 135, 420-426.

Dourish C.T. (1981) Behavioural effects of acute and chronic β-phenylethylamine administration in the rat: evidence for the involvement of 5-hydroxytryptamine. Neuropharmacol. 20, 1067-1072.

Dourish C.T. (1982a) An observational analysis of the behavioural effects of β-phenylethylamine in isolated and grouped mice. Prog. Neuro-psychopharmacol. and Biol. Psychiat. 6, 143-158.

Dourish C.T. (1982b) A pharmacological analysis of the hyperactivity syndrome induced by β-phenylethylamine in the mouse. Br. J. Pharmac. 77, 129-139.

Dourish C.T. and Cooper S.J. (1981) Effects of acute or chronic administration of low doses of a dopamine agonist on drinking and locomotor activity in the rat. Psychopharmacol. 72, 197-202.

Dourish C.T. and Dewar K.M. (1982) Behavioural variations in response to β-phenylethylamine administration in three rat strains. IRCS Med. Sci. 10, 56-57.

Dourish C.T., Greenshaw A.J. and Boulton A.A. (1983) Deuterium substitution enhances the effects of β-phenylethylamine on spontaneous motor activity in the rat. Pharmac. Biochem. Behav. (in press).

Fray P.J., Sahakian B.J., Robbins T.W., Koob G.F. and Iversen S.D. (1980) An observational method for quantifying the behavioural effects of dopamine agonists: contrasting effects of *d*-amphetamine and apomorphine. Psychopharmacol. 69, 253-259.

Green A.R., Hall J.E. and Rees A.R. (1981) A behavioural and biochemical study in rats of 5-hydroxytryptamine receptor agonists and antagonists, with observations on structure-activity requirements for the agonists. Br. J. Pharmac. 73, 703-719.

Jackson D.M. (1972) The effect of β-phenylethylamine upon spontaneous motor activity in mice: a dual effect on locomotor activity. J. Pharm. Pharmacol. 24, 383-389.

Jackson D.M. (1978) β-phenylethylamine: studies on the mechanism of its stimulant effects, in Noncatecholic Phenylethylamines, Part 1, Phenylethylamine: Biological Mechanisms and Clinical Aspects (Mosnaim A.D. and M.E. Wolf, eds), pp. 289-313. Marcel Dekker, New York.

Jackson D.M. and Jenkins O.F. (1983) Beta-phenylethylamine: some preliminary chronic studies, in Neurobiology of the Trace Amines (in press).

Jacobs B.L. (1976) Minireview: an animal behaviour model for studying central serotonergic synapses. Life Sci. 19, 777-786.

Janowsky D.S. and Davis J.M. (1974) Dopamine, psychomotor stimulants and schizophrenia: effects of methylphenidate and the stereo isomers of amphetamine in schizophrenia, in Advances in Biochemical Psychopharmacology, Vol. 12 (Usdin E., ed), pp. 317-338. Raven Press, New York.

Janssen P.A.J., Niemegeers C.J.E., Schellekens K.H.L. and Lenaerts F.M. (1967) Is it possible to predict the clinical effects of neuroleptic drugs (major tranquillizers) from animal data? Arzneim Forsch. 17, 841-854.

Jeste D.V., Doongaji D.R., Panjwani D., Datta M., Potkin S.G., Karoum F., Thatte S., Sheth A.S., Apte J.S. and Wyatt R.J. (1980) Cross cultural study of a biochemical abnormality in paranoid schizophrenia. Psychiat. Res. 3, 341-352.

Karoum F., Linnoila M., Potter W.Z., Chuang L.W., Goodwin F.K. and Wyatt R.J. (1982) Fluctuating high urinary phenylethylamine excretion rates in some bipolar affective disorder patients. Psychiat. Res. 6, 215-222.

Krsiak M.H., Steinberg H. and Stolerman I.P. (1970) Uses and limitations of photocell activity cages for assessing effects of drugs. Psychopharmacol. 17, 258-274.

Leysen J.E., Niemegeers C.J.E., Tollenaere J.P. and Laduron P.M. (1978) Serotonergic component of neuroleptic receptors. Nature 272, 168-171.

Mantegazza P. and Riva M. (1963) Amphetamine-like activity of β-phenylethylamine after a monoamine oxidase inhibitor in vivo. J. Pharm. Pharmacol. 15, 472-478.

Mogilnicka E. and Braestrup C. (1976) Noradrenergic influence on the stereotyped behaviour induced by amphetamine, phenylethylamine and apomorphine. J. Pharm. Pharmacol. 28, 253-255.

Mogilnicka E. and Klimek V. (1977) Drugs affecting dopamine neurons and yawning behavior. Pharmac. Biochem. Behav. 7, 303-305.

Moja E.A., Stoff D.M., Gillin J.C. and Wyatt R.J. (1976) Dose-response effects of β-phenylethylamine on stereotyped behaviour in pargyline-pretreated rats. Biol. Psychiat. 11, 731-742.

Moore K.E. (1978) Amphetamines: Biochemical and behavioral actions in animals, in Handbook of Psychopharmacology 11 (Iversen L.L., S.D. Iversen and S.H. Snyder, eds), pp. 41-98. Plenum Press, New York.

Munro A. (1978) Monosymptomatic hypochondrial psychosis manifesting as delusions of parasitosis. Arch. Dermatol. 114, 941-943.

Murphy D.L., Karoum F., Alterman I., Lipper S. and Wyatt R.J. (1983) Phenylethylamine and tyramine in patients with affective disorders: associations with clinical states and antidepressant drug treatment, in Neurobiology of the Trace Amines (in press).

Nielsen E.B., Lyon M. and Ellison G. (1983) Apparent hallucinations in monkeys during around-the-clock amphetamine for seven to fourteen days. J. Nerv. Ment. Dis. 171, 222-233.

Niemegeers C.J.E., Colpaert F.C., Leysen J.E., Awouters F. and Janssen P.A.J. (1983) Mescaline-induced head-twitches in the rat: an in vivo method to evaluate serotonin S_2 antagonists. Drug Devel. Res. 3, 123-135.

Niemegeers C.J.E. and Janssen P.A.J. (1979) Minireview: a systematic study of the pharmacological activities of dopamine antagonists. Life Sci. 24, 2201-2216.

Potkin S.G., Karoum F., Chuang L.-W., Cannon-Spoor H.E., Phillips, I. and Wyatt R.J. (1979) Phenylethylamine in paranoid chronic schizophrenia. Science 206, 470-471.

Randrup A. and Munkvad I. (1966) Dopa and other naturally occurring substances as causes of stereotypy and rage in rats. Acta Psychiat. Scand. (Suppl.) 191, 193-199.

Reynolds G.P. (1983) Phenylethylamine and schizophrenia - clinical and pharmacological results, in Neurobiology of the Trace Amines, (in press).

Reynolds G.P., Cowey L., Rossor M.N. and Iversen L.L. (1982) Thioridazine is not specific for limbic dopamine receptors. Lancet ii (8296), 499-500.

Robbins T.W. (1977) A critique of the methods available for the measurement of spontaneous motor activity, in Handbook of Psychopharmacology 7 (Iversen L.L., S.D. Iversen and S.H. Snyder, eds) pp. 37-82. Plenum Press, New York.

Silberman E.K. and Post R.M. (1980) The march of symptoms in psychotic decompensation. J. Nerv. Ment. Dis. 168, 104-110.

Sloviter R.S., Connor J.D. and Drust E.G. (1980) Serotonergic properties of β-phenylethylamine in rats. Neuropharmacology 19, 1071-1076.

Sloviter R.S., Drust E.G. and Connor J.D. (1978) Evidence that serotonin mediates some behavioural effects of amphetamine. J. Pharmacol. Exp. Ther. 206, 348-352.

Snyder S.H. (1973) Amphetamine psychosis: a "model" schizophrenia mediated by catecholamines. Am. J. Psychiat. 130, 61-67.

Snyder S.H., Banerjee S.P., Yamamura H.I. and Greenberg D. (1974) Drugs, neurotransmitters and schizophrenia. Science 184, 1243-1253.

Spano P.F., Biggo G., Casu M., Gessa G.L., Bareggi S.R., Grovini S. and Trabucchi M. (1978) Interaction of metergoline with striatal dopamine system. Life Sci. 23, 2383-2392.

Weinstock M., Weiss C. and Gitter S. (1977) Blockade of 5-HT receptors in the central nervous system by β-adrenoreceptor antagonists. Neuropharmacol. 16, 273-276.

Widelitz M.M. (1979) The effect of β-phenylethylamine on behavior and polyribosomes in the rat. Biochem. Biophys. Res. Commun. 86, 48-56.

Free Communications

EFFECTS OF TRYPTAMINE AND 5-HYDROXYTRYPTAMINE ON FOOD INTAKE IN THE RAT

Colin T. Dourish and Julie Broadbent

Psychiatric Research Division
Cancer and Medical Research Building
University of Saskatchewan
Saskatoon, Saskatchewan S7N OWO

INTRODUCTION

It is well established that peripherally administered 5-hydroxytryptamine (5-HT) has profound effects on feeding and drinking in rodents (Bray and York, 1972; Rezek and Novin, 1975; Kikta *et al*., 1981). Since 5-HT does not penetrate the blood-brain barrier (Oldendorf, 1971), these effects are presumably mediated by peripheral 5-HT receptors located in the kidney, liver and gut (Pollock and Rowland, 1981; Kikta *et al*., 1981). When injected IP, 5-HT produces a dose-dependent decrease in feeding in hungry rats, an effect which does not appear to be secondary to sensorimotor impairment or gastrointestinal distress (Pollock and Rowland, 1981).

Recently, tryptamine (the non-hydroxylated form of 5-HT) has been reported to modulate the physiological and behavioural effects of brain 5-HT. In electrophysiological studies, the depressant responses of single neurones to iontophoretically applied 5-HT were profoundly enhanced by a weak concurrent application of tryptamine, which by itself did not alter the baseline cell firing rate (Jones and Boulton, 1980). Similarly, in behavioural studies the effects of intracerebroventricular administration of 50 μg 5-HT or 500 μg tryptamine (a large dose since tryptamine is metabolized very rapidly by monoamine oxidase, Durden and Philips, 1980) alone on the spontaneous motor activity of rats was not significantly different from control.

However, the injection of 5-HT plus tryptamine at these doses produced a stereotyped syndrome of headweaving, forepaw padding and splayed hindlimbs (Dourish and Greenshaw, 1983).

The aim of the present study was to examine the effects of peripherally administered tryptamine on food and water intake in the rat and to compare tryptamine's effects with those of peripherally administered 5-HT.

EXPERIMENT 1

Methods

Eighty male Wistar rats (Charles River, Canada, Montreal) weighing 200-250 g were used. The animals were housed and tested in individual metal cages (dimensions 30x20x12 cm) which were equipped with paper towel bedding material to allow recovery of spilled food during testing. The rats were habituated to the cages for a minimum of 3 days prior to the start of testing. During this time food (standard chow pellets) and water (from a 50 ml calibrated bottle) were available ad libitum. Twenty-four hours prior to testing the animals were deprived of food. Water remained available ad libitum.

In this study, dose-response curves for peripherally administered 5-HT and tryptamine were determined for food and water intake during 60 min. Each rat acted as its own control and was tested twice, once in the drug condition and once in the saline (control) condition. There was a four day interval between successive tests. The order of administration of saline and drug injections was counter balanced in order to control for order effects, and a separate group of rats was tested at each drug dose. Tryptamine hydrochloride (Sigma Chemical Co.) was injected at doses of 5, 10, 20, 40, 80 and 160 mg/kg; 5-hydroxytryptamine creatinine sulphate (Sigma Chemical Co.) was injected at doses of 0.5, 1.C, 2.0 and 4.0 mg/kg; n = 8 rats tested at each drug dose. Drugs were dissolved in 0.9% saline vehicle (which also served as a control solution) and injected in a volume of 3 ml/kg. Immediately after injection each rat was given a weighed amount of food and water. The amount of food and water consumed during 60 min was measured to the nearest 0.1 g (including a correction for spillage). The intake of each animal under the drug condition was expressed as a percentage of the animal's intake after saline injection.

Results and Discussion

It is evident from Figure 1 that 5-HT produced a dose related decrease in the food intake of hungry rats. This was confirmed by analysis of variance [$F(4,35) = 10.31$, $p < 0.01$]. Multiple comparisons (Dunnett, $\alpha = 0.05$) revealed that food intake was significantly decreased by 1.0, 2.0 and 4.0 mg/kg 5-HT. A maximal inhibition of feeding was

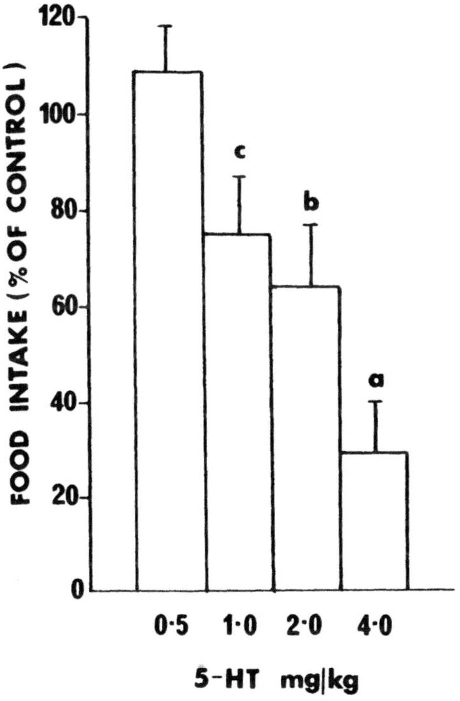

Fig. 1. Effects of 5-HT on food intake in hungry rats. Letters refer to significant differences determined by Dunnett's test between drug and control conditions. a = $p<0.01$; b = $p<0.025$; c = $p<0.05$.

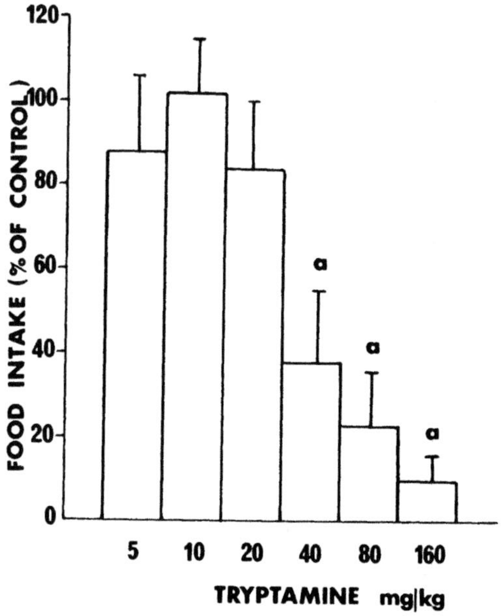

Fig. 2. Effects of tryptamine on food intake in hungry rats. Details are as in Fig. 1.

observed at 4.0 mg/kg 5-HT where intake was reduced to 30% of control.

Similarly, tryptamine produced a dose-dependent decrease in food intake (Figure 2). Analysis of variance [$F(6,48) = 9.20$, $p < 0.01$] followed by multiple comparisons revealed that food intake was significantly decreased by 40, 80 and 160 mg/kg tryptamine. A 160 mg/kg dose of tryptamine virtually abolished feeding (Figure 2). In contrast to the effects observed on feeding neither tryptamine [$F(6,48) = 0.88$, $p > 0.05$] nor 5-HT [$F(4,35) = 1.88$, $p > 0.05$] had any significant effect on water intake during the test.

These data confirm a number of previous findings that peripherally administered 5-HT reduces food intake in hungry rats (Bray and York, 1972; Pollock and Rowland,

1981). Previous studies have also reported that 5-HT increases water intake in water replete rats (Kikta et al., 1981) with a maximal hyperdipsia being evident 2-4 hr post-injection. Our failure to observe any hyperdipsic effect of 5-HT in the present study may be explained by the measurement of water intake 1 hr post-injection. Most interestingly, our data illustrate that tryptamine possesses anorectic properties at doses of 40-160 mg/kg. This is the first report of an effect of this amine on ingestive behaviour and since tryptamine readily crosses the blood-brain barrier (Oldendorf, 1971), it is unclear whether the anorexia observed is of peripheral or central origin. Consequently, a further study was carried out in an attempt to address this question, by examining the effect of tryptamine on 5-HT anorexia when given in conjunction with 5-HT.

EXPERIMENT 2

Methods

Seventy-two male Wistar rats (200-250 g) were allocated to 9 groups which were matched for body weight. Each rat was injected twice prior to the 60 min feeding test. For the first injection the animals were allocated to 3 groups (n = 24 each), and were treated with saline, 0.5 and 2.0 mg/kg 5-HT creatinine sulphate respectively, IP 5 min before the test. Within each group animals were divided amongst 3 groups (n = 8 each) and were treated with saline, 10 and 40 mg/kg tryptamine HCl, IP immediately before the test. Drugs were dissolved in 0.9% saline and administered in a volume of 1 ml/kg. All other procedures were as described for Experiment 1.

Results and Discussion

Tryptamine 10 mg/kg had no effect on food intake, whereas tryptamine 40 mg/kg significantly decreased feeding, thus confirming the results of Experiment 1. The data appear to provide some evidence for an interaction of 5-HT and tryptamine on food intake. A dose of 10 mg/kg tryptamine plus 2.0 mg/kg 5-HT significantly decreased food intake compared to control (Dunnett, $p < 0.05$) whereas 10 mg/kg tryptamine or 2.0 mg/kg 5-HT alone were without effect. Similarly, 40 mg/kg tryptamine plus 0.5 or 2.0 mg/kg 5-HT produced a larger decrement in feeding than 40 mg/kg tryptamine alone. However, these differences were not statistically significant due to large within-group variation, and, therefore, further studies are required to substantiate this finding.

CONCLUSION

In conclusion, the present data demonstrate that 5-HT and tryptamine reduce food intake when administered peripherally to hungry rats. It seems likely that the amines may act via a common peripheral receptor population located in the liver and gut to produce their anorectic effects.

ACKNOWLEDGEMENTS

We thank Dr. A.A. Boulton for encouragement and the provision of facilities and Saskatchewan Health for continuing financial support.

REFERENCES

Bray G.A. and York D.A. (1972) Studies on food intake in genetically obese rats. Am. J. Physiol. 223, 176-179.

Dourish C.T. and Greenshaw A.J. (1983) Effects of intraventricular tryptamine and 5-hydroxytryptamine on spontaneous motor activity in the rat. Res. Comm. Psychol. Psychiat. Behav. 8, 1-9.

Durden D.A. and Philips S.R. (1980) Kinetic measurements of the turnover rates of phenylethylamine and tryptamine *in vivo* in the rat brain. J. Neurochem. 34, 1725-1732.

Jones R.S.G. and Boulton A.A. (1980) Tryptamine and 5-hydroxytryptamine: actions and interactions on cortical neurones in the rat. Life Sci. 27, 1849-1856.

Kikta D.C., Threatte R.M., Barney C.C., Fregley M.J. and Greenleaf J.E. (1981) Peripheral conversion of L-5-hydroxytryptophan to serotonin induces drinking in rats. Pharmac. Biochem. Behav. 14, 889-893.

Oldendorf W.H. (1971) Brain uptake of radiolabelled amino acids, amines and hexoses after arterial injection. Am. J. Physiol. 221, 1629-1639.
Pollock J.D. and Rowland N. (1981) Peripherally administered serotonin decreases food intake in rats. Pharmac. Biochem. Behav. 15, 179-183.
Rezek M. and Novin D. (1975) The effects of serotonin on feeding in the rabbit. Psychopharmacologia 43, 255-258.

5HT INVOLVEMENT IN TRYPTAMINE INDUCED BEHAVIOUR IN MICE

Jane Irons, C.M. Robinson and C.A. Marsden

Department of Physiology and Pharmacology
Medical School, Queen's Medical Centre
Clifton Boulevard, Nottingham NG7 2UH

INTRODUCTION

Several studies have shown that tryptamine (T) administered to rodents pretreated with an MAO inhibitor produces characteristic behaviour (e.g. Clineschmidt and Lotti 1974, Marsden and Curzon 1978, Luscombe et al 1982). Depletion of 5HT decreases this response indicating involvement of neuronal 5HT (Foldes and Costa 1975, Marsden and Curzon 1978). Other studies however have suggested T may act directly on 5HT receptors while another possibility is the existence of specific T receptors in brain (Kellar and Cascio 1982). Recent behavioural studies have supported the concept of distinct 5HT and T receptors; for example T infused into rat hypothalamus causes hyperthermia while 5HT produces hypothermia (Cox et al 1981). Alternatively T could modulate 5HT mediated transmission via an action at 5HT receptors (Jones and Boulton 1980). The present study investigated the behavioural effects of administered T in the mouse with a view to elucidating the role of endogenous 5HT and 5HT receptors in the response.

METHODS

Drugs were administered i.p. dissolved in 0.9% saline except for p-chlorophenylalanine (PCPA) which was prepared as a suspension. Mice (male CS-1, 20-25g) were placed singly into observation boxes 30 min prior to injection and given tranylcypromine (TCP) 10 mg/kg 30 min before T (2.5 - 10.00 mg/kg). 5HT antagonists were given 30 min before TCP while PCPA (400 mg/kg) was given at 24 h intervals for 2 or 3 days and then the mice left for 24 h before TCP + T injection. The behavioural parameters scored for 30 or 45 min post T injection were head twitches (number of distinct involuntary head movements); forepaw clonus (number); rearing (number of episodes with both forepaws off the ground); locomotor activity (% of time). All behavioural observations were made between 14.00-17.00 h.

For the tryptophan (TP) deficient diet experiments the mice were maintained on one of three diets for 14 days. (a) Normal 41B diet containing 0.2% TP; (b) TP deficient maize diet containing 0.07% TP; (c) the maize diet with 0.2% TP added. Endogenous 5HT release from hypothalamic slices was measured under fixed volume conditions (Robinson and Marsden, this volume). 5HT levels, released and in brain, were assayed by HPLC with electrochemical detection.

RESULTS

T (2.5 - 10 mg/kg) given to TCP pretreated mice produced a dose related increase in head twitches, forepaw clonus and locomotor activity without a significant effect on rearing. In all subsequent studies the dose of T was 5 mg/kg. PCPA administration decreased the behavioural effects of administered T; however the degree of reduction of specific parameters was dependent on the degree of 5HT depletion. Whereas forepaw clonus was markedly reduced when the reduction of 5HT was only 20%; greater reductions (35-45%) were required before either head twitches or locomotor activity were decreased. In contrast, although mice maintained on a TP deficient diet showed reductions in brain 5HT (-41% $P < 0.01$, n=6) the head twitch response was increased compared to the controls (TCP + T, normal diet). This increase was prevented by additions of TP to the deficient diet (Table 1) which also reversed the fall in 5HT. In another group mice were given PCPA (400 mg/kg) 48 and 24 h before TCP + T. 5HT was decreased in all diet groups by 40-65% with the T response in the normal diet group reduced by 56% and no significant enhancement of T-induced head twitches observed in the TP deficient group (Table 1).

Table 1

Effect of dietary TP on T induced head-twitches in mice

Diet	T	Head Twitches Control	Head Twitches PCPA
Normal	-	1.3 ± 0.5	0
Normal	+	23.0 ± 1.3*	9.8 ± 2.0†
Low TP	+	45.0 ± 2.0*×	13.0 ± 1.4†
Low TP + TP	+	21.0 ± 3.4*	8.4 ± 1.0†

Mice maintained on the different diets were either given saline or PCPA (400 mg/kg x 2) and 24 h after the second injection given TCP (10 mg/kg) then either saline or T (5 mg/kg) 30 min later and head twitches counted for 45 min. Results given as mean ± SEM (n=6) * $P < 0.01$ cf group 1 (control) × $P < 0.01$ cf group 2 (control) † $P < 0.01$ cf appropriate control group.

5HT release from hypothalamic slices from mice kept on the normal diet was significantly increased (+30%) by T (10^{-7}M) added to the incubation system. T (10^{-7}M) also increased 5HT release from slices obtained from mice kept on the TP deficient diet or the deficient diet with added TP (+58% and +62% respectively). Basal release (no T added) however was significantly reduced (-47%) in the TP deficient slices compared to control and addition of TP to the diet reversed this effect.

Effects of 5HT receptor antagonists and an uptake blocker

Both cyproheptadine (3 mg/kg) and metergoline (1.5 mg/kg), at doses previously shown to correspond to approximate ED_{50} for antagonism of 5HT receptor induced responses in mice, produced comparable inhibition of the T induced head twitch and locomotor activity. Cyproheptadine however was more effective at antagonising forepaw clonus than metergoline (Table 2) Paroxetine (3 mg/kg), a 5HT uptake inhibitor, almost abolished forepaw clonus without significantly altering T induced head twitches or locomotor activity (Table 2).

Table 2

Effect of metergoline (1.5 mg/kg), cyproheptadine (3 mg/kg) and paroxetine (3 mg/kg) on T induced behaviour

Drug	H.Twitches	F.Clonus	L.Activity
Metergoline (6)	-80%	-18%	-73%
Cyproheptadine (6)	-62%	-57%	-56%
Paroxetine (6)	+ 3%	-83%	-10%

Results given as % change from control. Drugs given 30 min pre TCP.

DISCUSSION

The present results with the mouse confirm the involvement of neuronal 5HT in T mediated behaviour but suggest that some parameters (i.e. forepaw clonus) are highly dependent on the presence of 5HT while others (head twitches and locomotor activity) are less dependent. Furthermore metergoline potently inhibited the less 5HT dependent behaviour but was less active against forepaw clonus, while another 5HT antagonist, cyproheptadine, prevented all three types of behaviour, suggesting some differentiation in terms of receptors between the various components of the behavioural response. Depletion of 5HT by inhibition of TP hydroxylase reduced the behavioural effects of T; in contrast lowering brain 5HT by maintaining mice on a low TP diet enhanced the effects of T. The *in vitro* release results show that T was able to release 5HT from slices of both normal and TP deficient brains although basal release was reduced in the latter compared with the former. One explanation for these results is that in the chronic dietary experiments a prolonged decrease in brain 5HT and/or T (as TP is the precursor of T) results in altered responsiveness of the receptors involved in the T behavioural effects. This is supported by the fact that the enhanced response to T did not occur when the TP deficient diet was supplemented with TP in which situation no decrease in either brain 5HT or basal 5HT release was observed. There remains the need to explain the difference between the effects of depleting brain 5HT with PCPA or low TP diet on the T response. Other studies have shown that depleting 5HT by only 40% completely prevented T induced myoclonus in guinea-pigs (Luscombe et al 1982). Inhibition of TP hydroxylase could preferentially deplete a readily releasable pool of 5HT. Alternatively PCPA or a metabolite could have a direct influence on 5HT release.

In summary the results confirm that the behavioural effects of T involve 5HT neurones, but that some components of behavioural response are relatively less dependent on 5HT and so support the concept of T acting via stimulation on more than one receptor type.

We thank the Wellcome Trust for financial support and Beechams PLC and Carlo Erba Ltd. for generous supplies of paroxetine and metergoline.

REFERENCES

Clineschmidt B V and Lotti V J (1974) Indoleamine antagonists: Relative potencies as inhibitors of tryptamine and 5-hydroxytryptophan evoked responses. Br.J.Pharmac.50 311-313

Cox B, Lee T F and Martin D (1981) Different hypothalamic receptors mediate 5-hydroxytryptamine and tryptamine-induced core temperature changes in the rat. Br.J.Pharmac. 72 477-482

Foldes A and Costa E (1975) Relationship of brain monoamine and locomotor activity in rats. Biochem. Pharmac. 24 1617-1621

Jones R S G and Boulton A A (1980) Tryptamine and 5-hydroxytryptamine. Actions and interactions on cortical neurones in the rat. Life Sci. 27 1849-1856

Kellar K J and Cascio C S (1982) [^{3}H]-Tryptamine: high affinity binding sites in rat brain. Eur. J.Pharmac. 78 475-478

Luscombe G, Jenner P and Marsden C D (1982) Tryptamine-induced myoclonus in guinea-pigs pretreated with a monoamine oxidase inhibitor indicates pre- and post-synaptic actions of tryptamine upon central indoleamine systems. Neuropharmacology 21 1257-1265

Marsden C A and Curzon G (1978) The contribution of tryptamine to the behavioural effects of L-tryptophan in tranylcypromine-treated rats. Psychopharmacology 57 71-76

HYPODIPSIC EFFECTS OF β-PHENYLETHYLAMINE, PHENYLETHANOLAMINE, N-METHYLPHENYLETHYLAMINE AND d-AMPHETAMINE: A TEMPORAL ANALYSIS

J. Broadbent, A.J. Greenshaw and A.A. Boulton

Psychiatric Research Division
Cancer and Medical Research Building
University of Saskatchewan
Saskatoon, Saskatchewan S7N OWO

INTRODUCTION

The trace amine phenylethylamine (PE) is an endogenous constituent of mammalian brain tissue (Durden *et al*, 1973). d-Amphetamine and PE are structurally similar, PE lacking only the α-methyl group of amphetamine. Administration of amphetamine results in a marked attenuation of food and water intake in many species (Bizzi *et al*, 1970; Holtzman and Jewett, 1971; Soulairac and Soulairac, 1970). Whereas PE has been reported to induce hypophagia (Dourish and Boulton, 1981), several reports have indicated that PE has no effect on fluid consumption (Dourish and Boulton, 1981; 1979; Thornton, 1980). No previous studies, however, have been specifically directed to an analysis of the effects of PE on water intake. The present study, therefore, attempted to reveal any hypodipsic effects of PE by employing a temporal analysis of licking within the test period. In addition to PE and d-amphetamine, the PE derivatives, phenylethanolamine (PEOH) and N-methylphenylethylamine (NMPE) were also included in this study.

METHODS

Male Wistar rats (Charles River, Montreal) weighing 200-250 g at the start of the experiment were used. The animals were group housed with food freely available in the home cages. Water was available according to the experimental protocol. Four groups of rats were used (PEOH, NMPE and d-amphetamine n=8, PE n=5). Subjects were allowed access to water for two 30 min periods per day. The first period was the test session in which animals could gain water by licking at a drinking spout recessed into one wall of an operant chamber, (Coulborn Inst. E10-10). Licks were recorded during each min of the session by an infra-red photosensor linked to a microcomputer (CBM 4016). The total volume of fluid consumed was measured to the nearest 0.1 ml from a 50 ml burette supplying water to the spout. Following the test period, water was available for a further 30 min in a separate cage. After a ten day training period, the effects of PE.HCl, PEOH.HCl or NMPE.HCl were assessed. Each subject received doses of 3.24 to 41.7 mg/kg of one compound in a mixed order. These were administered two min before the test session. d-Amphetamine sulphate was given 5 min prior to testing in doses of 0.15 to 5.0 mg/kg. These logarithmic dose-response curves were chosen to allow a detailed dose-response assessment. All drugs were dissolved in 0.9% saline and administered IP in a volume of 1 ml/kg. At least three days separated drug trials. Equivalent injections of 0.9% saline were given on these non-drug days.

RESULTS

The effects of each compound on overall water intake are displayed in Table 1. Repeated measures ANOVA revealed that each compound significantly decreased water intake [PE, $F(6,24) = 34.84$, $p<0.01$; PEOH, $F(6,42) = 7.20$, $p<0.01$; NMPE, $F(6,42) = 4.36$, $p<0.01$; d-amphetamine, $F(6,42) = 19.56$, $p<0.01$]. Further analysis with Dunnett's multiple comparison test ($\alpha = 0.05$) indicated that significant decreases occurred at doses of 25.0 and 41.7 mg/kg for PE, PEOH and NMPE. For d-amphetamine differences from control were significant at 1.25, 2.5 and 5.0 mg/kg.

The effects of these compounds on the temporal distribution of licking are displayed in Figs. 1 and 2. For statistical analysis, the 30 minute session was divided into two 15 min periods. The total number of licks in each period was then subjected to ANOVA. Effects of PE (Fig. 1a) were significant for both periods [1-15 min, $F(6,24) =$

TABLE 1

MEAN FLUID INTAKE (ml ± S.E.)

		Dose mg/kg					
Compound	Control	3.24	5.4	9.0	15.0	25.0	41.7
PE	13.9±0.3	13.8±1.0	14.3±0.7	11.8±0.6	12.4±0.4	9.8±0.8*	4.58±0.8*
PEOH	13.6±0.9	13.3±1.0	14.3±1.0	15.3±0.9	13.8±5.7	7.7±2.3*	4.1 ±2.0*
NMPE	12.0±0.6	11.5±1.0	11.1±0.7	11.3±0.9	9.7±0.4	8.0±0.8*	8.9 ±0.9*

		Dose mg/kg					
Compound	Control	0.15	0.31	0.625	1.25	2.5	5.0
d-AMPH	12.9±0.9	13.3±1.7	11.7±1.2	11.4±0.8	7.24±2.3*	2.76±1.6*	0*

The effects of PE, PEOH and NMPE and of d-amphetamine on fluid intake. Test sessions were of 30 minutes duration (*$p<0.05$).

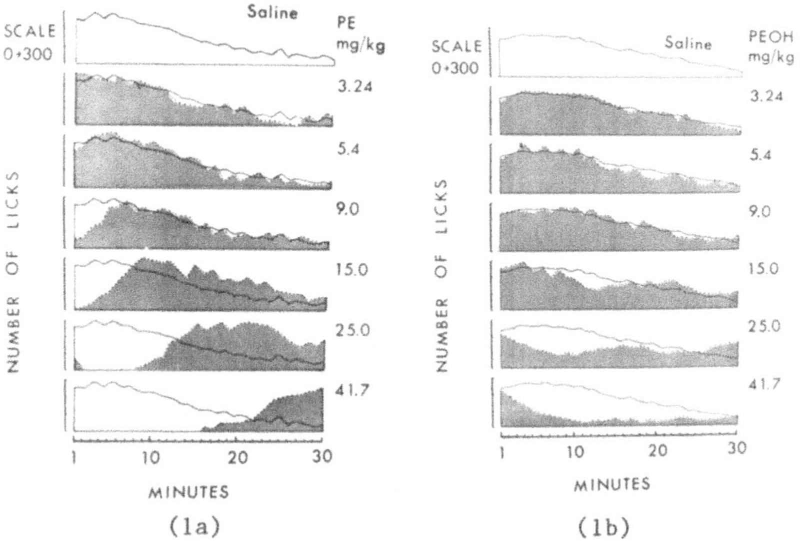

Fig. 1a and b - The effects of saline and of each dose of PE (a) and PEOH (b) on the temporal analysis of licking. Running averages of the number of licks emitted by the groups of rats during each minute of the test sessions are shown. For ease of comparison the performance of each group after saline (solid line - upper panel) is superimposed on each groups' performance after drug administra tion (dotted line - lower panels).

51.61, $p<0.01$; 16-30 min, $F(6,24) = 28.99$, $p<0.01$] at 9.0, 15.0, 25.0 and 41.7 mg/kg in the first period but only at 15.0, 25.0 and 41.7 mg/kg in the second period. Significant effects on water intake were only seen at 25.0 and 41.7 mg/kg (Table 1) suggesting that the initial decrease in licking at 15.0 mg/kg was compensated for later in the session. Administration of PEOH (Fig. 1b) resulted in decreases in local lick rate between one and 15 min [$F(6,42) = 8.60$, $p<0.01$] at 25.0 and 41.7 mg/kg and between 16 and 30 min [$F(6,42) = 4.39$ $p<0.01$] at 41.7 mg/kg. Significant changes in the licking pattern after NMPE [1-15 min, $F(6,42) = 4.76$, $p<0.01$; 16-30 min, $F(6,42) = 4.71$, $p<0.01$] were shown to be due to decreases in the first half of the session at 25.0 and 41.7 mg/kg, followed by increases in the second half of the session. The small effects of NMPE

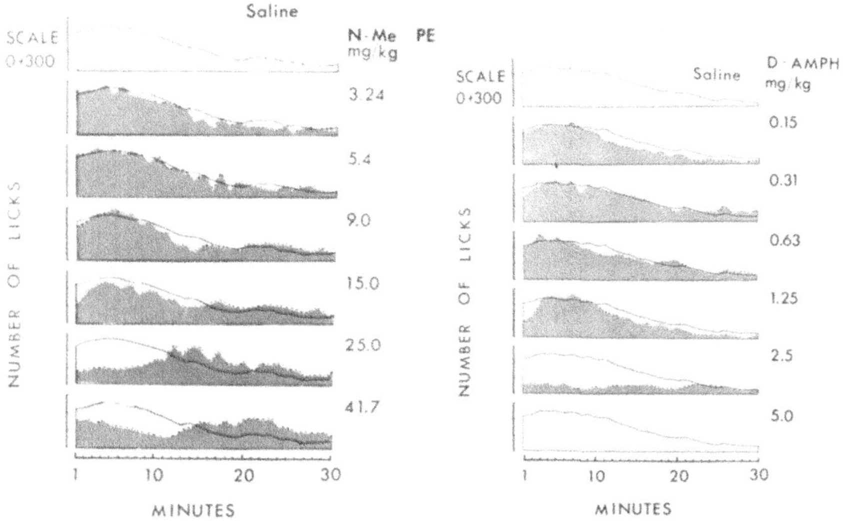

(2a) (2b)

Fig. 2a and b. The effects of saline and of each dose of NMPE (a) and d-amphetamine (b) on the temporal analysis of licking. Details as for Fig. 1.

on overall fluid intake (Table 1) can therefore be explained in terms of these opposing changes in licking rate (Fig. 2a).

d-Amphetamine administration (Fig. 2b) caused a sustained suppression of licking which was found to be significant in both halves of the session [1-15 min $F(6,42) = 16.83$, $p<0.01$; 16-30 min $F(6,42) = 3.43$, $p<0.01$]. These effects occurred at 1.25, 2.5 and 5.0 mg/kg in the first period and also in the second 15 min after 5.0 mg/kg.

DISCUSSION

The present results indicate that PE, PEOH and NMPE all share the hypodipsic properties of d-amphetamine. Differences in the relative potency and pharmacodynamic profiles of these compounds were, however, revealed by the temporal analysis of licking. The effective doses of each compound used in the present study appeared to induce changes in spontaneous motor activity. The observed hypodipsia may, therefore, be a secondary effect attributable to incompatability of the drug-induced motor response with drinking. Consistent with this proposal, and in contrast to anorexic effects of PE (Dourish and Boulton, 1981), the time course of the present hypodipsic effect of PE appears to parallel the time course of the syndrome of stereotyped behaviour induced by PE (Dourish et al, 1983).

ACKNOWLEDGMENT

We thank Saskatchewan Health and the MRC of Canada for continuing financial support.

REFERENCES

Bizzi A., Bonaccorsi A., Jespersen S, Jori A. and Garratini S. (1970) Pharmacological studies on amphetamine and fenfluramine, in Amphetamines and Related Compounds (Costa E. and Garratini S., eds) pp 577-595, Raven Press, New York.

Dourish C.T. and Boulton A.A. (1981) The effects of acute and chronic administration of β-phenylethylamine on food intake and body weight in rats. Prog. Neuro-psychopharmacol. 5, 411-414.

Dourish C.T., Greenshaw A.J. and Boulton A.A. Deuterium substitution selectively enhances the behavioural response to β-phenylethylamine in rats. Pharmac. Biochem. Behav. (In press).

Durden D.A., Philips S.R. and Boulton A.A. (1973) Identification and distribution of β-phenylethylamine in the rat. Can. J. Biochem. 51, 995-1002.

Holtzman S.G. and Jewett R.E. (1971) The role of brain norepinephrine in the anorexic effects of dextroamphetamine and monoamine oxidase inhibitors in the rat. Psychopharmacologia 22, 157-161.

Soulairac A. and Soulairac M-L. (1970) Effects of amphetamine-like substances and L-Dopa on thirst, water intake and duiresis, in Amphetamine and Related Compounds (Costa E. and Garratini S., eds) pp 819-837, Raven Press, New York.

Thornton E.W. (1980) Effect of β-phenylethylamine on social dominance in rat dyads during competition for water. IRCS Medical Science: Psychology and Psychiatry 8, 860-861.

TASTE-DEPENDENT EFFECTS OF LOW DOSES OF β-PHENYLETHYLAMINE AND D-AMPHETAMINE ON DRINKING IN THE RAT

Suzanne Turkish[1], C.T. Dourish[2] and S.J.Cooper[1]

[1]Psychology Department, University of Birmingham, Birmingham B15 2TT, U.K.
[2]Psychiatric Research Division, CMR Building, University of Saskatchewan, Canada S7N OWO.

INTRODUCTION

β-Phenylethylamine (PE) is an endogenous trace amine which has been identified in human, rat and other mammalian brain tissue (Boulton and Juorio, 1982). PE is structurally related to d-amphetamine (AMPH), lacking only the AMPH α-methyl group. A functional relationship between the two has been proposed. Thus it has been suggested that PE may be a major mediator for the central actions of AMPH (Borison et al., 1975).

One of the most well known of AMPH's behavioural effects is a depression of feeding and drinking responses. In addition to its anorectic effect, AMPH also reduces water consumption in water-deprived rats and mice (e.g. Dobrzanski and Doggett, 1976a; Stolerman and D'Mello, 1978). An anorectic effect of PEA has been reported (Dourish and Boulton, 1981), and PEA has also been shown to reduce water consumption in water-deprived rats (Cooper and Dourish, 1983). Interestingly, at low dose levels, AMPH can produce a paradoxical increase in food consumption (e.g. Dobrzanski and Doggett, 1976b; Winn et al., 1982).

Our first aim was to compare the possible effects of PE and AMPH on drinking responses in water-deprived rats, over wide dose ranges. A second aim was to determine whether or not taste factors can modulate the effects of PE and/or AMPH on drinking behaviour.

METHODS

Animals used in the study were 360 adult, male hooded rats (General strain), which were bred in the Psychology Department, University of Birmingham. They were housed individually in stainless steel cages with free access to food pellets (Diet 41B, Heygate and Sons, U.K.). They were maintained on a 12 hr light- 12 hr dark cycle, with room temperature kept constant at 21°C. The animals weighed 250-350 g at testing.

All animals were first adapted to a water-deprivation schedule of 24 hr water-deprivation on alternate days. Drinking tests were run at the end of a deprivation period. In the AMPH experiment, 180 animals were allocated at random to 3 fluid-condition groups: water, 0.005 M sodium saccharin, and 0.9% sodium chloride solution, respectively. Each group was presented with its respective fluid for a 30 min period immediately after the 24 hr deprivation period. Food was removed from the home cage during the fluid presentation period. Animals were thoroughly familiarised with the procedure before drug trials were begun. Within each fluid condition, the animals were allocated at random to 6 injection conditions: 0.03, 0.1, 0.3, 1.0 and 3.0 mg/kg AMPH and the corresponding vehicle (isotonic saline), respectively. Injections were given intraperitoneally (I.P.) 15 min before the start of the drinking test. During the test, the volume of liquid consumed was determined at 2 min intervals throughout the 30 min test period.

In the PE experiment, 180 animals were similarly divided according to the same 3 fluid conditions, familiarised with the procedures, and were then assigned to 6 injection conditions: 3.125, 6.25, 12.5, 25 and 50 mg/kg PE and the corresponding vehicle (distilled water) respectively. Injections were administered I.P. immediately before the test. Volume of liquid consumed was determined at 2 min intervals throughout the 30 min drinking test which followed after the 24 hr water deprivation period.

The drinking data were analysed using analysis of variance procedures. Drug doses and fluid conditions were independent factors. For the repeated measures factor (time course), scores were divided into five 6 min intervals.

RESULTS

d-Amphetamine. There were highly significant overall main effects for drug dose, fluid condition, and time-course of drinking [$F(5,162)=22.6$; $F(2,162)=27.0$; $F(4,648)=288.9$, respectively]. The time-course of drinking followed the usual pattern of a high initial level of consumption and a progressive decline in drinking in the remainder of the test period. During the first 6 min interval of high intake, there was a significant drug effect [$F(5,162)=9.08$, $P<0.001$], due to significant depressions in fluid consumption after injection of 3.0 mg/kg AMPH. However, there was not a significant fluid condition effect, or drug x fluid interaction. The results for the second 6 min period were different (Fig. 1). Here, there were significant drug, and fluid condition effects, and drug x fluid interaction [$F(10,162)=2.09$, $P<0.05$]. It is clear that water consumption was reduced in a dose-dependent manner by AMPH. But there was a differential effect in the case of saccharin solution and of saline drinking. Saccharin solution and saline consumption were elevated above control levels, after injection of the very low dose of AMPH, 0.03 mg/kg. In contrast, intake of both solutions was markedly depressed at the highest dose used, 3 mg/kg. Later time intervals in the drinking test will not be considered here, except to note that intakes fell to very low levels as the animals became satiated.

β-Phenylethylamine. There were highly significant overall main effects for PE dose, fluid condition, and time-course of drinking [$F(5,162)=22.0$; $F(2,162)=37.8$; $F(4,648)=280.0$, respectively]. In the first 6 min interval of the drinking test, there was a significant drug effect [$F(5,162)=63.8$, $P<0.001$], due to a depression of drinking at 25 mg/kg, and a virtual abolition of drinking at 50 mg/kg. There were no significant differences between fluid conditions. Results for the second 6 min period are shown in Fig. 2. For this interval, there were significant drug and fluid effects, and a significant drug x fluid interaction [$F(10,162)=2.30$, $P<0.05$]. Intake in each of the three fluid conditions was entirely suppressed by the largest dose of PE, 50 mg/kg. Some increases in consumption were, however, in evidence after injection of low doses of PE. The stimulant effect was most marked in the case of saccharin

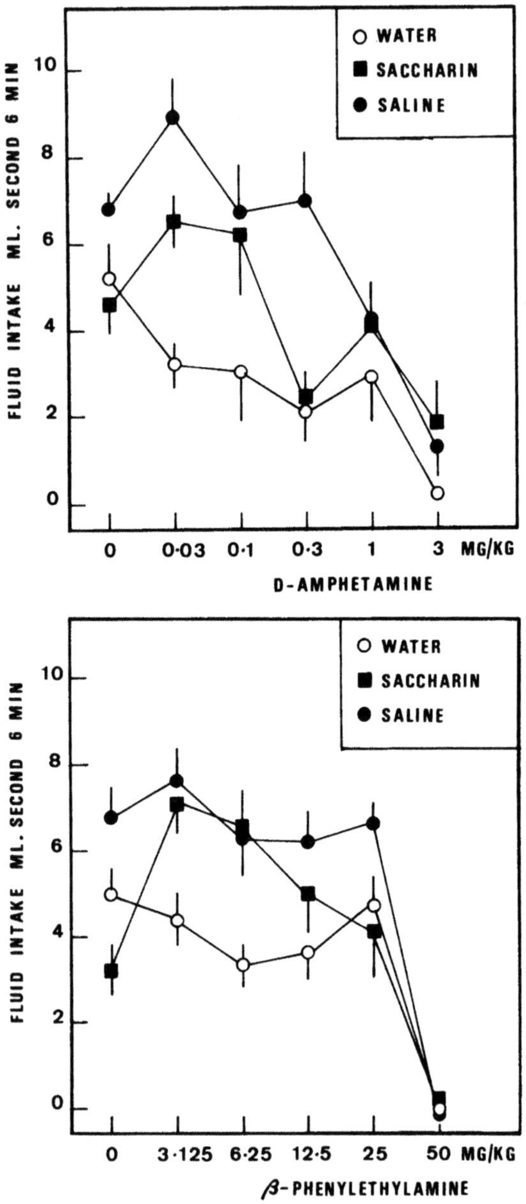

Fig. 1 (top) Effects of AMPH on fluid intake in second 6 min interval. Fig. 2 (bottom) Data for PE in same interval.

solution drinking. There was a modest rise in saline consumption following injection of 3.125 mg/kg PE. Water consumption was never increased by PE.

DISCUSSION

Effects of PE and AMPH on fluid consumption in water-deprived rats were found to depend on (i) drug dosage, (ii) flavour condition, (iii) time-course of drinking. There were notable similarities between the effects of PE and AMPH. During the first 6 min interval, the only effect of PE and AMPH was the pronounced hypodipsia produced by the largest doses of each drug. As control levels of consumption fell during the second 6 min interval, a novel stimulant effect of PE and AMPH became apparent. When very low doses of PE and AMPH were administered, there was clear evidence for drug-induced enhancement of drinking, which depended on the taste condition. Saccharin solution drinking was particularly affected; both PE and AMPH stimulated saccharin solution drinking at low dose levels (Figs. 1 and 2). Saline drinking was increased by AMPH. In constrast, there was no instance where water consumption was increased by either AMPH or PE.

ACKNOWLEDGEMENTS

D-amphetamine was generously supplied by Smith, Kline and French, Ltd. S.T. is a recipient of a postgraduate studentship from the M.R.C. (U.K.).

REFERENCES

Borison R.L., Mosnaim A.D. and Sabelli H.C. (1975) Brain 2-phenylethylamine as a major mediator for the central actions of amphetamine and methylphenidate. Life Sci. 17, 1331-1344.

Boulton A.A. and Juorio A.V. (1982) Brain trace amines, in Handbook of Neurochemistry, Vol. 1. (Lajtha A., ed.) pp. 189-222. Plenum Press, New York.

Cooper S.J. and Dourish C.T. (1983) Hypodipsia, stereotypy and hyperactivity induced by β-phenylethylamine in the water-deprived rat. Pharmacol. Biochem. Behav.

(submitted for publication)

Dobrzanski S. and Doggett N.S. (1976a) On the relation between hypodipsia and anorexia induced by (+)-amphetamine in the mouse. J. Pharm. Pharmacol. 28, 922-924.

Dobrzanski S. and Doggett N.S. (1976b) The effects of (+)-amphetamine and fenfluramine on feeding in starved and satiated mice. Psychopharmacology 48, 282-286.

Dourish C.T. and Boulton A.A. (1981) The effects of acute and chronic administration of β-phenylethylamine on food intake and body weight in rats. Prog. Neuro-psychopharmacol. 5, 411-414.

Stolerman I.P. and D'Mello G.D. (1978) Amphetamine-induced hypodipsia and its implications for conditioned taste aversion in rats. Pharmacol. Biochem. Behav. 8, 333-338.

Winn P., Williams S.F. and Herberg L.J. (1982) Feeding stimulated by very low doses of d-amphetamine administered systemically or by microinjection into the striatum. Psychopharmacology 78, 336-341.

β-PHENYLETHYLAMINE AND d-AMPHETAMINE: DIFFERENTIAL POTENCY IN THE CONDITIONED TASTE AVERSION PARADIGM

A.J. Greenshaw and C.T. Dourish

Psychiatric Research Division
Cancer and Medical Research Building
University of Saskatchewan
Saskatoon, Saskatchewan S7N OWO

INTRODUCTION

Previous studies have demonstrated that d-amphetamine is an effective stimulus for the induction of conditioned taste aversion (CTA) learning. In this paradigm aversion is defined as reduced intake of a flavoured liquid, the ingestion of which has been followed by the administration of drugs or certain other stimuli.

In the present experiments we have compared the relative CTA-inducing potency of d-amphetamine and of β-phenylethylamine (PE). Amphetamine is an extremely potent compound in terms of its ability to induce a CTA (Booth *et al.*, 1977; Sanger *et al.*, 1980). PE, a biogenic amine present in mammalian tissues in trace amounts (Durden *et al.*, 1973), is structurally similar to amphetamine, lacking only the α-methyl group in the alkyl side chain.

A substantial body of evidence indicates that the effects of PE resemble those of amphetamine pharmacologically (Fuxe *et al.*, 1967), electrophysiologically (Sabelli *et al.*, 1973) and behaviourally (Stein, 1964; Huang and Ho, 1974; Dourish and Boulton, 1981). A recent study has, however, demonstrated that PE is a very weak stimulus for CTA learning (Greenshaw and Dourish, 1983). In view of the structural and functional similarities of PE and d-amphetamine described earlier this result is surprising. It has been proposed that duration of action may be an important factor in determining potency in the CTA paradigm (Goudie and Dickins, 1978). PE is known to have a very short half life in mammalian tissue (Wu and Boulton, 1975) and this factor is an important consideration for the interpretation of the lack of potency of PE in this context.

In an attempt to address this issue the present study examined the possible CTA inducing effects of PE in the presence of monoamine oxidase (MAO) inhibitors. These compounds (in the present study pargyline and l-deprenyl respectively) protect the PE molecule from the normally rapid breakdown by MAO, thereby shifting the dose-response curve for PE to the left and greatly increasing the duration of action of PE.

EXPERIMENT 1

Method

Male Wistar rats (Charles River, Montréal) served as subjects. The animals weighed 200 to 250 g at the time of testing. They were housed in groups under a 12 h light/dark cycle (lights on 6 a.m.) at a temperature of 20 ± 1°C. Food was continuously available in the home cages. Water was available according to the experimental procedure.

The subjects were water deprived for 23 h and then individually given access to two water bottles in experimental test cages (30x18x12 cm) for a period of 30 min per day, for three days. The volume of water consumed by each animal on each day was measured to the nearest 0.5 ml. After the third drinking session the animals were randomly assigned to groups of six. These groups were matched for daily water intake. The matched groups were allocated to various treatment conditions: saline/saline; saline/2.5 mg/kg PE.HCl (Sigma); saline/75 mg/kg pargyline.HCl (Sigma); 2.5 mg/kg PE/75 mg/kg pargyline.HCl; 2.5 mg/kg d-amphetamine sulphate (Smith Kline and French). On the fourth test day each animal was given an injection of pargyline or saline prior to other injection treatments. After a 90 min period each animal was then given access to a 0.1% solution of sodium saccharin in two 100 ml bottles for a period of 30 min. Immediately after the 30 min period of exposure to saccharin each animal was injected with saline, PE.HCl, or d-amphetamine sulphate according to the group allocation described above. The drugs were injected IP in a 0.9% saline vehicle at a volume of 1 ml/kg. The injections of 0.9% saline alone were by the same route at an equivalent volume. For each of the subsequent four days the animals were exposed to two-bottle choice retention tests in which they were simultaneously given access to one bottle containing water and one containing 0.1% saccharin. This procedure allowed an assessment of any change in the animals' preference for saccharin while enabling the subjects to maintain their normal level of fluid intake.

RESULTS

The animals' saccharin intake on each of the four retention trials was converted to a percentage of their

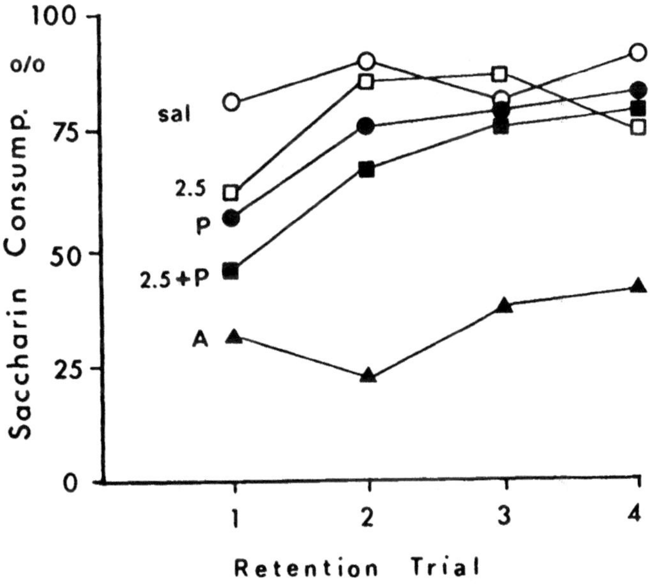

Fig. 1. The effect on subsequent saccharin preference of 2.5 mg/kg of PE and of 75.0 mg/kg of pargyline (P) alone and in combination. These effects are displayed in relation to that of the saline (sal) vehicle and of 2.5 mg/kg of d-amphetamine (A).

total fluid intake on each of these days. The average saccharin consumption (as a percentage of total intake) is displayed for each group, on each of the four retention trials, in Figure 1.

A two-way analysis of variance (treatment x trials) with repeated measures on the trial factor revealed a significant effect of treatments [$F (4,25) = 7.41$, $p < 0.05$]. There was a significant effect of trials [$F (3,75) = 7.75$, $p < 0.05$] but no significant interaction. Multiple comparisons (Dunnett $\alpha = 0.05$) conducted between the group means for each retention trial revealed a significant difference between the amphetamine group and the saline group on each of the four trials. The 2.5 mg/kg PE + 75 mg/kg pargyline group also exhibited significantly less saccharin preference than the saline group on day 1, however, no other comparisons of drug vs saline groups yielded significant results. These data suggest that PE is not a potent agent for eliciting CTA learning even when its duration of action is increased by MAO inhibition.

As different MAO inhibitors selectively alter the metabolism of biogenic amines, a second experiment was carried out using a different irreversible MAO inhibitor, deprenyl. For this experiment deprenyl was chosen because it is reported to preferentially elevate PE levels in rat brain (Philips and Boulton, 1979).

EXPERIMENT 2

Method

The subjects, as described for Experiment 1, were adapted to the 23 h water deprivation schedule under the conditions described earlier. After three baseline sessions the matched groups were injected with 4 mg/kg deprenyl or saline. Their water consumption was monitored on the following day. On the fifth day the animals were given access to saccharin in two bottles for 30 min. The animals were then injected with 2.5 or 5.0 mg/kg PE or with saline according to the prior group allocations: saline/saline; deprenyl/saline; saline/2.5 mg/kg PE; saline/5.0 mg/kg PE; deprenyl/2.5 mg/kg PE; deprenyl/5.0 mg/kg PE. The injection protocol was as for Experiment 1.

On each of the following three days the animals were exposed to two-bottle choice trials to assess the groups' relative saccharin/water preference.

RESULTS

Deprenyl pretreatment had no effect on baseline water intake. As described for Experiment 1, each animal's saccharin consumption is presented as a percentage of total fluid intake in Figure 2. An analysis of variance with repeated measures revealed no significant effects of treatments [F (5,30) = 2.50, $p < 0.05$]. There was a significant effect of trials [F (2,60) = 38.53, $p < 0.05$] but no significant interaction.

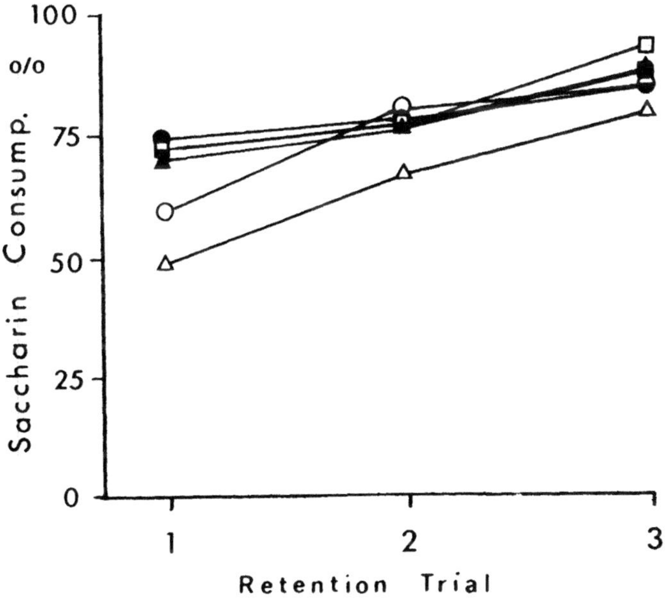

Fig. 2. The effects on subsequent saccharin preference of 2.5 and 5.0 mg/kg of PE (△–△ , □–□ respectively) and of 4.0 mg/kg of l-deprenyl (●–●) alone or in combination (filled triangles and squares respectively) compared to that of the saline vehicle (○–○).

DISCUSSION

The present results extend and confirm the previous report that PE and d-amphetamine have markedly different potencies in the CTA paradigm (Greenshaw and Dourish, 1983). With this study it is apparent that the differential potency of the two compounds may not be simply attributed to pharmacodynamic differences. This result, which does not support the duration of action hypothesis of CTA learning, is consistent with another recent report from this laboratory in which deuterium substitution was used to prolong the duration of action of PE (Dourish *et al.*, 1983).

The reason for the relative lack of aversive stimulus properties of PE remains unclear. It is, however, noteworthy that other pharmacologically and structurally similar compounds do have very different aversive stimulus properties (see Greenshaw and Dourish, 1983). It seems likely, therefore, that a consideration of structure-activity relationships may provide some insight in the present context. In any event the distinct difference between effects of PE and of d-amphetamine in CTA learning may preclude the proposal that PE may simply be an endogenous equivalent to amphetamine in the functional sense (Wyatt, 1978).

ACKNOWLEDGEMENTS

We thank Dr. A.A. Boulton for his advice and encouragement and Saskatchewan Health and the Canadian MRC for continuing financial support and Dr. P.H. Yu who kindly provided the deprenyl.

REFERENCES

Booth D.A., Pilcher C.W.T., D'Mello G.D., Stolerman I.P. (1977) Comparative potencies of amphetamine, fenfluramine and related compounds in taste aversion experiments in rats. *Br. J. Pharmacol.* *61*, 664-677.

Dourish C.T. and Boulton A.A. (1981) The effects of acute and chronic administration of β-phenylethylamine on food intake and body weight in rats. *Prog. Neuro-psychopharmacol.* *4*, 411-414.

Dourish C.T., Greenshaw A.J. and Boulton A.A. (1983) Deuterium substitution enhances the effects of β-phenylethylamine on spontaneous motor activity in the rat. Pharmacol. Biochem. Behav. (in press).

Durden D.A., Philips S.R. and Boulton A.A. (1973) Identification and distribution of β-phenylethylamine in the rat. Can. J. Biochem. 51, 995-1002.

Fuxe K., Grobecker H. and Jonsson J. (1967) The effects of β-phenylethylamine on central and peripheral monoamine containing neurons. Eur. J. Pharmacol. 2, 202-207.

Goudie A.J. and Dickins D.W. (1978) Nitrous oxide induced conditioned taste aversion in rats: the duration of drug exposure and its relation to the taste aversion/self administration paradox. Pharmacol. Biochem. Behav. 9, 587-592.

Greenshaw A.J. and Dourish C.T. (1983) Differential aversive stimulus properties of β-phenylethylamine and of d-amphetamine in rats. Psychopharmacol. (in press).

Huang J.T. and Ho B.T. (1974) The effect of pretreatment with iproniazid on the behavioural activities of β-phenylethylamine in rats. Psychopharmacol. 35, 77-81.

Philips S.R. and Boulton A.A. (1979) The effect of monoamine oxidase inhibitors on some arylalkylamines in rat striatum. J. Neurochem. 33, 159-167.

Sabelli H.C., Mosnaim A.D. and Vazquez A.J. (1973) A comparison of the functional roles of norepinephrine, dopamine and phenylethylamine in the central nervous system. Acta Physiol. Pol. 24, 33-40.

Sanger D.J., Greenshaw A.J., Thompson I.P. and Mercer J.D. (1980) Learned taste aversion to saccharin induced by orally consumed d-amphetamine. Pharmacol. Biochem. Behav. 13, 31-36.

Stein L. (1964) Self-stimulation of the brain and the central stimulant action of amphetamine. Fed. Proc. 23, 836-850.

Wu P.H. and Boulton A.A. (1975) Metabolism, distribution and disappearance of injected β-phenylethylamine in the rat. Can. J. Biochem. 53, 42-50.

Wyatt R.J. (1978) Is there an endogenous amphetamine? A testable hypothesis of schizophrenia, in The Nature of Schizophrenia (L.C. Wynne, R.L. Cromwell, S. Matthysse, eds.) pp 116-125, Wiley & Sons, New York.

LOCOMOTOR STIMULANT EFFECT OF β-PHENYLETHYLAMINE-NALOXONE MIXTURES

Colin T. Dourish[1] and Steven J. Cooper[2]
[1]Psychiatric Research Division, CMR Building, University of Saskatchewan, Canada S7N OWO.
[2]Psychology Department, University of Birmingham, Birmingham B15 2TT, U.K.

INTRODUCTION

The opiate receptor antagonist, naloxone (NAL), has been shown to block the locomotor stimulant effects of low doses of d-amphetamine (AMPH) (Hitzemann et al., 1982; Holtzman, 1974). Likewise, NAL has been reported to antagonize AMPH-induced ipsilateral turning in rats lesioned unilaterally with 6-hydroxydopamine in the substantia nigra (Dettmar et al., 1978). Recent biochemical evidence favours the view that NAL selectively blocks dopamine (DA) activity in the nucleus accumbens (Applegate et al., 1982). It seems probable therefore that DA synapses in the nucleus accumbens may serve as the substrate for the AMPH-NAL interactions which affect behaviour.

β-Phenylethylamine (PE) is an endogenous trace amine which has been identified in human, rat and other mammalian brain tissue (Boulton and Juorio, 1982). PE is structurally related to AMPH, lacking only the AMPH α-methyl group. Behaviourally, there are important similarities between the effects of PE and AMPH (e.g. Turkish et al., 1983). But there are also striking dissociations in their behavioural effects (e.g. Greenshaw and Dourish, 1983).

The aim of the present study was to determine the nature of possible interactions between PE and NAL in the rat. Measures of general activity, ambulation, rearing and behavioural stereotypy were taken to assess the degree of specificity of any interaction between the two drugs.

METHODS

Animals used in the study were 120 adult, male Sprague-Dawley rats (Charles River, Canada), weighing 200-250 g. Each rat was injected twice prior to the 30 min activity test. For the first injection, the animals were allocated at random to 5 groups (n=36 each), and were treated with 0, 1.0, 3.0 and 10.0 mg/kg NAL hydrochloride respectively. The injection was given by subcutaneous route, 15 min before testing. Within each group, animals were divided amongst 6 groups (n=6 each), and were treated with 0, 6.25, 12.5, 25 or 50 mg/kg PE hydrochloride respectively. The injection was given by intraperitoneal route, immediately before testing. Isotonic saline served as vehicle.

Animals were placed individually in large plexiglas cages (40 cm square, 23 cm high), positioned in automatic activity recording devices (Opto-Varimex Minor, Columbus Instruments). A logic circuit in these devices allowed a distinction to be made between whole-body ambulation, on the one hand, and total horizontal activity, on the other. Vertical activity (rearing and jumping) was recorded by interruption of photobeams 15 cm above the cage base. Automatic activity counts were registered and counted using an on-line microprocessor/Apple II plus microcomputer system. In addition, two observers recorded the presence and intensity of various behaviours, associated with PE-induced behavioural stereotypy, on a 5-point scale (see Dourish, 1981, 1982 for further details). Each rat was tested once only, and was naive to the apparatus.

Because PE has a short duration of action, automatic activity scores were analysed for the first 15 min of the test period. Individual scores were transformed ($\sqrt{Y + \frac{1}{2}}$), before being submitted to a 2-way analysis of variance. Individual group comparisons were made according to the Newman-Keuls procedure. Similarly, the behavioural ratings were analysed for the first 15 min, using appropriate nonparametric statistics.

RESULTS

NAL (1.0-10.0 mg/kg), given alone, produced no significant effects on the three automatic activity scores. PE (6.25-50 mg/kg), given alone, also had little effect on spontaneous activity of the rats. As Fig. 1 shows, however, the two drugs in combination produced a striking stimulation of ambulatory activity. PE (at doses 12.5-50 mg/kg), in combination with NAL (10.0 mg/kg) induced significantly greater activity than when PE was administered alone. Some activation of ambulation was also evident for the combination of 3.0 mg/kg NAL and 50 mg/kg PE.

A similar picture emerged in the case of total horizontal activity. PE (at doses 12.5-50 mg/kg), in combination with NAL (10.0 mg/kg) stimulated significantly greater activity than when PE was administered alone. Significant activation also occurred with 3 mg/kg NAL in combination with 50 mg/kg PE. The mutual potentiation between PE and NAL affected only horizontal activity, and did not produce any stimulation of vertical activity.

Higher doses of PE induced behavioural stereotypy in the rats, particularly at 50 mg/kg. NAL (3.0 and 10.0 mg/kg) antagonized the splayed hindlimbs induced by PE (50 mg/kg), and reduced the hyperreactivity associated with that dose of PE, but had little effect on induced headmovements.

DISCUSSION

These data show, for the first time, a striking mutual potentiation between PE and NAL in their effects on total horizontal and ambulatory activity. The results indicate that intense behavioural action can occur as a consequence of PE action coupled with putative opiate receptor antagonism. The activation appeared to be selective for horizontal activity, since vertical activity remained unaffected by the PE-NAL combinations. Since the data were obtained using naive animals, it is uncertain at the moment, whether a similar mutual potentiation would be obtained in animals that were familiar with the test environment.

Behaviourally, the stimulation may have arisen in one (or both) of two ways. Firstly, it is possible that PE-

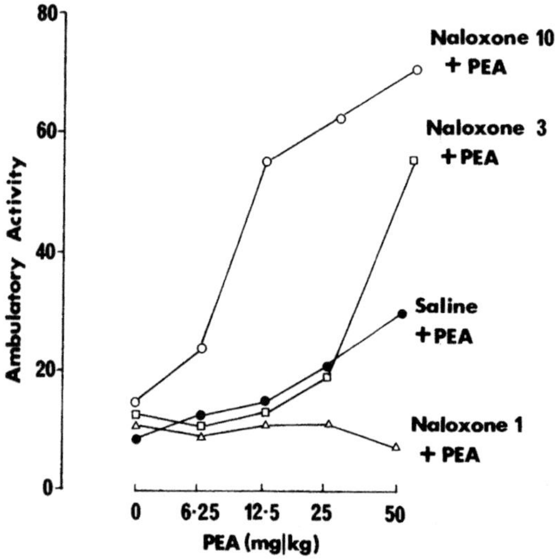

Fig. 1. Mutual potentiation between naloxone and β-phenylethylamine affecting ambulatory activity in naive male rats. Each point represents mean of 6 animals.

induced behavioural stereotypy may interfere with, or obscure, horizontal general activity associated with PE action. Antagonism of certain components of PE-induced stereotypy by NAL may therefore, according to this hypothesis, have led to a release of suppressed horizontal activity. But the fact that NAL in combination with 12.5 mg/kg PE produced behavioural stimulation (Fig. 1), tends to argue against this interpretation. PE at 12.5 mg/kg does not typically induce behavioural stereotypy in rats (e.g. Cooper and Dourish, 1983). Alternatively, PE and NAL in combination may have acted to produce a general increase in arousal, and directly affected horizontal activity. However, it is not clear why vertical activity was not also enhanced, as a further consequence of general arousal. A further possibility is that PE and NAL in combination act anxiolytically. Together, they may have reduced the fear/anxiety provoked by a novel environment, and thereby released horizontal exploratory activity.

So far, there appears to be no evidence in the rat that NAL potentiates the behavioural stimulant effects of AMPH. Instead, evidence consistently shows that NAL antagonizes the effects of AMPH. On the basis of interaction studies with NAL, therefore, there is little basis at present for drawing a comparison between PE and AMPH.

ACKNOWLEDGEMENTS

This work was supported by the Department of Health, Province of Saskatchewan. We thank Dr. A.A. Boulton for his enthusiastic support and provision of facilities. Naloxone was generously supplied by Endo Laboratories.

REFERENCES

Applegate C.D., Kuczenski R. and Leith N.J. (1982) Naloxone-amphetamine interactions: biochemical effects in dopamine terminal regions. Soc. Neurosci. Absts. 8, 121.

Boulton A.A. and Juorio A.V. (1982) Brain trace amines, in Handbook of Neurochemistry, Vol. 1 (Lajtha A., ed), pp. 189-222. Plenum Press, New York.

Cooper S.J. and Dourish C.T. (1983) Hypodipsia, stereotypy and hyperactivity induced by β-phenylethylamine in the water-deprived rat. Pharmacol. Biochem. Behav. (submitted for publication).

Dettmar P.W., Cowan A. and Walter D.S. (1978) Naloxone antagonizes behavioural effects of d-amphetamine in mice and rats. Neuropharmacology 17, 1041-1044.

Dourish C.T. (1981) Behavioural effects of acute and chronic β-phenylethylamine administration in the rat: evidence for the involvement of 5-hydroxytryptamine. Neuropharmacology 20, 1067-1072.

Dourish, C.T. (1982) A pharmacological analysis of the hyperactivity syndrome induced by β-phenylethylamine in the mouse. Br. J. Pharmac. 11, 129-139.

Greenshaw A.J. and Dourish C.T. (1983) Differential aversive stimulus properties of β-phenylethylamine and of d-amphetamine. Psychopharmacology (in press)

Hitzemann R., Curell, J., Ham D. and Loh H. (1982) Effects of naloxone on d-amphetamine- and apomorphine-induced behavior. Neuropharmacology 21, 1005-1011.

Holtzman S.G. (1974) Behavioral effects of separate and combined administration of naloxone and *d*-amphetamine. *J. Pharmacol. Exp. Ther.* 189, 51-60.

Turkish S., Dourish C.T. and Cooper S.J. (1983) Taste-dependent effects of low doses of -phenylethylamine and *d*-amphetamine on drinking in the rat, in *Neurobiology of the Trace Amines* (Boulton A.A., Baker G.B., Dewhurst W.G. and Sandler M., eds.), Humana Press, Clifton, New Jersey, in press.

E. CLINICAL STUDIES

THE ORIGIN, DRUG INTERACTION, URINE, PLASMA AND CSF CONCENTRATIONS OF PHENYLACETIC ACID IN NORMAL AND PSYCHIATRIC SUBJECTS

Farouk Karoum, E. Fuller Torrey, Dennis L. Murphy and Richard Jed Wyatt

Adult Psychiatry Branch, NIMH, Intramural Research Program, Saint Elizabeths Hospital, Washington, D.C. 20032

INTRODUCTION

Phenylacetic acid (PAA) is the major deaminated metabolite of phenylethylamine (PEA) in mammals. Its potential importance is highlighted by recent reports of elevated PEA excretion in some chronic paranoid schizophrenic patients (Potkin et al., 1979; Jeste et al., 1981) and in a subgroup of bipolar depressed psychotic patients (Karoum et al., 1982). Whether PEA is directly responsible for psychosis as suggested by the PEA hypothesis of schizophrenia (Fischer et al., 1975; Sandler and Reynolds, 1976; Wyatt et al., 1977) requires further investigation.

Although most urine PEA is derived from phenylalanine decarboxylation, its daily output is apparently not affected by diet or short term change in phenylalanine ingestion (Karoum et al., 1980). Most PEA produced in the body, therefore, appears to be endogenously regulated. Because of its high turnover and efficient metabolism PEA concentration in urine, plasma, CSF and brain is normally very low and hence its detection is below the sensitivity of many analytical procedures. Combined gas-chromatography mass-spectrometry (Karoum et al., 1979) or high resolution mass-spectrometry (Durden et al., 1973) are the current techniques of choice for the assay of PEA in biological media. The practical limitations of assaying PEA in biological materials in psychiatric research can theorotically be circumvented by assaying the rate of production of PAA. As described in this paper on a practical level, measuring PAA does not accurately reflect PEA metabolism.

In this communication we report on the excretion of PAA in schizophrenia and depression, the effects of PEA ingestion, and monoamine oxidase and dopadecarboxylase inhibitors on PAA excretion. We also report on the plasma and CSF concentrations of PAA in schizophrenic and depressed patients.

MATERIAL AND PATIENTS

Twenty-four hour urines were collected in receptacle containing 5 ml 10% ethylenediamine tetracetate (EDTA). After measuring total urine volume, aliquots were stored at -15°C until analyzed.

Urines were collected from 15 paranoid and 14 nonparanoid chronic schizophrenic patients and 31 young adult volunteers matched for age and gender. At least two urine collections were obtained from each subject. The controls and schizophrenic patients were maintained on normal diet. We also analyzed baseline urine collections from 11 unipolar depressed patients (these patients were maintained on low tyramine diet for at least three day before urine collection) and from another 10 depressed patients before and during the fourth week of chronic treatment with clorgyline (20-30 mg/day) or pargyline (50-125 mg/day). These 10 patients had been drug-free for a minimum of three weeks and an average of 26 days before the baseline urine samples were collected.

The schizophrenic patients met DSM III and Research Diagnostic Criteria (RDC) (Spitzer et al., 1978) for schizophrenia. Paranoid and nonparanoid subtype categorization was determined by semistructured interview and historical review (Potkin et al., 1978). The unipolar depressed patients met DSM III and RDC for major depression, were chronically ill and non-endogenous in terms of their symptoms pattern. The 10 depressed patients from the monoamine oxidase (MAO) inhibitor studies, met RDC criteria for major affective disorder (Sptizer et al., 1978) and had Hamilton depression scale scores in the moderately severe range (31.2 ± 4.3) on the 24 item scale (Lipper et al., 1979).

The schizophrenic and manic depressed patients studied for plasma and CSF concentrations PAA were all diagnosed according to DSM III for schizophrenia and bipolar disorder. These patients as well as the controls came from the same location in Western Ireland (County Roscommon, Maye and Galway). The patients were on a variety of medication while under investigation.

METHODS

Phenylacetic acid was assayed by a gas-chromatographic mass-fragmentographic method as previously described (Martin et al., 1979). A model 4000 Finnigan gas-chromatograph quadrupole mass-spectrometer was used. Phenylalanine and tyrosine were measured as previously reported (Karoum and Neff, 1982a). For the assay of urine PAA packed 3% SE 54 column was used (Martin et al., 1979). As for plasma and CSF concentration of PAA, a fused silica capillary column (30 m) bonded wall coated with SPB-5 (methylphenylvinylsilicone gum) was used (Supelco Inc., Bellefonte, PA.).

For the assay of plasma free PAA, an equal volume of acetone is added to a plasma aliquot (0.5 ml) containing 0.5 μg of $^2H_2$-PAA, mixed centrifuged and the clear supernatant transferred to a 10 ml glass tube. After the addition of 2N HCl (0.5 ml) and ethylacetate (5 ml), this mixture is vortex mixed, centrifuged and the organic phase (4 ml) separated. This extraction is repeated once more and after pooling, the organic phase is evaporated under reduced pressure (taking care not to over evaporate the residue), the residue is dissolved in methanol (0.3 ml) and transferred to a 1 ml microflex tube (Kontes Glass Co., Evanston, Ill.). After evaporation to dryness under a gentle stream of N_2, PAA is converted to its pentafluoropropionate ester (Martin et al., 1979). Excess esterification reagents are evaporated under N_2 to about 2 μl volume, followed by the addition of 15 μl of ethyl acetate. One or 2 μl of this solution is injected into the gas-chromatographic column.

For the assay of total plasma PAA containing $^2H_2$-PAA, 500 ng and $^2H_7$-PAG, ng, is deproteinized with acetone and the clear supernatant extracted with ethyl acetate exactly as described for free PAA. Both PAA and PAG and their deuterated isomers are extracted into ethyl acteate. The ethyl acetate extract is evaporated *in vacuo* and the dried residue mixed with 0.7 ml 6 N HCl. The acid solution is transferred into a 10 ml glass tube and heated at 100°C for 45 min to hydroxylze the PAG. After heating, 0.4 ml of 10 N NaOH is added to the hydrolyzed PAG inorder to neutralize most of the strong acid solution and the solution analyzed for total PAA as described for free PAA.

Assay of Free PAA in CSF

Cerebrospinal fluid, 0.5 ml, is added to 1 ml 1 N HCl followed by the addition of 100 ng of each of $^2H_2$-PAA and $^2H_7$-

PAA. The solution is mixed with ethyl acetate to extract the free PAA as described for free plasma PAA.

Assay of Total PAA in CSF

Cerebrospinal fluid, 0.5 ml, is added to 1 ml 10 N HCl followed by the addition of 100 ng of each of $^2H_2$-PAA and $^2H_7$-PAG. The mixture is heated at 100°C for 45 minutes, cooled in ice and then partially neutralized by adding 0.8 ml of 10 N NaOH. The mixture is then treated as described above for the assay of free PAA in CSF.

Quantification

Phenylacetic acid in both plasma and CSF is measured by simultaneously monitoring the following pairs of fragments: m/z 91 and 93 or 98, m/z 268 and 270, and m/z 268 and 275. The three sets of results are then evaluated for the presence of interferring substances by comparing the ratios of m/z 91 to 268, 93 to 270 and 98 to 275 for each sample and by observing the presence of peaks emerging close to the PAA or its deuterated isomers. Inconsistent ratios of a particular fragment will disqualify its use for quantification. The concentration of PAA is calculated by comparing the peak heights of the deuterated PAA internal standard with those corresponding to the appropriate fragments derived from PAA.

STUDIES AND RESULTS

Study I: PAA Excretion in Schizophrenia and Depression

Five schizophrenic patients were studied while stabilized on neuroleptics and again when off neuroleptics for at least two weeks. No effects of neuroleptic treatment on the excretion of free, conjugated or total PAA (NS, paired t-test) were observed. There also was no statistical difference in PAA excretion between 20 male and 11 female controls. Therefore, no further consideration was given to the patients' neuroleptic status or gender. Furthermore, no correlation could be established between urine 24 hr volume and total PAA excretion in the controls and patient groups studied (Pearson's r less than 0.3 for the controls and depressed patients and less than 0.1 for the schizophrenics).

The excretions of free, total and conjugated PAA in normal control subjects and chronic schizophrenic patients, subtyped as paranoid and nonparanoid, are summarized in Table I. As a group, chronic schizophrenic patients appear to excrete less total and conjugated PAA than the control subjects ($p < 0.01$, independent student t-test). When the patients were differentiated into paranoid and nonparanoid subtypes, only the nonparanoid patients had decreased total and conjugated PAA urinary excretion compared with controls (one-way analysis of variance, $p < 0.02$; Newman-Keules comparison, $p < 0.05$).

TABLE I

PHENYLACETIC ACID (PAA) EXCRETION (mg/ 24 hr) IN NORMAL CONTROLS AND CHRONIC SCHIZOPHRENIC PATIENTS

	FREE PAA		TOTAL PAA		CONJUGATED PAA	
Description	Mean ± SEM (n)	Range	Mean ± SEM (n)	Range	Mean ± SEM (n)	Range
Normal Controls	9.8 ± 1.5 (31)	3-45	162 ± 19 (31)	40-636	150 ± 19 (31)	43-591
Unipolar Depressed			149.2 ± 17 (21)	32-385		
Chronic Schizophrenics	11.0 ± 2.8 (23)	1-58	100 ± 12* (29)	6-288	69 ± 8* (23)	5-158
Paranoid chronic schizophrenics	14 ± 5 (11)	1-58	117 ± 17 (15)	25-288	83 ± 11+ (11)	24-158
Non-paranoid chronic schizophrenics	8.2 ± 3 (12)	1-29	82 ± 18+ (14)	6-269	64 ± 18+ (12)	5-244

*$p < .01$ by independent t-test comparing normal controls with all chronic schizophrenic patients.

+ One way analysis of variance for the three groups, $p < 0.02$, non-paranoid schizophrenics are significantly lower than normal controls, $p < 0.05$ by Newman-Keules comparisons.

N.B. The controls and schizophrenic patients were maintained on normal diet while the unipolar patients were restricted to tyramine low diet for at least three days before urine collection.

Total PAA excretion in 11 baseline unipolar depressed patients was not significantly different from the controls. In another study (Karoum et al., 1982b), a subgroup of these same unipolar patients were found to have no change in their PAA excretion after over four weeks treatment with deprenyl.

The scattergrams of the individual values of total PAA are illustrated in Figure 1. Seventy-five percent (23 of 31) of the control subjects had 24 hr total PAA urinary excretion of more than 100 mg in contrast with 29% of the nonparanoid (4 of 14 patients) and 53% of the paranoid (8 of 15 patients) schizophrenic patients.

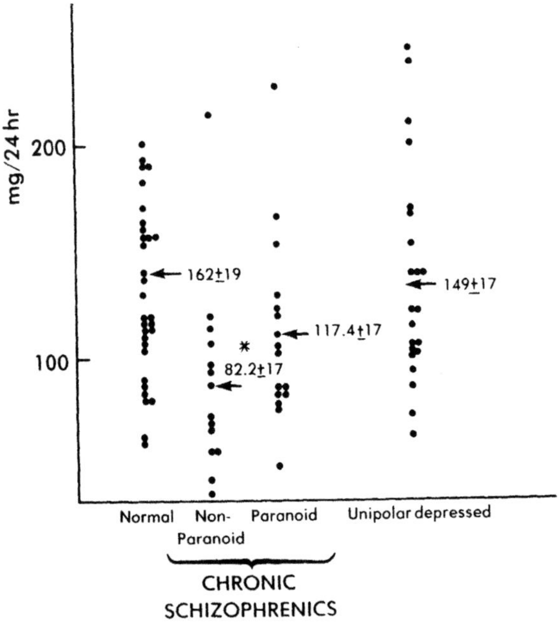

Fig. 1. Scattergrams of total PAA excretion in normal, schizophrenic and depressed subjects.
*=one way analysis of variance for the three groups, $p < 0.02$.

TABLE 2

PHENYLACETIC ACID (PAA) EXCRETION (mg/24 hrs) AFTER ORAL PEA ADMINISTRATIONS

Description	Free PAA	Total PAA	Conjugated PAA	Percentage change in total PAA
Patient 1, Baseline	20.5	280	259	
Patient 2, Baseline	16.1	199	183	
Patient 1 after 200 mg PEA, oral	57.7	478	421	70%
Patient 1 after 400 mg PEA, oral	60.2	754	694.0	169%
Patient 2 after 800 mg PEA, oral	84.9	1052.0	967.0	428%
Patient 2 after 1600 mg PEA, oral	227.3	1947	1719	878%

Study II: The Metabolism of Orally Administered PEA

In order to evaluate the metabolism of PEA to PAA we analzyed urine from four depressed patients after several doses of oral PEA. The individual results are summarized in Table 2. Phenylethylamine administered orally (200 to 1600 mg) was almost quantitatively metabolized to PAA.

Study III: Effects of Monoamine Oxidase Inhibitors (Pargyline and Clorgyline) on PEA Metabolism

This study was undertaken primarily to determine the contribution of endogenous PEA to the total urine excretion of PAA by metabolism through monoamine oxidase.

Two 24-hr urines were collected from a group of 10 depressed patients before and during the fourth week of treatment with either oral pargyline or clorgyline in a crossover study. The results obtained for the total PAA excretion are shown graphically in Figure 2.

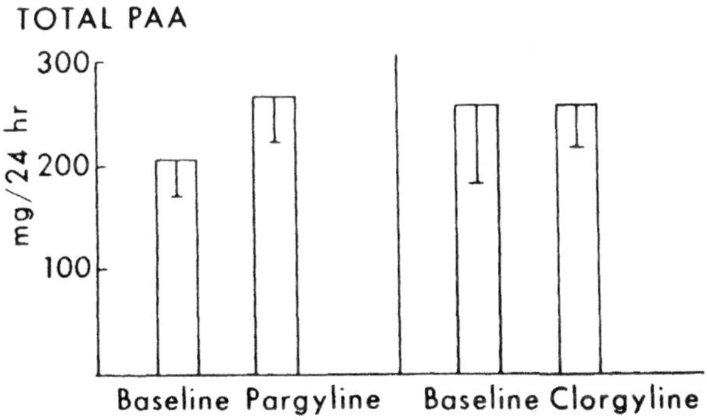

Fig. 2. Histograms of total PAA excretion in 10 depressed patients before and during four weeks oral pargyline (50-125 mg/day) or clorgyline (20-30 mg/day). Notice neither MAO inhibitor significantly changed PAA excretion.

Neither pargyline nor clorgyline significantly changed the mean 24-hr total PAA excretion, while PEA excretion was increased from two-fold (after clorgyline) to 100-fold (after pargyline) (Karoum et al., 1980). These results indicate that although PEA is rapidly metabolized to PAA via Type B MAO, the contribution of endogenous PEA to total urine PAA excretion is small. Therefore, urinary PAA is primarily derived from a metabolic pathway not involving the intermediate metabolism of PEA by MAO.

Study IV: The Effect of Chronic Carbidopa Administration on PAA Excretion

To further investigate the contribution of endogenous PEA towards total urine PAA, we measured total PEA and PAA excretion serially in two schizophrenic patients given 100 mg/24-hr oral carbidopa (peripheral dopa decarboxylase inhibitor) for three weeks. The excretion of PEA in all the urines collected after the second day of carbidopa administration was markedly reduced (Karoum et al., 1980). Carbidopa treatment failed to produce a consistent reduction in PAA excretion. These results indirectly support the conclusion made from Study III, that most urine PAA is not derived from PEA.

Study V: The Contribution of Phenylalanine to the Production of PAA

In this study 24-hr collections were gathered from a single schizophrenic patient (a high PEA excretor) while being maintained on a low phenylalanine diet (total daily phenylalanine intake not more than 700 mg) for four weeks. Total PAA excretion was gradually but consistently reduced, reaching a plateau after two weeks (total PAA was reduced from 105 to about 50 mg/24-hrs).

We also analyzed urines from two schizophrenic patients and five normal subjects after a 100 mg/kg oral phenylalanine load. As shown in Table 3, total PAA excretion did not consistently increase in all the urines that were analyzed.

TABLE 3

TOTAL PHENYLACETIC ACID (PAA) URINE EXCRETION (mg)
IN THE FOUR HOURS FOLLOWING ORAL PHENYLALANINE INGESTION+

	Total PAA	
Description	Baseline	After Phenylalanine
Chronic Schizophrenic	33	17
Chronic Schizophrenic	27	82
Normal Control	36	11
Normal Control	31	381
Normal Control	13	6
Normal Control	10	13
Normal Control	50	44
Mean ± SEM	28.6 ± 5.2	30.1 ± 10.2*

+L-phenylalanine was given at a dose of 100 mg/kg suspended in apple juice. Urine collection for 4 hr commenced immediately.

*The two means are not significantly different by paired t-test.

Although the number of subjects we have studied is small, the results, nevertheless, are interesting. The low, but consistent, reduction in PAA excretion observed in the one patient while on a low phenylalanine diet suggests that PAA in urine is at least partially derived from dietary phenylalanine. The failure of a single phenylalanine load, on the other hand, to elevate urine total PAA excretion during the four hours after loading, suggests that the metabolism of phenylalanine to PAA is controlled by a mechanism that responds slowly to changes in blood phenylalanine concentrations. Since a continuously high concentration of blood phenylalanine will stimulate PAA production (e.g., as in phenylketonuria (PKU)) (Blau, 1970) and prolonged low phenylalanine diet will induce a reduction of PAA excretion (Study V), a slow adaptation process of an unknown nature is apparently involved in the metabolism of phenylalanine to PAA.

Study VI: PAA Concentration in Plasma and CSF in Psychiatric Patients

The concentrations of PAA in the plasma and CSF from a group of manic depressed and schizophrenics were measured and compared with those from normal volunteers. These results are summarized in Table 4. As shown, there appear to be no difference in plasma and CSF concentrations of PAA betweeen genders. There is a tendency for PAA to be elevated in our patients, but only the paranoid schizophrenic and the bipolar depressed patients had significant elevations in their plasma and CSF of PAA. But, since most of the patients studied were on a variety of medications, our results need confirmation from the analysis of plasma and CSF from nonmedicated patients.

TABLE 4

CEREBROSPINAL FLUID (CSF) AND PLASMA CONCENTRATIONS OF PHENYLACETIC ACID (PAA) IN BIPOLAR DEPRESSED AND SCHIZOPHRENIC PATIENTS[a]

Description (Gender) (n)	CSF Total PAA	Plasma Total PAA
Normal Controls (M) (11)	39.8 ± 4.8	300 ± 58
Normal Controls (F) (8)	46.2 ± 4.1	281 ± 44
Normal Controls (M & F) (19)	42.5 ± 3.1	287 ± 38
Bipolar (Manic) Depressed Patients (4M & 4F) (8)	102.2 ± 15.9*	554 ± 123*
Paranoid Schizophrenic Patients (M) (10)	65.5 ± 10.5*	448 ± 45*
Undifferentiated Schizophrenic Patients (M) (22)	75.0 ± 15.5	337 ± 35
Schizoaffective Patients (M) (8)	59.3 ± 5.8	443 ± 86
Schizoaffective Patients (F) (7)	54.9 ± 3.2	381 ± 36

[a]The control subjects age ranged between 19-49 yrs with a mean and SEM of 28.65 ± 1.8 yrs. The bipolar and schizophrenic patients age ranged between 21 and 65 with a mean and SEM of 36.0 ± 1.56 yrs.

GENERAL DISCUSSION

As can be deduced from Table I and from previous reports, PAA is excreted in human urine primarily as a glutamine conjugate (James et al., 1972; Martin et al., 1979). Seakins (1971) has suggested that the majority of urinary PAA originated through the decarboxylation of phenylalanine to PEA in the gastrointestinal tract followed by its subsequent metabolism to PAA. Our data do not support this view; neither chronic ingestion of carbidopa (Study IV) nor a single load of phenylalanine (Study V) significantly changed PAA excretion. Furthermore, monoamine oxidase inhibitors (Study III) also failed to reduce PAA excretion (Figure 2), thus providing additional support for our conclusion that the contribution of gut PEA towards urinary PAA is small. In the one subject studied while on a low phenylalanine diet (Study V), total PAA excretion began to decrease only after one week (mean excretion, pretreatment, after one week, and after two weeks of low phenylalanine were 103, 49, and 22 mg/24-hr, respectively), which together with the results of studies III and IV suggests that most of urinary PAA is derived from phenylalanine via pathways not involving the formation of PEA as an intermediate.

Our interest in PAA excretion in schizophrenia stems from recent reports of high PEA excretion in some paranoid schizophrenic patients (Potkin et al., 1979; Jeste et al., 1981). The present studies were specifically undertaken to assess the usefulness of evaluating urine PAA as an index of changes in PEA metabolism and turnover. As is evident from Figure 2 and Studies III and IV, changes in PEA production cannot faithfully be represented by changes in PAA urinary output. This is because the contribution of PEA towards total excretion of PAA is apparently very small.

The relatively small amounts of PEA excreted (as compared with PAA) in man, however, does not exclude its possible importance as a naturally-occurring amphetamine-like substance. Potential central effects of PEA do not necessarily require production of milligram amounts. Neurotransmitters, supposedly, exert their influence at the synapse in very small quantities. Phenylethylamine is endogenously produced, has a fast turnover rate (Boulton, 1976; Karoum et al., 1980) and is efficiently metabolized (see Study II). In this context, it is worth mentioning that PEA is present in the rat brain at concentrations higher than those of several other non-catecholic phenylethylamines (e.g., p-, m-, and o-tyramine, p-octopamine, and phenylethanolamine (Boulton, 1976; Karoum et al., 1979)). Although the role of these latter non-catecholic amines is unclear in vertebrates, some of them (e.g., p-octopamine) may fulfill the criteria for a neurotransmitter in invertebrates (Saavedra et al., 1974).

PAA Excretion in Depression

The normal PAA excretion observed in the unipolar depressed patients (Table I) suggests there is a normal peripheral phenylalanine metabolism in this illness. As reported previously the urinary excretion of other biogenic amines (e.g., tyramine and octopamine) metabolites in depression also appears to be similar to normal subjects (Karoum et al., 1982b).

PAA Excretion in Schizophrenia

The low PAA excretion observed in our schizophrenic patients cannot be related to decreased PEA deamination (see Figure I), since MAO inhibition failed to reduce PAA excretion. The low PAA excretion may be attributed to other effects such as the patient's illness and hospitalization, patients intake of medication, incomplete urine collections, diet and an abnormality in phenylalanine metabolism via its transamination pathway. These effects require separate consideration.

Neuroleptics have been reported to inhibit d-amino acid oxidase (an enzyme that utilizes FAD as a co-factor) in vitro (Gabay and Harris, 1967). This enzyme can deaminate d-phenylalanine to phenylpyruvic acid, which after decarboxylation gives rise to PAA. Inhibition of d-amino acid oxidase by neuroleptics does not appear to have played an important role in the alteration of PAA in our patients, because neuroleptics failed to change PAA excretion in the five schizophrenic patients studied while they were on and off neuroleptics (Study I).

Another explanation for our findings may be attributed to reduced long term intake of phenylalanine or protein and/or deficiencies in any of the cofactors involved in the transamination of phenylalanine, its oxidative deamination, or in the decarboxylation of phenylpyruvic acid to PAA. Low phenylalanine or protein intake or dietary cofactor deficiency are not likely to be important, since our patients had normal balanced diets and were on daily vitamins and mineral supplements. Further a comparison of urinary excretion of phenylalanine and tyrosine between the controls and schizophrenic patients revealed no statistical differences in the output of these amino acids.

Assuming that the above observations are both confirmed and not artifactual in nature, we can offer two possible explanations for the observed low PAA excretion in schizophrenic patients. The first, a reduction of phenylalanine transamination and/or its oxidative deamination, and the second a decrease in phenylpyruvic acid decarboxylation (see Figure 3 for pathway).

Abnormal Phenylalanine Metabolism in Schizophrenia

The concept of reduced phenylalanine metabolism in some schizophrenic patients is an attractive hypothesis that needs direct confirmation. Chronic elevation in phenylalanine secondary to its reduced peripheral transamination could slowly lead to increased brain phenylalanine concentration in a manner comparable, but at a relatively slower rate, to the situation seen in phenylketonuria (PKU). In this connection, there have been a number of epidemiological studies suggesting an increase incidence of schizophrenia among both homozygotes and heterozygotes for PKU (Wyatt et al., 1977). Since the excretion of PAA is highly elevated in PKU (Blau, 1970) we therefore would expect to find at least a tendency for moderate increase rather than a decrease in PAA excretion in our schizophrenic patients, if one assumes some are heterozygotes for PKU.

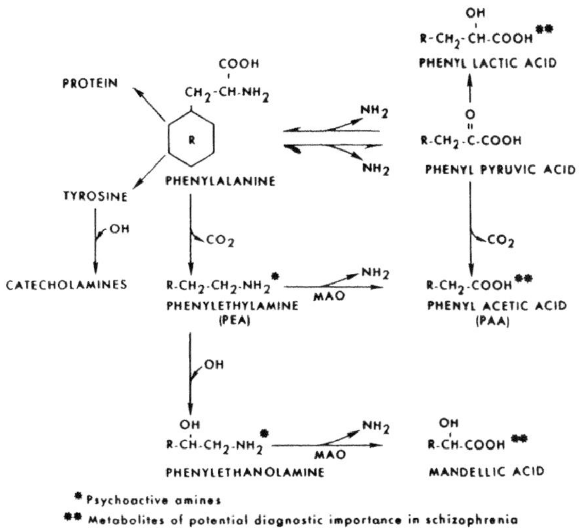

Fig. 3. A schematic and simplified metabolic pathway of phenylalanine.

To the extent that the reduced PAA excretion found in some schizophrenic patients reflects a decrease in phenylalanine transamination or oxidative deamination (assuming a normal dietary intake of phenylalanine), additional phenylalanine may be available to other relatively minor pathways. Included among these are p-hydoxylation to p-tyrosine and decarboxylation to PEA. Since PEA easily crosses the blood brain barrier (Karoum et al., 1982c), any increase in peripheral PEA, either secondary to reduced peripheral MAO activity or to excessive phenylalanine availability, would lead to an increase brain concentration of PEA. This increase in brain PEA might precipitate the symptoms of schizophrenia.

Plasma and CSF Concentration of PAA in Psychiatric Illness

A tendency for raised CSF concentration of PAA was observed in our bipolar depressed and schizophrenic patients especially the paranoid subtype. This latter is consistent with the report of Sandler et al. (1978) who reported elevated PAA in the CSF of a group of undifferentiated schizophrenics. In addition to CSF, we also found a tendency for raised plasma PAA in schizophrenia and manic depression. Because most of our patients were medicated, the results reported here should be considered tentative.

Our failure to observe a similar raise in urine PAA in our paranoid schizophrenic patient as we did for plasma and CSF of similar patients is difficult to explain without further investigation. As things presently stand, plasma and CSF estimation of PAA appears parodoxically to be better associated with psychosis than does urine PAA. Further, since PEA urinary excretion is apparently not directly associated with urine PAA and since high urine PEA is temporarily associated with psychosis (Karoum et al., 1982), it appears that abnormal PEA production is directly related to raised urine PEA excretion and raised plasma and CSF concentration of PAA.

In conclusion, urinary PAA excretion was found to be decreased in some schizophrenic patients but not in depression. A number of explanations for this low urinary PAA excretion were explored. Among these is the possibility of an abnormal phenylalanine metabolism. In contrast to urine, plasma and CSF concentrations of PAA appear to be elevated in paranoid schizophrenics. Manic depressives also showed a tendency for raised plasma and CSF concentrations of PAA.

REFERENCES

Blau, K. (1970) Aromatic acid excretion in phenylketonuria: Analysis of the unconjugated aromatic acid derived from phenylalanine. Clin. Chim. Acta 27, 5-18.

Boulton, A.A. (1976) Identification, distribution, metabolism and function of meta and para tyramine, phenylethylamine and tryptamine in brain. Adv. Biochem. Psychopharmacol. 15, 57-67.

Diagnostic and Statistical Manual of Mental Disorder (1980) Third Edition, American Psychiatric Association, Washington.

Durden, D.A., Philips, S.R. and Boulton, A.A. (1973) Identification and distribution of β-phenylethylamine in rat. Can. J. Biochem. 51, 995-1002.

Fischer, E. (1975) The phenylethylamine hypothesis of thymic hemeostatis. Biol. Psychiatry 10, 667-673.

Gabay, S. and Harris, S.R. (1967) Studies of flavin-adenine dinucleotide-requiring enzymes and phenothiazines-III inhibition kinetics with highly purified D-amino-acid oxidase. Biochem. Pharmacol. 16, 803-812.

James, M.O., Smith, R.L., Williams, R.T. and Reidenberg, M. (1972) The conjugation of phenylacetic acid in man, subhuman priamte and some non-primate species. Proc. Royal Soc. Lond. (B Biol. Sci.) 182, 25-35.

Jeste, D.V., Doonagaji, D.R., Panjwani, D., Datta, M., Potkin, S.G., Karoum, F., Thatte, S., Sheth, A.S., Apte, J.S. and Wyatt, R.J. (1981) Cross-cultural study of a biochemical abnormality in paranoid schizophrenia. Psychiatry Res. 3, 341-352.

Karoum, F., Nasrallah, H., Potkin, S., Chuang, L., Moyer-Schwing, J., Phillips, I. and Wyatt, R.J. (1979) Mass fragmentography of phenylethylamine, m- and p-tyramine and related amines in plasma, cerebrospinal fluid, urine and brain. J. Neurochem. 33, 201-212.

Karoum, F., Potkin, S.G., Murphy, D.L. and Wyatt, R.J. (1980) Quantitation and metabolism of phenylethylamine and tyramine's three isomers in humans, in Non-Catecholic Phenylethylamine (Mosnaim, A. and Wolf, M., eds), Part 2, pp. 177-191. Marcel Dekker Inc.

Karoum, F. and Neff, N.H. (1982a) Quantitative gas chromatography-mass spectrometry (GC-MS) of biogenic amines: Theory and practice, in Modern Methods in Pharmacology (Spector, S. and Bock, N., eds), pp. 39-54. Alan R. Liss, Inc., New York.

Karoum, F., Chuang, L.-W., Eisler, T., Calne, D.B., Liebowitz, M.R., Quitkin, F.M., Klein, D.R. and Wyatt, R.J. (1982b) Metabolism of (-) deprenyl to amphetamine and methamphetamine may be responsible for deprenyl's therapeutic benefit: A biochemical assessment. Neurology 32, 503-509.

Karoum, F., Speciale, S.G. Jr., Chuang, L.-W. and Wyatt, R.J. (1982c) Selective effects of phenylethylamine on central catecholamines: A comparative study with amphetamine. J. Pharmacol. Ther. 223, 432-439.

Karoum, F., Linniola, M., Potter, W.Z., Chuang, L.-W., Goodwin, F.K. and Wyatt, R.J. (1982d) Fluctuating high urinary phenylethylamine excretion rates in some bipolar affective disorder patients. Psych. Res. 6, 215-222.

Lipper, S., Murphy, D.L., Slater, S. and Buchsbaum, M.S. (1979) Comparative behavioral effects of clorgyline and pargyline in man: A preliminary evaluation. Psychopharmacology 62, 123-128.

Martin, M.E., Karoum, F. and Wyatt, R.J. (1979) Phenylacetic acid excretion in man. Anal. Biochem. 99, 283-287.

Potkin, S.G., Cannon, H.E., Murphy, D.L. and Wyatt, R.J. (1978) Are paranoid schizophrenics biologically different from other schizophrenics? New Engl. J. Medicine 298, 61-66.

Potkin, S.G., Karoum, F., Chuang, L., Cannon-Spoor, H.E., Phillips, I. and Wyatt, R.J. (1979) Phenylethylamine in paranoid chronic schizophrenia. Science 206, 470-471.

Potkin, S.G., Cannon-Spoor, H.E., DeLisi, L.E., Neckers, L.M. and Wyatt, R.J. (1983) Plasma phenylalanine, tyrosine and tryptophan in schizophrenia. Arch. Gen. Psychiatry 40, 749-754.

Saavedra, J.M., Brownstein, M.J., Carpenter, D.O. and Axelrod, J. (1974) Octopamine: Presence in single neurons of Aplysia suggests neurotransmitter function. Science 185, 364-365.

Sandler, M. and Reynolds, G.P. (1976) Does phenylethylamine cause schizophrenia? Lancet I, 70-71.

Sandler, M., Ruthven, C.R.J., Goodwin, B.L., King, G.S., Pettit, B.R., Reynolds, G.P., Tyrer, S.P., Weller, M.P. and Hirsh, S.R. (1978) Raised cerebrospinal fluid phenylacetic acid concentration: Preliminary support for the phenylethylamine hypothesis of schizophrenia. Comm. Psychopharm. 2, 199-202.

Seakins J.W.T. (1971) The determination of urinary phenylacetylglutamine as phenylacetic acid. Studies on its origin in normal subjects and children with cystic fibrosis. Clin. Chim. Acta 35, 121-131.

Spitzer, R.L., Endicott, J., Robins, E. (1978) Research Diagnostic Criteria (RDC) for a Selected Group of Functional Disorders, Biometrics Research, New York State Psychiatric Institute, New York.

Wyatt, R.J., Gillin, J.C., Stoff, D.M., Moja, E.A. and Tinklenburg, J.R. (1977) β-phenylethylamine (PEA) and the neuropsychiatric disturbances, in Neuroregulators and Psychiatric Disorders (Usdin, E., Barchas, J. and Hamburg, D., eds), pp. 31-45. Oxford University Press, New York.

THE CATABOLISM OF TRACE AMINES
IN SOME PSYCHIATRIC DISORDERS

[1]P.H. Yu, [1]B.A. Davis, [2]R.D. Bowen, [3]S. Wormith, [3]D. Addington and [1]A.A. Boulton

[1]Psychiatric Research Division, CMR Building
University of Saskatchewan, Saskatoon, Saskatchewan
[2]Department of Psychiatry, University of Saskatchewan
Saskatoon, Saskatchewan
[3]Regional Psychiatric Center, Saskatoon, Saskatchewan

INTRODUCTION

Trace amines possess an exceedingly high turnover rate (Durden and Philips, 1980; Wu and Boulton, 1974, 1975). They possess several routes of catabolism such as oxidative deamination, N-acylation, conjugation with sulfate, carbohydrates and amino acids as well as non-specific methylation. With little doubt oxidative deamination by monoamine oxidase (MAO) is the major degradative pathway and it has received much attention in the past decade. Many investigations on human platelet MAO have been reported (see reviews: Sandler *et al.*, 1981; Fowler *et al.*, 1982). Recently we have become interested in a group of agoraphobic patients. Agoraphobia has now been widely accepted as a distinct diagnostic syndrome. It is a phobic disorder characterized by pervasive fear of being alone or away from home and is often associated with recurrent panic attack episodes (Hallam, 1978). Benzodiazapine, an effective anxiolytic drug, is not effective in the treatment of this disorder (Sheehan *et al.*, 1980). MAO inhibitors and tricyclic antidepressants are, however, quite effective in ameliorating the panic attack and decreasing the frequency of the attacks even in the absence of major depressive illness (see review by Phol *et al.*, 1982). Such treatment indicates that biogenic amines may be involved.

In addition, we have also investigated the catabolism of trace amines in human psychopaths. The available evidence suggests that human aggression is genetically linked and that biological factors are involved (Hutchings and Melnick, 1975). Amphetamine decreases the fighting behavior of dominant rats. It facilitates, however, aggression in the subordinate animals (Miczek, 1974; Welch and Welch, 1971). Amphetamine and the related compound, methylphenidate, are used in the treatment of hyperactive children (Arnold *et al.*, 1973). Paradoxically amphetamine abuse may often cause homicidal violence (Ellinwood, 1971); this may be related to the amphetamine-induced paranoid psychosis. β-phenylethylamine (PE) has been proposed to be an "endogenous amphetamine," and is thus becoming an important subject for research in mental disorders.

Since direct measurement of trace amines in blood plasma is rather difficult and furthermore does not reflect the turnover rate of the amines, we have, instead, estimated both the enzyme involved in the catabolism and the acid metabolite formed. We have done this in patients (depressive, agoraphobic), and violent and non-violent prison inmates as well as in the healthy controls.

METHODS

Assay of platelet MAO activity. Platelets and plasma were prepared as previously described (Yu *et al.*, 1982). MAO activity was determined radioenzymatically using ^{14}C-labelled *p*-tyramine (*p*-TA), PE and tryptamine (T) as substrates (Yu *et al.*, 1982).

Determination of plasma trace acids. As previously described, a gas chromatographic-mass spectrometric (GC-MS) procedure utilising suitably deuterated trace acids as internal standards was used for the analysis of both unconjugated (U) and conjugated (C) phenylacetic acid (PAA), *p*-hydroxyphenylacetic acid (*p*-HPA) and *m*-hydroxyphenylacetic acid (*m*-HPA) (Davis *et al.*, 1982).

Research subjects. A group of 32 patients (mean age 34.03 ± 10.53) diagnosed as agoraphobic and 22 depressed patients (mean age 35.17 ± 12.50) without apparent agoraphobic symptoms, admitted to University Hospital, Saskatoon, were studied. Diagnoses were made according to

DSM III criteria. The diagnoses and psychological ratings were undertaken blind of the biochemical investigation and *vice versa*.

Two studies concerning human aggressiveness have been carried out. In the first (Study I) 103 individuals (age 26.8 ± 7.4) from the Regional Psychiatric Center (Saskatoon), and in the second (Study II), 26 inmates (age 30.8 ± 8.7) from a maximum security penitentiary volunteered to participate. All subjects were male. They have been grouped into violent (V), semi-violent (SV) and non-violent (NV) categories according to their court records and in-prison disciplinary reports.

All research subjects fasted overnight before blood samples were taken between 8 and 9 a.m.

RESULTS AND DISCUSSION

As can be seen in Table 1, platelet MAO activity in the agoraphobics is significantly higher than that of healthy control subjects (Control group II); it was also significantly higher than in matched non-agoraphobic depressed patients (Control group I).

Some recent reports have suggested that the dimension of social introversion and extraversion may be related to platelet MAO activity (Adler *et al*., 1980; Gattaz and Beckman, 1981; Demisch *et al*., 1982). Briefly, platelet MAO positively correlates to Wiggins social maladjustment scores. These scores have been described as meaning roughly that high scores are characterized with the parameters socially bashful, shy, embarrassed, self-conscious and socially reserved. Such personality characteristics are commonly found in agoraphobic patients. Low MAO activity is often associated with high social activity, sensation seeking, positive affect, hypomania, high ego strength and suicidal attempts. High platelet MAO activity seems to be associated with self-control, passiveness, withdrawal, uncommunicativeness and less social activity.

It has been reported that the depressed patients who show the greatest overall improvement after treatment with MAO inhibitors are most often characterized by clinical "features" such as psychic anxiety, somatic anxiety,

TABLE 1

Platelet MAO activity in agoraphobic patients and control subjects

MAO activity towards[1]	Agoraphobics (n=29)	Control I[2] (n=17)	Control II[2] (n=64)
p-Tyramine	3.57[3]	2.78	2.59
β-Phenylethylamine	13.58[3]	10.33	9.17
Tryptamine	1.15	0.82	0.90

[1]Results are expressed as means of platelet MAO activity (nmol/mg/h). [2]Control group I are non-agoraphobic depressed patients; Control group II are healthy volunteers with no apparent psychiatric disorders. [3]Values significantly different (by ANOVA) from Control groups I and II, $p<0.05$.

hypochondriasis, socialphobia and agoraphobia. The association of high platelet MAO in agoraphobic patients with panic attacks may possibly be related to the effective therapy with MAO inhibitors and warrant further study.

The biochemical basis of the elevation of the platelet MAO activity in these patients is not yet clear although our preliminary studies indicate that both enzyme concentration and substrate selectivity have been changed.

In addition to enzyme activity, we have also found that unconjugated PAA and *p*-HPA in plasma were significantly reduced (Table 2). This reduction, however, was not specific for agoraphobia since reduction of unconjugated PAA was also observed in the depressed patients. A reduction in the urinary excretion of PAA in depressed patients has recently been reported (Sabelli *et al*., 1983) and in fact these authors have proposed that urinary PAA excretion could be used as a state marker for the diagnosis of some forms of unipolar major depressive disorders. We should not overlook, however, that the urinary PAA is perhaps not solely derived by oxidative deamination of PE; furthermore it is not yet clear whether transamination of phenylalanine

TABLE 2

Unconjugated trace acid levels in the plasma of agoraphobic and control subjects

Trace acid	Agoraphobic (n=29)	Control I (n=17)	Control II (n=64)
PAA	107.3	82.0	132.8
m-HPA	11.9	14.7	19.2
p-HPA	62.7*	53.8*	83.7

Results are expressed in ng/ml. Only the mean is shown. See legend to Table 1 for further details. * Significantly different (by ANOVA) from Control group II, $p < 0.05$.

or p-tyrosine may contribute quantitatively to the urinary levels of the trace acids.

Positive correlations between Hamilton scores and platelet MAO activities have been claimed. In our studies, the short form of the Beck Scale for assessment of depression and the STAI scale for measurement of anxiety were utilised. Although platelet MAO activity towards PE tends to be correlated positively to the overall scores of both STAI and Beck scales, it did not reach statistical significance.

The platelet MAO activity and the conjugated and unconjugated trace acid levels in the plasma of violent offenders from Study I are shown in Table 3. Enzyme activity was reduced in the violent group with respect to all three substrates. Sandler et al. (1978) have reported that a group of incarcerated offenders with a history of violent crime exhibited elevated unconjugated PAA levels compared to offenders convicted of non-violent crimes and with no history of aggression. In this study we were unable to confirm this finding (Table 3). Instead, we found that unconjugated p-HPA and conjugated PAA were significantly lower in the violent group. These findings are at least in partial agreement with the results from an animal study (Dourish et al., 1982). In aggressive mice (induced by

TABLE 3

Platelet MAO activity and conjugated (C) and unconjugated (U) plasma trace acid levels in violent (V), semi-violent (SV) and non-violent (NV) prisoners in Study I

	V(n=35)	SV(n=30)	NV(n=38)
Platelet MAO			
p-tyramine	2.2	2.4	2.7
β-phenylethylamine	8.4	8.5	9.8
Tryptamine	0.5*	0.6	0.7
PAA (U)	136.1	119.7	126.9
PAA (C)	207.4**	335.3	361.3
m-HPA (U)	9.6	7.8	9.5
m-HPA (C)	1.3	0.8	1.1
p-HPA (U)	64.5*	78.5	99.5
p-HPA (C)	9.8	17.6	13.0

*$p < 0.05$, **$p < 0.02$ derived from ANOVA.
Subjects in Study I were 103 inmates from the Regional Psychiatric Centre in Saskatoon. Units are as described in Tables 1 and 2.

isolation) both PAA and *p*-HPA in blood plasma, urine and the brain were lower than in non-aggressive (i.e. grouped) mice. It would be interesting to know whether these biochemical changes in mice are related to their aggressiveness or whether they are a consequence of the isolation.

The human study (Study I) was complicated by the fact that most of the inmates were categorized as psychiatric patients and some of them had received drug treatment although MAO inhibitors were not included. Because of this we undertook Study II in which "non-psychiatric" inmates were assessed. In this second study subjects with a history of psychotic mental illness or who were being prescribed any psychotropic drugs were screened out. Their frequency of conviction and injury ratings in the community as well as in prison are illustrated in Table 4. It is

TABLE 4

Comparison of the violent incidences and injury rating of the violent (V) and non-violent (NV) groups of prisoners in Study II

	V(n=16)	NV(n=10)
Violence in the community		
Violence incidence	3.2 ± 1.9*	0.9 ± 1.0
Injury rating	3.4 ± 1.5**	1.0 ± 1.3
In prison disturbance		
Violence incidence	0.9 ± 0.09	0
Injury rating	0.9 ± 0.89	0

*$p < 0.01$; **$p < 0.001$.
Subjects in Study II were 26 prisoners from a maximum security penitentiary in Saskatchewan.

TABLE 5

Platelet MAO activity and conjugated and unconjugated plasma trace acid levels in violent (V) and non-violent (NV) prisoners in Study II

	V(n=16)	NV(n=10)	Significance (V versus NV)
Platelet MAO activity			
p-tyramine	2.14	2.70	p<0.05
β-phenylethylamine	7.46	9.79	p<0.01
Tryptamine	0.47	0.73	p<0.02
PAA (U)	102.63	124.62	
PAA (C)	262.19	263.78	
m-HPA (U)	9.64	6.72	
m-HPA (C)	0.90	1.70	p<0.05
p-HPA (U)	56.24	41.15	p<0.01
p-HPA (C)	19.19	21.59	

quite clear that the V group was significantly more violent than the NV group.

In the biochemical analyses (Table 5), the reduction in platelet MAO activity in the V group, towards the three substrates, was consistent with the first study. This reduction is not due to drug effect. Unconjugated and conjugated PAA, however, were not different between the two groups. p-HPA in the V group increased in comparison to the first study. It is difficult to interpret this paradoxical finding at this time.

We were also interested in the correlation, if any, between platelet MAO activities and plasma trace acid levels. As can be seen from Fig. 1, there was no correlation between plasma total PAA levels and platelet MAO activity towards PE. Neither was the platelet MAO activity correlated with the conjugated acid or the unconjugated acid levels. There was also no correlation between pHPA level in the plasma and platelet MAO activity towards p-TA. This indicates that other factors besides oxidative deamination are involved in the formation of the trace acids in plasma.

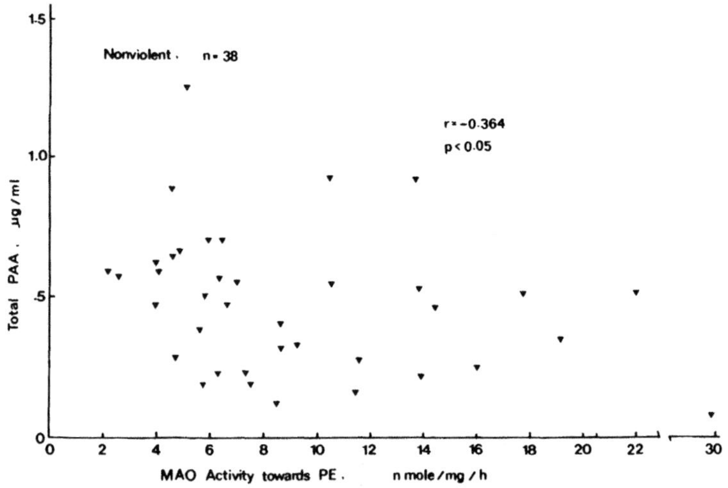

Fig. 1. Correlation between platelet MAO activity towards PE and the plasma PAA level in the non-violent offenders from aggression Study I.

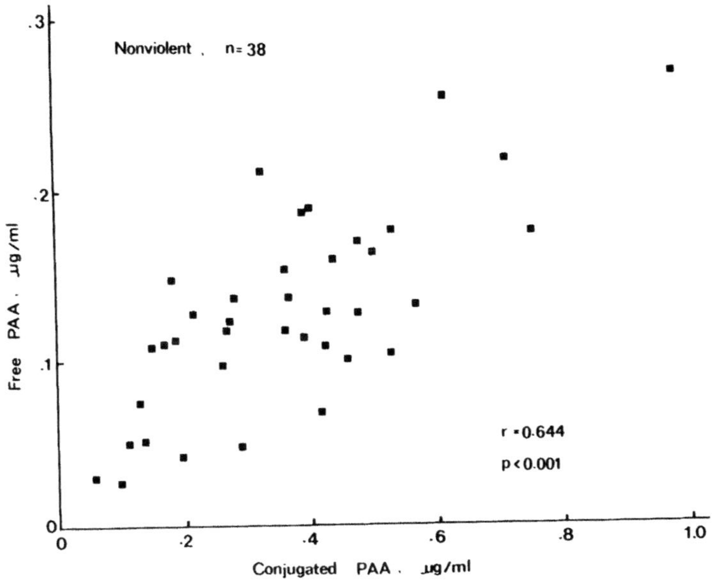

Fig. 2. Correlation between unconjugated and conjugated PAA levels in plasma of the non-violent offenders from aggression Study I.

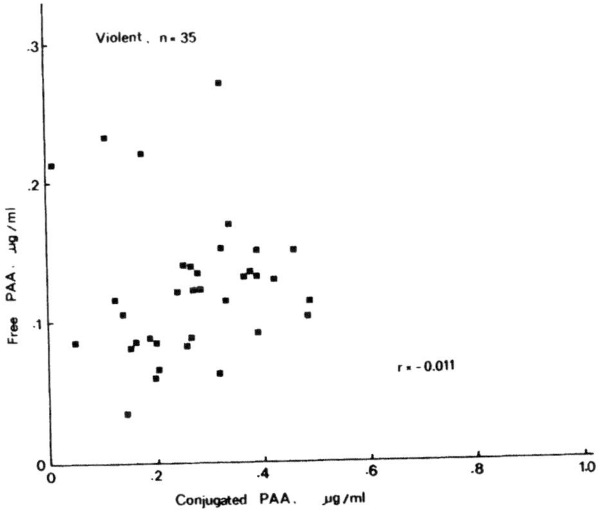

Fig. 3. Correlation between unconjugated and conjugated PAA levels in plasma of the violent offenders from aggression Study I.

A positive correlation between unconjugated and conjugated PAA, however, was found (Fig. 2) among the individuals in the NV group from aggression Study I. In the V group, however, this correlation did not hold (Fig. 3), perhaps indicating a defect in the conjugation system in some of the violent inmates.

We conclude that the catabolism of trace amines in mental disorder is quite complicated. Analysis of both enzyme activity and trace acid metabolites seems necessary. Whether any of the observed changes can be used as biological markers for a particular diagnostic group remains to be established.

ACKNOWLEDGEMENT

We thank the Saskatchewan Health, the MRC of Canada and the Canadian Penitentiary Service for their continuing financial support.

REFERENCES

Adler S.A., Gottesman I.I., Orsulak P.J., Kizuka P.P. and Schildkraut J.J. (1980) Platelet MAO activity: Relationships to clinical and psychometric variables. Schizophrenia Bull. 6, 226-231.

Arnold L.E., Kirileuk V., Corson S.A. and Corson O.L. (1973) Levo-amphetamine and dextro-amphetamine: differential effect on aggression and hyperkinesis in children and dogs. Am. J. Psychiat. 130, 165-170.

Davis B.A., Durden D.A. and Boulton A.A. (1982) Plasma concentrations of *p*- and *m*-hydroxyphenylacetic acid and phenylacetic acid in humans: gas chromatographic-high resolution mass spectrometric analysis. J. Chromatog. and Biomed. Appli. 230, 219-230.

Dourish C.T., Davis B.A., Dyck L.E., Jones R.S.G. and Boulton A.A. (1982) Alteration in trace amine and trace acid concentrations in isolated aggressive mice. Pharmacol. Biochem. and Behav. 17, 1291-1294.

Durden D.A. and Philips S.R. (1980) Kinetic measurements of the turnover rates of phenylethylamine and tryptamine *in vivo* in the rat brain. J. Neurochem. 34, 1725-1732.

Demisch L., Georgi K., Patzke B., Demisch K. and Bochnik H.J. (1982) Correlation of platelet MAO activity with introversion: a study on a German rural population. Psychiat. Res. 6, 303-311.

Ellinwood E.H. (1971) Assault and homicide associated with amphetamine abuse. Am. J. Psychiat. 127, 1170-1175.

Fowler C.J., Tipton K.F., Mackay V.P. and Youdim B.H. (1982) Human platelet monoamine oxidase - a useful enzyme in the study of psychiatric disorders? Neurosci. 7, 1577-1594.

Gattaz W.F. and Beckmann H. (1981) Platelet MAO activity and personality characteristics: a study in schizophrenic patients and normal individuals. Acta Psychiat. Scand. 63, 479-485.

Hallam R.S. (1978) Agoraphobia: a critical review of the concept. Br. J. Psychiat. 133, 314-319.

Hutchings B. and Mednick S.A. (1975) Registered male criminal adoptees and biological parents of registered male criminal adoptees, in Genetic Research in Psychiatry (Fieve R.R., Rosenthal D. and Brill H., eds.), pp 105-116, Johns Hopkins University Press, Baltimore.

Miczek K.A. (1974) Inter species aggression in rats: effects of d-amphetamine and chlordiazepoxide. Psychopharmacol. 39, 279-301.

Pohl R., Berchou R. and Rainey J.M. (1982) Tricyclic antidepressants and monoamine inhibitor in the treatment of agoraphobia. J. Clin. Psychopharm. 2, 399-407.

Sabelli H.C., Fawcett J., Gusovsky F., Javaid J. and Edwards J. (1983) Urinary phenyl acetate: a diagnostic test for depression. Science 220, 1187-1188.

Sheehan D.V., Ballenger J. and Jacobsen G. (1980) Treatment of endogenous anxiety with phobic, hysterical, and hypochondrial symptoms. Arch. Gen. Psychiat. 37, 51-59.

Sandler M., Reveley M.A. and Glover V. (1981) Human platelet monoamine oxidase activity in health and disease: a review. J. Clin. Path. 34, 292-302.

Welch A.S. and Welch B. (1971) Isolation, reactivity and aggression: evidence for an involvement of brain catecholamine and serotonin, in The Psychology of Aggression and Defeat (Eleftherion B.E. and Scott J.P., eds.), pp 91-142, Plenum Press, New York.

Wu P.H. and Boulton A.A. (1974) Distribution, metabolism and disappearance of intraventricularly injected p-tyramine in the rat. Can. J. Biochem. 52, 374-381.

Wu P.H. and Boulton A.A. (1975) Metabolism, distribution and disappearance of injected β-phenylethylamine in the rat. Can. J. Biochem. 53, 42-50.

Yu P.H., Bowen R., Carlson K., O'Sullivan K. and Boulton A.A. (1982) A comparison of some biochemical properties of platelet monoamine oxidase in some mentally disordered and healthy individuals. Psychiat. Res. 6, 107-121.

TYRAMINE AND DEPRESSIVE ILLNESS

M. Sandler, Susan M. Bonham-Carter and P.L. Walker
Bernhard Baron Memorial Research Laboratories and Institute of Obstetrics and Gynaecology, Queen Charlotte's Maternity Hospital, London W6 0XG, U.K.

Abstract

We have demonstrated repeatedly that patients with depressive illness show a consistent and highly significant decrease in urinary output of conjugated tyramine compared with control subjects when challenged with an oral tyramine load. This finding has now been confirmed by others who have demonstrated that the biochemical deficit is more severe in patients whose depression is endogenous. Although the disturbance is most prominent for tyramine degradation, we have identified a similar impairment of dopamine and octopamine metabolism in these patients. The conjugation deficit persists even when the patients are clinically recovered; thus, it may be considered as a marker for *predisposition* to the illness, rather than of the depressive state itself and used as such as a predictor for vulnerability. In a small population study, we were able to identify a significantly higher proportion of patients with a lifetime history of depressive illness in low conjugated tyramine excretors, compared with high excretors. The nature of the biochemical lesion is still unknown; platelets from depressed patients possess their full complement of the sulphate-conjugating enzyme, phenolsulphotransferase.

The past twenty five years has witnessed a more or less general acceptance of biological psychiatry, with the monoamines as one of the main struts of this new discipline. Indeed, our knowledge of the catecholamines and 5-hydroxytryptamine in relation to central nervous system function is now prodigious. However, the group of monoamines covered by the umbrella term "trace amines" has been almost totally neglected. These compounds possess no specific storage mechanism and, in consequence, are present in brain tissue in minute concentration, despite a brisk turnover. The first compilation on "Trace amines and the brain" (Usdin & Sandler, 1976) went some way towards bringing this blind spot into view. The present volume is intended to continue the process and monitor any advances made in a long-dormant area during the past seven years. In this connection, we wish to present our recent observations on the role of tyramine in depressive illness.

In the 1950's, tyramine was best known as a substrate for the *in vitro* assay of monoamine oxidase (MAO). Even when it emerged that the presence of this amine in human urine must point to its endogenous generation, the physiological implications were barely discussed (Oates, 1961). Experimentally, the pharmacology of tyramine was somewhat better served. The classical studies of Burn & Rand (1958) demonstrated that the pressor action of tyramine is indirect, resulting from a secondary release of noradrenaline. These findings set the stage for the discovery in the early 1960's of the "cheese effect", the hypertensive crisis supervening when patients under treatment with MAO-inhibiting drugs ingest tyramine-containing foodstuffs (Blackwell, 1963). Somewhat later, tyramine became implicated as a substance responsible for the initiation of headache attacks in patients suffering from so-called dietary migraine (Hanington, 1967). As a result of this development, my colleagues and I began to investigate the metabolism of tyramine in man.

Apart from the major inactivation pathway, involving oxidative deamination by MAO, it had been well recognised (Richter, 1940; Richter & MacIntosh, 1941; Beyer & Shapiro, 1945; Häggendal, 1963; Resnick, 1963) but lost sight of that when monoamines are given orally, they are degraded by conjugation to an extent depending on the nature of the amine (Morgan et al., 1969; Hengstmann et al., 1975).

We established that normal subjects metabolise an oral tyramine load to the extent of about 10-15% but that patients with dietary migraine appear to have some impairment of this conjugation process (Youdim et al., 1971). Other workers reported similar findings (Smith et al., 1971). With hindsight, it seems not unlikely that our self-selected patient group of that time suffered from some degree of depressive illness; indeed, the term "masked depression" has been used for just such a group of patients with headache as a depressive equivalent (Lopez Ibor, 1972). Whatever the clinical status of these subjects, our data led us to further experiments which resulted in the identification of a group of patients with depressive illness who had a highly significant decrease in tyramine-conjugating ability (Sandler et al., 1975). At that time we interpreted this finding in terms of a generalized increase in MAO activity, with a resulting deficit in transmitter monoamine at the synaptic cleft. Indeed, this hypothesis received some support from our subsequent direct demonstration of increased platelet MAO activity in depressive illness (Reveley et al., 1981). However, this observation does not solve the problem completely because, as is well known, the platelet contains solely MAO B (Donnelly & Murphy, 1977) and we still have no readily available direct probe of *in vivo* MAO A activity. Any increase of MAO action overall might be expected to lead to an enhanced urinary excretion of the major oxidatively deaminated metabolite of tyramine, *p*-hydroxyphenylacetic acid, in affected subjects. If anything, however, there tends to be a *deficit* in urinary output of this metabolite in affected subjects compared with controls after oral tyramine challenge (Bonham-Carter et al., 1978a).

Our first set of investigations in depressive patients was followed by observations on another more severely afflicted group (Bonham-Carter et al., 1978a). We again noted a highly significant decrease in tyramine-conjugating ability. Whilst certain clinical features suggestive of delayed gastrointestinal absorption are present in depressive illness, indirect experiments failed to reveal any gut malfunction sufficient to account for our findings (Bonham-Carter et al., 1978b). Although the particular tyramine conjugate we were investigating was known to be a sulphate, either substantially or completely (Mullen & Smith, 1971), we were unable to find any evidence of sulphate deficiency in our group of depressive patients (Bonham-Carter et al., 1980a).

The deficit was clearly shown to persist, even after clinical recovery (Bonham-Carter et al., 1978a). This finding was important, for it was the first indication we had had of the biochemical lesion being a trait - rather than a state- marker and, accordingly, a possible predictor of vulnerability to depression. We therefore decided to perform a prospective study of the phenomenon.

The so-called "four day blues" is a relatively mild, semi-physiological phenomenon, a transient and self-limiting bout of depression supervening shortly after childbirth in many women. Not to be confused with it, although the two may shade into each other, is post-partum depression proper which is viewed for most workers as a manifestation of depressive illness itself. Females prone to depression have a considerably greater chance of experiencing an episode in the period following childbirth than at any other time in their lives (Kendell et al., 1976). Even mild degrees of this illness must be taken seriously, for they may well have a disruptive effect on the mother-child relationship, leading to an increase in infant morbidity, if not mortality (H. Caplan, in preparation). Thus, if we were able to identify prospectively those women who are more likely to develop the illness than others, by virtue of some pre-existing vulnerability, this would enable appropriate social and psychiatric support to be given and would mark an important step forward. In an attempt to identify such women prospectively (Bonham-Carter et al., 1980b), we screened a group of patients attending an antenatal clinic to see whether the oral tyramine challenge test we had been using (100mg of tyramine orally followed by measurement of tyramine sulphate urinary output over a timed period) would identify those more likely to become depressed after childbirth. This pilot study demonstrated clearly that patients with low output values had a significantly higher lifetime incidence of depressive illness compared with those with high values. None of these patients, it must be stated, was depressed at the time of tyramine loading although two of them subsequently developed depressive illness in the puerperium of the particular pregnancy under investigation.

Despite the success of this pilot experiment, it has not been practicable, up to the present time, to carry out a more extensive study because the oral tyramine loading test is tedious and cumbersome to administer, both for the clinician and the patient, requiring timed urine collection over a number of hours. We are now trying to modify it by measuring plasma concentration of conjugated tyramine with time, i.e. performing a "tyramine tolerance test".

It seemed important to try to establish whether the conjugation deficit, whatever its mechanism, is specific for tyramine or is a manifestation of a more generalized defect. Whilst our findings (Bonham-Carter, S.M., Walker, P.L. & Sandler, M., in preparation) are preliminary, they indicate that in depressive illness, both *p*-octopamine and dopamine sulphation, at least, are impaired in a manner similar to that of tyramine. In fact, this deficit was almost overlooked, for the time curve of gastrointestinal absorption of these two amines is rather longer than that of tyramine.

The actual mechanism of the conjugation deficit has proved peculiarly difficult to unravel. To test our original hypothesis of an increase in MAO activity would require a readily available source of MAO A but none has so far been identified. In the near future, however, we hope to obtain gut biopsy material from recovered depressives, who are able to give informed consent. As mentioned earlier, these patients still manifest with a conjugation deficit, even during clinical remission. Small gut mucosa possesses high activity of MAO A (Squires, 1972; Elsworth et al, 1978). We would also, with the same clinical material, be able to test directly the possibility of some defect existing of the sulphate conjugating enzyme, phenolsulphotransferase (PST), an obvious theoretical cause of deficient sulphation.

Human PST has recently become the centre of great interest in mental health research (see Sandler & Usdin, 1981). Identification of its activity in the human platelet (Hart et al,., 1979; Anderson & Weinshilboum, 1980; Rein et al., 1981) has made direct clinical studies possible for the first time. Very recently, my colleagues and I have been able to show that PST exists in two forms which we have termed M and P enzymes (Rein et al., 1982). They appear to be under separate genetic control (Bonham-Carter et al., 1981; Reveley et al., 1983). Specific substrates for the M form include the monoamines and many of their metabolites whist the P form degrades a number of exogenous phenols of various kinds (Bonham-Carter et al., 1983; Sodha et al., 1983). These forms differ in tissue distribution, inhibitor specificity, thermostability and substrate preference. Disappointingly, careful investigation has revealed no deficit of either form of the enzyme in platelets from depressive patients (Bonham-Carter et al., 1981). Sulphate conjugation of an orally administered monoamine predominantly takes place in the small gut mucosa (Nose & Lipmann, 1958); however, final proof must lie in analysis of the gut biopsy material we hope to obtain shortly.

ATP is essential for the generation of 3'-phosphoadenosine 5'-phosphosulphate (PAPS: "active sulphate") from which sulphate is transferred by the action of PST to a phenolic acceptor (see Sandler & Usdin, 1981). We also measured the concentration of ATP, therefore, in plasma from depressives and controls (Bonham-Carter et al., 1981). Again, no deficit was apparent. Before completing this line of experimentation, however, we need to measure the plasma concentration of active sulphate itself. If this were to prove normal, some explanation based on impairment of sulphate uptake would have to be our fallback position.

The whole problem is fraught with unknowns. It is not yet clear whether all clinical groups of affective disorder are involved equally. Nor do we know whether the existence of an affective component in other psychiatric illnesses, such as panic/anxiety disorder and alcoholism, will be expressed as a greater or lesser degree of the deficit. Even the detailed pharmacokinetics of tyramine and the size of dose necessary to demonstrate the effect are relatively unknown. Although we must assume that our findings point to a trait-dependent deficit, extensive family studies are obviously required to uphold such an interpretation. Even so, it is of great interest that a tyramine -conjugation deficit in depression has very recently been confirmed by Harrison et al. (1982). In corroborating our findings, they showed that, after oral tyramine challenge, both unipolar and bipolar endogenous depressives put out significantly lower amounts of conjugated tyramine compared with normals: patients with atypical depression did not differ significantly in their tyramine excretion pattern from normal controls. Whatever the mechanism of the metabolic defect we have observed turns out to be, we can still use it empirically as a trait marker, much as others have employed decreased platelet 5-hydroxytryptamine uptake (Coppen et al., 1978; Tuomisto et al., 1979), alteration in α_2-adrenoceptor density (Charney et al., 1981), altered ^{3}H-impramine binding (Langer et al., 1981) and abnormalities in the dexamethasone suppression test (Carroll et al., 1976) without properly understanding their mechanism.

Although the direct assay of platelet PST was developed by our group to provide the answer to the problem of this tyramine conjugation deficit, in the event it failed to do so. Even so, platelet PST assay has now been performed in other conditions and interesting data have emerged. Thus, concordant twins have been shown to manifest a higher degree of correlation of platelet PST activity than discordant twins: there was no difference between normal and schizophrenic twin pairs (Reveley et al., 1983). Patients with Parkinson's disease have been shown to possess higher platelet PST activity than normals but this finding, in fact, was later shown to stem from an adaptive enzyme increase in response to L-dopa administration. Parkinsonians not under treatment with this drug and those on a lower dosage regimen had a proportionately smaller rise in activity (Glover et al., 1983).

One of the most interesting findings to emerge from these studies, a significant deficit of PST P in patients with dietary migraine compared with normal and non-dietary migraine, was also the first demonstration of an objective physicochemical change to be demonstrated in patients belonging to this clinical group (Littlewood et al., 1982). It may well be that such a finding reflects a similar deficit of the corresponding enzyme in the gut wall; there may thus be insufficient activity there to detoxicate noxious dietary phenols which, as a consequence, gain access to the circulation and initiate a headache attack. The situation, however, may be even more complex for our very recent data (Littlewood, J., Glover, V., Sandler, M., in preparation) show clearly that red wine, a notorious migraine-triggering agent, inhibits PST P powerfully, whilst white wine does so to a lesser extent. Thus, any natural tendency to a low PST P activity may be reinforced by the action of a foodstuff containing a presumably phenolic inhibitor. Here again, we are left with further questions because it turns out that tea also contains a powerful PST inhibitor - although tea is not generally considered to be a migraine-triggering agent. It may well be that a mixture of phenolic compounds is responsible for intitiating a headache attack in predisposed subjects, some of which inhibit PST P directly whilst others overwhelm the weakened enzymatic barriers to gain access to the circulation and trigger a headache episode. Although these findings appear remote from the original tyramine conjugation deficit, their emergence is a direct consequence of that observation!

It is not only this present set of findings which implicate tyramine in the pathogenesis of depressive illness. There are other pointers too. Ghose et al., (1975), for example, noted an enhanced pressor sensitivity to intravenous tyramine in patients suffering from depressive illness. We ourselves have demonstrated an apparent decrease in endogenous output of the major metabolite of tyramine, *p*-hydroxyphenylacetic acid, in depression, and this biochemical lesion was combined with a decreased output of the major metabolite of octopamine, *p*-hydroxymandelic acid (Sandler et al., 1979a), whilst a decreased concentration of the major metabolite of phenylethylamine, phenylacetic acid, both free and conjugated, was identified in cerebrospinal fluid from these patients (Sandler et al., 1979b). All in all, there may be a better case for the existence of a tyramine deficit in depressive illness than of noradrenaline or 5-hydroxytryptamine!

References

Anderson, R.J., Weinshilboum, R.M. (1980) Phenolsulphotransferase in human tissue: radiochemical enzymatic assay and biochemical properties. *Clin. Chim. Acta* 103: 79-90.

Beyer, K.H., Shapiro, S.H. (1945) Excretion of conjugated epinephrine related compounds. *Am. J. Physiol.* 144: 321-330.

Blackwell, B. (1963) Tranylcypromine. *Lancet* ii: 414.

Bonham-Carter, S.M., Glover, V., Sandler, M., Gillman, P.K., K.P. Bridges, P.K. (1981). Human platelet phenolsulpho-

transferase: separate control of the two forms and activity range. *Clin. Chim. Acta* 117: 333-334.
Bonham-Carter, S.M., Goodwin, B.L., Sandler, M., Gillman, P.K., Bridges, P.K. (1980a). Decreased conjugated tyramine output in depression: the effect of oral L-cysteine. *Br. J. Clin. Pharmac.* 10: 305-308.
Bonham-Carter, S.M., Rein G., Glover, V., Sandler, M., Caldwell, J. (1983). Human platelet phenolsulphotransferase M and P: substrate specificities and correlation with *in vivo* sulphoconjugation of paracetamol and salicylamide. *Br. J. Clin. Pharmac.* 15: 323-330.
Bonham-Carter, S.M., Reveley, M.A., Sandler, M., Dewhurst, J., Little, B.C., Hayworth, J., Priest, R.G. (1980b). Decreased urinary output of conjugated tyramine is associated with lifetime vulnerability to depressive illness. *Psychiat. Res.* 3: 13-31.
Bonham-Carter, S.M., Sandler, M., Goodwin, B.L. Sepping, P., Bridges, P.K. (1978a). Decreased urinary output of tyramine and its metabolites in depression. *Br. J. Psychiat.* 132: 125-132.
Bonham-Carter, S.M., Sandler, M., Sepping, P., Bridges, P.K., (1978b). Decreased conjugated tyramine outputs in depression: gastrointestinal factors. *Br. J. Clin. Pharmac.* 5: 269-272.
Burn, J.H., Rand, M.J. (1958). The action of sympathomimetic amines in animals treated with reserpine. *J. Physiol.* 144: 314-336.
Carroll, E.J., Curtis, G.C., Mendels, J.M. (1976). Neuroendocrine regulation in depression. 11. Discrimination of depressed from nondepressed patients. *Arch. Gen. Psychiat.* 33: 1051-1058.
Charney, D.S., Heninger, G.R., Sternberg, D.E., Redmond, D.E., Leckman, J.F., Maas, J.W., Roth, R.J. (1981). Presynaptic adrenergic receptor sensitivity in depression. *Arch. Gen. Psychiat.* 38: 1334-1340.
Donnelly, C.H., Murphy, D.L. (1977). Substrate- and inhibitor-related characteristics of human platelet monoamine oxidase. *Biochem. Pharmac.* 26: 753-858.
Elsworth, J.D., Glover, V., Reynolds, G.P., Sandler, M., Lees, A.J., Phuapradit, P., Shaw, K.M., Stern, G.M., Kumar, P. (1978). Deprenyl administration in man: a selective monoamine oxidase B inhibitor without the "cheese effect". *Psychopharmacology* 57: 33-38.
Ghose, K., Turner, P., Coppen, A. (1975) Intravenous tyramine pressor response in depression. *Lancet* i: 1317-1318.

Glover, V., Lees, A.J., Ward, C., Stern, G.M., Sandler, M., (1983) Platelet phenolsulphotransferase activity in Parkinson's disease. *J. Neural Transmiss., in the press.*

Häggendal, J. (1963) The presence of conjugated adrenaline and noradrenaline in human blood plasma. *Acta physiol. Scand.* 59: 255-260.

Hanington, E. (1967) Preliminary report on tyramine headache. *Br. Med. J.* 2: 550-551.

Harrison, W., Cooper, T., Quitkin, F., Liebowitz, M., McGrath, P., Stewart, J., Klein, D. (1982) Tyramine excretion test in depressive illness. *Abstr. ACNP Mtg., Puerto Rico,* p. 56.

Hart, R.F., Renskers, K.J., Nelson, E.B., Roth, J.A. (1979) Localization and characterization of phenol sulphotransferase in human platelets. *Life Sci.* 24: 125-130.

Hengstmann, J.H., Konen, W., Konen, C., Eichelbaum, M., Dengler, H.J. (1975). Bioavailability of m-octopamine in man related to its metabolism. *Eur.J.Clin.Pharmac.* 8: 33-39.

Kendell, R.E., Wainwright, S., Hailey, A., Shannon, B. (1976). The influence of childbirth on psychiatric morbidity. *Psychol. Med.* 6: 297-302.

Langer, S.Z., Zarifian, E., Briley, M., Raisman, R., Sechter, D. (1981). High-affinity binding of ^{3}H-imipramine in brain and platelets and its relevance to the biochemistry of affective disorders. *Life Sci.*29:211-220.

Littlewood, J., Glover, V., Sandler, M., Petty, R., Peatfield, R., Rose, F.C. (1982) Platelet phenolsulphotransferase deficiency in dietary migraine. *Lancet* i: 983-986.

Lopez Ibor, J.J. (1972). Masked depression and depressive equivalents. In: Depressive Illness, Diagnosis, Assessment, Treatment. *Ed.P. Kielholz, Hans Huber, Bern,* pp. 38-44.

Morgan, C.D., Ruthven, C.R.J., Sandler, M. (1969). The quantitative assessment of isoprenaline metabolism in man. *Clin. Chim. Acta* 26: 381-386.

Mullen, P.E., Smith, I. (1971). Tyramine metabolism and migraine: a metabolic defect. *Br.J.Pharmac.*41: 413-414P.

Nose, Y., Lipmann, F. (1958). Separation of steroid sulfokinases. *J.Biol.Chem.* 233: 1348-1351.

Oates, J.A.(1961). Measurement of urinary tryptamine, tyramine & serotonin. In: Methods in Medical Research, Vol. 9, *Ed.J.H. Quastel, Year Book Medical Publishers,*

Inc., Chicago, pp. 169-174.
Rein, G., Glover, V., Sandler, M.(1981). Sulphate conjugation of biologically active monoamines and their metabolites by human platelet phenolsulphotransferase. *Clin.Chim.Acta* 111: 247-256.
Rein, G., Glover, V., Sandler, M., (1982). Multiple forms of phenolsulphotransferase in human tissues: selective inhibition by dichloronitrophenol. *Biochem.Pharmac.* 31: 1893;1897.
Resnick, O.(1963). The metabolism of orally ingested epinephrine in man. *Life Sci.* Pt. 9: 629-636.
Reveley, M.A., Glover, V., Sandler, M., Coppen, A.(1981). Increased platelet monoamine oxidase activity in affective disorders. *Psychopharmacology* 73: 257-260.
Reveley, A.M., Bonham-Carter, S.M., Reveley, M.A., Sandler, M. (1983). A genetic study of platelet phenolsulphotransferase activity in normal and schizophrenic twins. *J.Psychiat.Res., in the press.*
Richter, D., (1940). The inactivation of adrenaline *in vivo* in man. *J.Physiol.Lond.* 98: 361-371.
Richter, D., MacIntosh, F.C.(1941). Adrenaline ester. *Am.J.Physiol.* 135: 1-5.
Sandler, M., Bonham-Carter, S.M., Cuthbert, M.F., Pare, C.M.B. (1975). Is there an increase in monoamine oxidase activity in depressive illness? *Lancet* i: 1045-1049.
Sandler, M., Ruthven, C.R.J., Goodwin, B.L., Coppen, A. (1979a). Decreased cerebrospinal fluid concentration of free phenylacetic acid in depressive illness. *Clin.Chim.Acta* 93: 169-171.
Sandler, M., Ruthven, C.R.J., Goodwin, B.L., Reynolds, G.P., Rao, V.A.R., Coppen, A. (1979b). Deficient production of tyramine and octopamine in cases of depression. *Nature, Lond.* 278: 357-358.
Sandler, M., Usdin, E. (1981) (Eds.). Phenolsulphotransferase in Mental Health Research. *Macmillan, Basingstoke.*
Smith, I., Kellow, A.H., Mullen, P.E., Hanington, E.(1971). Dietary migraine and tyramine metabolism. *Nature, Lond.* 230: 246-248.
Sodha, R.J., Glover, V., Sandler, M. (1983). Phenolsulphotransferase in human placenta. *Biochem. Pharmac.* 32: 1655-1657.
Squires, R.F.(1972). Multiple forms of monoamine oxidase in intact mitochondria as characterized by selective inhibitors and thermal stability: a comparison of eight mammalian species. *In: Monoamine Oxidases - New Vistas.* Eds. E. Costa, M. Sandler, Raven Press,

New York, pp. 355-370.

Tuomisto, J., Tukiainen, E., Ahlfors, U.G. (1979). Decreased uptake of 5-hydroxytryptamine in blood platelets from patients with endogenous depression. *Psychopharmacology* 65: 141-148.

Usdin, E., Sandler, M. (1976) (Eds.). Trace amines and the brain. *Dekker, New York.*

Youdim, M.B.H., Bonham-Carter, S., Sandler, M., Hanington, E., Wilkinson, M. (1971). Conjugation defect in tyramine-sensitive migraine. *Nature, Lond.* 230: 127-128.

PHENYLETHYLAMINE, TYRAMINE AND OTHER TRACE AMINES IN PATIENTS WITH AFFECTIVE DISORDERS: ASSOCIATIONS WITH CLINICAL STATE AND ANTIDEPRESSANT DRUG TREATMENT

D.L. Murphy, F. Karoum, I. Alterman,
S. Lipper and R. J. Wyatt
National Institute of Mental Health
Bethesda, Maryland 20205 U.S.A.

ABSTRACT

In 22 patients with major affective disorders, phenylethylamine (PE) excretion tended to be higher and more variable in 12 females (25 $\pm$ 39 μg/d) than in 10 males (8 $\pm$ 1 μg/day). In a subgroup of 15 patients studied intensively as part of a treatment trial with the MAO-inhibitors, clorgyline and pargyline, pretreatment PE excretion was not consistently associated with differences in severity of depression, anxiety, activation or hypomania, although almost all correlations with psychopathology were in the negative direction, including several [self-rated dysphoria (r=-.55), anger (r=-.49), functional deficit (r=-.59)] which were statistically significant. Excretion of m-tyramine (m-TA) also exhibited generally negative associations with psychopathology, including self-rated depression (r=-.52) and anxiety (r=-.59), while hypomania was positively correlated (r=.72). Significant cross-correlations among PE, m-TA, and p-TA were not present at baseline, but MAO-inhibitor treatment led to intercorrelated increases in these trace amines, as well as phenylethanolamine, o-TA and platelet MAO activity changes. Increased PE excretion was associated with significantly decreased self-rated depression and anxiety during MAO-inhibitor treatment. Some baseline cardiovascular measures were negatively correlated with PE excretion, including systolic blood pressure, pulse and plasma norepinephrine concentrations, while baseline p-TA excretion was significantly correlated with pulse increases and systolic and diastolic blood pressure reductions produced by standing. These associations suggest that further study of behavioral connections with PE and m-TA excretion and of cardiovascular interactions with p-TA and PE might be of interest. Our data could be considered consistent with an amended hypothesis relating phenylethylamine to a cluster of depression-related symptoms including dysphoria and anger, but not depressed mood alone.

INTRODUCTION

Monoamine metabolism abnormalities in patients with affective disorders have been the focus of an intensive research effort during the past 15 years. Most of this effort has been directed towards the major amine neurotransmitters, especially norepinephrine, serotonin, and, to a lesser extent, dopamine. In the absence of any clear-cut evidence of a single neuroamine abnormality in all depressed patients, recent hypotheses have emphasized heterogeneity in depression subtypes, possible abnormalities in the balance among neurotransmitter systems themselves, as well as altered neuroamine interactions with neuroendocrine and neuropeptide systems.

In contrast to the studies of noradrenergic and serotonergic function in depression, there have been only sporadic suggestions, and fewer investigations, regarding the possible involvement of the trace amines in the affective disorders. Fischer and Heller's (1972) original suggestion of a deficit of phenylethylamine in depression was subsequently elaborated into a phenylethylamine hypothesis for the affective disorders (SABELLI & MOSNAIM, 1974; SABELLI et al., 1978). Evidence of decreased urinary phenylethylamine excretion in some depressed patients has been found (FISHER & HELLER, 1972; SABELLI & MOSNAIM, 1974; SABELLI *et al*., 1978; FISHER *et al*., 1972; SPATZ & SPATZ, 1978), but methodologic problems limit these reports. Reductions in cerebrospinal fluid concentrations of phenylacetic acid, the major metabolite of phenylethylamine, have been observed in depressed patients (SANDLER *et al*., 1979a). Reduced tyramine formation has also been suggested to be associated with depression (SANDLER *et al*., 1979b; BONHAM CARTER *et al*., 1980), and the question of an accumulation of octopamine or other false transmitter trace amines raised (MURPHY, 1972; BALDESSARINI & FISCHER, 1978).

Two books (MOSNAIM & WOLF, 1978, 1980) and several reviews (SABELLI *et al*., 1978; BALDESSARINI & FISCHER, 1978; JONES, 1982) have considered some of the evidence for the involvement of phenylethylamine, tyramine, tryptamine and other trace amines in psychiatric and neurologic disorders. There has been particular interest in the possibility that the amphetamine-like properties of phenylethylamine or the hallucinogenic properties of some derivitives of the trace amines might be implicated in schizophrenia, particularly its paranoid form (POTKIN *et al.*, 1979; JESTE *et al.*, 1980), or mania (FISCHER & HELLER, 1972; SABELLI & MOSNAIM, 1974; SABELLI *et al.*, 1978).

In our present studies of affective disorder patients, we have attempted to ask two major questions: (A), In a heterogeneous group of depressed patients, are there any clinical characteristics of the patients which are associated with differences in phenylethylamine or tyramine excretion?; and (B), When phenylethylamine and tyramine are increased by the administration of monoamine oxidase (MAO)-inhibiting antidepressants, are there any behavioral or biological consequences which are correlated with the changes in trace amines?

METHODS

Free urinary phenylethylamine, m-tyramine and p-tyramine were determined in 24-hour urine samples from 22 hospitalized patients (10 males, 12 females) with RDC-defined major depressive disorder. All patients were maintained on a diet limiting monoamines, monoamine precursors and caffeine and remained off all drugs for a minimum 2-week period before 2 to 6 24-hour urine samples were obtained for baseline trace amine measurements using mass fragmentography assays previously described (KAROUM, *et al.*, 1979). The 2 to 6 day urine samples were averaged to obtain a mean value for each patient.

A subgroup of 15 of these patients participated in studies using the MAO-inhibiting antidepressants clorgyline (7 patients, 25-30 mg/day for 4 weeks) or pargyline (8 patients, 75-105 mg/day for 4 weeks) following a 3-week baseline placebo period (LIPPER *et al.*, 1979; MURPHY *et al.*, 1981). These two drugs are phenylethylamine derivatives and are structurally similar, both acting to inhibit MAO via their acetylenic groups. Clorgyline acts selectively on MAO type A, while pargyline, at the doses studied, affects both MAO-A and MAO-B (LIPPER *et al.*, 1979; MURPHY *et al.*, 1981). These 15 patients were studied intensively with a battery of rating scales including the Hamilton depression scale, the NIMH self-rating scale, the Profile of Mood States (POMS) and the Minnesota Multiphasic Personality Inventory (MMPI) given in the week immediately before drug treatment began; the Hamilton and NIMH scales

Table 1. Trace Amine Excretion in 22 Depressed Patients

	Trace Amines (μg/24h)		
	Phenylethylamine	p-Tyramine	m-Tyramine
All Males (N = 10)	8 ± 1	483 ± 72	56 ± 9
All Females (N = 12)	25 ± 39	433 ± 88	49 ± 8

were given again in the fourth week of MAO-inhibitor treatment. Urine collections and cardiovascular measures were also obtained at the same times.

RESULTS

A. Phenylethylamine Excretion and Baseline Behavioral and Biological Measures

While there was no significant association with age, there was a trend towards a sex difference in phenylethylamine excretion among the 22 depressed patients studied (Table 1). Female patients had higher mean urinary phenylethylamine concentrations than did the males, but this difference was primarily attributable to two women patients with values over 100 μg/24h, while none of the other female or male patients had mean values higher than 20 μg/24h.

These two patients were not distinguished by diagnostic subtype or any general clinical characteristics, although a more detailed study of these individuals is underway. Neither of these patients entered the clorgyline or pargyline trials, one because her symptoms began to improve spontaneously during the placebo period and she was discharged without drug treatment, and the other because her agitation and psychotic depressive state worsened and required neuroleptic treatment.

The 15 depressed patients who were studied with multiple behavioral and biological measures during their off-drug, diet-controlled baseline period had mean urinary phenylethylamine values which ranged from 2.7 to 13.6 μg/day (a 5-fold range), and averaged 8.8 ± 0.8 μg/day.

Table 2. Correlations between Baseline Phenylethylamine Excretion and Ratings of Psychopathology in Depressed Patients

Urinary Phenylethylamine versus:	Correlation Coefficient (r)
Activation - Vigor - Hypomania	
NIMH Activation Scale	-.05
POMS Vigor	.01
MMPI-Ma (hypomania)	.17
Anxiety - Tension - Dysphoria	
NIMH Anxiety Scale	-.44
NIMH Dysphoria Scale	-.55*
POMS Tension	-.39
POMS Anger	-.49*
Depression	
NIMH Scale	-.39
POMS	-.31
MMPI-D (depression)	-.53*
Hamilton	-.26
Other	
NIMH Functional Deficit Scale	-.59*
MMPI-MF (masculine-feminine)	-.64**

*$p < .05$; **$p < .01$

As indicated in Table 2, although phenylethylamine has been suggested to be an "endogenous amphetamine" and possesses some stimulant and locomotor activity-enhancing properties in animals, there was no evidence whatsoever for any association between phenylethylamine excretion in these 15 patients and the NIMH activation scale ratings--a scale which has been found to be highly sensitive to the stimulant effects of d- and l-amphetamine in a comparable group of depressed patients (VAN KAMMEN & MURPHY, 1975). Ratings on two similar scales, POMS vigor and MMPI hypomania, were also not correlated with phenylethylamine excretion (Table 2).

In regard to the other behavioral measures, the most striking observation was that almost all measures of psychopathology were inversely correlated with phenylethylamine excretion, with the trends in the direction that lower phenylethylamine values were associated with higher ratings for anxiety, dysphoria, anger and depression (Table 2). Only some of these correlations attained statistical significance, but nonetheless there seemed to be an overall pattern involving some consistency among these different ratings which could be considered in accord with the suggested association between reduced phenylethylamine and some depression-related symptomatology.

Among the biological measures, there was only a weak, non-significant correlation between platelet MAO activity and phenylethylamine excretion (r = -.34, Table 3).

Table 3. Correlations between Baseline Phenylethylamine Excretion and Biological Measures in Depressed Patients

Urinary Phenylethylamine versus:	Correlation Coefficient (r)
<u>Platelet MAO Activity</u>	-.34
<u>Urinary Amines and Metabolites</u>	
p-Tyramine	-.43
m-Tyramine	.14
3-methoxy, 4-hydroxy phenylglycol (MHPG)	-.44
<u>Cardiovascular Measures - Recumbent</u>	
Systolic Pressure	-.60*
Pulse	-.30
Plasma Norepinephrine	-.52*
<u>Cardiovascular Measures - Standing</u>	
Systolic Pressure	-.55*
Pulse	-.62*
Plasma Norepinephrine	-.63*

*$p < .05$

Reducing MAO activity by MAO-inhibiting antidepressant administration raises phenylethylamine excretion in animals and humans, and if endogenous variations in platelet MAO activity reflected functional differences in whole-body MAO-B activity, a higher negative correlation might have been predicted.

There were several internally consistent correlations between phenylethylamine excretion and blood pressure, pulse and plasma norepinephrine concentrations measured both in a recumbent position and after five minutes of standing, with patients with lower phenylethylamine generally tending to have higher systolic blood pressures, pulses, plasma norepinephrine and urinary MHPG concentrations (Table 3). This finding is not directly consistent with phenylethylamine's known pharmacologic properties as a catecholamine releasing agent (KAROUM et al., 1982a). Correlational relationships of this type, however, may reflect multiple interacting and independent factors, including the general finding that depressed patients studied on our research units have elevated plasma norepinephrine concentrations (LAKE et al., 1982).

B. Relationships between Changes in Phenylethylamine Excretion and Changes in Behavioral and Biological Measures During MAO-Inhibitor Administration

We have reported previously that in the larger sample of patients evaluated in the efficacy trials of pargyline and clorgyline, clorgyline produced substantial antidepressant effects, while pargyline treatment led to improvement in a few patients, worsening in others and, overall, was significantly less effective than clorgyline on all therapeutic outcome measures (LIPPER et al., 1979; MURPHY et al., 1981). Phenylethylamine excretion was increased less than two-fold during clorgyline administration, but over 80-fold by pargyline (Table 4). This would seem strong evidence that MAO-inhibitor antidepressant efficacy is not due to increased phenylethylamine availability. In fact, fairly robust associations between noradrenergic system changes in these patients and changes in behavioral ratings in response to clorgyline suggest that clorgyline's efficacy may be linked to noradrenergic neurotransmitter system changes (MURPHY et al., 1981; ROY et al., submitted).

Correlations between trace amine excretion changes and rating scale changes would not be expected to yield

Table 4. Changes in the Urinary Excretion of Phenylethylamine, m- and p- Tyramine during Pargyline and Clorgyline Administration

	Amine Excretion (μg/24h)			
	Placebo	Pargyline	Placebo	Clorgyline
	(N = 8)		(N = 7)	
Phenylethylamine	8 ± 1	680** ± 162	7 ± 1	14 ± 4
m-Tyramine	30 ± 6	114* ± 33	32 ± 7	56 ± 19
p-Tyramine	330 ± 40	905*** ± 103	360 ± 45	453 ± 46

(Means ± S.E.M.; *$p < .05$; **$p < .01$; ***$p < .001$)

reliable results when examined in the clorgyline- and pargyline-treated subgroups separately, because of the small number of patients in each treatment subgroup. Under the premise that if trace amine changes were possibly associated with any therapeutic effects, side effects or biological changes resulting from MAO inhibition, such associations might emerge from the examination of such changes in a correlation matrix combining both of these related drugs, we examined our data in this fashion.

As indicated in Table 5, there again was no evidence of an association between increased phenylethylamine and increased behavioral activation. There were moderate positive correlations between increased phenylethylamine excretion and improvement in self-rated depression as well as improvement in anxiety on the NIMH scales. This finding was not supported by similar observer-rated Hamilton depression scale changes, however. Because of the many positive associations reported elsewhere (MURPHY et. al., 1981; ROY _et al._, submitted) between noradrenergic system changes and antidepressant effects with clorgyline treatment, a very tenuous suggestion might relate the weak, but positive correlation between the urinary phenylethylamine and urinary MHPG changes (Table 5) as supportive of some interaction between phenylethylamine and norepinephrine and

Table 5. Correlations during MAO-Inhibitor Treatment between Phenylethylamine Excretion Changes and Behavioral and Biological Changes

Urinary Phenylethylamine Changes versus:	Correlation Coefficient (r)
Activation Increases	
NIMH Activation Scale	-.07
Depression Improvement	
NIMH Scale	.58*
Hamilton Scale	.15
Anxiety Improvement	
NIMH Anxiety Scale	.48
Urinary Amine and Metabolite Changes	
p-Tyramine	.76**
m-Tyramine	.53*
o-Tyramine	.62*
Phenylethanolamine	.47
MHPG	.39

*$p < .05$; **$p < .01$

clinical response. However, a confirmatory pattern of similar associations with, for example, plasma norepinephrine or other cardiovascular measures, did not emerge.

Positive correlations were consistently present between phenylethylamine increases produced during MAO inhibitor treatment and increases in the excretion of the other trace amines (Table 5). As expected, all of these trace amine changes were fairly highly correlated with platelet MAO activity reductions (Table 6).

Table 6. Biological Changes Associated with Platelet MAO Activity Changes during MAO-Inhibitor Treatment

Platelet MAO Activity Changes versus Changes in:	Correlation Coefficient (r)
Phenylethylamine Excretion	-.95**
p-Tyramine Excretion	-.75**
m-Tyramine Excretion	-.62*
o-Tyramine Excretion	-.69**
Phenylethanolamine Excretion	-.52*
MHPG Excretion	-.03

*p < .05; **p < .01

C. m-Tyramine and p-Tyramine Excretion and Behavioral and Biological Measures

p-Tyramine and m-tyramine excretion did not exhibit either sex or age differences among the 22 depressed patients studied (Table 1).

In the 15 intensively studied patients for whom the battery of behavioral, cardiovascular and biological measures were available, p-tyramine excretion was not correlated at a statistically significant level with any behavioral measure (Table 7). Of interest, p-tyramine excretion was fairly strongly related to one cluster of cardiovascular measures. While neither baseline recumbent nor standing blood pressures or pulses were correlated with p-tyramine, the fall in systolic and diastolic pressure and the increase in pulse produced by the patients going from a recumbent position to standing for five minutes were significantly associated with baseline p-tyramine excretion (Table 7).

Table 7. Correlations between Baseline p-Tyramine Excretion and Behavioral and Biological Measures

Urinary p-Tyramine versus:	Correlation Coefficient (r)
Behavioral Measures	N.S.
Cardiovascular Responses to Standing	
Systolic Pressure Change	-.64**
Diastolic Pressure Change	-.71**
Pulse Change	.63*
Platelet MAO Activity	.58*

*p < .05; **p < .01; N.S., non-significant

Table 8. Correlations between Baseline p-Tyramine Excretion and Behavioral and Biological Measures

Urinary p-Tyramine versus:	Correlation Coefficient (r)
Behavioral Measures	
NIMH Depression Scale	-.52*
NIMH Anxiety Scale	-.59*
MMPI-MA (hypomania)	.72**
MMPI-PD (psychopathology)	.54*
MMPI-HS (hypochondrias)	.59*
Biological Measures	N.S.

*p < .05; **p < .01; N.S., non-significant

m-Tyramine excretion, in contrast, exhibited no significant correlations with any cardiovascular or other biological measures, but was inversely associated with depression and anxiety ratings on the NIMH scale (Table 8).

m-Tyramine also exhibited positive associations with several MMPI scales, including hypomania. There is some conceivable coherence to these findings, because the MMPI has both "trait" and "state" components in its scales.

In response to treatment with the MAO-inhibitors, m-tyramine and p-tyramine both increased (Table 4), but the increases were apparently not proportionate to each other, and did not exhibit a significant positive correlation, even though both amines became significantly positively correlated with phenylethylamine excretion during drug treatment.

DISCUSSION

Urinary phenylethylamine excretion in most of the 22 hospitalized depressed patients included in our study was virtually identical to that found in 15 normals [7.1 ± 0.5 µg/day, range 5.7 - 11.2, in 8 males, 7 females combined (KAROUM *et al.*, 1979)] and also to similar values in 32 normals (POTKIN *et al.*, 1979) studied using the same assay as that in the present investigation (KAROUM *et al.*, 1979).

Two depressed women, however, had unusually high urinary phenylethylamine excretion. Karoum and coworkers have previously reported high, fluctuating urinary phenylethylamine excretion in five women with bipolar affective disorder and a characteristic symptom pattern which differentiated them from six other patients with affective disorders (KAROUM *et al.*, 1982b). As noted above, our two patients with high phenylethylamine values were not included in our correlational studies, and may comprise a depressed patient subgroup requiring further investigation. Among non-affective disorder patients, a significantly higher incidence of elevated (> 10 µg/day) phenylethylamine excretion has also been found in some schizophrenic patients (12 out of 31 patients, 10 of 12 having paranoid schizophrenia) compared to controls (3 out of 32 individuals) (POTKIN *et al.*, 1979).

Severe stress associated with an initial parachuting experience has recently been reported to be associated with marked short-term increases in urinary phenylethylamine (PAULOS & TESSEL, 1982). At present, we can only conclude that altered phenylethylamine excretion appears not to be generally associated with severe depressions requiring hospitalization in either men or women. More studies in normals, as well as in patients, especially in relation to gender differences and behavioral and biological indicators of stress are clearly needed.

In our correlational studies in a subgroup of intensively-investigated depressed patients, we found some indications of possible associations between both phenylethylamine excretion and m-tyramine excretion (but not p-tyramine excretion) and several psychopathological elements associated with depression. These findings could be considered consistent with an amended hypothesis relating severity of a cluster of depression-related symptoms including dysphoria, anxiety, and anger (but not depressed mood alone) with a phenylethylamine deficit state. There was, however, no indication of any associations between behavioral activation and phenylethylamine, either in the baseline state or when phenylethylamine was markedly increased by MAO-inhibitor treatment. This data fairly strongly argues against a simple, direct "endogenous amphetamine" interpretation of phenylethylamine's behavioral properties. We have also concluded that the antidepressant effects of MAO-inhibitors are not likely to be mediated by increased phenylethylamine itself or associated changes in phenylethylamine or the tyramines, primarily on the basis of the lack of change in these measures during treatment with the MAO-A selective inhibitor, clorgyline which had marked antidepressant effects without altering phenylethylamine (MURPHY _et al_., 1983).

The associations between m-tyramine excretion and several possible "trait" measures from the MMPI in depressed patients seem of interest for further study, particularly since m-tyramine excretion in the depressed patients (53 ± 7 µg/d) was lower than that previously reported in normals (87 ± 8 µg/d) studied with the same assay methodology (KAROUM _et al_., 1979). However, as in the case of phenylethylamine, matched normal controls need to be studied under the same conditions as the patients before satisfactory conclusions can be drawn about such possible differences. In regard to the significant correlations between baseline p-tyramine and pulse and blood pressure responses to standing, further study of p-tyramine excretion in relation to postural stress may also be of interest in regard to certain cardiovascular disorders such as orthostatic hypotension.

REFERENCES

BALDESSARINI R.J. & FISCHER J.E. (1978) Trace amines and alternative neurotransmitters in the central nervous system. Biochem. Pharmcol. 27, 621-626.

BONHAM CARTER S.M., REVELEY M.A., SANDLER M., DEWHURST S. J., LITTLE B.C., HAYWORTH J. & PRIEST R.G. (1980) Decreased urinary output of conjugated tyramine is associated with lifetime vulnerability to depressive illness. Psychiatry Res. 3, 13-21.

FISCHER B. & HELLER B. (1972) Phenethylamine as a neurohumoral agent in brain. Behav. Neuropsychiat. 4, 8-11.

FISCHER E., SPATZ H., SAAVEDRA J.M., REGGIANI H., MIRO A.H. & HELLER B. (1972) Urinary elimination of phenethylamine. Biol. Psychiatry 5, 139-147.

JESTE D.V., DOONGAJI D.R., PANJWANI D., DATTA M., POTKIN, S.G., KAROUM F., THATTE S., SHETH A.S., APTE J.S. & WYATT R.J. (1980) Cross-cultural study of a biochemical abnormality in paranoid schizophrenia. Psychiatry Res. 3, 341-352.

JONES R.S.G. (1982) Tryptamine: A neuromodulator or neurotransmitter in mammalian brain? Prog. Neurobiol. 19, 117-139.

KAROUM F., NASRALLAH H., POTKIN S., CHUANG L., MOYER-SCHWING J., PHILLIPS I. & WYATT R.J. (1979) Mass fragmentography of phenylethylamine, m- and p-tyramine and related amines in plasma, cerebrospinal fluid, urine and brain. J. Neurochem. 33, 201-212.

KAROUM F., SPECIALE S.G. JR., CHUANG L.-W. & WYATT R.J. (1982a) Selective effects of phenylethylamine on central catecholamines: A comparative study with amphetamine. J. Pharmacol. Ther. 223, 432-439.

KAROUM F., LINNIOLA M., POTTER W.Z., CHUANG L.-W., GOODWIN, F.K. & WYATT R.J. (1982b) Fluctuating high urinary phenylethylamine excretion rates in some bipolar affective disorder patients. Psychiatry Res. 6, 215-222.

LAKE C.R., PICKAR D., ZIEGLER M.G., LIPPER S., SLATER S. & MURPHY D.L. (1982) High plasma norepinephrine levels in patients with major affective disorder. Am. J. Psychiatry 139, 1315-1318.

LIPPER S., MURPHY D.L., SLATER S. & BUCHSBAUM M.S. (1979) Comparative behavioral effects of clorgyline and pargyline in man: A preliminary evaluation. Psychopharmacology 62, 123-128.

MOSNAIM A.D. & WOLF M.E. (1978) Noncatecholic Phenylethylamines - Part 1 - Phenylethylamine: Biological Mechanisms and Clinical Aspects, Marcel Dekker, Inc., New York.

MOSNAIM A.D. & WOLF M.E. (1980) Noncatecholic Phenylethylamines - Part 2 - Phenylethanolamine, Tyramines, and Octopamine. Marcel Dekker, Inc., New York.

MURPHY D.L. (1972) Amine precursors, amines and false neurotransmitters in depressed patients. Am. J. Psychiatry 129, 141-148.

MURPHY D.L., LIPPER S., PICKAR D., JIMERSON D., COHEN R.M., GARRICK N.A., ALTERMAN I.S. & CAMPBELL I.C. (1981) Selective inhibition of monoamine oxidase type A: Clinical antidepressant effects and metabolic changes in man, in Monoamine Oxidase Inhibitors. The State of the Art (Youdim M.B.H. & Paykel E.S., eds.) pp. 189-205. John Wiley & Sons, New York.

MURPHY D.L., COHEN R.M., SIEVER L.J., ROY B., KAROUM F., WYATT R.J., GARRICK N.A. & LINNOILA M. (1983) Clinical and laboratory studies with selective monoamine-oxidase-inhibiting drugs: Implications for hypothesized neurotransmitter changes associated with depression and antidepressant drug effects, in Monoamine Oxidase and Its Selective Inhibitors: New Concepts in Therapy and Research (Beckmann H. & Riederer P., eds.) pp. 287-303. Basel, Karger.

PAULOS M.A. & TESSEL R.E. (1982) Excretion of β-phenethylamine is elevated in humans after profound stress. Science 215, 1127-1129.

POTKIN A.G., KAROUM F., CHUANG L.-W., CANNON-SPOOR H.E., PHILLIPS I. & WYATT R.J. (1979) Phenylethylamine in paranoid chronic schizophrenia. Science 206, 470-471.

ROY B.F., MURPHY D.L., LIPPER S., SIEVER L.J., ALTERMAN I.S. JIMERSON D., LAKE C.R. & COHEN R.M. (submitted) Cardiovascular effects of the selective monoamine oxidase-inhibiting antidepressant clorgyline: Correlations with clinical responses and changes in catecholamine metabolism.

SABELLI H.C. & MOSNAIM A.D. (1974) Phenylethylamine hypothesis of affective behavior. Am. J. Psychiatry 131, 695-699.

SABELLI H.C., BORISON R.L., DIAMOND B.I., HAVDALA H.S. & NARASIMHACHAR N. (1978) β-Phenylethylamine and brain function. Biochem. Pharmacol. 27, 1729-1730.

SANDLER M., RUTHVEN C.R., GOODWIN B.L. & COPPEN A. (1979a) Decreased cerebrospinal fluid concentration of free phenylacetic acid in depressive illness. Clin. Chem. Acta 93, 169-171.

SANDLER M., RUTHVEN C.R.J., GOODWIN B.L. & REYNOLDS G.P. (1979b) Deficient production of tyramine and octopamine in cases of depression. Nature 278, 357-358.

SPATZ H. & SPATZ N. (1978) Urinary and brain phenylethylamine levels under normal and pathological conditions, in Noncatecholic Phenylethylamines - Part 1 - Phenylethylamine: Biological Mechanisms and Clinical Aspects (Mosnaim A.D. & Wolf M.E., eds.) pp. 447-474. Marcel Dekker, New York.

VAN KAMMEN D.P. & MURPHY D.L. (1975) Attenuation of the euphoriant and activating effects of d- and l-amphetamine by lithium carbonate treatment. Psychopharmacologia 44, 215-224.

PHENYLETHYLAMINE AND SCHIZOPHRENIA - CLINICAL AND PHARMACOLOGICAL RESULTS.

GAVIN P. REYNOLDS

MRC Neurochemical Pharmacology Unit, Brain Tissue Bank,
Department of Neurological Surgery and Neurology,
Addenbrooke's Hospital, Hills Road,
Cambridge CB2 2QQ.

ABSTRACT

Evidence is reviewed which suggests phenylethylamine may contribute to the aetiology of schizophrenia, in particular the paranoid subgroup of the disease. Supporting pharmacological results are assessed and some new results indicating a specific action for phenylethylamine on 5-hydroxytryptamine receptors in vitro are presented. This information is discussed in terms of other current neurotransmitter hypotheses of schizophrenia.

INTRODUCTION

The phenylethylamine (PE) hypothesis of schizophrenia stems primarily from the structural and pharmacological similarities between PE and amphetamine (α-methyl-PE) and the resemblance of the symptoms of amphetamine overdose to paranoid schizophrenic psychosis. Since this hypothesis was first proposed (Sandler and Reynolds, 1976), much effort has been put into the search for a neurochemical aetiology of schizophrenia, although conclusive results have yet to emerge. However, in that time our understanding of the pharmacology of PE, particularly in relation to other 'classical' neurotransmitters (some of which have also been implicated in schizophrenia), has substantially increased. In addition, studies have indicated changes in PE function in schizophrenia. Thus it seems timely to review this information and reassess the possible role of PE in schizophrenia, particularly considering how it might be relevant to (or at least consistent with) other neurochemical hypotheses of the disease.

PHENYLETHYLAMINE AND SCHIZOPHRENIA - RELEVANT CLINICAL FINDINGS

There are certainly indications that there is an increased concentration of PE in schizophrenic patients. The urinary excretion of the amine by paranoid schizophrenics is apparently elevated (Reynolds, 1977) and the effect appears to be specific to this disease subgroup (Potkin et al., 1979). However such studies are confounded by artifactual problems associated with institutionalisation of the patients and neuroleptic drug treatment. For example, the latter has been found to affect monoamine oxidase activity (DeLisi et al., 1981) which might serve to elevate PE levels. The reported decrease in plasma phenylacetic acid in some schizophrenics (Davis et al., 1982) is not inconsistent with such an interpretation.

CSF concentrations should reflect brain metabolism more closely and with a far smaller contribution from the gut. One recent study indicated that although there is no general abnormality in CSF concentrations of PE (Beckmann et al., 1982), four of the 13 untreated schizophrenics exhibited CSF concentrations of PE which were well above the normal range, an effect which may be associated with paranoid psychosis in these patients. On the other hand, PE concentrations in those schizophrenics receiving neuroleptic treatment were in all cases indistinguishable from control values. The concentration of PAA, the major metabolite of PE, has been variously reported to be increased (Sandler et al., 1978a) or decreased (Beckmann et al., 1982) in schizophrenic CSF, although medication, diagnostic and sex differences between the groups may have contributed to this discrepancy.

Post-mortem brain tissue is being increasingly widely used to study the neurochemistry of neuropsychiatric disorders. We have measured PAA in brain tissue taken post-mortem from patients diagnosed as schizophrenics (Figure 1) and obtained results reminiscent of the reported PE concentrations in CSF. While no overall difference between schizophrenia and control values is observed, a few patients exhibited greatly increased PAA concentrations, in one case

Figure 1. PAA concentrations in human caudate nucleus

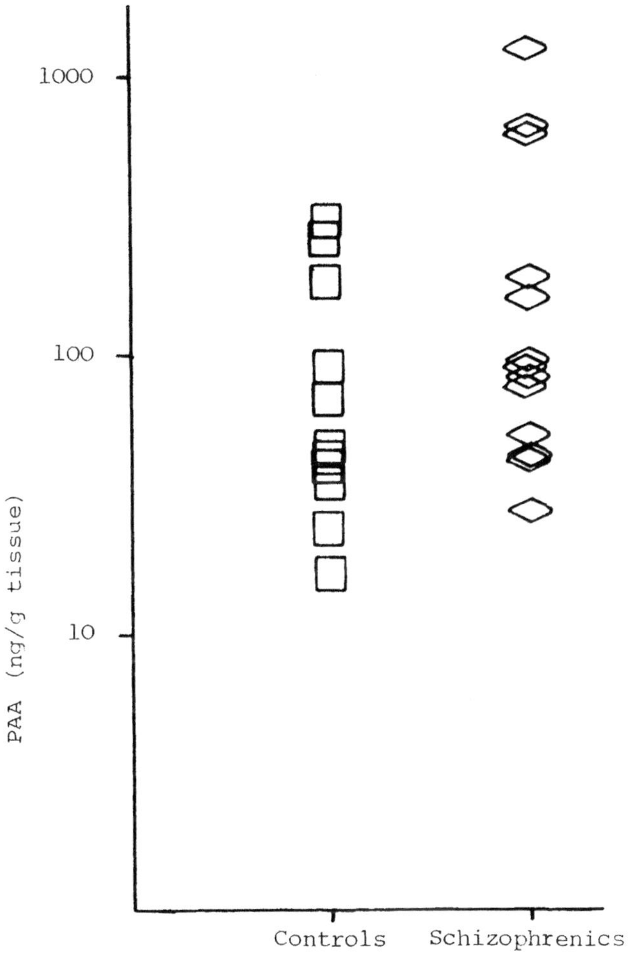

Free PAA was measured using a method based on that of Fellows et al. (1978) and adapted for brain tissue.

by a factor of over ten times. These results were apparently independent of medication (one patient with high PAA had not received neuroleptic drug treatment) or cause of death.

As yet a definitive conclusion from these reports is unwarranted, although there are strong indications that a small proportion of schizophrenic patients, possibly a paranoid subgroup, may have an abnormality in PE metabolism. That this is a reflection of grossly aberrant phenylalanine metabolism is unlikely since a recent study (Reveley and Reveley, 1982) indicated the lack of any misdiagnosed phenylketonuric patients in a large number of schizophrenic patients. Thus a more subtle abnormality is indicated.

AMPHETAMINE AND SCHIZOPHRENIA

How does our understanding of the pharmacology of PE contribute to an assessment of the possible dysfunction of PE in schizophrenia? Before attempting to answer this, it is perhaps relevant to point out how developments in PE pharmacology have closely paralleled, or followed, work on amphetamine. The psychosis which follows high dose and/or chronic amphetamine has, as already mentioned, served as model and guide in our search to understand paranoid schizophrenic psychosis. Originally, dopamine was especially implicated in schizophrenia as this transmitter was clearly involved in the behavioural effects of amphetamine. Stereotyped behaviour induced by amphetamine in animals was found to be dependent on intact presynaptic dopamine function and is inhibited by the antipsychotic drugs which block dopamine D_2 receptors. However, changes in the concentrations of dopamine and its metabolites in post-mortem brain tissue from schizophrenic patients are not consistently found, and the reported increase in their dopamine receptors is probably an effect of neuroleptic medication (e.g., Crow et al., 1981; MacKay et al., 1982). Recently attention has (re)turned to the other monoamine neurotransmitters, noradrenaline (Hornykiewicz, 1982) and 5-hydroxytryptamine (5-HT) (Trulson and Jacob, 1979) each of which are also involved in some components of amphetamine-induced behaviour. In particular, the effects of large doses or chronic administration of amphetamine to animals induces effects referred to as the '5-HT behavioural syndrome' (Peroutka et al., 1981). Implicated in this syndrome are the 5-HT receptors, to which many neuroleptic drugs and hallucinogenic compounds such as LSD bind with high affinity. However the relative antipsychotic potencies of neuroleptics parallel their affinities to D_2 receptors better than to $5\text{-}HT_2$ sites in human brain (Reynolds, 1983). Thus 5-HT has not yet replaced dopamine in the forefront of research into the neurochemical aetiology of schizophrenia.

PHENYLETHYLAMINE PHARMACOLOGY

Pharmacological investigations of PE have shown the amine to be similar to amphetamine, in having effects mediated by dopamine, noradrenaline and 5-HT transmission. Thus the biphasic stimulation of locomotor activity in mice has been thought to involve dopaminergic and noradrenergic systems, including a direct action (possibly via a metabolite) on post-synaptic dopamine receptors (Jackson, 1975). A more rigorous investigation has been performed by Dourish (1982) who has assessed the importance of the three amine neurotransmitters in the various behavioural effects of high dose PE in mice. 5-HT is particularly implicated in many of these behaviours. In fact other authors (Sloviter et al., 1980) have suggested that PE (but not amphetamine) acts directly as an agonist at 5-HT receptors in producing this behavioural syndrome. It has been argued that PE administration to animals provides a better model for schizophrenia than does amphetamine on the basis of PE (but not amphetamine) stereotypy being preferentially inhibited by those neuroleptics which have fewer extrapyramidal side effects (Borison et al., 1977). However to view these drugs as being specific for limbic dopamine receptors is probably fallacious (Reynolds et al., 1982) and, as pointed out by Dourish and Boulton (1982), the preferential inhibition of PE-induced behaviour by clozapine more likely reflects its relatively high affinity for $5\text{-}HT_2$ receptors. This tends to negate the argument for PE providing a better model for schizophrenia, since neuroleptic antagonism of the PE-induced behaviour does not correlate so well with relative antipsychotic potency.

PHENYLETHYLAMINE AND 5 HT RECEPTORS

In order to investigate whether there is an in vitro biochemical correlate of the relationship between PE-induced behaviour and 5-HT systems, in particular the proposed agonist action of PE on 5-HT receptors, we have studied the effect of PE and some related compounds on the specific binding of radiolabelled 5-HT to preparations of human post-mortem brain tissue. The results (table 1) show that PE is over ten times more able to displace 5-HT from its binding sites than D-amphetamine. This is consistent with the differences in the pharmacological effects of the two compounds which have been discussed above and which suggest PE to be a 5-HT agonist (Sloviter et al., 1980a), although this experiment is unable to distinguish agonist and antagonist effects.

The greater effect of the L-isomer over that of D-amphetamine is notable and suggests that while this interaction may be important in hyperreactivity in mice, where the startle response shows an analogous isomeric effect (Handley and Thomas, 1979). However, the 5-HT behaviour syndrome is more readily elicited by the D-isomer, probably via 5-HT release (Sloviter et al., 1980b).

There is evidence consistent with an interaction between PE and 5-HT systems in vivo in man. The observations by Sandler et al (1978b) that PAA is increased in the plasma of 'aggressive psychopaths', parallel the reports of abnormalities of central 5-HT metabolism as indicated by decreased CSF concentrations of 5-HIAA in such subjects (Brown et al., 1979). Whether this is a causal relationship or not is, of course, unclear.

Table 1. Inhibition of Specific 5-HT Binding to Human Brain Tissue

	K_I (M)
Phenylethylamine	8.0×10^{-6}
D-amphetamine	1.1×10^{-4}
L-amphetamine	3.0×10^{-5}

Values are means calculated from IC_{50} values of 3-5 displacement experiments with each concentration assayed in triplicate. (^3H)5-HT was used at 4 nM; 10^{-6} M 5-HT defined specific binding.

CONCLUDING DISCUSSION

To develop the observations and results discussed here into a comprehensive hypothesis for the neurochemical aetiology of schizophrenic psychosis is beyond the scope of this short report. However, it is perhaps worth summarising the findings and attempting to integrate them in the more classical hypotheses for neurotransmitter dysfunction in schizophrenia.

Results suggest that an increase in PE concentrations may occur in the CNS of some schizophrenic patients, particularly those exhibiting paranoid psychotic symptoms. This seems most likely to have a direct effect of 5-HT binding sites, which represent the 5-HT_1 subgroup of receptors. While it is the 5-HT_2 receptors which are antagonised by many neuroleptic drugs, a close interaction of the receptors in 5-HT function has been proposed whereby 5-HT inhibition and excitation are mediated by 5-HT_1 and 5-HT_2 receptors respectively (Peroutka et al., 1981). However, since the cell bodies of dopaminergic pathways receive substantial innervation from 5-HT systems, a PE-induced 5-HT dysfunction may well affect dopaminergic activity. Thus it is not necessary to invoke an anti-psychotic action of neuroleptics involving sites other than post-synaptic dopamine receptors.

REFERENCES

Beckmann H., Reynolds G.P., Sandler M., Waldmeier P., Lauber J., Riederer P. and Gattaz W.F. (1982) Phenylethylamine and phenylacetic acid in CSF of schizophrenics and healthy controls. Arch. Psychiatr. Nervenkr. 232, 463-471.

Borison R.L., Havdala H.S. and Diamond I. (1977) Chronic phenylethylamine stereotypy in rats: a new animal model for schizophrenia. Life Sci. 21, 117-122.

Brown G.L., Ballinger G.A.C., Minichiello M.D. and Goodwin F.K. (1979) Human aggression and its relationship to cerebrospinal fluid 5-hydroxyindole acetic acid, 3-methoxy 4-hydroxyphenylglycol and homovanillic acid. In Psychopharmacology of aggression (ed. M. Sandler) 131-148, Raven Press, New York.

Crow T.J., Owen F., Cross A.J., Ferrier N., Johnstone E.C., McCreadie R.M., Owens D.G.C. and Poulter M. (1981) Neurotransmitter enzymes and receptors in post-mortem brain in schizophrenia: evidence that an increase in D_2 dopamine receptors is associated with the type I syndrome. In Transmitter biochemistry of human brain tissue (eds. Ruderer P. and Usdin E.) 85-96, Macmillan London.

Davis B.A., Yu P.H., Carlson K., O'Sullivan K. and Boulton A.A. (1982) Plasma levels of phenylacetic acid m- and p-hydroxyphenylacetic acid and platelet monoamine oxidase activity in schizophrenic and other patients. Psychiatr. Res. 6, 97-105.

DeLisi L.E., Wise C.D., Bridge T.P., Rosenblatt J.E., Wagner R.L., Morihisa J., Karson C., Potkin S.G. and Wyatt R.J. (1981) A probably neuroleptic effect on platelet monoamine oxidase in chronic schizophrenic patients. Psychiatry Research 4, 95-107.

Dourish C.T. (1982) A pharmacological analysis of the hyperactivity syndrome induced by β-phenylethylamine in the mouse. Brit. J. Pharmacol. 77, 129-139.

Dourish C.T. and Boulton A.A. (1982) Putative PEA receptors? Psychiatr. Res. 7, 387.

Fellows L.E., King G.S., Pettit B.R., Goodwin B.L., Ruthven C.R.J. and Sandler M. (1978) Phenylacetic acid in human cerebrospinal fluid and plasma: selected ion monitoring assay. Biomed. Mass. Spectr. 5, 508-511.

Handley S.L. and Thomas K.V. (1979) Potentiation of startle response by d- and l-amphetamine. Psychopharmacol. 64, 105-111.

Hornykiewicz O. (1982) Brain catecholamines in schizophrenia - a good case for noradrenaline. Nature 299, 484-486.

Jackson D.M. (1975) β-phenylethylamine and locomotor activity in mice. Arzneim. Forsch. 25, 622-626.

Mackay A.V.P., Iversen L.L., Rossor M., Spokes E., Bird E., Arregui A., Creese I. and Snyder S.H. (1982) Increased brain dopamine and dopamine receptors in schizophrenia. Arch. Gen. Psychiatr. 39, 991-997.

Peroutka S.J., Lebowitz R.M. and Snyder S.H. (1981) Two distinct serotonin receptors with different physiological functions. Science 212, 827-829.

Potkin S.G., Karoum F., Chuang L-W., Cannon-Spoor H.E., Phillips I. and Wyatt R.J. (1979) Phenylethylamine in paranoid chronic schizophrenia. Science 206, 470-471.

Reveley A.M. and Reveley M.A. (1982) Screening for adult phenylketonuria in psychiatric inpatients. Biol. psychiatry 17, 1343-1345.

Reynolds G.P. (1977) PhD Thesis, University of London.

Reynolds G.P. (1983) (^{3}H)-Ketanserin binding to 5-HT_2 receptors in human brain. Brit. J. Pharmacol. 273P.

Reynolds G.P., Cowey L., Rossor M. and Iversen L.L. (1982) Thioridazine is not specific for limbic dopamine receptors. Lancet ii 499-500.

Sandler M. and Reynolds G.P. (1976) Does phenylethylamine cause schizophrenia? Lancet i 70-71.

Sandler M., Ruthven C.R.J., Goodwin B.L., King G.S., Pettit B.R., Reynolds G.P., Tyrer S.P., Weller M.P. and Hirsch S.R. (1978a) Raised cerebrospinal fluid phenylacetic acid concentration: preliminary support for the phenylethylamine hypothesis of schizophrenia? Commun. Psychopharmacol. 2, 199-202.

Sandler M., Ruthven C.R.J., Goodwin B.L., Field H. and Matthews R. (1978b) Phenylethylamine overproduction in aggressive psychopaths. Lancet ii 1269-1270.

Sloviter R.S., Connor J.D. and Drust E.G. (1980a) Serotonergic properties of β-phenylethylamine in rats. Neuropharmacol. 19, 1071-1074.

Sloviter R.S., Damiano B.P. and Connor J.D. (1980b) Relative potency of amphetamine isomers in causing the serotonin behavioural syndrome in rats. Biol. Psychiatr. 15, 789-796.

Trulson M.E. and Jacobs B.L. (1979) Long-term amphetamine treatment decreases brain serotonin metabolism: implications for theories of schizophrenia. Science 205, 1295-1297.

URINARY EXCRETION OF TRYPTAMINE IN COMPARISON TO NORMETANEPHRINE AND BETA-PHENYLETHYLAMINE IN HUMAN VOLUNTEERS AFTER SUBCHRONIC TREATMENT WITH DIFFERENT MONOAMINE OXIDASE INHIBITORS

Bieck*, P.R., Nilsson*, E., Schick*, C., Waldmeier**, P.C., and Lauber**, J.

*Human Pharmacology Institute CIBA-GEIGY GmbH, D-74 Tuebingen, Federal Republic of Germany

**Research Department, Pharmaceuticals Division, CIBA-GEIGY Ltd., CH-4002 Basle, Switzerland

Abstract: Measurement of urinary tryptamine excretion has been used extensively in the past as indicator for monoamine oxidase (MAO) inhibition in man. Basal tryptamine excretion (ug/g creatinine) differs between sexes ($p < 0.001$). The values obtained ($\bar{x} \pm SD$) were: 66 ± 26 in 13 males (n = 90) and 112 ± 42 in 9 females (n = 60). The higher basal tryptamine excretion and the larger tryptamine/indole acetic acid ratios in females, together with the greater effect of MAO inhibitors suggest a lower basal MAO-A activity in women. Different hormonal influences could explain the large interindividual and intraindividual seasonal variation of tryptamine excretion in females. By comparison of urinary tryptamine, normetanephrine and beta-phenylethylamine during subchronic treatment with three different MAO inhibitors it could be shown that tryptamine is a MAO-A substrate. MAO-A inhibition by the new reversible selective inhibitor CGP 11 305 A (150mg/d) caused parallel increases of tryptamine (320%) and normetanephrine (360%), but did not change urinary phenylethylamine. Selective MAO-B inhibition with deprenyl (10mg/d) enhanced phenylethylamine excretion (2600%), but did not affect tryptamine. During treatment with tranylcypromine (20mg/d), the unspecific inhibitor of both MAO-A+B, all three metabolites increased significantly (between 440% and 720%).

INTRODUCTION

Sjoerdsma et al. demonstrated already in 1959 that a rise in urinary tryptamine (T) excretion results from monoamine oxidase (MAO) inhibition in man. In spite of the extensive use of urinary T as a sensitive index of MAO inhibition in tissues (Levine and Sjoerdsma, 1962) it is unknown which of the two forms (A and/or B) of the (MAO) is responsible for T deamination in man. MAO-A deaminates noradrenaline (NA) but not phenylethylamine (PE), and is preferentially inhibited by the selective MAO-A inhibitor drug clorgyline (for review see Youdim and Finberg, 1983). As a result the urinary concentration of the O-methylated metabolite normetanephrine (NMN) increases substantially (LINNOILA et al., 1982). MAO-B prefers PE as substrate and is selectively inhibited by deprenyl (for review see Youdim and Finberg, 1983). A 20 to 90fold rise in PE excretion was found after 10mg/d for one week in healthy volunteers (Elsworth et al., 1978). We tried to establish the substrate specificity of T in man by comparing the effects of selective and non-selective MAO inhibitor drugs on the urinary concentrations of NMN, PE and T. Preliminary results on T excretion have already been reported (Nilsson and Schick, 1983).

MATERIALS AND METHODS

Subjects

25 healthy volunteers of both sexes, aged between 22 and 44 years participated in these studies. None of the volunteers took any other

medication for one week prior to and during the studies. During the experiments no tyramine-containing foodstuffs or alcohol were permitted. All volunteers were ambulatory and carried out their usual activities.

Dosage

In all studies there were initial control periods of 7 days (d) and between treatments drug free intervals of 14d. In study I - II, the subjects received weekly increasing oral doses of the monoamine oxidase inhibitors or placebo (P) in a randomized cross-over design. CGP 11 305 A (study I): 50mg, 75mg, 100mg and 150mg. Deprenyl (study II): 5mg, 10mg, 15mg and 20mg. In study III, the subjects received in an open pilot trial increasing oral doses of 10mg, 15mg, 20mg and 25mg tranylcypromine (TCP). In study IV, placebo and CGP 11 305 A (100mg/d) were taken orally for one week in randomized order. Third treatment was tranylcypromine for one week (20 mg/d p.o.).

Analytical methods

24h urines were collected using as a preservative sodium disulfite (40ml of a 10% solution per 24h urine). Aliquots were kept frozen until assayed. Creatinine was measured by the Jaffe reaction with autoanalyzers (SMA 12 and Beckman Creatinine Analyzer 2).

Tryptamine was either measured fluorometrically by a modification of the method of Sjoerdsma et al. (1959) or by GC according to Baker et al. (1979). The validity and comparability of the fluorometric method was proved by cross-checking with GC. Addition of the MAO inhibitor drugs to urine did not interfere.

Indole-3-acetic acid (IAA) was analyzed by HPLC with UV-detection at 280 nm. Details will be reported (Nilsson and Hengen, 1983).

Normetanephrine was determined by HPLC with electrochemical detection after acid hydrolysis

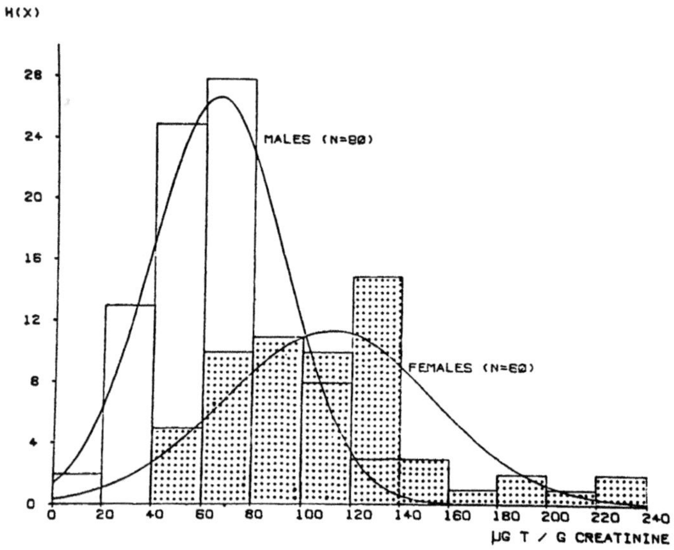

Fig. 1. Sex-dependent frequency distribution of urinary tryptamine (ug/g creatinine). H(x) = number of samples (n). Blanc columns: 13 males; dotted columns: 9 females.

and purification by a combination of solvent extraction/ion exchange procedures (Waldmeier et al., 1983b). Beta-phenylethylamine was analyzed by capillary GCMS (Carlo Erba Fractovap 2150 coupled to a Finnigan 4000 mass spectrometer). (For details see Lauber and Waldmeier, 1983).

Calculations

All 24h urine values of amine metabolites were calculated per g creatinine and are given as such throughout the paper. If not stated otherwise the values are means ± SD. The molar T/IAA ratio was obtained from (ug T/ug IAA) x 100 x correction factor (for the differing molecular weights). Statistical differences were tested

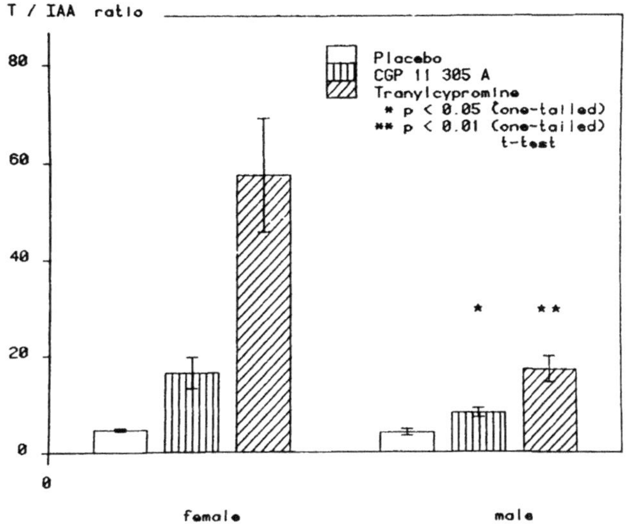

Fig. 2. Urinary T/IAA ratio in 3 female and in 3 male subjects. Mean $\pm$ SEM. Comparison for sex-difference with one-tailed t-test.

using the t-tests of Student, Student-Welch and Dunnett and variance analysis (random-block-factor-design).

RESULTS

Urinary T Excretion in Unmedicated Subjects

Sex difference in urinary T excretion
In pooling the urinary T excretion data of all untreated volunteers participating in the studies we could compare the values of 13 males (n=90) and of 9 females (n=60) (Fig. 1). The figure shows a histogram of our data. The means were 66 $\pm$ 26 (males) and 112 $\pm$ 42 (females), resp. In Fig. 2 the effect of MAO inhibition on the T/IAA ratio is shown. Values for females and

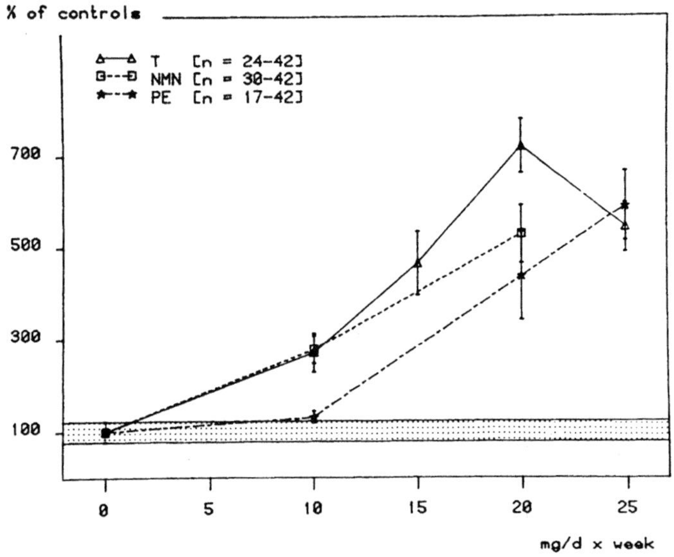

Fig. 3. Effect of subchronic treatment with weekly increasing doses of tranylcypromine on urinary T, NMN and PE in 6 subjects. Values in % of controls. Mean ± SEM.

males were considered separately. It can be seen that the effect of the two MAO inhibitors is much more marked in females compared to males. For each sex the increase during MAO inhibition compared to placebo is significant.

Urinary T Excretion during Subchronic Therapy with Non-selective and Selective MAO Inhibitors

Non-selective MAO-A+B inhibition. During the control period of the tranylcypromine study the volunteers excreted in the average 77 ± 38 ug T. Urinary T increased in a dose-dependent manner up to 550 ± 280 ug during the 20mg TCP dosage. 4 d after drug discontinuation pretreatment values were reached again.

Fig. 3 shows the increase of urinary metabo-

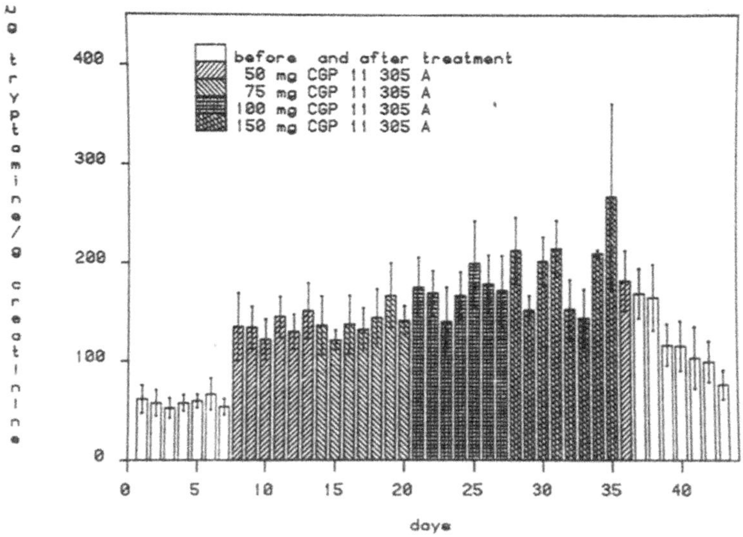

Fig. 4. Urinary tryptamine (ug/g creatinine) during subchronic treatment with weekly increasing doses of the selective MAO-A inhibitor CGP 11 305 A in 6 subjects. Mean ± SEM.

lites in % of controls. The dotted area represents the largest SD of the control period. NMN and PE excretion increased parallel to T. During the week with the dosage 20mg TCP/d values of 440% (PE), 530% (NMN), and 720% (T) were reached.

Selective MAO-A inhibition. During administration of 50mg CGP 11 305 A urinary T approximately doubled in the average from 59 ± 25 to 136 ± 56 ug. There was a further dose-dependent increase up to 320% of controls during the highest dose of 150mg CGP 11 305 A (Fig. 4). All treatment periods were significantly different from controls ($p<0.01$; Dunnett's t-test). About one week after stopping therapy T excretion reached control levels.

Fig. 5 shows the increase of urinary meta-

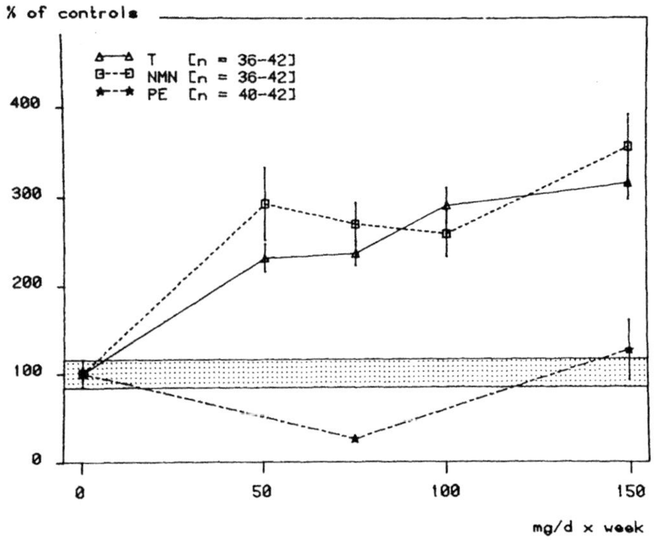

Fig. 5. Effect of subchronic treatment with weekly increasing doses of CGP 11 305 A. Parameters as in Fig. 3.

bolites in % of controls. Urinary NMN parallels that of T and goes up to 360%. Urinary PE was below controls during the 75mg dosage and remained unchanged from controls during the highest dosage of 150mg/d CGP 11 305 A.

Selective MAO-B inhibition. Urinary T during subchronic therapy with deprenyl increased only minimally with the higher dosage of 15mg and 20mg/d. Maximal weekly excretion was 134 ± 88 ug compared to control values of 81 ± 30 ug.

Fig. 6 depicts T and the simultaneously measured urinary PE excretion. During the week with 10mg/d deprenyl PE reached 2600% of controls. T was unchanged (and NMN was not measured).

Ratio T/IAA. Fig. 7 shows the pathways of tryptophan metabolism. Urinary T in man is derived entirely from decarboxylation of tryptophan

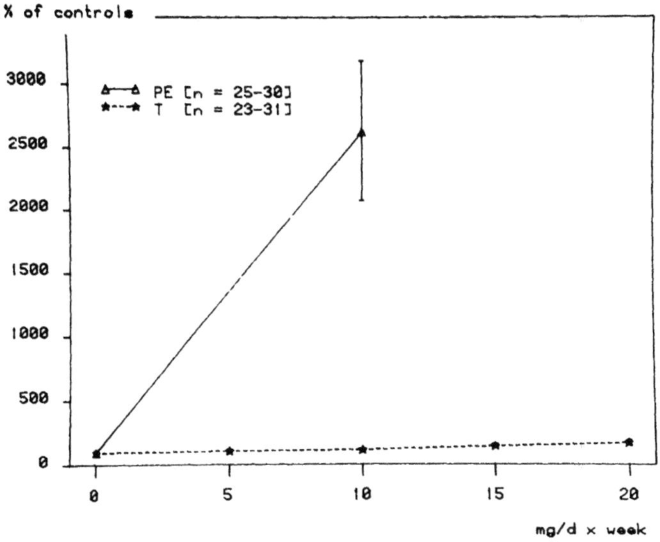

Fig. 6. Effect of subchronic treatment with weekly increasing doses of deprenyl on urinary T and PE in 6 subjects. Values in % of controls.

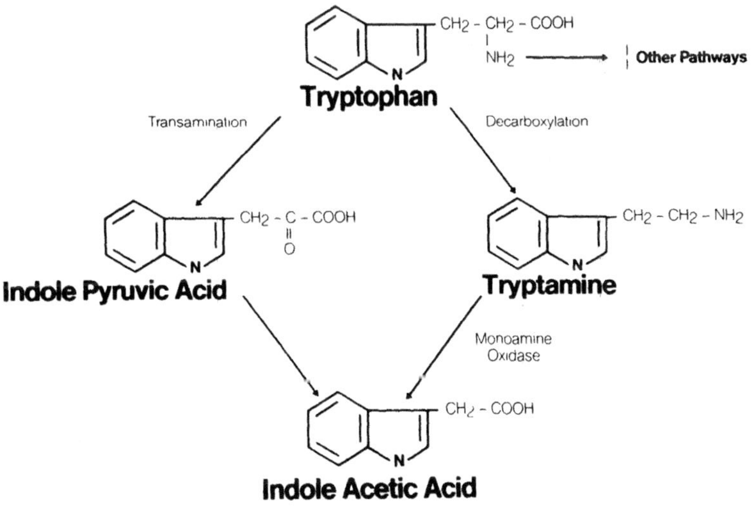

Fig. 7. Pathways of tryptophan metabolism.

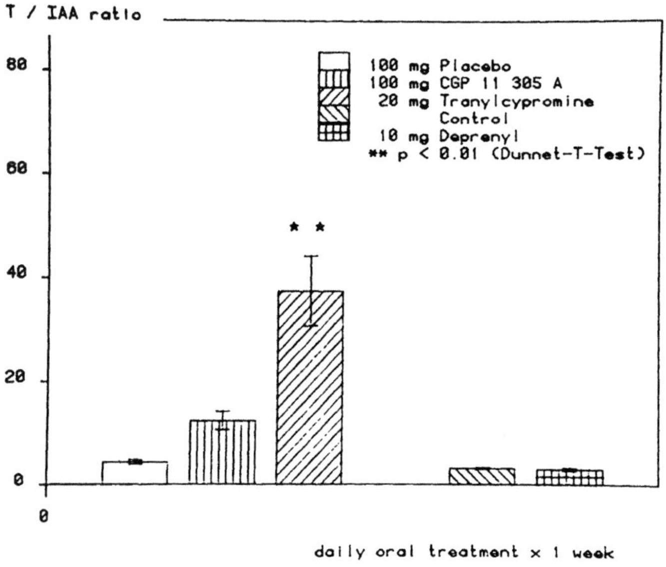

Fig. 8. Urinary T/IAA ratio during treatment with 3 different MAO inhibitors. Left: one week administration of placebo, CGP 11 305 A and tranylcypromine to 6 subjects. Right: two weeks administration of deprenyl to 6 subjects (1.week: 5mg; urine collection during 2.week: 10mg/d). Mean ± SEM of 40 collection days (average).

in the kidney (Rodnight, 1961). Indole acetic acid, the primary metabolite of T is derived up to 30% also along other pathways (Weissbach et al., 1959). Under conditions of relatively constant dietary intake of tryptophan (up to 5g/d) the measurement of urinary T excretion is a reliable index of MAO inhibition. The estimation of the T/IAA ratio in urine was found to be almost equivalent to T alone (LaBrosse et al., 1964). Only under such conditions as tryptophan loading it is a more sensitive index of MAO inhibition.

Fig. 8 shows the urinary T/IAA ratio during subchronic therapy for only one week with the 2 MAO inhibitors CGP 11 305 A (100mg/d) and tranylcypromine (20mg/d). (This Fig. corresponds to Fig. 2, but the values of all six subjects are pooled). It can be seen that there is an increase of the ratio from 4 to 12 and 38, respectively. The latter is statistically significant. In contrast, 10mg deprenyl/d did not change the T/IAA ratio at all.

DISCUSSION

Our finding of a significant difference in urinary T excretion between females and males has not been described before. Fluctuations of ovarian sex hormones have been associated with fluctuations in MAO activity in different tissues of the rat (Holzbauer and Youdim, 1973) and estrogens seem to have an antidepressant effect on women in the postmenopause. In rat brain estradiol reduces preferentially the activity of MAO-A (Luine and McEwen, 1977). The fluctuations in female sex hormones, therefore, could easily explain the much larger urinary variation of the MAO-A substrate T in females (Fig. 1). If one assumes a relationship between T excretion rate and activity of renal MAO-A the T/IAA ratio should be higher in females. This was actually found, but the difference between 4.7 (3 females) and 4.2 (3 males) is not significant. The formation of IAA from metabolic pathways other than those involving T could be important here. However, after MAO inhibition the T/IAA ratio is enhanced significantly more in females than in males (Fig. 2). This supports the assumption of a basically lower renal MAO-A activity in women.

The classification in two functional forms of MAO does not necessarily apply to enzymes in different species or to different tissues of a given species (for review see Fowler et al., 1978). Mostly from in vitro experiments it has been claimed that oxidation of T is solely by MAO B (Houslay and Tipton, 1974) or by both A and B forms of the MAO (White and Tansik, 1979). Some in vivo data in animals suggest T to be a type A MAO substrate (Fuller and Roush, 1972). In man the measurement of amine metabolic endproducts in cerebrospinal fluid, plasma and urine is an established means for assessment of the effect of MAO inhibitors (Murphy et al., 1983). The simultaneously measured urinary excretion rates of the major noradrenaline metabolite NMN and of PE together with T during subchronic treatment with 2 selective and one non-selective MAO inhibitor enabled us to identify T as a prevalent MAO-A substrate, at least in the human kidney. Supported is this by the unchanged T/IAA ratio after selective MAO-B inhibition with deprenyl. The only minimal and insignificant T increase with higher deprenyl doses than 10mg/d are in favour of its relatively selective action also during subchronic therapy. It has been stated that higher doses of deprenyl also inhibit MAO-A (Elsworth et al., 1978). Such an assumption would be consistent with own findings after challenge with intravenous tyramine of deprenyl treated subjects (Cremer et al., 1983). The 26fold rise of urinary PE after 10mg/d deprenyl subchronically compares well with the 20 to 90fold increase described in the literature (Elsworth et al., 1978.)

The different extent of enhanced T excretion after tranylcypromine (720%) and CGP 11 305 A (up to 390%) is explained by the irreversible and reversible MAO inhibition, respectively. This conclusion is supported also by the extent and duration of the potentiated intravenous tyramine pressor sensitivity (Bieck and Antonin, 1982; Bieck et al., 1983).

Our data on T excretion also confirm earlier studies in animals (Waldmeier et al., 1983a) and in man showing that CGP 11 305 A is a relatively specific and reversible MAO-A inhibitor (Bieck et al., 1983 and Waldmeier et al., 1983b). From the results of these studies it is proposed to use urinary T as a clinical index of selective MAO-A inhibition.

Acknowledgements: We thank for the assistance of Ms. M. Huegler, J. Fischer, C. Gerster and K. Kress.

REFERENCES

Baker, G.B., Calverley, D.G., Dewhurst, W.G. and Martin I.L. (1979) A sensitive gas chromatographic technique for quantification of urinary tryptamine. Br. J. Pharmacol. 67, 469 P.

Bieck, P.R. and Antonin, K.H. (1982) Monoamine oxidase inhibition by tranylcypromine: assessment in human volunteers. Eur. J. Clin. Pharmacology 22, 301 - 308.

Bieck, P., Antonin, K.H., and Jedrychowski, M. (1983) Monoamine oxidase inhibition in healthy volunteers by CGP 11 305 A, a new specific inhibitor of MAO-A. Mod. Probl. Pharmacopsychiat. 19, 53 - 62.

Cremer, G., Antonin, K.H. and Bieck, P. (1983) Monoamine oxidase inhibition in healthy volunteers by (-)deprenyl (D). Naunyn-Schmiedeberg's Arch. Pharmacol. 322, Suppl. R 132.

Elsworth, J.D. (1978) Deprenyl andministration in man: A selective monoamine oxidase B inhibitor without the "cheese effect". Psychopharmacology 57, 33 - 38.

Fowler, C.J., Callingham, B.A., Mantle, T.J. and Tipton, K.F. (1978) Monoamine oxidase A and B: a useful concept? Biochem. Pharmacol. 27, 97 - 101.

Fuller, R.W. and Roush, B.W. (1972) Substrate-selective and tissue selective inhibition of monoamine oxidase. Arch. Int. Pharmacodyn. Ther. 198, 270 - 276.

Holzbauer, M. and Youdim, M.B.H. (1973) The oestrus cycle and monoamine oxidase activity. Br. J. Pharmacol. 48, 600 - 608.

Houslay, M.D. and Tipton, K.F. (1974) A kinetic evaluation of monoamine oxidase activity in rat liver mitrochondrial outer membranes. Biochem. J. 139, 645 - 652.

LaBrosse, E.H., Kopin, I.J., Felix Jr., W.R. and Westlake, R.J. (1964) Urinary tryptamine and indole-3-acetic acid excretion by schizophrenic patients: use of the tryptamine/indole acetic ratio as an index of monoamine oxidase inhibition. J. Psychiat. Res. 2, 185 - 197.

Lauber, J. and Waldmeier, P.C. (1983) In prepara-

tion. Biol. Med. 109, 225 - 227.
Levine, R.J and Sjoerdsma, A. (1962) Monoamine oxidase activity in human tissues and intestinal biopsy specimens. Proc. Soc. Exp. Biol. Med. 109, 225 - 227.
Linnoila, M., Karoum, F., Potter, W.Z. (1982) Effect of low-dose clorgyline on 24-hour urinary monoamine excretion in patients with rapidly cycling bipolar affective disorder. Arch. Gen. Psychiatry. 39, 513 - 516.
Luine, V.N. and McEwen, B.S. (1977) Effect of oestradiol on turnover of type A monoamine oxidase in brain. J. Neurochem. 28, 1221 - 1226.
Murphy, D.L., Cohen, R.M., Siever, L.J., Roy, B., Karoum, F., Wyatt, R.J., Garrick, N.A. and Linnoila, M. (1983) Clinical and laboratory studies with selective monoamine-oxidase-inhibiting drugs. Mod. Probl. Pharmacopsychiat. 19, 287 - 303.
Nilsson, E. and Hengen, N. (1983) Measurement of indole-3-acetic acid in human urine by HPLC-UV. In preparation.
Nilsson, E. and Schick, Ch. (1983) Effect of different monoamine oxidase (MAO) inhibitors on human urinary tryptamine excretion. Naunyn-Schmiedeberg's Arch. Pharmacol. 322, Suppl. R 133.
Rodnight, R. (1961) Body fluid indoles in mental illness. Int. Rev. Neurobiol. 3, 251 - 292.
Sjoerdsma, A., Oates, J.A., Zaltzman, P., Udenfriend, S. (1959) Identification and assay of urinary tryptamine: application as an index of monoamine oxidase inhibition in man. J. Pharmacol. Exp. Ther. 126, 217 - 222.
Waldmeier, P.C., Baumann, P.A., Delini-Stula, A., Bernasconi, R., Sigg, K., Buech, O., Felner, A.E. (1983a) Characterization of a new, short-acting and specific inhibitor of type A monoamine oxidase. Mod. Probl. Pharmacopsychiat. 19, 31 - 52.
Waldmeier, P.C., Antonin, K.H., Feldtrauer, J.J. Grunenwald, Ch., Paul, E., Lauber, J. and Bieck, P. (1983b) Urinary excretion of o-methylated catecholamines, and phenylethylamine in human volunteers treated with

tranylcypromine and CGP 11 305 A. Eur. J. Clin Pharmacology, in print.

Weissbach, H., King, W., Sjoerdsma, A. and Udenfriend, S. (1959) Formation of indole-3-acetic acid and tryptamine in animals. J. Biol. Chem. 234, 81 - 86.

White, H.L. and Tansik, R.L. (1979) Characterization of multiple substrate binding sites of MAO, in Monoamine oxidase: structure, function and altered functions (Singer, Van Korff, Murphy, eds.) pp. 129 - 144. Academic Press, New York.

Youdim, M.B.H. and Finberg, J.P.M. (1983) Monoamine oxidase inhibitor antidepressants, in Psychopharmacology Part I: Preclinical Psychopharmacology (GrahameSmith, Cowen, eds.) pp. 38 - 70. Excerpta Medica, Amsterdam.

Free Communications

p-HYDROXYPHENYLACETIC ACID CONCENTRATION IN THE CSF OF PATIENTS WITH NEUROLOGICAL AND PSYCHIATRIC DISORDERS

K. Kobayashi, Y. Imazu and T. Shohmori

Institute for Neurobiology, Okayama University Medical School, Okayama, Japan

In recent years substantial evidence suggesting that *p*-tyramine (*p*-TA) acts as a neurotransmitter or neuromodulator in the brain has been presented (Boulton, 1978; Boulton and Juorio, 1979; Juorio and Jones, 1981). Although the physiological roles of this amine in the central nervous system (CNS) are as yet unknown, it has been claimed that *p*-TA may be involved in the etiology of certain mental disorders (Boulton and Juorio, 1979; Boulton, 1980). However, there is little clinical evidence for this proposal. *p*-Hydroxyphenylacetic aid (*p*-HPAA), a major metabolite of *p*-TA in the brain (McQuade et al., 1981), has been detected and quantitated in the cerebrospinal fluid (CSF)(Karoum et al., 1975; 1977), but a comparison of *p*-HPAA concentrations in the CSF of patients with various CNS disorders has not yet been made. We have recently developed a sensitive high-performance liquid chromatographic (HPLC) method for the determination of *p*-HPAA in CSF (Kobayashi et al., 1982). In this communication we describe CSF levels of *p*-HPAA in patients with various neurological and psychiatric disorders with special reference to schizophrenia.

MATERIALS AND METHODS

The first 3-4 ml of CSF obtained by lumbar puncture from 202 patients, aged from 3 months to 77 years, with a variety of neurological and psychiatric disorders was used. After collecting the next 4-5 ml of CSF for routine diagnostic purposes, a second sample of CSF (8 to 10th ml) was obtained from 14 patients. The patients were sampled without regard to medication. The diagnosis of psychiatric disorders was made according to the DSM III criteria. The schizophrenic group (mean age, 31 years; range, 16-54 years) consisted of 12 disorganized, 3 catatonic, 8 paranoid, one residual and 2 undifferentiated patients. A group of 17 non-neurological patients (mean age, 49 years; range, 22-80 years) served as the adult control. In most cases lumbar puncture was done in the lateral decubitus position between 9 and 10 a.m. Samples were immediately frozen and kept at -20°C until assayed. All investigations were performed with patients' or their parents' informed consent.

p-HPAA in the CSF was measured by a HPLC method reported elsewhere (Kobayashi et al., 1982). Using this method p-HPAA can be determined concurrently with the dopamine metabolites, 3,4-dihydroxyphenylacetic acid (DOPAC) and homovanillic acid (HVA). Student's *t*-test was used for the statistical analysis.

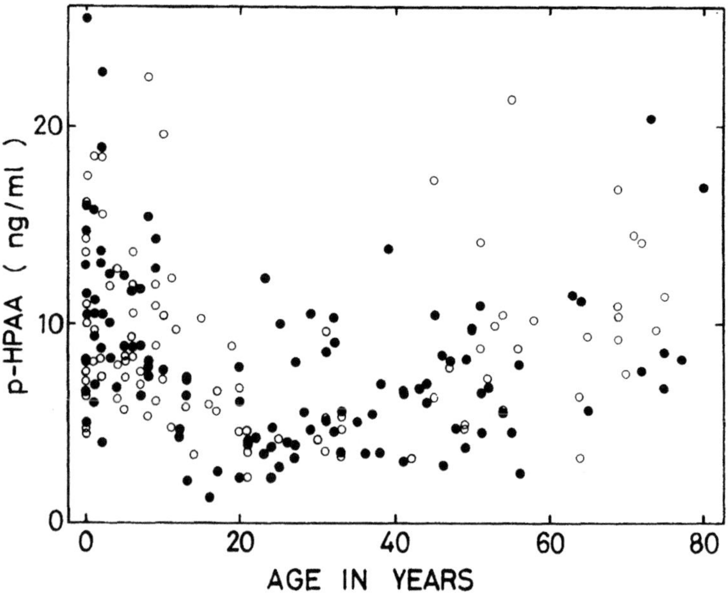

FIG. 1. Values of p-HPAA in the lumbar CSF plotted against age. ●, male; ○, female.

RESULTS

A U-shaped curve was obtained in a total of 219 cases when the values of p-HPAA in the lumbar CSF were plotted as a function of age as shown in Fig. 1. There were no sex differences in the CSF p-HPAA levels. Since the patients under 15 years old had a significantly higher mean p-HPAA value (9.9 ± 0.4 ng/ml, n=105; p<0.001) than the higher age group (7.2 ± 0.4 ng/ml, n=114), p-HPAA values between CNS disorders were analyzed only in the adult group of patients over 16 years old.

p-HPAA levels in the CSF of controls and patients with various CNS disorders are summarized in Table 1, and the values for the five diagnostic subtypes of schizophrenia and related disorders are shown in Fig. 2. Although there is a considerable overlap in the distribution of CSF p-HPAA concentrations between control and schizophrenic subjects, the mean value in schizophrenic patients was significantly lower than in controls. Among the schizophrenic subtypes only the disorganized schizophrenic patients showed a sig-

TABLE 1. p-HPAA concentration in the lumbar CSF of patients with CNS disorders compared with controls

	n	p-HPAA (ng/ml)	
Controls	17	8.5 ± 1.1	(4.2-20.5)
Schizophrenic disorders	26	5.3 ± 0.5*	(2.3-10.5)
Acute paranoid disorder	2	3.0 ± 0.7	(2.3- 3.6)
Atypical psychosis	4	4.0 ± 0.3	(3.4- 4.6)
Affective disorders	9	5.4 ± 0.9	(2.6-10.5)
Anxiety disorders	4	7.2 ± 2.3	(1.4-12.3)
Somatoform disorders	3	10.0 ± 4.2	(3.6-17.8)
Organic mental disorders	5	8.8 ± 1.9	(3.3-14.1)
Epilepsy	15	5.7 ± 0.6	(2.3-10.3)
Parkinson's disease	5	8.9 ± 1.5	(5.7-14.5)
Cerebrovascular accident	14	10.9 ± 1.1	(6.9-21.4)
Pseudobulbar palsy	3	9.0 ± 1.5	(6.1-11.2)
Migraine	3	5.7 ± 0.6	(4.8- 6.8)
Others	4	9.4 ± 2.0	(5.7-14.5)

Values are mean ± SEM with the ranges in parentheses.
* Significantly different from controls at p<0.01.

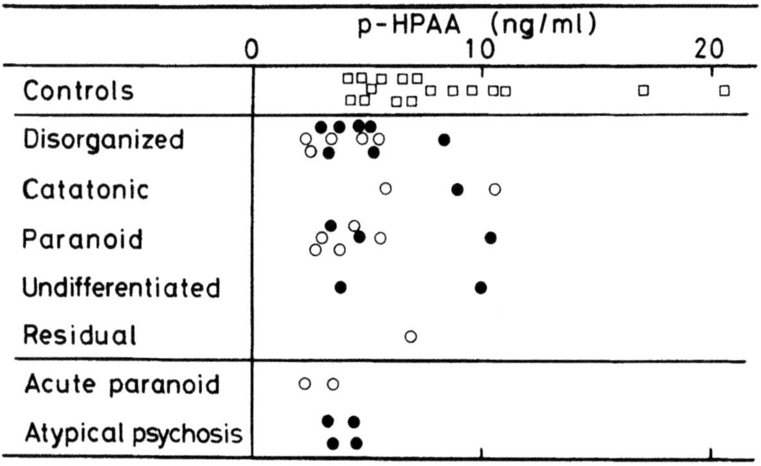

FIG. 2. Distribution of p-HPAA levels in the lumbar CSF of controls and patients with diagnostic subtypes of schizophrenia and related disorders. ○, unmedicated patients; ●, medicated patients.

nificantly lower mean value compared with controls ($p<0.01$). Since there were no significant differences in the CSF p-HPAA levels between medicated and unmedicated (for at least 2 weeks) patients as shown in Fig. 2, the low mean p-HPAA concentration is probably not a result of drug treatment, at least in this schizophrenic population.

The concentration of p-HPAA in the patients with acute paranoid disorder and atypical psychosis was lower than in controls, but this difference failed to reach statistical significance probably due to the small number of subjects. The patients with somatoform disorders and cerebrovascular accident showed somewhat, but not significantly, higher p-HPAA values compared with controls. Though there were no differences between the mean values of controls and epileptics, the patients with tonic-clonic seizures had a significantly lower mean value than controls ($p<0.05$).

An apparent increase in p-HPAA levels from the first 3-4 ml (I) to 8-10th ml (II) of CSF was found in 14 samples examined in this study. The mean ratio between II and I was 1.22 (range, 1.04-1.58). There was a low but significant correlation between p-HPAA and HVA concentrations in the CSF ($r=0.47$, $p<0.001$).

DISCUSSION

As was also found in our preliminary study (Kobayashi et al., 1982), schizophrenic patients, especially of the disorganized type, showed a significantly lower mean value of the CSF p-HPAA than controls in the present investigation. Karoum et al. (1975) have reported that p-HPAA concentrations were 9.9 ± 3.5 ng/ml in the lumbar CSF of patients with a variety of psychiatric illnesses (their diagnoses not shown) and 22.8 ± 3.6 ng/ml in the ventricular CSF obtained at the time of neurosurgical procedures. Subsequently, the mean p-HPAA concentration in the lumbar CSF of 12 unmedicated schizophrenic patients has been reported to be 6.6 ± 0.8 ng/ml by the same author (1977). However, since this value was obtained from CSF collected after 16 ml of CSF had been withdrawn, it is difficult to assess it as a baseline level of p-HPAA in schizophrenic patients in that a concentration gradient for p-HPAA was reported by Karoum et al. (1975) and also found in the spinal CSF obtained in the present study.

The biochemical and physiological significance of the low concentration of p-HPAA in schizophrenic patients needs careful consideration for several reasons. One point is that the origin of p-HPAA in the CSF has not been thoroughly elucidated yet. The existence of a concentration gradient of p-HPAA suggests the central origin of this metabolite in the lumbar CSF. However, the relative contributions of the brain and spinal cord to the concentration of p-HPAA is not known. Secondly, the effects of neuroleptics on p-HPAA in the CSF can not be ruled out, although they are unlikely because there were no differences in p-HPAA values between medicated and unmedicated schizophrenic patients in this study. In experimental animals it has been reported that neuroleptics produced a marked reduction in the striatal concentration of p-TA (Juorio, 1977; Boulton, 1978). Another point is that the age distribution of control and schizophrenic groups was not identical in this study. Since older patients tend to have a higher p-HPAA value as can be seen in Fig. 1, age differences between groups may affect the result. Finally, there has been no direct evidence suggesting that p-TA influences neural functions and, subsequently, human or animal behaviors. Final interpretation of the results obtained in this study must await further experimentation to clarify these points. Despite these difficulties, however, the present data do point to the possibility of a decreased concentration or metabolism of p-TA in the brain of schizophrenic patients.

Acknowledgement. We are grateful to Ms. Fumiko Sato for her skillful technical assistance.

REFERENCES

Boulton A.A. (1978) The tyramines: functionally significant biogenic amines or metabolic accidents? Life Sci. 23, 659-672.

Boulton A.A. and Juorio A.V. (1979) The tyramines: are they involved in the psychoses? Biol. Psychiat. 14, 413-419.

Boulton A.A. (1980) The properties and potential function of some brain trace amines. Prog. Clin. Biol. Res. 39, 291-303.

Juorio A.V. (1977) Effect of chlorpromazine and other antipsychotic drugs on mouse striatal tyramines. Life Sci. 20, 1663-1668.

Juorio A.V. and Jones R.S.G. (1981) The effect of mesencephalic lesions on tyramine and dopamine in the caudate nucleus of the rat. J. Neurochem. 36, 1898-1903.

Karoum F., Gillin J.C., Wyatt R.J. and Costa E. (1975) Massfragmentography of nanogram quantities of biogenic amine metabolites in human cerebrospinal fluid and whole rat brain. Biomed. Mass Spectrom. 2, 183-189.

Karoum F., Bunney W., Gillin J.C., Jimerson D., Van Kammen D. and Wyatt R.J. (1977) Effect of probenecid on the concentration of the lumbar cerebrospinal fluid acidic metabolites of tyramine, octopamine, dopamine and norepinephrine. Biochem. Pharmacol. 26, 629-632.

Kobayashi K., Koide Y. and Shohmori T. (1982) Determination of p-hydroxyphenylacetic acid in cerebrospinal fluid by high-performance liquid chromatography with electrochemical detection. Clin. Chim. Acta 123, 161-168.

McQuade P.S., Juorio A.V. and Boulton A.A. (1981) Estimation of the p and m isomers of hydroxyphenylacetic acid in mouse brain by a gas chromatographic procedures: their regional distribution and the effects of some drugs. J. Neurochem. 37, 735-739.

NONCATECHOLIC PHENYLETHYLAMINES AND MAO ACTIVITY IN DIABETES AND MIGRAINE

A.D. Mosnaim, M.E. Wolf, and S. Diamond

Pharmacol. and Psychiat., Chgo.Med.Sch. and

Diamond Headache Clinic, Chicago, IL.

INTRODUCTION

Human studies have resulted in extensive and many times conflicting reports of alterations in plasma and platelet monoamine oxidase (MAO) activity, Vmax and Km in a variety of disease states (1,2). A partial list of these conditions include affective bipolar disorders, schizophrenia, essential hypertension, iron deficiency anemia, specific developmental dislexia, migraine headache and insulin-dependent (Type I) diabetes (2). Kinetic parameters of this enzyme also appear to be significantly modified by a number of drug treatments (3).

The general observation that emotional crises precipitate and aggravate diabetic conditions (4), the involvement of serotonin, catecholamines and noncatecholic phenylethylamine (NCPEAs) in both brain mechanisms and insulin secretion (2,5-7), the regulatory role played by MAO on the levels of these amines (2), together with earlier data indicating a direct and/or indirect hypoglycemic effect of MAO inhibitors (7), led us to investigate a possible role for the MAO/NCPEAs system in the pathophysiology of diabetes. Similarly, reports of decreased platelet MAO activity during a migraine episode (8) and the known ability of phenylethylamine (PEA) and p-tyramine (TRM) rich foods to trigger such a headache in vulnerable patients (9), raised our interest to study the MAO/PEA system in this condition.

EXPERIMENTAL

Blood was obtained from controlled Type I diabetics, free from concurrent illness and not receiving other drugs besides insulin, mature-onset diabetics (Type II), free from other diseases and controlled by either diet or hypoglycemic drugs, and from age- and sex-matched controls. For the migraine study, blood was drawn (as early as possible in an acute attack and 1 to 3 weeks after the headache episode) from a group of female (20-40 years of age) classical and non-classical migraineurs and from drug-free, sex-, and age-matched controls. All patients had suffered migraine for several years and most of them were under pain medication when participating in this study.

Samples were analyzed in a double-blind fashion for platelet MAO activity and plasma PEA levels; in some specimens we also determined plasma TRM (o-, m-, and p-isomers) and phenylethanolamine (OHPEA) concentration. In brief, MAO activity (platelet from fresh blood used as enzyme source) was determined using p-methoxybenzylamine (spectrophotometrically) and m-iobenzylamine and p-TRM (spectrophotometrically), as substrates (1). Amines were analyzed both by gas-liquid chromatography (GLC) and GLC-mass spectrometry techniques (for details see Ref. 2,10).

RESULTS

When compared to sex- and age-matched controls Type I diabetic patients showed a statistically significant decrease in platelet MAO activity towards p-TRM, p-methoxybenzylamine (Table 1), and m-iodobenzylamine. Results obtained from a group of Type II diabetics, either on diet (n=3) or hypoglycemic drugs (n=4), were consistently in the lower end of the range showed by matched controls (n=10), but failed to reach statistical significance [30±0.6, 31±1.4 and 37.5±5.1 (p-TRM) and 156±2.8, 166±5.5 and 187±17.3 nmoles/h/10^9 platelets SD (p-methoxybenzylamine), respectively]. As previously (10) reported Type I diabetics have significantly higher mean PEA plasma levels than matched controls. We failed, however to show differences in plasma concentration of either p-TRM or

TABLE 1. PLASMA[1] CONCENTRATION OF NONCATECHOLIC PHENYLETHYLAMINES AND PLATELET MAO ACTIVITY IN TYPES I and II DIABETICS AND SEX- AND AGE-MATCHED CONTROLS.

Subjects	Age	[4]PEA	[5]PEA	[4]OHPEA	[4]TRM	[3]Monoamine Oxidase Activity p-tyramine	p-methoxy benzylamine
Males							
Control 1	34	0.1	0.6	0.1	0.38	28.3	127.8
Control 2	36	0.4	1.1	0.1	0.43	30.1	136.4
Control 3	34	0.3	0.8	0.5	0.10	27.4	131.7
Control 4	25	0.3	0.7	0.1	0.54	30.8	128.4
Controls	(25-36)	0.3±0.1	0.8±0.1	0.2±0.1	0.36±0.1	29.1±0.8	131.1±1.9
Controls[6]	(18-34)	—	—	—	—	31.0±4.2(n=19)	130±11.9(n=18)
Patient 1	32	6.4	14.6	0.1	0.13	11.2	80.6
Patient 2	35	0.4	1.3	0.3	1.82	16.6	92.5
Patient 3	25	3.3	7.1	0.5	2.27	4.1	15.9
Patient 4	20	0.9	3.6	0.9	0.11	5.3	56.1
Patient 5	28	3.8	7.8	0.5	0.55	10.0	62.3
Patient 6	31	0.6	4.8	0.1	0.51	8.3	73.2
Patient 7	34	0.7	3.9	0.1	0.66	9.1	53.8
Patients	(20-34)	2.3*±0.9	6.0*±1.7	0.4†±0.1	0.86†0.3	9.2±°1.6	62.1°±9.3
Females							
Controls[6]	(18-34)	—	—	—	—	43.0±3.4(n=23)	181.0±9.7(n=24)
Patient 1	19	—	—	—	—	6.6	39.5
Patient 2	19	—	—			6.2	19.3
Patient 3	23	—	—	—	—	7.3	110.5
Patient 4	30	—	—	—	—	21.3	123.7
Patients 1-4	(19-30)	—	—	—	—	10.4±3.6*	73.2±25.8*

[1]ng/ml plasma ± SEM; [2] see Results and Discussion for details; [3]nmol/h/10^9 platelets ± SD, equivalennt to 10^{-18} per single platelet; [4]determined by GLC-MS; neither o-nor m-TRM could be detected, levels less than 0.050ng/ml; [5]determined by GLC; individual values represent mean of 2-5 measurements from each biological sample; [6] Ref. 1; —values not available. Student's t-test $^*p<0.05$, $p>0.05$, $^{o}p<0.01$.

OHPEA (10); neither m- nor o-TRM could be detected (less than 0.050 ng/ml) (Most of these results have been already reported in detail, Refs. 2 and 10).

As shown in Table 2 there is no statistically significant difference in mean platelet MAO activity towards either p-TRM or p-methoxybenzylamine when comparing a group of migraineurs either during the attack or one to three weeks after the migraine episode, with sex- and age-matched controls. Similarly, no significant differences could be found in mean plasma PEA levels among the groups studied.

In both studies PEA plasma values, obtained by GLC and GLC-MS, followed a similar trend with the GLC results however, being consistently higher (for details see Ref. 10). The GLC procedure used ^{3}H-PEA and the GLC-MS technique deuterated amine isomers as internal standards to correct for recovery.

DISCUSSION

There are several conflicting studies reporting changes in MAO activity in diabetic patients. While serum MAO has been found to be increased (11) or remain unchanged (12) in Type I and Type II diabetics, the activity of the platelet enzyme is statistically significantly decreased in Type I and is not affected in Type II patients (2,10). Animal studies indicate that rat brain MAO activity is decreased after experimental alloxan diabetes but soon recovers to control values (13). Results obtained from a relatively small number of Type I patients indicate that, as a group, increased levels of plasma PEA accompany the observed decreased platelet MAO activity; these patients however, showed no differences in the plasma concentration of OHPEA and p-TRM. Similarly, there is a lack of statistically significant correlation between the observed decrease in platelet MAO and increase plasma PEA for individual patients, indicating, perhaps, that they happen

TABLE 2. PLATELET MAO ACTIVITY AND PLASMA PHENYLETHYLAMINE LEVEL IN MIGRAINE PATIENTS AND SEX- AND AGE-MATCHED CONTROLS

SUBJECT	MAO ACTIVITY (nmole/hr/10^9 platelets±S.D.)							
	p-Tyramine		p-Methoxybenzylamine		Phenylethylamine (ng/ml±SEM)			
	M	AM	M	AM	M[1]	AM[1]	M[2]	AM[2]
Controls	[3]43.0±3.4(n=23)		[3]181.0±9.7(n=24)		0.4±0.1(n=4)		1.7±0.2(n=1)	
1	44.3	—	169.3	183.4	0.234	0.280	1.5	2.1
2	33.4	40.2	183.4	171.7	—	—	1.3	1.8
3	44.9	35.4	174.8	—	0.349	0.116	2.1	1.6
4	47.6	44.3	161.6	183.4	—	—	2.1	—
5	38.0	43.7	189.9	167.9	—	—	1.3	2.1
6	38.9	43.0	183.4	176.5	—	—	—	1.4
7	41.4	35.8	177.8	183.4	0.259	0.471	1.7	0.9
8	43.2	38.4	—	167.6	—	—	1.7	2.1
9	39.4	54.3	182.5	175.9	0.643	0.259	1.9	1.1
10	40.9	50.9	181.4	177.5	—	—	1.9	1.3
(1-10)	41.2±4.0	42.9±6.4	178.2±8.5	176.4±6.3	0.37±0.2	0.28±0.1	1.7±0.3	1.6±0.4

[1]Determined by gas-liquid chromatography-mass fragmentography technique, [2]Determined by gas-liquid chromatography, M, during migraine, AM, 1-3 weeks after migraine episode, —Values not available, [3]Pers. Comm. Dr. E.A. Zeller and Ref. 1.

independently in Type I patients.

The initial report of Sandler *et al.* (8 and Ref. therein cited) of a "transitory deficit of platelet MAO activity" during a classical migraine episode has been followed by a large number of studies examining the possible use of the kinetic characteristics of this enzyme as a biological marker of this and other forms of headache. Sandler's observations and the dominant role of platelet MAO (Type B) in the degradation of amines (*p*-TRM and PEA) believed to play a precipitating role in the triggering of dietary migraine led us to study the possible role of the platelet MAO/circulating PEA system in the pathophysiology of this condition. Our results, which should be viewed with great caution as they represent only a small number of subjects, fail to show statistically significant differences in either platelet MAO activity or plasma PEA levels in a group of migraneurs during a migraine episode when compared to values obtained from the same patients when migraine-free (1-3 weeks after the attack) and with a group of sex- and age-matched controls.

It should be noted that the different studies herein cited usually differ in enzyme preparation techniques, MAO substrates used and enzyme assay procedures. It is clear that further work in this area is needed to clarify the role, if any, of the MAO/noncatecholic phenylethylamine system in diabetes and/or migraine (14).

ACKNOWLEDGEMENTS

We wish to thank Dr. F. Karoum for GLC-MS analysis of PEA in plasma samples and to Drs. E.A. Zeller, O.H. Callaghan, F. Karoum, S.P. Singh and S. Huprikar for fruitful discussions. This work was supported, in part, by the National Migraine Foundation and by the School of Graduate and Postdoctoral Studies, University of Health Sciences/The Chicago Medical School.

REFERENCES

1. Zeller E. A., Boshes B., Arbit J., Bieber M., Blonsky E. R., Dolkart M., and Huprikar S. V. (1976) Molecular biology of neurological and psychiatric disorders. 1. Effects of Parkinsonism, age, sex and L-DOPA on platelet monoamine oxidase. *J. Neural Transmission* **39**, 63-77.

2. Mosnaim A. D., Wolf M. E., Huprikar S., Singh S. P., and Zeller E. A. (1979) Reduced monoamine oxidase activity in blood platelets from insulin-dependent diabetics. Diabetes **28**, 455-456.
3. Friedhoff A. J., Miller J. C., and Weisenfreund J. (1978) Human platelet MAO in drug-free and medicated schizophrenic patients. Amer. J. Psychiat. **135**, 952-955.
4. Kimball C. P. (1971) Emotional and psychosocial aspects of diabetes mellitus. Med. Clin. North Am. **55**, 1007-1018.
5. Mosnaim A. D. and Wolf M. E. (eds.) (1978) Non-catecholic Phenylethylamines, Part 1: Phenylethylamine: Biological Mechanisms and Clinical Aspects. Marcel Dekker, New York.
6. Mosnaim A. D. and Wolf M. E. (eds.) (1980) Non-catecholic Phenylethylamines, Part 2: Phenylethanolamine, Tyramines and Octopamine. Marcel Dekker, New York.
7. Feldman J. and Lebovitz H. J. (1971) The nature of the interaction of amines with the pancreatic beta cell to influence insulin secretion. J. Pharmacol. Exp.Ther. **179**, 56-65.
8. Sandler M. (1977) Transitory platelet monoamine oxidase deficit in migraine: Some reflections. Headache **17**, 153-158.
9. Diamond S. and Medina J. L. (1980) Review article: Current thoughts on migraine. Headache **20**, 208-212.
10. Mosnaim A. D., Karoum F., Zeller A., Callaghan O. H., Singh S. P., and Wolf M. E. (1982) Platelet monoamine oxidase activity and plasma levels of non-catecholic phenylethylamines in insulin-dependent diabetic subjects. Clin. Chim. Acta **126**, 237-242.
11. Tryding N., Nilsson S. E., Tufvesson G., Berg R., Carlstrom S., Elmfors B., and Nilsson J. E. (1969) Physiological and pathological influences on serum monoamine oxidase level: Effect of age, sex, contraceptive steroids and diabetes mellitus. Scand. J. Clin. Lab. Invest. **23**, 79-84.
12. Billett E., Hanley S., Heptinstall S., and Mayer R. J. (1983) Monoamine oxidase activity in diabetes. Diabetes **32**, 130-133.
13. Mayanil C. S. K., Kazmi S. M. I., and Baquer N. Z. (1982) Changes in monoamine oxidase activity in rat brain during alloxan diabetes. J. Neurochem. **38**, 179-183.

14. Wolf M. E. and Mosnaim A. D. (1983) Phenylethylamine in neuropsychiatric disorders. Gen. Pharmacol. **14**, 385-390.

PHENYLETHYLAMINE AND TARDIVE DYSKINESIA

Marion E. Wolf, M.D. and Aron D. Mosnaim, Ph.D.

North Chicago Veterans Administration Medical Center and The University of Health Sciences/ The Chicago Medical School.
VAMC, North Chicago, Illinois 60064

The involvement of dopaminergic nigrostriatal mechanisms in the pathophysiology of tardive dyskinesia (TD) is well documented (Klawans, 1973). There is also growing evidence, supported by physiological, behavioral, pharmacological and clinical experimentation to implicate the non-catecholic phenylethylamines, e.g., phenylethylamine (PEA), octopamines, tyramines and phenylethanolamine, in brain dopaminergic mechanisms (Mosnaim and Wolf, 1978). Furthermore, recent research work would suggest a role for PEA in the pathophysiology of certain types of TD. (Wolf and Mosnaim, 1983a).

Several investigators have noted that affective disorder (AD) patients are more vulnerable to the development of TD (Kane et al., 1980 ; Rosenbaum et al., 1977; Wolf et al., 1982). Our work would suggest that TD in affective disorder patients constitutes a particular clinical entity characterized by development of a syndrome at an earlier age after relatively less exposure to neuroleptic agents than schizophrenic patients. Affective disorder patients with TD also differ from schizophrenic subjects with TD in their cognitive functions. (Wolf et al., 1983b).

Thus in the present study, plasma PEA levels were measured both in schizophrenic patients with and without TD and AD patients with and without TD, aiming at an assessment of the possible role of PEA in certain TD subgroups.

METHODOLOGY

Fourteen male subjects, eight chronic schizophrenics and six bipolar disorder patients had thorough psychiatric and neurologic examinations. Patients were diagnosed according to the Research Diagnostic Criteria (Spitzer et al., 1975) and evaluated for TD using the diagnostic criteria of Jeste et al., (1981). The severity of the movement disorder was assessed by the Simpson Rating Scale (Simpson et al., 1979). After a two week psychotropic drug washout period, blood was drawn for determination of plasma PEA levels which was measured by a gas liquid-mass fragmentography technique (Karoum et al., 1979).

RESULTS

The results obtained indicate that schizophrenic patients with and without TD did not differ in their plasma PEA levels. Bipolar patients with TD had however, statistically reduced plasma PEA levels in comparison to bipolar patients without TD (See Table 1).

PLASMA PHENYLETHYLAMINE LEVELS IN PSYCHIATRIC PATIENTS

Patient Group	PEA (ng/ml) ($\bar{x} \pm$ SD)
Schizophrenics with TD (n=4)	0.38 ± 0.19
Schizophrenics without TD (n=4)	0.24 ± 0.17
Bipolar Disorders with TD (n=3)	*0.11 ± 0.04
Bipolar Disorders without TD (n=3)	0.30 ± 0.04

*Student's t test $p < 0.05$ as compared to bipolar patients without TD.

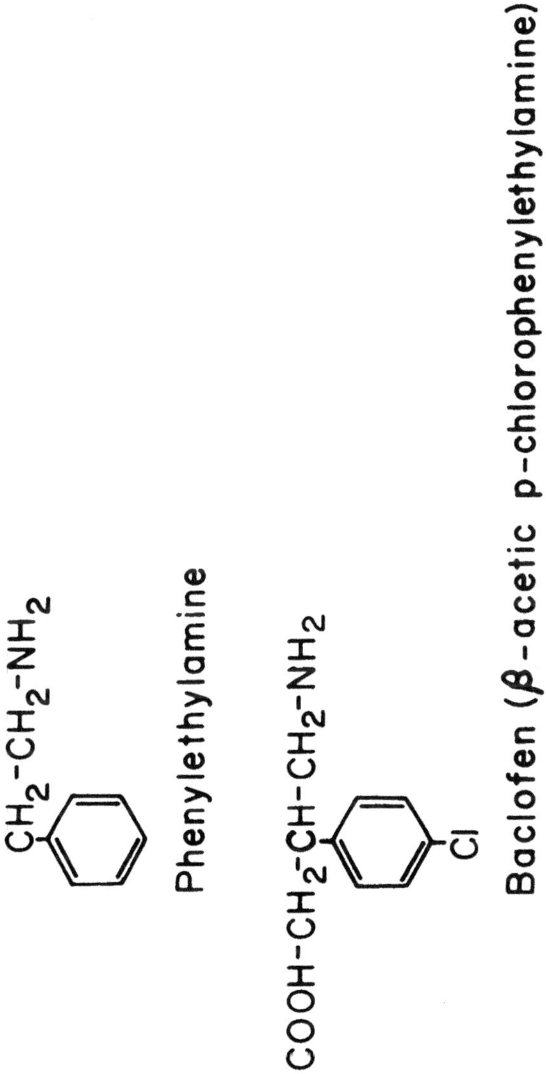

FIG 1. Chemical Formulae of Phenylethylamine and Baclofen

DISCUSSION

Potkin et al., (1978;1979) found that chronic paranoid schizophrenic patients had reduced platelet MAO activity and increased urinary PEA excretion, and more recently this group of investigators have reported that the subgroup of schizophrenic patients with low platelet MAO activity was characterized also by the presence of paranoid features and tardive dyskinesia (Jeste et al., 1982). Our preliminary data indicated that schizophrenic patients with TD had higher mean PEA values than schizophrenics without TD, but this difference did not achieve statistical significance. However, it is possible that if a larger number of subjects were evaluated or only paranoid schizophrenic subjects were included in the study, that the differences in the means may reach levels of significance.

Although our data must be viewed within the limitations imposed by the small number of subjects, they suggest that bipolar disorder patients with TD have lower plasma PEA levels than bipolar subjects without TD. These preliminary findings suggest that a deficit of PEA may be of pathophysiological importance in certain TD groups. Further support for this view is given by results of a selective therapeutic response to Baclofen, β-acetic p-chlorophenylethylamine (See Figure 1), by a group of patients with trunkal dyskinesia (Wolf et al., 1983c, Wolf et al., 1983d). Since Baclofen could be viewed as a PEA derivative, it has been postulated that this drug exerts therapeutic effects only in those patients presenting deficits of PEA, which may be reflected in decreased plasma PEA levels as observed in the bipolar TD patients in this study. Further work in this area is in progress.

ACKNOWLEDGEMENTS

We are grateful to Dr. Karoum for phenylethylamine analysis. This work was supported, in part, by the Illinois Department of Mental Health, Research and Development Grant #8324-01.

REFERENCES

Jeste D.V., Necker L.M., Wagner R.L., Wise D.C., Staub R.A. Rogol A., Potkins S., Bridge P.T., and Wyatt R.J., (1981) Lymphocyte monoamineoxidase and plasma prolactin and growth hormone in tardive dyskinesia J. Clin. Psychiat. 42: 75-77.

Jeste D., Kleinman J., Potkin G., Luchins D.J., and Weinberger D.R., (1982). Ex Uno Multi: Subtyping the Schizophrenic Syndrome. Biol. Psychiatry 17: 199-222.

Kane J., Struve F.A., Weinhold P.I, and Woerner M., (1980). Strategy for the study of patients at high risk for tardive dyskinesia. Am. J. Psychiat. 137: 1265-1267.

Karoum F., Nasrallah H., Potkin S., Chuang L., Moyer-Schwing J., Phillips I. and Wyatt R.J., (1979). Mass fragmentography of phenylethylamine, m- and p-tyramine and related amines in plasma, cerebrospinal fluid, urine and brain. J. Neurochem. 33: 201.

Klawans H.L., (1973). The pharmacology of tardive dyskinesia. Am. J. Psychiat. 130: 82-86.

Mosnaim A.D., and Wolf M.E., (1978) Non Catecholic Phenylethylamines. Part I. Phenylethylamine: Biological Mechanisms and Clinical Aspects. Marcel Dekker. Inc. New York.

Potkin S.G., Karoum F., Chuang L.W., Cannon Spoor H.E., Phillips J. and Wyatt R.J., Phenylethylamine in Paranoid Chronic Schizophrenia (1979). Science 206: 470-471.

Potkin S.G., Cannon E., Murphy D. and Wyatt R.J. (1978) New England J. Med. 298: 61-66.

Rosenbaum H.M., Niven R.G., and Hanson N., (1977). Tardive Dyskinesia: Relationship with primary affective disorder. Dis. Nerv. Syst. 38: 423-427.

Simpson G.M., Lee J.H., Zoubok K. and Gardos G.A., (1978). Rating Scale for Tardive dyskinesia Psychopharm 64: 171-179.

Spitzer R.L., Endicott J., Robins E.,Kuransky J. and Jurland B., (1975). Preliminary report of the reliability of research diagnostic criteria applied to psychiatric case records. In Predictability in psychopharmacology, preclinical and clinical correlations. Sudilovsky A,et al.,eds. p.1-47,Raven Press,N.Y.

Wolf M.E., and Mosnaim A.D., (1983a). Phenylethylamine in neuropsychiatric disorders. Comp. Biochem. and Physiol. In press.

Wolf M.E., Ryan J. and Mosnaim A.D., (1983b). Cognitive functions in tardive dyskinesia Psychol. Med. In press.

Wolf M., Almy G., Mathis P., and Mosnaim A.D. (1983c). Baclofen therapy in psychiatric patients. Clin. Pharm. and Therap. 33: 244.

Wolf M., Keener S., Mathis P. and Mosnaim A.D., (1983d). Phenylethylamine-like properties of Baclofen. Neuropsychol. In press.

Wolf M.E., Ryan J. and Mosnaim A.D. (1982). Organicity and tardive dyskinesia, Psychosomatics 23: 475-480.

Phenylethylamine, Phenylacetic Acid and Methionine Enkephalin Levels in Humans Following Profound Acute Stress

[1]Paulos,M.,[2]Mosnaim,A.,[2]Wolf,M.,and[3]Tessel,R.

[1]Du Pont Co., Wilmington,DE.,[2]Chicago Medical School, IL. and[3]Univ. of Kansas, KS.

The blood concentration of catecholamines reflects the balance between the basal sympathetic neuronal activity and adrenomedullary amine output, as influenced by a number of factors including age, stress (type and level) as well as drug effects and disease states, and the activity of the several enzymatic systems responsible for catecholamines degradation(1,2). Amine uptake by erythrocytes, platelets and other blood cells and binding to hemoglobin and plasma proteins further regulate the levels of free, circulating catecholamines(3). As could be expected from their similarities in chemical structure, neuroanatomical distribution, metabolic pathways of biosynthesis and degradation, physiological role and pharmacological effects, the circulating levels and blood distribution of the parent noncatecholic phenylethylamines, phenylethylamine (PEA), phenylethanolamine, *p*-octopamine and *p*-tyramine, share common regulatory mechanisms with the catecholamines(3-5).

Behavioral studies in animals and clinical observations, specially from the pathophysiology of paranoid schizophrenia, migraine headache and Type I diabetes(2,6) suggest that the monoamine oxidase - PEA system may be involved in the human reaction to psychological stress.

In a test of this hypothesis we have recently reported that following exposure to a profound acute stressor (initial parachuting experience) the urinary PEA excretion rate (ug/hr., urination was collected separately for 24 hr prior to and during the 24 hr. period following the jump) was makredly increased in each subject studied (n=8; range 2.6-15.8 fold above baseline)(7).

In the present, preliminary studies we have determined plasma phenylacetic acid (PAAc), the major metabolite of PEA(8), and methionine enkephalin (met-enk) in one individual (hourly for 10 hr.) following his first parachuting experience and urinary PEA excretion rate in one experienced jumper. The preliminary results here reported should encourage further work on the hypothesis that the physiological response to at least some forms of stress, involves changes in both the MAO/PEA and the met-enk systems.

MATERIALS AND METHODS

The protocol describing the criteria used in the selection of subjects and in the collection of blood and urine samples has been published in detail elsewhere (7,9.). Stress during the parachuting experience was operationally defined as an elevation in heart rate as followed by a Holter monitor (casette tape, continuous electrocardiogram recording), which was attached to each subject one hr. prior to jump, remained attached during the parachuting and was removed one hour after the subject had landed (9). Assay of urinary PEA was carried out using a gas chromatography-mass spectrometry technique (30m long x 0.32 mm I.D. fused silica capillary column, Ribermag(R) Quadrupole Mass Spectrometer Model R10-10; Riber Corp., Paris, France). Values reported for urinary PEA levels are corrected for losses in the amine extraction and derivatization procedures (deurated PEA; D_4-PEA, lateral chain, was used as internal standard for recovery)(9). Creatinine was assayed following standard procedures (9). Following its extraction from plasma samples PAAc levels were determined by GLC. Plasma samples (2ml) were added ^{3}H(G)-PAAc (as internal standard for recovery) and after adjusting pH to 2.0 (conc.HCl), PAAc was taken up into chloroform (4 ml x2). The acid was extracted into a pH 9.0 solution (0.1N NaOH), followed by pH adjustment (2.0, HCl) and its subsequent extraction into fresh chloroform (3 ml x 2). Chloroform was carefully evaporated and the residue dissolved in 100 ul of fresh solvent. This solution was placed on a

Quanta/Gram^{TM} cellulose TLC plate, along with a separate spot containing 5 ug 1-^{14}C-PAAc (10,000 dpm), and developed (ethanol [95%]: amonia [conc.], 20:1) for 2 hours at room t^{o}. After TLC radioscanning (radioactivity present in R_F 0.25-0.32 area), the section containing the standard PAAc was sprayed (fresh solution containing n-butanol: ethanol: water: aniline and glucose, 30:10:10:1 ml and 1 g, respectively) and heated in a ventilated oven at 115-120°C for 7-10 min. (PAAc gives an orange-brown color over a yellowish background, R_F 0.22-0.30). The area containing the plasma PAAc (as determined from the standard) was scrapped off the plate, extracted into chloroform and concentrated to 40 ul (N_2 blown, room t^{o}). After the addition of benzoic acid (5 ul containing 200 ng of the acid; gas-liquid chromatography [GLC] reference purposes), the solution was added BSA (3:1 v/v) and allowed to stand for 5 min. at room t^{o}. Samples were analyzed (not later than six hours after derivatization) in a Beckman GLC Model 65 equipped with a flame ionization detector. Aliquots were separated for counting (5-10 ul) and for injection (5-10 ul) into a 1.8 m length x 0.6 cm internal diameter coiled glass U-turn column system packed with 10% OV 17 on Diaport S (80-100 mesh) (Beckman Inst. Co.); initial column t^{o} was held at 100°C for 5 min., followed by programming (2.5^{o}C/min.) to final t^{o} of 250°C. The injection block and detector were maintained at 259°C and 310°C, respectively. The flame was maintained using hydrogen and air (42 ml/min. and 300 ml/min., respectively), and nitrogen was the carrier gas (80 ml/min.). PAAc concentration was estimated by comparing the average area under the peak response (5-10 ul injections; 2-5 determinations for each biological sample showing a variation range of 3to 7%) with a standard curve obtained by plotting areas under the peak response as a function of known amounts of PAAc (linear in the 5 ng-500 ng range). Studies on the completeness of the PAAc-BSA reaction and of the stability of the formed trimethylsilyl ester derivative (TMSP) in chloroform solution were carried out by adding BSA to chloroform solutions of 1-^{14}C-PAAc (100-1000 ng range). Aliquots were taken at 5 min., 60 min. and then hourly for ten hours and analyzed for unreacted labeled PAAc (TLC radioscanning)(for details see ref.10)

In each case the identify of the peak obtained from

plasma samples was confirmed as TMSP (R_T 10.4 min.) by comparing its retention time to that of the BSA derivative of benzoic acid (R_T 8.3 min.), and by analyzing a sample aliquot enriched with known amounts of standard TMSP. Values reported are corrected for percentage of ^3H(G)-PAAc derivatized (range 65-83%).

Levels of plasma met-enk were determined as published elsewhere (11). In brief, blood was collected in vacutainer tubes (containing EDTA) and immediately centrifuged (4^oC,800 xg for 15 min.). After extracting the pentapeptide (Octadecasilyl-silica column; Immunonuclear Corp., kit, Stillwater, Minnesota) the resultant methanol solution was evaporated to dryness and the sample reconstituted (1:1 v:v, bovine serum albumin-borate buffer pH 6.4). Two hundred ul aliquot was used for the radioimmunoassay (duplicate).

Levels of met-enk were calculated from a standard curve obtained for each set of samples by plotting the concentration of the standard peptide on the abcissa (nine values, range 0 to 2500 pg/ml) versus percent met-enk bound. The sensitivity of this assay has been reported as 8 pg, with a 50% displacement obtained by 60 pg of met-enk. The affinity constant of the antibody being 4×10^{12} L/M (11).

RESULTS

As seen in the Table, one hr. after parachuting the subject studied had significantly higher plasma met-enk than either the average or each of the individual values shown by a group of race-matched and comparable age controls. This high pentapeptide levels continued over the period of the experiment (10 hrs. after jumping). PAAc plasma levels showed a "delayed", significant increase at hr. 4 and 5, which postively correlated with changes in both systolic and diastolic blood pressure. There was no significant differences in the urinary PEA excretion rate in the experienced parachutist (ug/hr;for 24 hr before and then during the 24 hr following the jumping each urination was collected separately) when compared to its own baseline (non-stress conditions) baseline (for details see refs. 7,9).

DISCUSSION

There is ample evidence indicating that stressful events are accompanied by a variety of changes in

METHIONINE-ENKEPHALIN AND PHENYLACETIC ACID PLASMA LEVELS AND BLOOD PRESSURE IN ONE INDIVIDUAL AFTER AN ACUTE STRESS EXPERIENCE (PARACHUTING)

[1]Time (hrs.)	MET-ENK (pg/ml±SEM)	PAAc (ng/ml±SEM)	BP Systolic/Diastolic
1	85	99.4	138/86
2	86	111.3	124/80
3	52	*168.9	140/90
4	100	*193.1	138/100
5	80	125.6	122/80
6	69	115.7	122/80
7	93	124.3	114/72
8	78	109.7	116/78
9	115	117.8	120/82
(1-9)	84.3±6.0	129.5±10.3	-
[2]Controls	35.4±6.2 (n=10)	117.6±8.6 (n=31)	-

[1]After parachuting, [2]Matched for race and of comparable age.

*Student's t test $p<0.05$.

physiological mechanisms involving hormones and other neuroregulators (1,9). As the administration of amphetamine produced both a stress-like and a paronoid schizophrenia-like response in men we became interested in examining the possible role of the PEA metabolic pathway in the ethiology and response to stress (4,6). PEA, an amine conceptualized as an endogenous amphetamine, has been implicated as an ethiological factor in paranoid schizophrenia (6,12) and appears to mediate some of the effects produced by amphetamine (3). There is also some evidence suggesting an interaction of PEA and PEA derivatives with enkephalins in some of the behavioral and analgesic properties (11,see Mosnaim, Wolf and Zeller, this book) attributed to these substances. It is hoped that the preliminary results presented in this paper would encourage further work on the hypothesis that the physiological response to at least some forms of stress involves the phenylethylaminergic and enkephalinergic systems. At this moment we can only speculate on the possibility that both systems respond independently or in a concerted manner to the parachuting stress model.

ACKNOWLEDGEMENTS

We thank Mr. Rodger Wine for his expertise in met-enk determinations. This work was supported, in part, by the National Migraine Foundation.

REFERENCES

1. Da Prada M. (1980) Concentration, dynamics and functional meaning of catecholamines in plasma and urine. Trends Pharmac. Ther. **1**, 157.
2. Mosnaim A. D., Wolf M. E., Huprikar S., Singh S. P., and Zeller E. A. (1979) Reduced monoamine oxidase activity in blood platelets from insulin-dependent diabetic subjects. Diabetes **28**, 455-456.
3. Mason M., Wolf M. E., and Mosnaim A. D. (1983) Studies in the mechanism of phenylethylamine uptake by rabbit erythrocytes. Comp. Biochem. Physiol. (in press).
4. Mosnaim A. D. and Wolf M. E. (eds.) (1978) Non-catecholic Phenylethylamines, Part 1: Phenylethylamine:

Biological Mechanisms and Clinical Aspects. Marcel Dekker, New York.

5. Mosnaim A. D. and Wolf M. E. (eds.) (1980) Non-catecholic Phenylethylamines, Part 2: Phenylethanolamine, Tyramines and Octopamine. Marcel Dekker, New York.

6. Wolf M. E. and Mosnaim A. D. (1983) Phenylethylamine in neuropsychiatric disorders. Gen. Pharmacol. **14**, 385-390.

7. Paulos M. and Tessel R. E. (1982) Excretion of β-phenylethylamine is elevated in humans after profound stress. Science **215**, 1127-1129.

8. Fellows L. E., King G. S., Pettit B. R., Goodwin B. L., Ruthven C. R. J., and Sandler M. (1978) Phenylacetic acid in human cerebrospinal fluid and plasma: Selected ion monitoring assay. Biomed. Mass. Spectrom. **5**, 508-511.

9. Paulos M. (1981) Ph.D. Dissertation, University of Kansas, Lawrence, Kansas.

10. Madubuike P. (1975) Ph.D. Dissertation, University of Health Sciences/The Chicago Medical School, North Chicago, Illinois.

11. Mosnaim A. D., Wine R., Diamond S., and Wolf M. E. (1984) Proceed. of the IV Intern. Migraine Symp. (Marshall and Rose, eds.). Raven Press, New York.

12. Potkin S., Linnoila M., Potter W., Chuang L., Goodwin F.,and Wyatt R. (1982) Psychiat. Res. **6**, 215-222.

^{14}C-TRYPTAMINE BINDING IN PARKINSON'S DISEASE AND HEPATIC COMA

Elisabeth Kienzl, Peter Riederer,
Kurt Jellinger and Heinrich Noller [x)]

Ludwig Boltzmann-Institute of Clinical Neurobiology, Lainz Hospital, Vienna, and x) Institute for Physical Chemistry, Technical University, Vienna, Austria.

The serotonergic (5-HT) system seems to regulate at least in part sleep-wakefulness, body temperature, sympathetic activity, adrenomedullary and hormone secretions. However, functional changes in the central activity of 5-HT transmission are suggested to be partially modulated by tryptamine (T) (Jones, 1982a). The only way to synthesize brain T was found to be direct decarboxylation of L-tryptophan (TRP) by aromatic amino acid decarboxylase. Warsh et al. (1979) demonstrated a significant increase in brain T level followed by TRP plus MAO-inhibitor administration. Additional neuro- and electrophysiological data give evidence for the existence of post-synaptic receptors for T, independent of those for 5-HT (Tedeschi et al, 1959; Kellar and Cascia,1982). T neurons are thought to be located in the midbrain raphe nuclei and innervate several forebrain regions (Aghajanian and Asher,1971).

In membrane preparations of human post mortem frontal cortex we measured binding sites that are kinetically distinguishable and are referred to a high affinity (nM) or low affinity (μM) component on the basis of the relative magnitude of their apparent K_D-constants.

Binding characteristics in Parkinson's disease and hepatic coma reflect neuropathological manifestations. While in Parkinson's disease 5-HT is reduced in most of the brain areas (Bernheimer et al., 1961; Birkmayer et al.,1974), a substantial increase in 5-HT concentration and turnover has been measured in hepatic encephalopathy (Jellinger and Riederer,1977). 5-HT receptor binding showed nothing abnormal in Parkinson's disease (Rinne et al,1980) while a mild decrease of the receptor number was noted in the frontal cortex (Kienzl et al.,1981). However, a dramatic drop of 5-HT binding sites was measured in the frontal cortex in hepatic encephalopathy (Riederer et al., 1981).
If 5-HT transmission is modulated by T (Jones, 1982b) changes in T binding could be assumed to occur in Parkinson's disease and hepatic coma. Therefore, we have studied this problem in more detail.

MATERIAL AND METHODS

Brain homogenates of the medial frontal cortex were studied after dissection 2 - 10 hours post mortem. The neuropathological examination using routine histological methods did not show any abnormalities in 4 controls. Hepatic encephalopathy showed characteristic histological changes corresponding to cerebral edema and increase of pathological astroglia of the Alzheimer II-type.Furthermore, 8 subjects with Parkinson's disease (idiopathic [6] and vascular type [2]) according to clinical and neuropathologic criteria were studied. Clinical data are given in table 1.

Brain samples were stored frozen at -30°C. The preparation of tissue included homogenization with 10 volumes of 50 mM Tris-buffer, pH 7,4 containing 10 µM pargyline and 0,1 % ascorbic acid. The crude tissue homogenate suspension was then diluted 1:10 in the same buffer and centrifuged at 39,000 x g

for 18 min. at $4^{o}C$. The supernatant was discarded and the pellet washed and centrifuged again for 8 min., resuspended in buffer up to a protein concentration of 0,3 - 0,6 µg/ml. Aliquots (1 ml) were preincubated at $37^{o}C$ for 5 min. and then incubated for a further 7 min. with ^{14}C-T (50,3mCi/mmol; NEN) with varying concentrations between 0,2 - 6,0 µM. 1 mM unlabeled T has been added for the determination of unspecific binding. Preparation of crude synaptosomes was provided from unfrozen tissue, 3 hours post mortem. Frontal cortex was dissected and placed in 20 vol. cold 0,32 M Sucrose-Hepes buffer, homogenized and centrifuged for 4 min. at 1085 x g to remove nuclei and cellular debris. The supernatant was collected and centrifuged at 27ooo x g for 15 min. to sediment synaptosomes. Hypotonic shock provided the rupture of synaptosomes. 0,6-24 nM ^{3}H-T (39,5Ci mmol; NEN) was used in these experiments. 50 µM non radioactive T was added to define non-specific binding.
Separation of specific and unspecific radioactivity was performed on glass microfibre filters GF/B on a Millipore apparatus with 3 washes of 5 ml 50 mM Tris-HCl-buffer pH 7,4 each. Radioactivity retained on the filters was measured in a scintillation counter in a Rotiszint 22 - solution.

RESULTS

Kinetic data of the reversible, specific binding of ^{3}H-T and ^{14}C-T respectively to human brain homogenates and synaptosomes indicate the presence of two distinct saturable specific binding sites. The synaptosomal fraction reveals an additional 200 - 300 % increase in high affinity binding number thus indicating a predominance of T binding sites in neuronal structures.

The low affinity component for T in micromolar order revealed that brain dysfunction in the serotonergic system may effect the T-receptor interaction.

In Parkinson's disease Scatchard analysis of ^{14}C-T binding shows a significant increase in maximal binding capacity of about 200 % and a decrease in the affinity opposite to the effect of 5-HT binding characteristics. In hepatic coma the K_D-values remain unchanged and only a slight increase is observed in receptor density (table 1).
As a working hypotheses, the different relative strenghts of electron pair-donor and -acceptor sites of both compounds are suggested to their complementary physiological interactions at neuronal membranes. If ligand-receptor-binding is understood as catalytic effect, the weaker donor capacity of T (as the result of lacking the hydroxy-group in the 5th ring position) will cause the lower affinity as compared to 5-HT: T: K_D high affinity : 11,97 nM; 5-HT: K_D high affinity: 4,76 nM (controls) (fig.1, table 1)

TRYPTAMINE

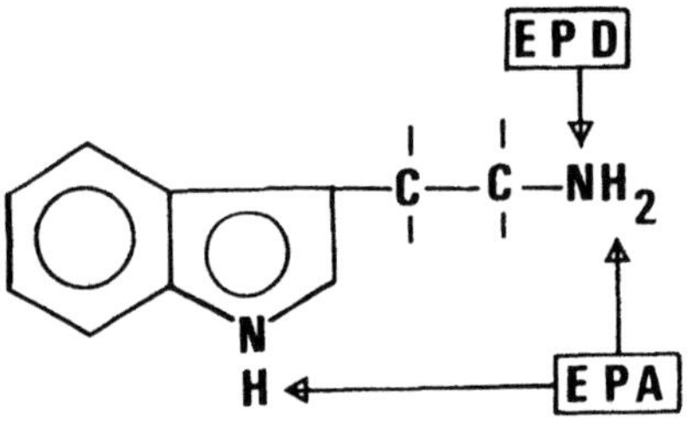

SEROTONIN

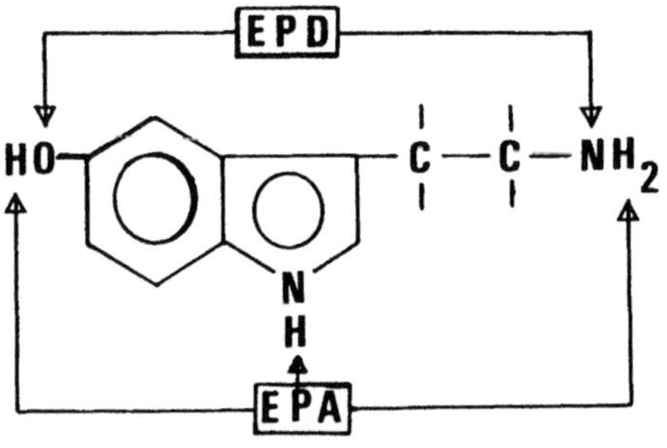

Fig.1: Electron pair donor (EPD) and electron pair acceptor (EPA) sites in tryptamine and serotonin molecules

Table 1

^{14}C-TRYPTAMINE BINDING IN HUMAN FRONTAL CORTEX

	N	F/M	AGE (yrs)	POST MORTEM TIME (hrs)	K_D [μM]	B_{MAX} [pmol/mg prot.]
LOW AFFINITY STATE						
CONTROLS	4	4/0	72,50 ± 9,4*)	6,0 ± 1,7	5,03 ± 0,32*)	71,98 ± 8,55*)
HEPATIC COMA	4	3/1	65,20 ± 8,4	3,7 ± 2,0	5,58 ± 0,66	93,58 ± 17,85
MORBUS PARKINSON	6	2/4	69,66 ± 9,8	7,1 ± 4,8	8,74 ± 4,14	165,79 ± 30,90
VASCULAR PARKINSONISM	2	1/1	84/72	20/10	5,34/13,67	202,14/184,86
HIGH AFFINITY STATE					K_D [nM]	B_{MAX} [fmol/mg prot.]
CONTROLS						
Homogenate: frozen tissue	2	2/0	75/59	7/6,5	3,86/6,40	59,86/85,89
Crude synaptosomes unfrozen tissue	2	2/0	75/59	7/6,5	12,84/11,11	180,17/183,14

*) DATA expressed in $\bar{x}$ ± SD

DISCUSSION

T can displace the postsynaptic binding of 5-HT in brain tissue and is a possible antagonist to physiological and pharmacological effects of 5-HT (Bennett and Snyder,1976; Whittaker and Seeman, 1978; Seeman et al.,1980). The dynamic changes of kinetic behaviour of T-binding in Parkinson's disease as well as in hepatic coma are paralleled by the significant decrease in 5-HT binding (Riederer et al.,1981). Hepatic encephalopathy is characterized by a regional increase of 5-HT levels and turnover, with preference for the brain-stem and parts of the limbic system. In Parkinson's disease the reduced B_{max} values and drops in K_D are accompanied by selective histopathological alterations, changes in cellular structure and an additional 40 - 50 % diminished 5-HT concentration in almost all parkinsonian brain regions (Bernheimer et al.,1961). The compensating effect seems to be an increase in tryptaminergic receptor activity in Parkinson's disease, while there are only mild changes in hepatic coma. Smith and Kellos (1969) reported the urinary excretion of abnormal high amounts of T in Parkinsonian patients while such data are lacking for human brain tissue. Tremor as well as the disturbance of body temperature regulation in Parkinsonism seem to be related to the imbalance in the T-5-HT system (Tedeschi et al.,1959; Cox et al.,1981). Furthermore, T seems to be involved in the hyperkinetic hehavioral syndrome observed in animals after administration of TRP and tranylcypromine (Marsden and Curzon, 1978). This has been related to an increase in T concentration affecting dopaminergic neuronal activity by enhancing the release of dopamine (DA) (Emis et al.,1981; Baker et al.,1980). Therefore, it may be speculated that dyskinesias may be, at least in part, correlated to changes in the 5-HT-T regulation both affecting dopaminergic activity in the striatum. This is further substantiated by findings showing a competition of neutral amino acids for brain uptake. L-Dopa therapy might reduce the uptake of other aromatic amino acids including TRP (Riederer,1980), while in hepatic encephalopathy the uptake of aromatic amino acids, expecially of TRP, is ėnhanced (Riederer et al.,1981). It is noteable that two patients with vascular parkinsonism - a condition with reduced nutritional supply to the brain - had B_{max}-values at the upper range of patients with Parkinson's disease, although no depigmentation of substantia nigra was noted. Therefore, T concentration might be changed under special pathological conditions as reflected by binding studies.

Further studies to define T as neurotransmitter and/or neuromodulator in human brain are essential to get more detailed insights into the pathophysiology of Parkinson's disease and metabolic disturbances.

REFERENCES

Aghajanian G.K. and Asher J.M. (1971) Histochemical fluorescence of raphe-neurones: Selective enhancement by tryptophan. Science 172, 1159-1161

Baker G.B., Hiob L.E., Martin I.L., Mitchell P.R., and Dewhurst W.G. (1980) Interactions of tryptamine analogues with 5-hydroxytryptamine and dopamine in rat striatum in vitro. Proc.West. Pharm.Soc. 23, 167-70

Bennet J.P. and Snyder S.H. (1976) Serotonin and lysergic acid diethylamide binding in rat brain membranes: Relationship to postsynaptic serotonin receptors. Mol.Pharmacol.12, 273-389

Bernheimer H., Birkmayer W. and Hornykiewicz O. (1961) Verteilung des 5-Hydroxytryptamins (Serotonin) im Gehirn des Menschen und sein Verhalten bei Patienten mit Parkinson-Syndrom. Klin. Wschr. 39, 1056-1059

Birkmayer W., Danielczyk W., Neumayer E., and Riederer P. (1974) Nucleus ruber and L-dopa psychosis: biochemical post mortem findings. J.Neural Transm. 35, 93-116

Cox B., Leet F. and Mardin D. (1981) Different hypothalamic receptors mediate 5-HT and tryptamine induced core temperature changes in the rat. Brit.J.Pharmac. 72, 447-482

Ennis C., Kemp J.D. and Cox B (1981) Characterization of inhibitory receptors that modulate dopamine release in the striatum. J.Neurochem. 36, 1515-1520

Jones R.S.G. (1982a) Tryptamine: A Neuromodulator or Neurotransmitter in Mammalian Brain? Progr. Neurobiol. 19, 117-139

Jones R.S.G. (1982b) Responses of cortical neurrones to stimulation of the nucleus raphe medianus. A pharmacological analysis of the role of indolamines. Neuropharm. 21, 511-520

Jellinger K. and Riederer P. (1977) Brain monoamines in metabolic (endotoxic) coma: A preliminary biochemical study in human post-mortem material. J.Neural Transm. 41, 275-286

Kellar K.J. and Cascio C.S. (1982) [^{3}H]-Tryptamine: High affinity binding sites in rat brain. Eur. J.Pharm. 78, 475-478

Kienzl E., Riederer P., Jellinger K., and Wesemann W. (1981) Transitional states of central serotonin receptors in Parkinson's disease. J.Neural Transm. 51, 113-122

Marsden C.A. and Curzon G. (1978) The role of tryptamine in the behavioral effects of tranylcypromine + L-tryptophan. Neuropharm. 18, 159-164

Riederer P. (1980) L-Dopa competes with tyrosine and tryptophan for human brain uptake. Nutr. Metab. 24, 417-423

Riederer P., Kruzik P., Kienzl E., Kleinberger G. Jellinger K., and Wesemann W. (1981) Central aminergic function and its disturbance by hepatic disease: The current status of L-valine pharmacotherapy in metabolic coma; in Transmitter Biochemistry of Human Brain Tissue Riederer P. and Usdin E. eds) pp 143-182. Macmillan Publishers LTD, London

Rinne U.K., Koskinen V., and Lönnberg P. (1980) Neurotransmitter receptors in the parkinsonian brain; in Parkinson's Disease - Current Progress. Problems and Management (Rinne U.K., Klingler M, Stamm G. eds.) p 93 Elsevier, North Holland Amsterdam

Seeman P., Westman K., Coscina D. and Warsh J.J. (1980) Serotonin receptors in hippocampus and frontal cortex. Eur.J.Pharmac. 66, 179-191

Smith I and Kellow A.H. (1969) Aromatic amines and Parkinson's disease. Nature 221, 1261

Tedeschi D.H., Tedeschi R.E. and Fellows E.J. (1959) The effects of tryptamine on the central nervous system including a pharmacological procedure for evaluation of iproniazid-like drugs. J.Pharm.Exp. Ther. 126, 223-232

Warsh J.J., Coscina D.V., Godse D.D. and Chan P.W. (1979) Dependence of brain tryptamine formation on tryptophan availability. J. Neurochem. 32 1191-1196

Whittaker P.M. and Seeman P. (1978) High affinity ^{3}H-serotonin binding to caudate: Inhibition by hallucinogens and serotonergic drugs. Psychopharm. 59, 1-5

BIOCHEMICAL EFFECTS AND ANTIDEPRESSANT MECHANISMS OF L-DEPRENYL

Michael R. Liebowitz M.D., Farouk Karoum Ph.D., Frederic M. Quitkin M.D., Jonathan Stewart M.D., Patrick McGrath M.D., Wilma Harrison M.D., Sharon O. Davies R.N., Donald F. Klein M.D.
From the Department of Psychiatry, College of Physicians and Surgeons, Columbia University and New York State Psychiatric Institute. Dr. Karoum is from the Adult Psychiatry Branch, National Institute of Mental Health, Washington, D.C.

INTRODUCTION

At low doses (e.g. 10 mg/day), L-deprenyl is thought to selectively inhibit the B isoenzyme of monoamine oxidase (MAO). The biochemical effects of L-deprenyl in depressed patients have, however, received little detailed study. It is unclear if L-deprenyl maintains MAO specificity when chronically administered in effective antidepressant dosages. Also, L-deprenyl has pharmacological effects other than MAO inhibition, such as blockade of norepinephrine (NE), reuptake (Knoll, 1976), inhibition of NE release (Knoll, 1976) and metabolism to methamphetamine and amphetamine (Reynolds *et al.*, 1978).

Detailed biochemical studies, carried out in conjunction with a clinical trial, were therefore indicated to clarify L-deprenyl's biochemical effects and antidepressant mechanisms in depressed patients.

METHODS

Subjects: The patient group included three males and

five females; a subset of these entered into a treatment study. Mean age was 37.9 years, with a range of 26-51. All patients met previously defined criteria for atypical depression (Liebowitz *et al.*, 1981; Quitkin *et al.*, 1982). In addition, all patients met Research Diagnostic Criteria (RDC) for Major Depression (Spitzer *et al.*, 1978).

Procedures: Patients were hospitalized for a 10 day placebo period (during which baseline biochemical samples were gathered) and their first week of active drug; after that, with one exception, they were treated as outpatients. All were on low tyramine diets throughout the study. The dosage schedule for placebo non-responders was 10 mg/day for four weeks, followed by 20 mg/day for the fifth week, and 30 mg/day for the sixth week. The one patient responding to 10 mg/day, however, was maintained at that dose.

Twenty-four hour urines were obtained from patients from 11 PM to 11 PM for two successive days at each collection point, and the results averaged. Urines were collected in plastic jars using 500 mg sodium metabisulfite as a preservative, refrigerated during collection, and stored at -20°C. Specimens were then shipped to Farouk Karoum, Ph.D. and Markku Linnoila, M.D., Ph.D. for analysis using procedures previously described (Karoum *et al.*, 1982; Petruccelli *et al.*, 1982). Urines with 24 hour creatinine levels of less than 250 mg, all of which had low volumes, were excluded as incomplete collections.

MAO determinations were carried out under the supervision of Morton Levitt, Ph.D. using a modification of the method of Murphy *et al.* (1977). Platelets and plasma MAO activities were assayed using ^{14}C-benzylamine as a substrate.

Biochemical Variables: The biochemical variables being reported and their interrelationships are shown on Table 1.

RESULTS

Deprenyl at 10 mg/day produced a significant increase in PE and p-TA outputs (see Table 2), while p-HPAA and MOPEG were significantly decreased. Mood changes did not contribute to the biochemical changes observed at this dosage level.

At L-deprenyl, 20 mg/day, the p-TA increase and the MOPEG decrease were maintained. The PE and p-HPAA effects seemed constant but significance was no longer maintained because of reduced sample size.

TABLE 1. Biochemical Variables.

Measured Substance	Interrelationship
Monoamines	
phenylethylamine (PE)	MAO B substrate
p-tyramine (p-TA)	MAO A and B substrate
dopamine (DA)	MAO A and B substrate
norepinephrine (NA)	MAO A substrate
Monoamine Metabolites	
p-hydroxyphenylacetic acid (p-HPAA)	deaminated p-TA metabolite
3-methoxytyramine (3-MTA)	aminated DA metabolite
3-methoxy-4-hydroxy-phenylglycol (MOPEG)	deaminated NA metabolite
5-hydroxyindoleacetic acid (5-HIAA)	deaminated serotonin (5-HT) metabolite
methamphetamine	aminated L-deprenyl metabolite
amphetamine	aminated L-deprenyl metabolite
platelet MAO	MAO B
plasma MAO	uncertain

TABLE 2. Biochemical Measures at Different Dosages of L-Deprenyl.

	Baseline (N=8)	10mg (N=6)	20 mg (N=4)	30 mg (N=2)
PE (μg)	4.70[a] (3.10)	122.37* (82.18)	173.52† (133.64)	221.55 (265.52)
p-TA (μg)	819.25 (384.98)	1466.17* (490.93)	1861.75** (513.50)	1153.00† (354.97)
p-HPAA (mg)	1.361 (.864)	.510* (.172)	.484 (.163)	.413 (.184)
MOPEG (mg)	2.541 (.982)	1.185* (.394)	.964** (.534)	.950 (.424)
AMPH (μg)	---- (-)	625.65 (387.03)	1607.75 (1022.53)	1725.00 (1755.04)
MAMPH (μg)	---- (-)	1593.33 (681.68)	3428.12 (2526.54)	3830.50 (4198.09)

NOTE: Correlated means t differences between baseline values and dosage level of L-deprenyl treatment.
[a]MEAN (S.D.); †.05 < p < .10; *p < .05; **p <. 01

With L-deprenyl, 30 mg/day, the findings were similar, but statistical comparisons were no longer significant due to the small number of patients. What remained was a trend for p-TA to be increased over baseline.

At no L-deprenyl dosage were significant changes from baseline seen for NA, DA, 3-MTA, 5-HIAA or creatinine. At six weeks, five of eight patients showed marked improvement. No baseline biochemical differences between subsequent responders and non-responders were found.

Stepwise increases in amphetamine and methamphetamine outputs were seen with increased doses of L-deprenyl (see Table 2). Platelet MAO activity was more than 95% inhibited within the first week on 10 mg/day, and remained so at all three doses of L-deprenyl.

DISCUSSION

It is clear that L-deprenyl is a MAO B inhibitor. This effect, however, is marked in less than a week and at low dosages (10 mg or less), while L-deprenyl appears to require four to six weeks and/or higher dosages (20-30 mg) to benefit most atypical depressives (Quitkin _et al._, unpublished paper). It is possible, however, that the effects of MAO B inhibition, such as PE elevation, are crucial, but take more than four weeks to exert their effect.

MAO A inhibition is not clearly found in our sample at any of the L-deprenyl dosages studied. MOPEG changes indicate that some process other than MAO B inhibition occurred at each L-deprenyl dose, but it is unclear whether this involved MAO A inhibition or a more specific disturbance of noradrenergic function or metabolism. The lack of statistically significant increase in 3-MTA or fall in 5-HIAA weigh against MAO A inhibition. Lack of simultaneous measures of other DA and NA metabolites, coupled with small sample size, however, make any conclusions on this issue tentative, especially for the higher doses of L-deprenyl.

The fall in MOPEG seen in our sample with L-deprenyl is consistent with NA reuptake inhibition, and is similar to values reported for six weeks of desipramine after doses producing a steady state plasma level of 100-150 ng/ml (Linnoila _et al._, 1982). Reuptake blockade could explain the reduced tyramine vulnerability reported for L-deprenyl dosages as high as 60 mg/day (Elsworth _et al._, 1978; Stern _et al._, 1978; Pickar _et al._, 1981).

L-deprenyl may also exert its antidepressant effects through its metabolism to methamphetamine and amphetamine. L-deprenyl, however, is thought to be converted to L-amphetamine, which has less euphoria-producing potency in healthy volunteers than D amphetamine (Smith and Davis, 1977). If amphetamine conversion and euphoria were a principal antidepressant mechanism, one would expect equal doses of D-deprenyl, which would presumably be converted to D-amphetamine, to be more active as an antidepressant. To date, the antidepressant potencies of L- and D-deprenyl have not, to our knowledge, been systematically compared. Direct measurements of the optical activity of the amphetamines generated from L-deprenyl have also not been made (Reynolds et al., 1978).

Finally, L-deprenyl may exert some or all of its antidepressant activity through other mechanisms. These might include direct effects on dopamine receptors, indirect sympathomimetic activity (Simpson, 1978), or direct interactions with recently identified CNS amphetamine receptors (Paul and Skolnick, 1982).

To summarize, while L-deprenyl 10-30 mg/day was found to significantly inhibit MAO B but not MAO A, further specificity studies are required, especially for what appears to be the more effective antidepressant dose range of 20 mg/day and above. In addition, amine reuptake blockade and/or conversion to methamphetamine and amphetamine may contribute to L-deprenyl's clinical and side effect profiles. Further studies are required to examine the biochemical mechanisms of L-deprenyl.

REFERENCES

Elsworth J. D., Glover V., Reynolds G. P., Sandler M., Lees A. J., Phuapradit P., Shaw K. M., Stern G. M., and Kumar P. (1978) Deprenyl administration in man: A selective monoamine oxidase B inhibitor without the 'cheese effect.' Psychopharmacology **57**, 33-38.

Karoum F., Chuang L.-W., Eisler T., Calne D. B., Liebowitz M. R., Quitkin F. M., Klein D. F., and Wyatt R. J. (1982) Metabolism of (-) deprenyl to amphetamine and methamphetamine may be responsible for deprenyl's therapeutic benefit: A biochemical assessment. Neurology **32**, 503-509.

Knoll J. (1976) Analysis of the pharmacological effects of selective monoamine oxidase inhibitors, in Monoamine Oxidase and Its Inhibition, Ciba Foundation Symposium 39 (new series), pp. 135-161. Elsevier/ Excerpta Medica/North-Holland.

Liebowitz M. R., Quitkin F. M., Stewart J. W., McGrath P. J., et al. (1981) Phenelzine and imipramine in atypical depression. Psychopharmacol. Bull. **17(3)**, 159-161.

Linnoila M., Karoum F., Calil H. M., Kopin I. J., and Potter W. Z. (1982) Alteration of norepinephrine metabolism with desipramine and zimelidine in depressed patients. Arch. Gen. Psychiatry **39**, 1025-1027.

Murphy D. L., Belmaker R., Carpenter W. T., et al. (1977) Monoamine oxidase in chronic schizophrenia: Studies of hormonal and other factors affecting enzyme activity. Br. J. Psychiatry **130**, 151-158.

Paul S. M. and Skolnick P. (1982) (+)-Amphetamine binding to rat hypothalamus: Relation to anorexic potency of phenylethylamines. Science **218**, 487-489.

Petruccelli B., Bakris G., Miller T., Korpi E., and Linnoila M. (1982) Quantitation of 5-hydroxytryptophan, serotonin and 5-hydroxyindoleacetic acid in human body fluids. Acta Pharmacol. Toxicol. **51**, 421-427.

Pickar D., Cohen R. M., Jimerson D. C., and Murphy D. L. (1981) Tyramine infusions and selective monoamine oxidase inhibitor treatment. Psychopharmacology **74**, 4-7.

Quitkin F. M., Schwartz D., Liebowitz M. R., Stewart J. W., et al. (1982) Atypical depressives: A preliminary report of antidepressant response, sleep patterns, and cortisol secretion, in Treatment of Depression: Old Controversies and New Approaches (Clayton P. J. and Barrett J. E., eds), pp. 253-263. Raven Press, New York.

Quitkin F. M., Liebowitz M. R., Stewart J. W., McGrath P. J., Harrison W., and Davies S. O. Deprenyl in atypical depressives: I. Clinical efficacy. (Unpublished paper)

Reynolds G. P., Elsworth J. D., Blau K., Sandler M., Lees A. J., and Stern G. M. (1978) Deprenyl is metabolized to methamphetamine and amphetamine in man. Letter to the Editors, Br. J. Clin. Pharmacol. **6**, 542-544.

Simpson L. L. (1978) Evidence that deprenyl, a type B monoamine oxidase inhibitor, is an indirectly acting sympathomimetic amine. Biochem. Pharmacol. **27**, 1591-1595.

Smith R. C. and Davis J. M. (1977) Comparative effects of d-amphetamine, l-amphetamine, and methylphenidate on mood in man. Psychopharmacology **53**, 1-12.

Spitzer R. L., Endicott J., and Robins E. (1978) Research diagnostic criteria: Rationale and reliability. Arch. Gen. Psychiatry **35**, 773-782.

Stern G. M., Lees A. J., and Sandler M. (1978) Recent observations on the clinical pharmacology of (-) deprenyl. Neural Transmission **43**, 245-251.

SUBJECT INDEX

GPSR Compliance

The European Union's (EU) General Product Safety Regulation (GPSR) is a set of rules that requires consumer products to be safe and our obligations to ensure this.

If you have any concerns about our products, you can contact us on ProductSafety@springernature.com

In case Publisher is established outside the EU, the EU authorized representative is:

Springer Nature Customer Service Center GmbH
Europaplatz 3
69115 Heidelberg, Germany

Batch number: 10755248

Printed by Printforce, the Netherlands